The Greek Alphabet

Alpha	A	α	Iota	I	ι	Rho	P	ρ
Beta	B	β	Kappa	K	κ	Sigma	Σ	σ
Gamma	Γ	γ	Lambda	Λ	λ	Tau	T	τ
Delta	Δ	δ	Mu	M	μ	Upsilon	Y	υ
Epsilon	E	ϵ	Nu	N	ν	Phi	Φ	ϕ
Zeta	Z	ζ	Xi	Ξ	ξ	Chi	X	χ
Eta	H	η	Omicron	O	o	Psi	Ψ	ψ
Theta	Θ	θ	Pi	Π	π	Omega	Ω	ω

Abbreviations for Units

A	ampere	lb	pound
Å	angstrom (10^{-10} m)	L	liter
atm	atmosphere	m	meter
Btu	British thermal unit	MeV	mega-electron volts
Bq	becquerel	Mm	megameter (10^6 m)
C	coulomb	mi	mile
°C	degree Celsius	min	minute
cal	calorie	mm	millimeter
Ci	curie	ms	millisecond
cm	centimeter	N	newton
dyn	dyne	nm	nanometer (10^{-9} m)
eV	electron volt	pt	pint
°F	degree Fahrenheit	qt	quart
fm	femtometer, fermi (10^{-15} m)	rev	revolution
		R	roentgen
ft	foot	Sv	seivert
Gm	gigameter (10^9 m)	s	second
G	gauss	T	tesla
Gy	gray	u	unified mass unit
g	gram	V	volt
H	henry	W	watt
h	hour	Wb	weber
Hz	hertz	y	year
in	inch	yd	yard
J	joule	μm	micrometer (10^{-6} m)
K	kelvin	μs	microsecond
kg	kilogram	μC	microcoulomb
km	kilometer	Ω	ohm
keV	kilo-electron volts		

Elementary Modern Physics

Elementary
Modern Physics

Paul A. Tipler

Worth Publishers

Elementary Modern Physics

Paul A. Tipler

Copyright © 1992 by Worth Publishers, Inc.

All rights reserved

Printed in the United States of America

Library of Congress Catalog Card Number: 91-68315

ISBN: 0-87901-569-1

Printing: 5   4   3   2   1   Year: 96   95   94   93   92

Development Editors: Valerie Neal and Steven Tenney

Design: Malcolm Grear Designers

Art Director: George Touloumes

Production Editors: Karen Landovitz, Elizabeth Mastalski

Production Supervisor: Sarah Segal

Layout: Patricia Lawson

Picture Research: Steven Tenney, John Schultz of PAR/NYC, and Lana Berkovich

Line Art: York Graphic Services and Demetrios Zangos

Composition: York Graphic Services

Printing and binding: R. R. Donnelley and Sons

Cover: Beam-fanning patterns produced by a crystal of barium titanate. Light beams from separate lasers are directed along the same path in the crystal. These laser beams create diffraction gratings inside the crystal, causing the beams to be deflected into the bright fan of multicolored ellipses shown on the screen. The bright spot is from the undeflected laser beams. (Related effects are discussed in more detail on pp. 220–221.) When using weak laser beams, with intensities of about $1W/cm^2$, the crystal takes a few seconds to respond to the light. If the $BaTiO_3$ crystal were perfect and pure, it would not bend the beams at all. The defects or impurities responsible for the effect are known to be present at a few parts per million, but their identity is unknown. (Photograph courtesy of Roger Cudney and Jack Feinberg, University of Southern California.)

Illustration credits begin on p. **IC-1**, and constitute an extension of the copyright page.

Worth Publishers

33 Irving Place

New York, NY 10003

Preface

This is a textbook for a one-semester or one-quarter course in modern physics that typically follows the introductory physics course for engineers and science majors. It provides an introduction to relativity, quantum physics, and the physics of atoms, molecules, solids, nuclei, and elementary particles, plus a concluding chapter on astrophysics and cosmology. Most of the material appeared earlier in the third edition of my introductory physics textbook *Physics for Scientists and Engineers*. To help the student review various topics from classical physics that are needed for the study of modern physics, two appendixes have been added: Appendix B reviews the properties of classical waves, and Appendix C reviews the Maxwell–Boltzmann distribution, the equipartition theorem, and the gaussian distribution function. In addition, a table of nuclear masses is given in Appendix G.

I wrote this book because I saw a need for a more elementary, more up-to-date, and more exciting book for students such as engineering, physics, chemistry, or life science majors who have had a course in introductory physics and want to know about the physics of the 20th century, including the exciting discoveries of the last several years.

Because of the widespread familiarity with my previous book *Modern Physics*, Second Edition (1978), some comparisons with that book may be in order. Both books begin with relativity, but here I have deleted treatment of kinetic theory, and have greatly reduced the space devoted to what is often called "old quantum theory." I have done so in order to devote more time to contemporary topics, including elementary particle theory and cosmology, and the many modern applications of physics, such as scanning tunneling microscopes, holograms, charge-couple devices, photorefractivity, superconductivity, SQUIDs, etc. In addition, this text is at a considerably lower mathematical level and is less detailed in its coverage than *Modern Physics*.

Although the coverage of modern physics is necessarily more descriptive than that of classical physics, the important ideas of modern physics are often introduced by the study of specific, simple problems. For example, the concepts of quantum numbers and degeneracy arise naturally in the study of the three-dimensional infinite square well in Chapter 3. Similarly, the Pauli exclusion principle arises from symmetry considerations in the study of two identical particles in a one-dimensional infinite square well in the same chapter. In Chapter 6 the infinite square well is again used to introduce the concept of Fermi energy needed to understand the electrical and thermal properties of solids.

Contemporary Applications

Many applications of modern physics are presented in full-color photographs with extended captions. This allows the student to see how modern physics is applied in contemporary technology without the need for extensive in-text development. In addition, three essays on fascinating new technologies have been written to enhance the text:

Scanning Tunneling Microscopy by Ellen D. Williams (Chapter 3, page 111)
Trapped Atoms and Laser Cooling by D. J. Wineland (Chapter 4, page 156)
SQUIDs by Samuel J. Williamson (Chapter 6, page 230)

Examples, Exercises, and Problems

The understanding of physics and the development of problem solving skills are enhanced by the extensive and integrated use of examples, in-text exercises, and graded sets of problems. Nearly all of the 64 worked examples are numerical and have been written to ensure correspondence between the examples and the end-of-chapter problems (especially those at the intermediate level). There are several numerical in-text exercises (with answers given immediately) that ask the student to perform a simple calculation that reinforces understanding.

Problems at the end of each chapter are grouped into three levels of difficulty. Level I are relatively easy single-step problems. They are keyed to the appropriate sections in the chapter so that the student can quickly find help if needed. Level II problems require a more sophisticated understanding and are not presented by section. Level III problems are the most challenging and will be of value primarily to advanced students. There are 288 Level I problems, 160 Level II, and 49 Level III.

Ease of Review

Several pedagogical features are included to aid students in their initial study and review. Important equations, laws, and tables are highlighted by a color screen. Margin heads are provided for quick reference. Key terms are introduced in **boldface** type, defined in the text, and listed in the review section of each chapter. Thought questions immediately follow some sections within each chapter. These include routine questions that can be easily answered from the preceding text as well as open-ended questions that can serve as a basis for classroom discussion.

Each chapter concludes with a summary of the important laws and results that were discussed, along with the equations that will be most useful in solving problems, a list of suggested further readings, a review section, and the graded sets of problems. The review section contains a list of learning objectives, the key terms that the student should be able to identify and define, and a set of true-false questions.

Acknowledgments

Many people have contributed to this book.

Ralph Llewellyn (University of Central Florida) wrote the exciting and informative chapter on astrophysics and cosmology (Chapter 9).

Many new and interesting end-of-chapter problems were provided by Howard Miles (Washington State University), John Russell (Southeastern Massachusetts University), and Grant Hart (Brigham Young University). Among them, they have also provided one of the independently worked sets of solutions for all of the problems in the text.

James Walker (Washington State University) prepared the answers listing at the end of the text, produced the complete answers listing and the elegant solutions that are published in the accompanying *Solutions Manual and Instructor's Resources*, and also offered many valuable suggestions for improving and clarifying the end-of-chapter problems.

Robin Macqueen (University of British Columbia) contributed the Suggestions for Further Reading for each chapter.

The accuracy of the numerical calculations in the examples and exercises has been expertly checked by Edward Brown (Manhattan College). Professor Brown also offered many helpful suggestions in his reviews of the end-of-chapter problems.

Ron Gautreau (New Jersey Institute of Technology) has written a *Study Guide* which includes key ideas, numbers and key equations, potential pitfalls, true and false questions with responses, questions and answers, and problems and solutions for each chapter.

The *Solutions Manual and Instructor's Resources* has been prepared by Robert Allen (Victor Valley Community College), John Davis (University of Washington), James Walker (Washington State University), and Vicki Williams (Pennsylvania State University). It provides a comprehensive selection of demonstrations, a film and video guide, critical-thinking questions, complete solutions, and a complete answers listing for all end-of-chapter problems.

Many other instructors have provided extensive and invaluable reviews. They have all made a deeply appreciated and fundamental contribution to the quality of this book, and I therefore wish to thank:

Ralph Baierlin, *Wesleyan University*

Walter Borst, *Texas Technological University*

Edward Brown, *Manhattan College*

Roger Clapp, *University of South Florida*

Miles Dresser, *Washington State University*

Manuel Gómez-Rodríguez, *University of Puerto Rico, Rio Piedras*

Harry Otteson, *Utah State University*

Larry Panek, *Widener University*

Malcolm Perry, *Cambridge University, United Kingdom*

Brooke Pridmore, *Clayton State College*

Robert Rundel, *Mississippi State University*

John Russell, *Southeastern Massachusetts University*

Jim Smith, *University of Illinois, Urbana-Champaign*

Richard Smith, *Montana State University*

Edward Thomas, *Georgia Institute of Technology*

Gianfranco Vidali, *Syracuse University*

Thad Zaleskiewicz, *University of Pittsburgh, Greensburg*

Finally, I would like to thank everyone at Worth Publishers for their help and encouragement, and particularly Steven Tenney, Valerie Neal, Betsy Mastalski, Karen Landovitz, George Touloumes, and Sarah Segal.

Berkeley, California
February 1992

About the Author

Paul Tipler was born in the small farming town of Antigo, Wisconsin, in 1933. He graduated from high school in Oshkosh, Wisconsin, where his father was Superintendent of the Public Schools. He received his BS at Purdue University in 1955 and his PhD at the University of Illinois in 1962, where he studied the structure of nuclei. He taught for one year at Wesleyan University in Connecticut while writing his thesis. He then moved to Oakland University in Michigan, where he was one of the original members of the physics department, playing a major role in developing the physics curriculum. During the next 20 years, he taught nearly all the physics courses and wrote the first and second editions of his widely used textbooks *Modern Physics* (1969, 1978) and *Physics* (1976, 1982). In 1982 he moved to Berkeley, California, where he now resides and where he wrote *College Physics* (1987) and the third edition of *Physics For Scientists and Engineers* (1991). In addition to physics, his interests include music, hiking, and camping, and he is an accomplished jazz pianist and poker player.

Contents

Chapter 1   **Relativity 2**
   1-1   Newtonian Relativity 3
   1-2   The Michelson–Morley Experiment 5
   1-3   Einstein's Postulates 8
   1-4   The Lorentz Transformation 9
   1-5   Clock Synchronization and Simultaneity 15
   1-6   The Doppler Effect 20
   1-7   The Twin Paradox 21
   1-8   The Velocity Transformation 24
   1-9   Relativistic Momentum 26
   1-10 Relativistic Energy 28
   1-11 General Relativity 34
          Summary 37
          Suggestions for Further Reading, Review, Problems 40

Chapter 2   **The Origins of Quantum Theory 47**
   2-1   The Origin of the Quantum Constant: Blackbody Radiation 49
   2-2   The Photoelectric Effect 50
   2-3   X Rays 54
   2-4   Compton Scattering 56
   2-5   Quantization of Atomic Energies: The Bohr Model 58
   2-6   Electron Waves and Quantum Theory 63
          Summary 68
          Suggestions for Further Reading, Review, Problems 69

Chapter 3   **Quantum Mechanics 74**
   3-1   The Electron Wave Function 75
   3-2   Electron Wave Packets 78
   3-3   The Uncertainty Principle 82
   3-4   Wave–Particle Duality 85
   3-5   The Schrödinger Equation 86
   3-6   A Particle in a Box 88
   3-7   A Particle in a Finite Square Well 94
   3-8   Expectation Values 98

xii   Contents

    3-9   Reflection and Transmission of Electron Waves: Barrier Penetration   **100**
    3-10   The Schrödinger Equation in Three Dimensions   **105**
    3-11   The Schrödinger Equation for Two Identical Particles   **106**
    Summary   **108**
    Essay Ellen D. Williams, *Scanning Tunneling Microscopy*   **111**
    Suggestions for Further Reading, Review, Problems   **115**

Chapter 4

Atoms   **120**

    4-1   Quantum Theory of the Hydrogen Atom   **121**
    4-2   The Hydrogen-Atom Wave Functions   **125**
    4-3   Magnetic Moments and Electron Spin   **129**
    4-4   The Stern–Gerlach Experiment   **131**
    4-5   Addition of Angular Momenta and the Spin-Orbit Effect   **133**
    4-6   The Periodic Table   **136**
    4-7   Optical and X-Ray Spectra   **141**
    4-8   Absorption, Scattering, and Stimulated Emission   **146**
    4-9   The Laser   **147**
    Summary   **155**
    Essay D. J. Wineland, *Trapped Atoms and Laser Cooling*   **156**
    Suggestions for Further Reading, Review, Problems   **163**

Chapter 5

Molecules   **167**

    5-1   Molecular Bonding   **168**
    5-2   Polyatomic Molecules   **175**
    5-3   Energy Levels and Spectra of Diatomic Molecules   **177**
    Summary   **186**
    Suggestions for Further Reading, Review, Problems   **187**

Chapter 6

Solids   **190**

    6-1   The Structure of Solids   **190**
    6-2   The Classical Free-Electron Theory of Metals   **195**
    6-3   The Fermi Electron Gas   **201**
    6-4   Quantum Theory of Electrical Conduction   **206**
    6-5   Band Theory of Solids   **209**
    6-6   Impurity Semiconductors   **213**
    6-7   Semiconductor Junctions and Devices   **214**
    6-8   Superconductivity   **221**

Summary 227

Essay Samuel J. Williamson, *SQUIDs* 230

Suggestions for Further Reading, Review, Problems 234

Chapter 7

Nuclei 238

7-1 Properties of Nuclei 239

7-2 Nuclear Magnetic Resonance 244

7-3 Radioactivity 246

7-4 Nuclear Reactions 252

7-5 Fission, Fusion, and Nuclear Reactors 255

7-6 The Interaction of Particles with Matter 263

Summary 269

Suggestions for Further Reading, Review, Problems 270

Chapter 8

Elementary Particles 276

8-1 Hadrons and Leptons 277

8-2 Spin and Antiparticles 280

8-3 The Conservation Laws 283

8-4 The Quark Model 286

8-5 Field Particles 289

8-6 The Electroweak Theory 290

8-7 The Standard Model 290

8-8 Grand Unification Theories 292

Summary 292

Suggestions for Further Reading, Review, Problems 293

Chapter 9

Astrophysics and Cosmology 297

9-1 Our Star, the Sun 298

9-2 The Stars 304

9-3 The Evolution of Stars 308

9-4 Cataclysmic Events 311

9-5 Final States of Stars 313

9-6 Galaxies 316

9-7 Gravitation and Cosmology 322

9-8 Cosmogenesis 324

Summary 329

Suggestions for Further Reading, Review, Problems 330

Appendix

A   Summary of Selected Mathematical Relations   **AP-1**

B   Properties of Classical Waves   **AP-5**

C   The Maxwell–Boltzmann Distribution, the Equipartition Theorem, and the Gaussian Distribution   **AP-13**

D   Numerical Data   **AP-18**

E   Conversion Factors   **AP-21**

F   Periodic Table of the Elements   **AP-23**

G   Properties of Nuclei   **AP-24**

Illustration Credits   **IC-1**

Answers to True–False and Odd-Numbered Problems   **A-1**

Index   **I-1**

Elementary Modern Physics

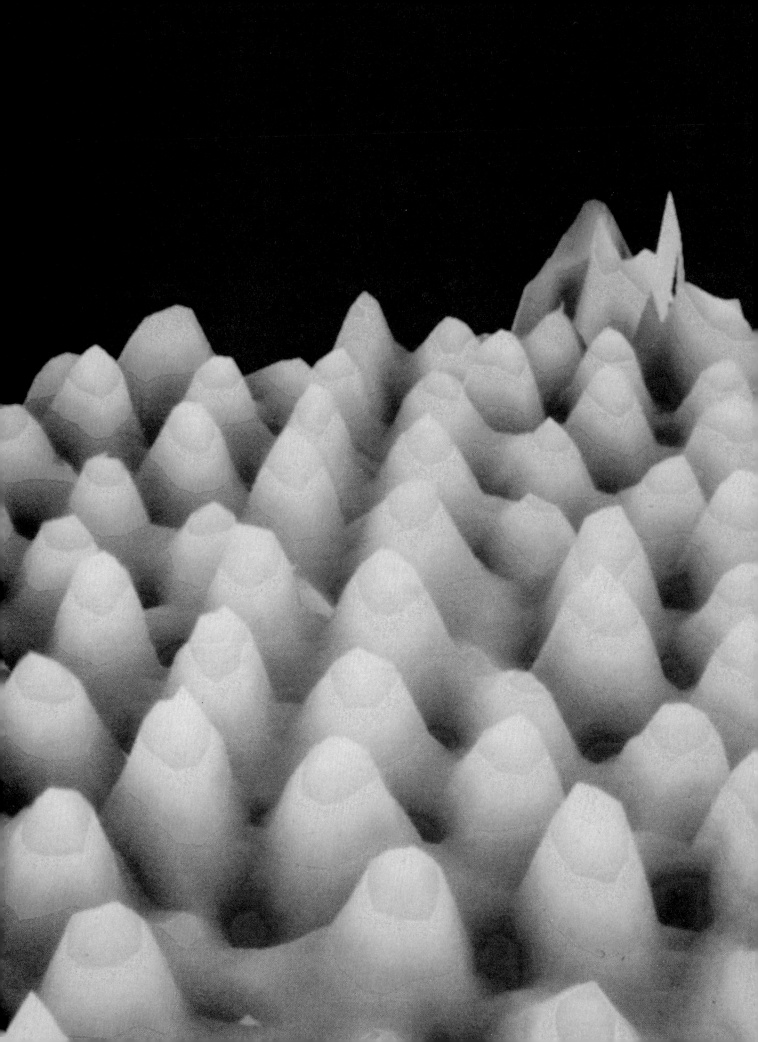

The surface of silicon, an important ingredient in many semiconductor devices, is shown here at a magnification of about 1,000,000,000. Individual atoms of a silicon crystal are seen as hills in this micrograph, which was obtained with a scanning tunneling microscope. The digital information collected by the microscope is plotted by a computer, which assigns false colors to accentuate the crystalline structure.

The spectacular successes of the laws of mechanics, electromagnetism, and thermodynamics as expressed by Newton, Maxwell, Carnot, and others led some to believe that there was little left to do in physics except apply these laws to various phenomena. Such optimism (or pessimism, depending on your point of view) was short-lived, for the late nineteenth and early twentieth centuries saw a rapid accumulation of discoveries and experiments which could not be explained in terms of the known laws generally referred to as classical physics. The breakdown in classical physics occurred in many different areas: the Michelson-Morley null result contradicted Newtonian relativity; the blackbody radiation spectrum and the measured heat capacities of solids and gases contradicted predictions of thermodynamics; the photoelectric effect, the Compton effect, and the spectra of atoms could not be explained by electromagnetic theory; and the exciting discoveries of radioactivity and x rays seemed to be outside the framework of classical physics. The development of the theories of relativity and quantum mechanics in the early twentieth century provided answers to all the puzzles listed above and many more. The application of these theories to such microscopic systems as atoms, molecules, nuclei, and elementary particles, and to macroscopic systems of gases, liquids, solids, and the cosmos itself, has given us a deep understanding of the workings of nature and has revolutionized our way of life.

Chapter 1

Relativity

Albert Einstein in 1916.

Albert Einstein was probably the best known scientist in the world. The well-publicized observation during the eclipse of 1919 of the deflection of a beam of light due to the gravitational attraction to the sun was in agreement with his prediction 3 years earlier from his theory of general relativity, and brought him instant fame. His special theory of relativity, published in 1905 with its predictions of time dilation, length contraction, the twin paradox, and the famous equation $E = mc^2$, which relates mass and energy, has fascinated both scientists and the general public. His explanation of the photoelectric effect (also published in 1905), in which he boldly predicted the quantization of energy in electromagnetic radiation, was instrumental in the development of quantum theory, and brought him the Nobel prize in physics in 1921. (We will study the photoelectric effect along with other origins of quantum theory in Chapter 2.)

In this chapter we will study relativity. The theory of relativity consists of two rather different theories, the special theory and the general theory. The special theory, developed by Einstein and others in 1905, concerns the comparison of measurements made in different inertial reference frames moving with constant velocity relative to one another. Its consequences, which can be derived with a minimum of mathematics, are applicable in a wide variety of situations encountered in physics and engineering. On the other hand, the general theory, also developed by Einstein and others around 1916, is concerned with accelerated reference frames and gravity. A thorough understanding of the general theory requires sophisticated mathematics, and the applications of this theory are chiefly in the area of gravitation. It is of great importance in cosmology, but it is rarely encountered in other areas of physics or in engineering. We will therefore concentrate on the special theory (often referred to as *special relativity*) and discuss the general theory only briefly in the last section of this chapter.

1-1 Newtonian Relativity

Newton's first law does not distinguish between a particle at rest and one moving with constant velocity. If there is no net external force acting, the particle will remain in its initial state—either at rest or moving with its initial velocity. Consider a particle at rest relative to you with no forces acting on it. According to Newton's first law, it will remain at rest. Now consider the same particle from the point of view of a second observer who is moving with constant velocity relative to you. From this observer's "frame of reference," both you and the particle are moving with constant velocity. Newton's first law holds also for him. (Note that if the second observer were accelerating relative to you, he would see the particle accelerating relative to him with no external forces acting on it. Thus Newton's first law would not hold for him.) How might we distinguish whether you and the particle are at rest and the second observer is moving with constant velocity, or the second observer is at rest and you and the particle are moving?

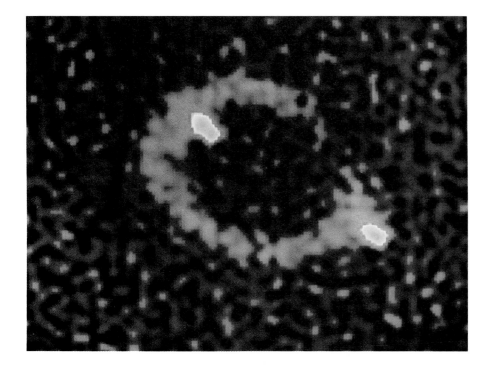

This ringlike structure of the radio source MG1131 + 0456 is thought to be due to "gravitational lensing," first proposed by Einstein in 1936, in which a source is imaged into a ring by a large massive object in the foreground.

Figure 1-1 A boxcar moving with constant velocity along a straight track. The reference frame S' is at rest relative to the car and is moving with speed V relative to S, which is at rest relative to the track. It is impossible to tell by doing mechanics experiments inside the car whether the car is moving to the right with speed V or the track is moving to the left with speed V.

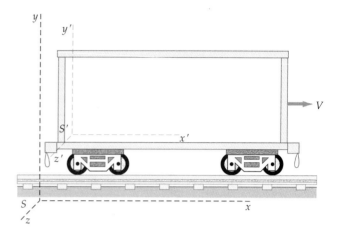

Let us consider some simple experiments. Suppose we have a train moving along a straight, flat track with a constant velocity V. (We assume there are no bumps or shakes in the motion.) Let us choose a coordinate system xyz with its x axis along the track as shown in Figure 1-1. It doesn't matter where along the track we choose the origin. For different choices, the position (relative to the origin) of the train and its contents will differ, but their velocities will be the same. A set of coordinate systems at rest relative to each other is called a **reference frame**. We will call the reference frame at rest relative to the track frame S. We now consider doing various mechanics experiments in a closed boxcar of the train. For these we choose a coordinate system at rest relative to the train. This coordinate system is in reference frame S', which is moving to the right with speed V relative to frame S. We note that a ball at rest in the train remains at rest. If we drop the ball, it falls straight down in frame S' with an acceleration g due to gravity. (Of course, when viewed in frame S, the ball moves along a parabolic path because it has an initial velocity V to the right.) No mechanics experiment that we can do—measuring the period of a pendulum or a body on a spring, observing the collisions between two bodies, or whatever—will tell us whether the train is moving and the track is at rest or the track is moving and the train is at rest. Newton's laws hold for reference frame S' as well as reference frame S.

A reference frame in which Newton's laws hold is called an **inertial reference frame**.

> All reference frames moving at constant velocity relative to an inertial reference frame are also inertial reference frames.

If we have two inertial reference frames moving with constant velocity relative to each other, such as S and S', there are no mechanics experiments that can tell us which is at rest and which is moving or if they are both moving. This result is known as the principle of **newtonian relativity**:

> Absolute motion cannot be detected.

This principle was well known by Galileo, Newton, and others in the seventeenth century. By the late nineteenth century, however, this view had changed. It was then generally thought that newtonian relativity was not valid and that absolute motion could be detected in principle by a measurement of the speed of light.

1-2 The Michelson–Morley Experiment

From our study of wave motion, we know that all mechanical waves require a medium for their propagation, and that the speed of such waves depends only on the properties of the medium. For example, the speed of sound waves in air depends on the temperature of the air. This speed is relative to still air. Motion relative to still air can indeed be detected. If you are moving relative to still air, you feel a wind.

It was therefore natural to expect that some kind of medium supports the propagation of light and other electromagnetic waves. This proposed medium was called the **ether**. As proposed, the ether had to have unusual properties. For example, it had to have great rigidity to support waves of such high velocity. (Recall that the velocity of waves on a string depends on the tension of the string and that of longitudinal sound waves in a solid depends on the bulk modulus of the solid.) Yet the ether could introduce no drag force on the planets, as their motion is fully accounted for by the law of gravitation. It was suspected that the ether was at rest relative to the distant stars, but this was considered to be an open question. It was therefore of considerable interest to determine the velocity of the earth relative to the ether. Experiments to do this were undertaken by Albert Michelson, first in 1881 and then again with Edward Morley in 1887 with greater precision. It was thought that a measurement of the speed of light relative to some reference frame moving through the ether would yield a result greater or less than c by an amount that depended on the speed of the frame relative to the ether and the direction of motion relative to the direction of the light beam. Thus, in 1881 Michelson set out to measure the speed of light relative to the earth and from this measurement to determine the velocity of the earth relative to the ether.

According to Maxwell's theory of electromagnetism, the speed of light and other electromagnetic waves is

$$c = \frac{1}{\sqrt{\epsilon_0 \mu_0}} = 3 \times 10^8 \text{ m/s}$$

where ϵ_0 and μ_0 are, respectively, the permittivity and permeability of free space. There is nothing in Maxwell's equations that tells us in what reference frame the speed of light will have this value, but the expectation was that this was the speed of light relative to its natural medium, the ether.

In the usual measurements of the speed of light, the time it takes for a light pulse to travel to and from a mirror is determined. Figure 1-2 shows a light source and a mirror a distance L apart. If we assume that both are moving with speed v through the ether, classical theory predicts that the light will travel toward the mirror with speed $c - v$ and back with speed $c + v$ (both speeds being relative to the mirror and the light source). The time for the total trip will be

$$t_1 = \frac{L}{c-v} + \frac{L}{c+v} = 2c\frac{L}{c^2 - v^2} = \frac{2L}{c}\left(1 - \frac{v^2}{c^2}\right)^{-1} \qquad 1\text{-}1$$

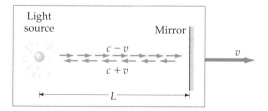

Figure 1-2 A light source and mirror moving with speed v relative to the "ether." According to classical theory, the speed of light relative to the source and mirror is $c - v$ toward the mirror and $c + v$ away from the mirror.

We can see that this differs from the time $2L/c$ by the factor $(1 - v^2/c^2)^{-1}$, which is very nearly equal to 1 if v is much less than c. We can simplify this expression for small values of v/c by using the binomial expansion

$$(1 + x)^n = 1 + nx + n(n-1)\frac{x^2}{2} + \cdots \simeq 1 + nx \qquad \text{1-2}$$

when x is much less than 1. Using $n = -1$ and $x = v^2/c^2$, Equation 1-1 becomes

$$t_1 \simeq \frac{2L}{c}\left(1 + \frac{v^2}{c^2}\right) \qquad \text{1-3}$$

The orbital speed of the earth about the sun is about 3×10^4 m/s. If we take this for an estimate of v we have $v = 3 \times 10^4$ m/s, $v/c = (3 \times 10^4 \text{ m/s})/(3 \times 10^8 \text{ m/s}) = 10^{-4}$, and $v^2/c^2 = 10^{-8}$. Thus the correction for the earth's motion is small indeed.

Michelson realized that although this effect is too small to be measured directly, it should be possible to determine v^2/c^2 by a difference measurement. For this measurement, he used a Michelson interferometer. In this experiment, one beam of light moves along the direction of the earth's motion and another moves perpendicular to that direction (Figure 1-3). The difference between the round-trip times of these beams depends on the speed of the earth and can be determined by an interference measurement. Let us assume that the interferometer is oriented such that the beam that strikes mirror M_1 is in the direction of the assumed motion of the earth. Equation 1-3 then gives the classical result for the round-trip time t_1 for the transmitted beam. The beam that reflects from the beam splitter and strikes mirror M_2 travels with some velocity **u** (relative to the earth) perpendicular to the earth's velocity. Relative to the ether, it travels with velocity **c** as shown in Figure 1-4. The velocity **u** (according to classical theory) is then the vector difference $\mathbf{u} = \mathbf{c} - \mathbf{v}$, as shown in the figure. The magnitude of **u** is $\sqrt{c^2 - v^2}$, so the round-trip time t_2 for this beam is

$$t_2 = \frac{2L}{\sqrt{c^2 - v^2}} = \frac{2L}{c}(1 - v^2/c^2)^{-1/2} \qquad \text{1-4}$$

Again using the binomial expansion, we obtain

$$t_2 \simeq \frac{2L}{c}\left(1 + \frac{1}{2}\frac{v^2}{c^2}\right) \qquad \text{1-5}$$

This expression is slightly different from that given for t_1 in Equation 1-3.

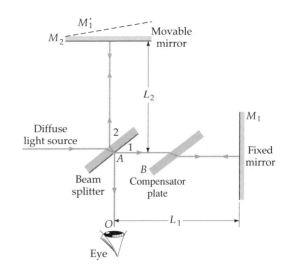

Figure 1-3 Michelson interferometer. The dashed line M_1' is the image of mirror M_1 in mirror A. The interference fringes observed are those of a small wedge-shaped film of air formed between the sources M_2 and M_1'. Assume that the light beam reflecting off mirror M_1 is parallel to the motion of the earth and that reflecting off mirror M_2 is perpendicular to the earth's motion. The interference of the two beams depends on the relative number of waves in each path, which depends in turn on the speed of the light beams relative to the earth. If the speed of light along the parallel path is different from that along the perpendicular path, the interference fringe pattern will shift when the interferometer is rotated through 90°.

The difference in these two times is

$$\Delta t = t_1 - t_2 \approx \frac{L}{c} \frac{v^2}{c^2} \quad \text{1-6}$$

This time difference is to be detected by observing the interference of the two beams of light.

Because of the difficulty of making the two paths of equal length to the precision required, the interference pattern of the two beams is observed and the whole apparatus is then rotated 90°. The rotation produces a time difference given by Equation 1-6 for each beam. The total time difference of $2\,\Delta t$ results in a phase difference $\Delta\phi$ between the two beams, where

$$\Delta\phi = 2\pi \frac{2c\,\Delta t}{\lambda}$$

and λ is the wavelength of the light. The interference fringes observed in the first orientation should thus shift by a number of fringes ΔN given by

$$\Delta N = \frac{\Delta\phi}{2\pi} = \frac{2c\,\Delta t}{\lambda} = \frac{2L}{\lambda} \frac{v^2}{c^2} \quad \text{1-7}$$

In Michelson's first attempt in 1881, L was about 1.2 m and λ was 590 nm. For $v^2/c^2 = 10^{-8}$, ΔN was expected to be 0.04 fringes. However, no shift was observed. In case the earth just happened to be at rest relative to the ether when the experiment was performed, the experiment was repeated six months later when the motion of the earth relative to the sun was in the opposite direction. Even though the experimental uncertainties were estimated to be of about the same magnitude as the expected fringe shift, Michelson reported the observation of no fringe shift as evidence that the earth did not move relative to the ether. In 1887, when he repeated the experiment with Edward W. Morley, he used an improved system for rotating the apparatus without introducing a fringe shift due to mechanical strains, and he increased the effective path length L to about 11 m by a series of multiple reflections. Figure 1-5 shows the configuration of the Michelson–Morley apparatus. For this attempt, ΔN was expected to be 0.4 fringes, about 20 to 40 times the minimum that could be observed. Once again, no shift was observed. The experiment has since been repeated under various conditions by a number of people, and no shift has ever been found.

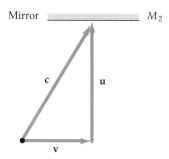

Figure 1-4 The light beam reflected from the beam splitter in a Michelson interferometer. The interferometer is moving to the right relative to the ether with velocity **v**, and the light beam moves perpendicular to mirror M_2 with velocity **u**. The velocity of light is **c** in the frame of the ether. Relative to the earth, in which the interferometer is fixed, the velocity of light is $\mathbf{u} = \mathbf{c} - \mathbf{v}$. The speed of light relative to the earth according to classical theory is then $u = (c^2 - v^2)^{1/2} = c(1 - v^2/c^2)^{1/2}$.

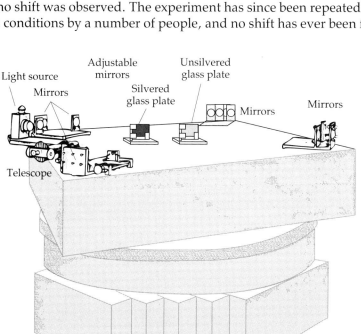

Figure 1-5 Drawing of Michelson and Morley's apparatus for their 1887 experiment. The optical parts were mounted on a sandstone slab 5 ft square, which was floated in mercury to reduce the strains and vibrations that had affected the earlier experiments. Observations could be made in all directions by rotating the apparatus in the horizontal plane.

In 1905, at the age of 26, Albert Einstein published a paper on the electrodynamics of moving bodies.* In this paper, he postulated that absolute motion cannot be detected by any experiment. (We will discuss the Einstein postulates in detail in the next section.) The null result of the Michelson–Morley experiment is therefore expected. We can consider the whole apparatus and the earth to be at rest. Thus no fringe shift is expected when the apparatus is rotated 90° since all directions are equivalent. Einstein did not set out to explain the results of the Michelson–Morley experiment. His theory arose from his considerations of the theory of electricity and magnetism and the unusual property of electromagnetic waves that they propagate in a vacuum. In his first paper, which contains the complete theory of special relativity, he made only a passing reference to the Michelson–Morley experiment, and in later years he could not recall whether he was aware of the details of this experiment before he published his theory.

1-3 Einstein's Postulates

The theory of special relativity can be derived from two postulates proposed by Einstein in his original paper in 1905. Simply stated, these postulates are

Einstein's postulates

Postulate 1. Absolute, uniform motion cannot be detected.

Postulate 2. The speed of light is independent of the motion of the source.

Postulate 1 is merely an extension of the newtonian principle of relativity to include all types of physical measurements (not just those that are mechanical). Postulate 2 describes a common property of all waves. For example, the speed of sound waves does not depend on the motion of the sound source. When an approaching car sounds its horn, the frequency heard increases according to the doppler effect, but the speed of the waves traveling through the air does not depend on the speed of the car. The speed of the waves depends only on the properties of the air, such as its temperature.

Although each postulate seems quite reasonable, many of the implications of the two together are quite surprising and contradict what is often called common sense. For example, one important implication of these postulates is that every observer measures the same value for the speed of light independent of the relative motion of the source and the observer. Consider a light source S and two observers, R_1 at rest relative to S and R_2 moving toward S with speed v, as shown in Figure 1-6a. The speed of light measured by R_1 is $c = 3 \times 10^8$ m/s. What is the speed measured by R_2? The answer is *not $c + v$*. By postulate 1, Figure 1-6a is equivalent to Figure 1-6b, in which R_2 is at rest and the source S and R_1 are moving with speed v. That is, since absolute motion cannot be detected, it is not possible to say which is really moving and which is at rest. By postulate 2, the speed of light from a moving source is independent of the motion of the source. Thus, looking at Figure 1-6b, we see that R_2 measures the speed of light to be c, just as R_1 does. This result is often considered as an alternative to Einstein's second postulate:

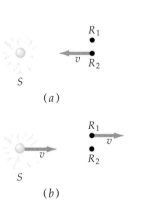

Figure 1-6 (a) A stationary light source S and a stationary observer R_1, with a second observer R_2 moving toward the source with speed v. (b) In the reference frame in which the observer R_2 is at rest, the light source S and observer R_1 move to the right with speed v. If absolute motion cannot be detected, the two views are equivalent. Since the speed of light does not depend on the motion of the source, observer R_2 measures the same value for that speed as observer R_1.

---

*_Annalen der Physik_, vol. 17, 1905, p. 841. For a translation from the original German, see W. Perrett and G.B. Jeffery (trans.), _The Principle of Relativity: A collection of Original Memoirs on the Special and General Theory of Relativity_ by H. A. Lorentz, A. Einstein, H. Minkowski, and W. Weyl, Dover, New York, 1923.

Postulate 2 (Alternate). Every observer measures the same value c for the speed of light.

This result contradicts our intuitive ideas about relative velocities. If a car moves at 50 km/h away from an observer and another car moves at 80 km/h in the same direction, the velocity of the second car relative to the first car is 30 km/h. This result is easily measured and conforms to our intuition. However, according to Einstein's postulates, if a light beam is moving in the direction of the cars, observers in both cars will measure the same speed for the light beam. Our intuitive ideas about the combination of velocities are approximations that hold only when the speeds are very small compared with the speed of light. Even in an airplane moving with the speed of sound, it is not possible to measure the speed of light accurately enough to distinguish the difference between the results c and $c + v$, where v is the speed of the plane. In order to make such a distinction, we must either move with a very great velocity (much greater than that of sound) or make extremely accurate measurements, as in the Michelson–Morley experiment.

1-4 The Lorentz Transformation

Einstein's postulates have important consequences for measuring time intervals and space intervals as well as relative velocities. Throughout this chapter we will be comparing measurements of the positions and times of events (such as lightning flashes) made by observers who are moving relative to each other. We will use a rectangular coordinate system xyz with origin O, called the S reference frame, and another system $x'y'z'$ with origin O', called the S' frame, that is moving with a constant velocity $\mathbf{V}$ relative to the S frame. Relative to the S' frame, the S frame is moving with a constant velocity $-\mathbf{V}$. For simplicity, we will consider the S' frame to be moving with speed V along the x axis in the positive x direction relative to S. Then, relative to S', the S frame is moving with speed V in the negative x' direction along the x' axis. In each frame, we will assume that there are as many observers as are needed who are equipped with measuring devices, such as clocks and metersticks, that are identical when compared at rest (see Figure 1-7).

We need many observers, for example, to determine the times of events. If one observer is distant from an event, then his time observation can be thrown off by the time it takes for the information about the event to travel to his location (such as the transit time for a light pulse). The observer can avoid such problems by recording only events *local* to him, and leaving other events to other observers at those locations. It's like having one official at the beginning of a racetrack and another at the end.

We will use Einstein's postulates to find the general relation between the coordinates x, y, and z and the time t of an event as seen in reference frame S and the coordinates x', y', and z' and the time t' of the same event as seen in reference frame S', which is moving with uniform velocity relative to S. We will consider only the simple case in which the origins are coincident at time $t = t' = 0$. The classical relation, called the **galilean transformation,** is

$$x = x' + Vt' \qquad y = y' \qquad z = z' \qquad t = t' \qquad \text{1-8}a$$

The inverse transformation is

$$x' = x - Vt \qquad y' = y \qquad z' = z \qquad t' = t \qquad \text{1-8}b$$

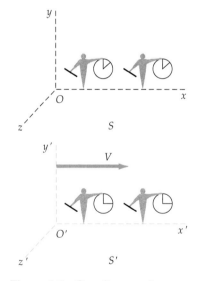

Figure 1-7 Coordinate reference frames S and S' moving with relative speed V. In each frame, there are observers with metersticks and clocks that are identical when compared at rest.

Galilean transformation

These equations are consistent with experimental observations as long as V is much less than c. They lead to the familiar classical addition law for velocities. If a particle has velocity $u_x = dx/dt$ in frame S, its velocity in frame S' is

$$u'_x = \frac{dx'}{dt'} = \frac{dx'}{dt} = \frac{dx}{dt} - V = u_x - V \qquad 1\text{-}9$$

If we differentiate this equation again, we find that the acceleration of the particle is the same in both frames:

$$a_x = du_x/dt = du'_x/dt' = a'_x$$

It should be clear that the galilean transformation is not consistent with Einstein's postulates of special relativity. If light moves along the x axis with speed c in S, these equations imply that the speed in S' is $u'_x = c - V$; rather than $u'_x = c$, which is consistent with Einstein's postulates and with experiment. The classical transformation equations must therefore be modified to make them consistent with Einstein's postulates. We will give a brief outline of one method of obtaining the relativistic transformation.

We assume that the relativistic transformation equation for x is the same as the classical equation (Equation 1-8a) except for a constant multiplier on the right side. That is, we assume the equation is of the form

$$x = \gamma(x' + Vt') \qquad 1\text{-}10$$

where γ is a constant that can depend on V and c but not on the coordinates. The inverse transformation must look the same except for the sign of the velocity:

$$x' = \gamma(x - Vt) \qquad 1\text{-}11$$

Let us consider a light pulse that starts at the origin of S at $t = 0$. Since we have assumed that the origins are coincident at $t = t' = 0$, the pulse also starts at the origin of S' at $t' = 0$. Einstein's postulates require that the equation for the x component of the wavefront of the light pulse is $x = ct$ in frame S and $x' = ct'$ in frame S'. Substituting ct for x and ct' for x' in Equations 1-10 and 1-11, we obtain

$$ct = \gamma(ct' + Vt') = \gamma(c + V)t' \qquad 1\text{-}12$$

and

$$ct' = \gamma(ct - Vt) = \gamma(c - V)t \qquad 1\text{-}13$$

We can eliminate either t' or t from these two equations and determine γ. We get

$$\gamma^2 = (1 - V^2/c^2)^{-1}$$

$$\gamma = \frac{1}{\sqrt{1 - V^2/c^2}} \qquad 1\text{-}14$$

(It is important to note that γ is always greater than 1 and that when V is much less than c, $\gamma \approx 1$.) The relativistic transformation for x and x' is therefore given by Equations 1-10 and 1-11 with γ given by Equation 1-14. We can obtain equations for t and t' by combining Equation 1-10 with the inverse transformation given by Equation 1-11. Substituting $x = \gamma(x' + Vt')$ for x in Equation 1-11, we obtain

$$x' = \gamma[\gamma(x' + Vt') - Vt] \qquad 1\text{-}15$$

which can be solved for t in terms of x' and t'. The complete relativistic transformation is

$$x = \gamma(x' + Vt') \qquad y = y' \qquad z = z' \qquad \text{1-16}$$

$$t = \gamma\left(t' + \frac{Vx'}{c^2}\right) \qquad \text{1-17}$$

Lorentz transformation

The inverse is

$$x' = \gamma(x - Vt) \qquad y' = y \qquad z' = z \qquad \text{1-18}$$

$$t' = \gamma\left(t - \frac{Vx}{c^2}\right) \qquad \text{1-19}$$

The transformation described by Equations 1-16 through 1-19 is called the **Lorentz transformation**. It relates the space and time coordinates x, y, z, and t of an event in frame S to the coordinates x', y', z', and t' of the same event as seen in frame S', which is moving along the x axis with speed V relative to frame S.

We will now look at some applications of the Lorentz transformation.

Time Dilation

An important consequence of Einstein's postulates and the Lorentz transformation is that the time interval between two events that occur at the same place in some reference frame is always less than the time interval between the same events that is measured in another reference frame in which the events occur at different places. Consider two events that occur at x'_0 at times t'_1 and t'_2 in frame S'. We can find the times t_1 and t_2 for these events in S from Equation 1-17. We have

$$t_1 = \gamma\left(t'_1 + \frac{Vx'_0}{c^2}\right)$$

and

$$t_2 = \gamma\left(t'_2 + \frac{Vx'_0}{c^2}\right)$$

so

$$t_2 - t_1 = \gamma(t'_2 - t'_1)$$

The time between events that happen at the *same place* in a reference frame is called **proper time** t_p. In this case, the time interval $\Delta t_p = t'_2 - t'_1$ measured in frame S' is proper time. The time interval Δt measured in any other reference frame is always longer than the proper time. This expansion is called **time dilation**:

$$\Delta t = \gamma \Delta t_p \qquad \text{1-20}$$

Time dilation

Example 1-1

Two events occur at the same point x'_0 at times t'_1 and t'_2 in frame S', which is traveling at speed V relative to frame S. What is the spatial separation of these events in frame S?

From Equation 1-16, we have

$$x_1 = \gamma(x_0' + Vt_1')$$

and

$$x_2 = \gamma(x_0' + Vt_2')$$

Then

$$x_2 - x_1 = \gamma V(t_2' - t_1')$$
$$= V(t_2 - t_1)$$

The spatial separation of these events in S is the distance a single point, such as x_0' in S', moves in S during the time interval between the events.

We can understand time dilation directly from Einstein's postulates without using the Lorentz transformation. Figure 1-8a shows an observer A' a distance D from a mirror. The observer and the mirror are in a spaceship that is at rest in frame S'. He explodes a flash gun and measures the time interval $\Delta t'$ between the original flash and his seeing the return flash from the mirror. Since light travels with speed c, this time is

$$\Delta t' = \frac{2D}{c}$$

We now consider these same two events, the original flash of light and the receiving of the return flash, as observed in reference frame S, in which observer A' and the mirror are moving to the right with speed V as shown in Figure 1-8b. The events happen at two different places x_1 and x_2 in frame S. During the time interval Δt (as measured in S) between the original flash and the return flash, observer A' and his spaceship have moved a horizontal distance $V \Delta t$. In Figure 1-8b, we can see that the path traveled by the light is longer in S than in S'. However, by Einstein's postulates, light travels with the same speed c in frame S as it does in frame S'. Since it travels farther in S at the same speed, it takes longer in S to reach the mirror and return. The time interval in S is thus longer than it is in S'. From the triangle in Figure 1-8c, we have

$$\left(\frac{c \Delta t}{2}\right)^2 = D^2 + \left(\frac{V \Delta t}{2}\right)^2$$

or

$$\Delta t = \frac{2D}{\sqrt{c^2 - V^2}} = \frac{2D}{c}\frac{1}{\sqrt{1 - V^2/c^2}}$$

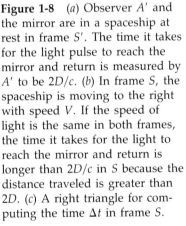
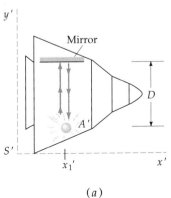
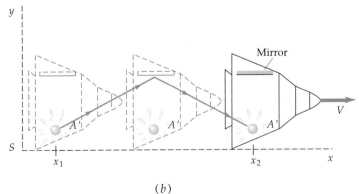

Figure 1-8 (a) Observer A' and the mirror are in a spaceship at rest in frame S'. The time it takes for the light pulse to reach the mirror and return is measured by A' to be $2D/c$. (b) In frame S, the spaceship is moving to the right with speed V. If the speed of light is the same in both frames, the time it takes for the light to reach the mirror and return is longer than $2D/c$ in S because the distance traveled is greater than $2D$. (c) A right triangle for computing the time Δt in frame S.

Using $\Delta t' = 2D/c$, we obtain

$$\Delta t = \frac{\Delta t'}{\sqrt{1 - V^2/c^2}} = \gamma \Delta t'$$

Example 1-2

Astronauts in a spaceship traveling away from the earth at $V = 0.6c$ sign off from space control, saying that they are going to nap for 1 hour and then call back. How long does their nap last as measured on earth?

Since the astronauts go to sleep and wake up at the same place in their reference frame, the time interval for their nap of 1 hour as measured by them is proper time. In the earth's reference frame, they move a considerable distance between these two events. The time interval measured in the earth's frame (using two clocks located at those events) is longer by the factor γ. With $V = 0.6c$, we have

$$1 - \frac{V^2}{c^2} = 1 - (0.6)^2 = 0.64$$

Then γ is

$$\gamma = \frac{1}{\sqrt{1 - V^2/c^2}} = \frac{1}{\sqrt{0.64}} = \frac{1}{0.8} = 1.25$$

The nap thus lasts for 1.25 hours as measured on earth.

Exercise

If the spaceship in Example 1-2 is moving at $V = 0.8c$, how long would a 1-hour nap last as measured on earth? (Answer: 1.67 h)

Length Contraction

A phenomenon closely related to time dilation is **length contraction**. The length of an object measured in the reference frame in which the object is at rest is called its **proper length** L_p. In a reference frame in which the object is moving, the measured length is shorter than its proper length. Consider a rod at rest in frame S' with one end at x'_2 and the other end at x'_1. The length of the rod in this frame is its proper length $L_p = x'_2 - x'_1$. Some care must be taken to find the length of the rod in frame S. In this frame, the rod is moving to the right with speed V, the speed of frame S'. The length of the rod in frame S is *defined* as $L = x_2 - x_1$, where x_2 is the position of one end at some time t_2, and x_1 is the position of the other end *at the same time* $t_1 = t_2$ as measured in frame S. Equation 1-18 is convenient to use to calculate $x_2 - x_1$ at some time t because it relates x, x' and t, whereas Equation 1-16 is not convenient because it relates x, x', and t':

$$x'_2 = \gamma(x_2 - Vt_2)$$

and

$$x'_1 = \gamma(x_1 - Vt_1)$$

Since $t_2 = t_1$, we obtain

$$x'_2 - x'_1 = \gamma(x_2 - x_1)$$

$$x_2 - x_1 = \frac{1}{\gamma}(x'_2 - x'_1) = \sqrt{1 - V^2/c^2}\,(x'_2 - x'_1)$$

Length contraction

$$L = \frac{1}{\gamma} L_p = \sqrt{1 - V^2/c^2}\, L_p \qquad 1\text{-}21$$

Thus the length of a rod is smaller when it is measured in a frame in which it is moving. Before Einstein's paper was published, Lorentz and FitzGerald tried to explain the null result of the Michelson–Morley experiment by assuming that distances in the direction of motion contracted by the amount given in Equation 1-21. This contraction is now known as the **Lorentz–FitzGerald contraction**.

Example 1-3

A stick that has a proper length of 1 m moves in a direction along its length with speed V relative to you. The length of the stick as measured by you is 0.914 m. What is the speed V?

The length of the stick measured in a frame in which it is moving with speed V is related to its proper length by Equation 1-21:

$$L = \frac{L_p}{\gamma}$$

Then

$$\gamma = \frac{L_p}{L} = \frac{1\text{ m}}{0.914\text{ m}} = \frac{1}{\sqrt{1 - V^2/c^2}} = 1.094$$

$$\sqrt{1 - V^2/c^2} = 0.914$$

$$1 - \frac{V^2}{c^2} = (0.914)^2 = 0.835$$

$$\frac{V^2}{c^2} = 1 - 0.835 = 0.165$$

$$V = 0.406c$$

An interesting example of time dilation or length contraction is afforded by the appearance of muons as secondary radiation from cosmic rays. Muons decay according to the statistical law of radioactivity:

$$N(t) = N_0\, e^{-t/\tau} \qquad 1\text{-}22$$

where N_0 is the original number of muons at time $t = 0$, $N(t)$ is the number remaining at time t, and τ is the mean lifetime, which is about 2 μs for muons at rest. Since muons are created (from the decay of pions) high in the atmosphere, usually several thousand meters above sea level, few muons should reach sea level. A typical muon moving with speed $0.998c$ would travel only about 600 m in 2 μs. However, the lifetime of the muon measured in the earth's reference frame is increased by the factor $1/\sqrt{1 - V^2/c^2}$, which is 15 for this particular speed. The mean lifetime measured in the earth's reference frame is therefore 30 μs, and a muon with speed $0.998c$ travels about 9000 m in this time. From the muon's point of view, it lives only 2 μs, but the atmosphere is rushing past it with a speed of $0.998c$. The distance of 9000 m in the earth's frame is thus contracted to only 600 m in the muon's frame as indicated in Figure 1-9.

It is easy to distinguish experimentally between the classical and relativistic predictions of the observation of muons at sea level. Suppose that we

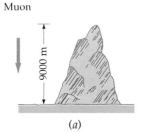

(a)

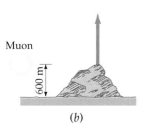

(b)

Figure 1-9 Although muons are created high above the earth and their mean lifetime is only about 2 μs when at rest, many appear at the earth's surface. (a) In the earth's reference frame, a typical muon moving at $0.998c$ has a mean lifetime of 30 μs and travels 9000 m in this time. (b) In the reference frame of the muon, the distance traveled by the earth is only 600 m in the muon's lifetime of 2 μs.

observe 10^8 muons at an altitude of 9000 m in some time interval with a muon detector. How many would we expect to observe at sea level in the same time interval? According to the nonrelativistic prediction, the time it takes for these muons to travel 9000 m is $(9000 \text{ m})/0.998c \approx 30 \ \mu\text{s}$, which is 15 lifetimes. Substituting $N_0 = 10^8$ and $t = 15\tau$ into Equation 1-22, we obtain

$$N = 10^8 \ e^{-15} = 30.6$$

We would thus expect all but about 31 of the original 100 million muons to decay before reaching sea level.

According to the relativistic prediction, the earth must travel only the contracted distance of 600 m in the rest frame of the muon. This takes only $2 \ \mu\text{s} = 1\tau$. Therefore the number of muons expected at sea level is

$$N = 10^8 \ e^{-1} = 3.68 \times 10^7$$

Thus relativity predicts that we would observe 36.8 million muons in the same time interval. Experiments of this type have confirmed the relativistic predictions.

Question

1. You are standing on a corner and a friend is driving past in an automobile. Both of you note the times when the car passes two different intersections and determine from your watch readings the time that elapses between the two events. Which of you has determined the proper time interval?

1-5 Clock Synchronization and Simultaneity

We saw in Section 1-4 that proper time is the time interval between two events that occur at the same point in some reference frame. It can therefore be measured on a single clock. However, in another reference frame moving relative to the first, the same two events occur at different places, so two clocks are needed to record the times. The time of each event is measured on a different clock, and the interval is found by subtraction. This procedure requires that the clocks be **synchronized.** We will show in this section that

> Two clocks that are synchronized in one reference frame are not synchronized in any other frame moving relative to the first frame.

A corollary to this result is

> Two events that are simultaneous in one reference frame are not simultaneous in another frame moving relative to the first.

(This is true unless the events and clocks are in the same plane and are perpendicular to the relative motion). Comprehension of these facts usually resolves all relativity paradoxes. Unfortunately, the intuitive (and incorrect) belief that simultaneity is an absolute relation is difficult to overcome.

Suppose we have two clocks at rest at points A and B a distance L apart in frame S. How can we synchronize these two clocks? If an observer at A looks at the clock at B and sets her clock to read the same time, the clocks will not be synchronized because of the time L/c it takes light to travel from one clock to another. To synchronize the clocks, the observer at A must set her clock ahead by the time L/c. Then she will see that the clock at B reads a time that is L/c behind the time on her clock, but she will calculate that the clocks are synchronized when she allows for the time L/c for the light to reach her. Any other observers except those equidistant from the clocks will see the clocks reading different times, but they will also calculate that the clocks are synchronized when they correct for the time it takes the light to reach them. An equivalent method for synchronizing two clocks would be for an observer C at a point midway between the clocks to send a light signal and for observers at A and B to set their clocks to some prearranged time when they receive the signal.

We now examine the question of **simultaneity.** Suppose A and B agree to explode flashguns at t_0 (having previously synchronized their clocks). Observer C will see the light from the two flashes at the same time, and since he is equidistant from A and B, he will conclude that the flashes were simultaneous. Other observers in frame S will see the light from A or B first, depending on their location, but after correcting for the time the light takes to reach them, they also will conclude that the flashes were simultaneous. We can thus define simultaneity as follows:

> Two events in a reference frame are simultaneous if light signals from the events reach an observer halfway between the events at the same time.

To show that two events that are simultaneous in frame S are not simultaneous in another frame S' moving relative to S, we will use an example introduced by Einstein. A train is moving with speed V past a station platform. We will consider the train to be at rest in S' and the platform to be at rest in S. We have observers A', B', and C' at the front, back, and middle of the train. We now suppose that the train and platform are struck by lightning at the front and back of the train and that the lightning bolts are simultaneous in the frame of the platform (S) (Figure 1-10). That is, an observer C on the platform halfway between the positions A and B, where the lightning strikes, sees the two flashes at the same time. It is convenient to suppose that the lightning scorches the train and platform so that the events can be easily located in each reference frame. Since C' is in the middle of the train, halfway between the places on the train that are scorched, the events are simultaneous in S' only if C' sees the flashes at the same time. However, the flash from the front of the train is seen by C' before the flash from the back of the train. We can understand this by considering the motion of C' as seen in frame S (Figure 1-11). By the time the light from the front flash reaches C', C' has moved some distance toward the front flash and some distance away from the back flash. Thus the light from the back flash has not yet reached C' as indicated in the figure. Observer C' must therefore conclude that the events are not simultaneous and that the front of the train was struck before the back. Furthermore, all observers in S' on the train will agree with C' when they have corrected for the time it takes the light to reach them.

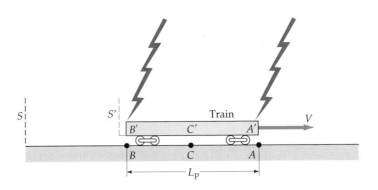

Figure 1-10 Simultaneous lightning bolts strike the ends of a train traveling with speed V in frame S attached to the platform. The light from these simultaneous events reaches observer C midway between the events at the same time. The distance between the bolts is $L_{p,\text{platform}}$.

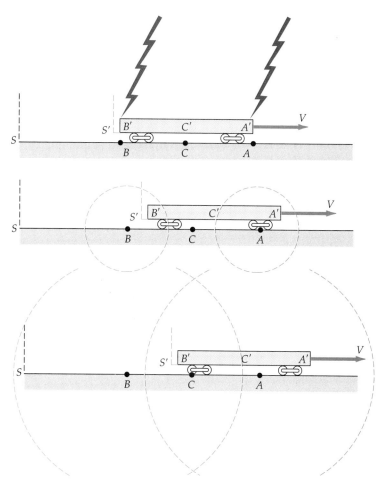

Figure 1-11 The light from the lightning bolt at the front of the train reaches observer C' at the middle of the train before that from the bolt at the back of the train. Since C' is midway between the events (which occur at the front and rear of the train), these events are not simultaneous for him.

Let $L_{p,\text{train}}$ be the proper length of the train, that is, the length of the train as measured in S' in which it is at rest. Also, let $L_{p,\text{platform}}$ be the proper length of the platform, that is, the distance between the scorch marks as seen in S. Since the scorch marks on the platform coincide with the front and back of the train at the instant (in S) that the lightning bolts strike, the distance between the scorch marks $L_{p,\text{platform}}$ equals the length of the train L_T as measured in frame S in which it is moving. This length is smaller than the proper length of the train because of length contraction; that is, $L_T = L_{p,\text{platform}} < L_{p,\text{train}}$.

Figure 1-12 shows the events of the lightning bolts as seen in the reference frame of the train (S') in which the train is at rest and the platform is moving. In this frame, the distance between the burns on the platform is contracted, so the platform is shorter than it is in S, and the train is at rest, so the train is longer than it is in S. When the lightning bolt strikes the front of the train at A', the front of the train is at point A, and the back of the train has not yet reached point B. Later, when the lightning bolt strikes the back of the train at B', the back has reached point B on the platform.

Figure 1-12 The lightning bolts of Figure 1-10 as seen in frame S' of the train. In this frame, the distance between A and B on the platform is less than $L_{p,\text{platform}}$, and the proper length of the train $L_{p,\text{train}}$ is longer than $L_{p,\text{platform}}$. The first lightning bolt strikes the front of the train when A' and A are coincident. The second bolt strikes the rear of the train when B' and B are coincident.

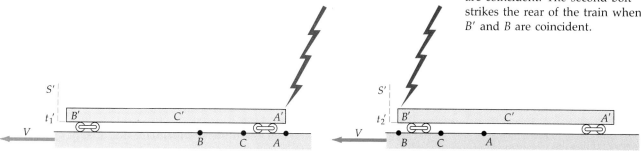

In reference frame S, the lightning bolts strike A and B simultaneously. Suppose there are clocks on the platform at A and B that are synchronized in frame S. From the point of view of frame S' attached to the train, the clocks and the platform are moving past the train. Lightning first strikes the front of the train, which is at point A, and some time later, lightning strikes the back of the train, which is now at point B. The moving clocks are thus not synchronized as seen from frame S'. If the clock at A reads 12:00 noon when the lightning bolt strikes A, the clock at B must read some time before 12:00 noon at that time. The clock at B reads 12:00 noon later when it reaches the back of the train and the lightning bolt strikes at B. Another way of saying this is that the clock at A leads the clock at B as seen in S'. In frame S', we will call the clock at A the "chasing" clock because in that frame the two clocks are moving in the negative x' direction with clock A at x_2 following, or chasing, clock B at x_1.

The time discrepancy of two clocks that are synchronized in frame S as seen in frame S' can be found from the Lorentz transformation equations. Suppose we have clocks at points x_1 and x_2 that are synchronized in S. What are the times t_1 and t_2 on these clocks as observed from frame S' at a time t'_0? From Equation 1-19, we have

$$t'_0 = \gamma \left(t_1 - \frac{Vx_1}{c^2} \right)$$

and

$$t'_0 = \gamma \left(t_2 - \frac{Vx_2}{c^2} \right)$$

Then

$$t_2 - t_1 = \frac{V}{c^2}(x_2 - x_1) \qquad 1\text{-}23$$

Note that the chasing clock (at x_2) leads the other (at x_1) by an amount that is proportional to their proper separation $x_2 - x_1$.

> If two clocks are synchronized in the frame in which they are at rest, they will be out of synchronization in another frame. In the frame in which they are moving, the chasing clock leads (shows a later time) by an amount
>
> $$\Delta t_s = L_p \frac{V}{c^2}$$
>
> where L_p is the proper distance between the clocks.

A numerical example should help clarify time dilation, clock synchronization, and the internal consistency of these results.

Example 1-4

An observer in a spaceship has a flash gun and a mirror (as in our time dilation example in Figure 1-8). The distance from the gun to the mirror is 15 light-minutes (written $15c\cdot\text{min}$) and the spaceship travels with speed $V = 0.8c$. The spaceship travels past a very long space platform that has two synchronized clocks, one at the position of the spaceship when the observer explodes the flash gun and the other at the position of the spaceship when the light returns to the gun from the mirror. Find the time intervals between the events (exploding the flash gun and receiving the return flash from the mirror) in the frame of the ship and in the frame of the platform. Find the distance traveled by the ship, and the amount by which the clocks on the platform are out of synchronization as viewed by the ship.

We will call the frame of the spaceship S' and that of the platform S. In the spaceship, the light travels from the gun to the mirror and back, a total distance $D = 30c \cdot \text{min}$. The time it takes for light to travel $30c \cdot \text{min}$ is

$$\Delta t' = \frac{D}{c} = \frac{(30c \cdot \text{min})}{c} = 30 \text{ min}$$

Since these events happen at the same place in the spaceship, the time interval is proper time:

$$\Delta t_p = D/c = 30 \text{ min}$$

During this time, the platform travels backwards past the ship a distance equal to the distance L' between the platform clocks measured in frame S':

$$L' = \Delta x' = V \Delta t' = (0.8c)(30 \text{ min}) = 24c \cdot \text{min}$$

In frame S, the time between the events is longer by the factor γ. Since $V/c = 0.8$, $1 - V^2/c^2 = 1 - 0.64 = 0.36$. The factor γ is thus

$$\gamma = \frac{1}{\sqrt{1 - V^2/c^2}} = \frac{1}{\sqrt{0.36}} = \frac{1}{0.6} = \frac{5}{3}$$

The time between the events as observed in frame S is therefore

$$\Delta t = \gamma \Delta t_p = \frac{5}{3}(30 \text{ min}) = 50 \text{ min}$$

During this time, the spaceship travels a distance in frame S equal to the proper distance between the platform clocks:

$$L_p = \Delta x = V \Delta t = (0.8c)(50 \text{ min}) = 40c \cdot \text{min}$$

Note that this distance is longer than the contracted distance between the clocks that is measured by observers in frame S' of the spaceship.

Observers on the platform would say that the spaceship's clock is running slow because it records a time of only 30 min between the events, whereas the time measured on the platform is 50 min.

Figure 1-13 shows the situation viewed from the spaceship in S'. The platform is traveling past the ship with speed $0.8c$. There is a clock at point x_1, which coincides with the ship when the flash gun is exploded, and another at point x_2, which coincides with the ship when the return flash is received from the mirror. We assume that the clock at x_1 reads 12:00 noon at the time of the light flash. The clocks at x_1 and x_2 are synchronized in S but not in S'. In S', the clock at x_2, which is chasing the one at x_1, leads by

$$\frac{L_p V}{c^2} = \frac{(40c \cdot \text{min})(0.8c)}{c^2} = 32 \text{ min}$$

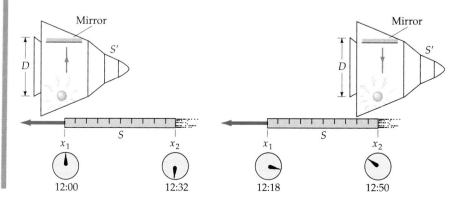

Figure 1-13 Example 1-4. Clocks on a platform as observed from the spaceship's frame of reference S'. During the time $\Delta t' = 30$ min it takes for the platform to pass the spaceship, the clocks on the platform run slow and tick off $(30 \text{ min})/\gamma = 18$ min. But the clocks are unsynchronized, with the chasing clock leading by $L_p V/c^2$, which for this case is 32 min. The time it takes for the spaceship to pass as measured on the platform is therefore 32 min + 18 min = 50 min.

> When the spaceship coincides with x_2, the clock there reads 12:50. The time between the events is therefore 50 min in S. Note that according to observers in S', this clock ticks off 50 min $-$ 32 min $=$ 18 min for a trip that takes 30 min in S'. Thus, observers in S' see this clock run slow by the factor 30/18 = 5/3.
>
> Every observer in one frame sees the clocks in the other frame run slow. According to observers in S, who measure 50 min for the time interval, the time interval in S' (30 min) is too small, so they see the single clock in S' run too slow by the factor 5/3. According to the observers in S', the observers in S measure a time that is too *long* despite the fact that their clocks run too slow because the clocks in S are out of synchronization. The clocks tick off only 18 min, but the second one leads the first by 32 min, so the time interval is 50 min.

Questions

2. Two observers are in relative motion. In what circumstances can they agree on the simultaneity of two different events?
3. If event A occurs before event B in some frame, might it be possible for there to be a reference frame in which event B occurs before event A?
4. Two events are simultaneous in a frame in which they also occur at the same point in space. Are they simultaneous in other reference frames?

1-6 The Doppler Effect

In the doppler effect for sound, the change in frequency for a given velocity V depends on whether it is the source or the receiver that is moving with that speed. Such a distinction is possible for sound because there is a medium (the air) relative to which the motion takes place, so it is not surprising that the motion of the source or the receiver relative to the still air can be distinguished. Such a distinction between motion of the source or receiver cannot be made for light or other electromagnetic waves in a vacuum. Therefore, the expressions for the doppler effect for sound cannot be correct for light. We will now derive the relativistic doppler-effect equations that are correct for light.

We will consider a source moving toward a receiver with velocity V, and we will work in the frame of the receiver. Let the source emit N electromagnetic waves. The first wave will travel a distance $c\,\Delta t_R$ and the source will travel a distance $V\,\Delta t_R$ in the time Δt_R measured in the frame of the receiver. The wavelength will be

$$\lambda' = \frac{(c\,\Delta t_R - V\,\Delta t_R)}{N}$$

The frequency f' observed by the receiver will therefore be

$$f' = \frac{c}{\lambda'} = \frac{c}{c - V}\frac{N}{\Delta t_R}$$

$$= \frac{1}{1 - V/c}\frac{N}{\Delta t_R}$$

If the frequency of the source is f_0, it will emit $N = f_0\,\Delta t_S$ waves in the time Δt_S measured by the source. Here Δt_S is the proper time interval (the first wave and the Nth wave are emitted at the same place in the source's reference frame). Times Δt_S and Δt_R are related by Equation 1-20 for time dilation, so $\Delta t_R = \gamma \Delta t_S$. Thus, when the source and receiver are moving toward one another we obtain

$$f' = \frac{1}{1 - V/c} \frac{f_0 \Delta t_S}{\Delta t_R} = \frac{f_0}{1 - V/c} \frac{1}{\gamma}$$

or

$$f' = \frac{\sqrt{1 - V^2/c^2}}{1 - V/c} f_0 = \sqrt{\frac{1 + V/c}{1 - V/c}} f_0 \qquad \text{approaching} \qquad 1\text{-}24a$$

This differs from our classical equation only in the time-dilation factor.

When the source and receiver are moving away from one another, the same analysis shows that the observed frequency is given by

$$f' = \frac{\sqrt{1 - V^2/c^2}}{1 + V/c} f_0 = \sqrt{\frac{1 - V/c}{1 + V/c}} f_0 \qquad \text{receding} \qquad 1\text{-}24b$$

It is left as a problem (see Problem 64) for you to show that the same results are obtained if the calculations are done in the reference frame of the source.

Example 1-5

The longest wavelength of light emitted by hydrogen in the Balmer series (see Chapter 2) has a wavelength of $\lambda_0 = 656$ nm. In light from a distant galaxy, this wavelength is measured to be $\lambda' = 1458$ nm. Find the speed at which the distant galaxy is receding from the earth.

If we substitute $f' = c/\lambda'$ and $f_0 = c/\lambda_0$ into Equation 1-24b, we obtain

$$\sqrt{\frac{1 - V/c}{1 + V/c}} = \frac{f'}{f_0} = \frac{\lambda_0}{\lambda'}$$

This equation is somewhat simplified if we use $\beta = V/c$. Then squaring the above equation and taking the reciprocal of each side, we obtain

$$\frac{1 + \beta}{1 - \beta} = \left(\frac{\lambda'}{\lambda_0}\right)^2 = \left(\frac{1458 \text{ nm}}{656 \text{ nm}}\right)^2 = 4.94$$

so

$$1 + \beta = 4.94 - 4.94 \beta$$

$$\beta = \frac{4.94 - 1}{4.94 + 1} = 0.663 = \frac{V}{c}$$

The galaxy is thus receding at a speed of $V = 0.663c$. The shift towards longer wavelengths of light from distant galaxies that are receding from us is called the **redshift**.

1-7 The Twin Paradox

Homer and Ulysses are identical twins. Ulysses travels at high speed to a planet beyond the solar system and returns while Homer remains at home. When they are together again, which twin is older, or are they the same age? The correct answer is that Homer, the twin who stays at home, is older. This problem, with variations, has been subject of spirited debate for decades, though there are very few who disagree with the answer.* The problem is a paradox because of the seemingly symmetric roles played by the twins with the asymmetric result in their aging. The paradox is resolved when the asymmetry of the twins' roles is noted. The relativistic result conflicts with common sense based on our strong but incorrect belief in absolute simultaneity. We will consider a particular case with some numerical magnitudes that, though impractical, make the calculations easy.

Let planet P and Homer on earth be at rest in reference frame S a distance L_p apart, as illustrated in Figure 1-14. We neglect the motion of the earth. Reference frames S' and S'' are moving with speed V toward and away from the planet, respectively. Ulysses quickly accelerates to speed V, then coasts in S' until he reaches the planet, where he stops and is momentarily at rest in S. To return he quickly accelerates to speed V toward earth and then coasts in S'' until he reaches earth, where he stops. We can assume that the acceleration times are negligible compared with the coasting times. We use the following values for illustration: $L_p = 8$ light-years and $V = 0.8c$. Then $\sqrt{1 - V^2/c^2} = 3/5$ and $\gamma = 5/3$.

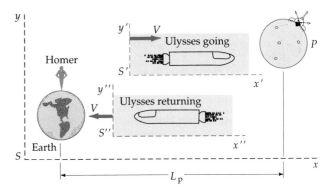

Figure 1-14 The twin paradox. The earth and a distant planet are fixed in frame S. Ulysses coasts in frame S' to the planet and then coasts back in frame S''. His twin Homer stays on earth. When Ulysses returns, he is younger than his twin. The roles played by the twins are not symmetric. Homer remains in one inertial reference frame, but Ulysses must accelerate if he is to return home.

It is easy to analyze the problem from Homer's point of view on earth. According to Homer's clock, Ulysses coasts in S' for a time $L_p/V = 10$ y and in S'' for an equal time. Thus Homer is 20 y older when Ulysses returns. The time interval in S' between Ulysses' leaving earth and his arriving at the planet is shorter because it is proper time. The time it takes to reach the planet by Ulysses' clock is

$$\Delta t' = \frac{\Delta t}{\gamma} = \frac{10 \text{ y}}{5/3} = 6 \text{ y}$$

Since the same time is required for the return trip, Ulysses will have recorded 12 y for the round trip and will be 8 y younger than Homer upon his return.

From Ulysses' point of view, the distance from the earth to the planet is contracted and is only

$$L' = \frac{L_p}{\gamma} = \frac{8 \text{ light-years}}{5/3} = 4.8 \text{ light-years}$$

At $V = 0.8c$, it takes only 6 y each way.

The real difficulty in this problem is for Ulysses to understand why his twin aged 20 y during his absence. If we consider Ulysses as being at rest and Homer as moving away, Homer's clock should run slow and measure

---

*A collection of some important papers concerning this paradox can be found in *Special Relativity Theory, Selected Reprints*, American Association of Physics Teachers, New York, 1963.

only 3/5(6) = 3.6 y. Then why shouldn't Homer age only 7.2 y during the round trip? This, of course, is the paradox. The difficulty with the analysis from the point of view of Ulysses is that he does not remain in an inertial frame. What happens while Ulysses is stopping and starting? To investigate this problem in detail, we would need to treat accelerated reference frames, a subject dealt with in the study of general relativity and beyond the scope of this book. However, we can get some insight into the problem by having the twins send regular signals to each other so that they can record the other's age continuously. If they arrange to send a signal once a year, each can determine the age of the other merely by counting the signals received. The arrival frequency of the signals will not be 1 per year because of the doppler shift. The frequency observed will be given by Equations 1-24a and 1-24b. Using $V/c = 0.8$ and $V^2/c^2 = 0.64$, we have for the case in which the twins are receding from each other

$$f' = \frac{\sqrt{1 - V^2/c^2}}{1 + V/c} f_0 = \frac{\sqrt{1 - 0.64}}{1 + 0.8} f_0 = \frac{1}{3} f_0$$

When they are approaching, Equation 1-24 gives $f' = 3f_0$.

Consider the situation first from the point of view of Ulysses. During the 6 y it takes him to reach the planet (remember that the distance is contracted in his frame), he receives signals at the rate of $\frac{1}{3}$ signal per year, and so he receives 2 signals. As soon as he turns around and starts back to earth, he begins to receive 3 signals per year. In the 6 y it takes him to return he receives 18 signals, giving a total of 20 for the trip. He accordingly expects his twin to have aged 20 years.

We now consider the situation from Homer's point of view. He receives signals at the rate of $\frac{1}{3}$ signal per year not only for the 10 y it takes Ulysses to reach the planet but also for the time it takes for the last signal sent by Ulysses before he turns around to get back to earth. (He cannot know that Ulysses has turned around until the signals begin reaching him with increased frequency.) Since the planet is 8 light-years away, there is an additional 8 y of receiving signals at the rate of $\frac{1}{3}$ signal per year. During the first 18 y, Homer receives 6 signals. In the final 2 y before Ulysses arrives, Homer receives 6 signals, or 3 per year. (The first signal sent after Ulysses turns around takes 8 y to reach earth, whereas Ulysses, traveling at $0.8c$, takes 10 y to return and therefore arrives just 2 y after Homer begins to receive signals at the faster rate.) Thus Homer expects Ulysses to have aged 12 y. In this analysis, the asymmetry of the twins' roles is apparent. When they are together again, both twins agree that the one who has been accelerated will be younger than the one who stayed home.

The predictions of the special theory of relativity concerning the twin paradox have been tested using small particles that can be accelerated to such large speeds that γ is appreciably greater than 1. Unstable particles can be accelerated and trapped in circular orbits in a magnetic field, for example, and their lifetimes can then be compared with those of identical particles at rest. In all such experiments, the accelerated particles live longer on the average than those at rest, as predicted. These predictions have also been confirmed by the results of an experiment in which high-precision atomic clocks were flown around the world in commercial airplanes, but the analysis of this experiment is complicated due to the necessity of including gravitational effects treated in the general theory of relativity.*

---

*The details of this experiment can be found in J. C. Hafele and Richard E. Keating, "Around-the-world Atomic Clocks: Predicted Relativistic Time Gains" and "Around-the-world Atomic Clocks: Observed Relativistic Time Gains," *Science*, July 14, 1972, p. 166.

1-8 The Velocity Transformation

We can find how velocities transform from one reference frame to another by differentiating the Lorentz transformation equations. Suppose a particle has velocity $u'_x = dx'/dt'$ in frame S', which is moving to the right with speed V relative to frame S. Its velocity in frame S is

$$u_x = \frac{dx}{dt}$$

From the Lorentz transformation equations (Equations 1-16 and 1-17), we have

$$dx = \gamma(dx' + V\,dt')$$

and

$$dt = \gamma\left(dt' + \frac{V\,dx'}{c^2}\right)$$

The velocity in S is thus

$$u_x = \frac{dx}{dt} = \frac{\gamma(dx' + V\,dt')}{\gamma\left(dt' + \dfrac{V\,dx'}{c^2}\right)} = \frac{\dfrac{dx'}{dt'} + V}{1 + \dfrac{V}{c^2}\dfrac{dx'}{dt'}} = \frac{u'_x + V}{1 + Vu'_x/c^2}$$

If a particle has components of velocity along the y or z axes, we can use the same relation between dt and dt', with $dy = dy'$ and $dz = dz'$, to obtain

$$u_y = \frac{dy}{dt} = \frac{dy'}{\gamma\left(dt' + \dfrac{V\,dx'}{c^2}\right)} = \frac{dy'/dt'}{\gamma\left(1 + \dfrac{V}{c^2}\dfrac{dx'}{dt'}\right)} = \frac{u'_y}{\gamma\left(1 + \dfrac{Vu'_x}{c^2}\right)}$$

and

$$u_z = \frac{u'_z}{\gamma\left(1 + \dfrac{Vu'_x}{c^2}\right)}$$

The complete relativistic velocity transformation is

Relativistic velocity transformation

$$u_x = \frac{u'_x + V}{1 + Vu'_x/c^2} \qquad \text{1-25}a$$

$$u_y = \frac{u'_y}{\gamma(1 + Vu'_x/c^2)} \qquad \text{1-25}b$$

$$u_z = \frac{u'_z}{\gamma(1 + Vu'_x/c^2)} \qquad \text{1-25}c$$

The inverse velocity transformation equations are

$$u'_x = \frac{u_x - V}{1 - Vu_x/c^2} \qquad \text{1-26}a$$

$$u'_y = \frac{u_y}{\gamma(1 - Vu_x/c^2)} \qquad \text{1-26}b$$

$$u'_z = \frac{u_z}{\gamma(1 - Vu_x/c^2)} \qquad \text{1-26}c$$

These equations differ from the classical and intuitive result $u_x = u'_x + V$, $u_y = u'_y$ and $u_z = u'_z$ because the denominators in Equations 1-25 and 1-26 are not equal to 1. When V and u'_x are small compared with the speed of light c, $\gamma \approx 1$ and $Vu'_x/c^2 \ll 1$. Then the relativistic and classical expressions are the same.

Example 1-6

A supersonic plane moves with speed 1000 m/s (about 3 times the speed of sound) along the x axis relative to you. Another plane moves along the x axis at speed 500 m/s relative to the first plane. How fast is the second plane moving relative to you?

According to the classical formula for combining velocities, the speed of the second plane relative to you is 1000 m/s + 500 m/s = 1500 m/s. If we assume that you are at rest in the S reference frame and that the first plane is at rest in the S' frame, which is moving at $V = 1000$ m/s relative to S, the second plane has velocity $u'_x = 500$ m/s in S'. The correction term for u_x in the denominator of Equation 1-25a is then

$$\frac{Vu'_x}{c^2} = \frac{(1000)(500)}{(3 \times 10^8)^2} \approx 5 \times 10^{-12}$$

This correction term is so small that the classical and relativistic results are essentially the same.

Example 1-7

Work Example 1-6 if the first plane moves with speed $V = 0.8c$ relative to you and the second plane moves with the same speed $0.8c$ relative to the first plane.

In this case, the correction term is

$$\frac{Vu'_x}{c^2} = \frac{(0.8c)(0.8c)}{c^2} = 0.64$$

The speed of the second plane in frame S is then

$$u_x = \frac{0.8c + 0.8c}{1 + 0.64} = 0.98c$$

This is quite different from the classically expected result of $0.8c + 0.8c = 1.6c$. In fact, it can be shown from Equation 1-25 that if the speed of an object is less than c in one frame, it is less than c in all other frames moving relative to that frame with a speed less than c. We will see in Section 1-10 that it takes an infinite amount of energy to accelerate a particle to the speed of light. The speed of light c is thus an upper, unattainable limit for the speed of a particle having mass. (Massless particles, such as photons, always move at the speed of light.)

Example 1-8

Light moves along the x axis with speed $u_x = c$. What is its speed in S'?

From Equation 1-26a, we have

$$u'_x = \frac{c - V}{1 - Vc/c^2} = \frac{c(1 - V/c)}{1 - V/c} = c$$

as required by Einstein's postulates.

Question

5. The Lorentz transformation for y and z is the same as the classical result: $y = y'$ and $z = z'$. Yet the relativistic velocity transformation does not give the classical result $u_y = u'_y$ and $u_z = u'_z$. Explain.

1-9 Relativistic Momentum

We have seen in previous sections that Einstein's postulates require important modifications in our ideas of simultaneity and in our measurements of time and length. Perhaps more importantly, they also require modifications in our concepts of mass, momentum, and energy. In classical mechanics, the momentum of a particle is defined as the product of its mass and its velocity, $\mathbf{p} = m\mathbf{u}$, where $\mathbf{u}$ is the velocity. In an isolated system of particles, with no net force acting on the system, the total momentum of the system remains constant.

In this section we will see from a simple thought experiment that the classical expression for momentum, $\mathbf{p} = m\mathbf{u}$, is just an approximation. That is, this quantity is not conserved in an isolated system. We consider two observers: observer A in reference frame S and observer B in frame S'. Each observer has a ball of mass m. The two balls are identical when compared at rest. Each observer throws his ball vertically with a speed u_0 such that it travels a distance L, makes an elastic collision with the other ball, and returns. Figure 1-15 shows how the collision looks in each reference frame. Classically, each ball has vertical momentum of magnitude mu_0. Since the vertical components of the momenta are equal and opposite, the total vertical component of momentum is zero before the collision. The collision merely reverses the momentum of each ball, so the total vertical momentum is zero after the collision.

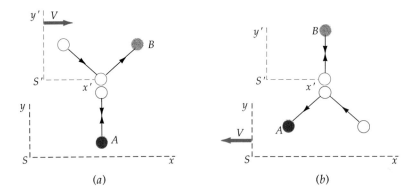

Figure 1-15 (a) Elastic collision of two identical balls as seen in frame S. The vertical component of the velocity of ball B is u_0/γ in S if it is u_0 in S'. (b) The same collision as seen in S'. In this frame, ball A has vertical component of velocity equal to u_0/γ.

Relativistically, however, the vertical components of the velocities of the two balls as seen by either observer are not equal and opposite. Thus, when they are reversed by the collision, classical momentum is not conserved. Consider the collision as seen by A in frame S. The velocity of his ball is $u_{Ay} = +u_0$. Since the velocity of B's ball in frame S' is $u'_{Bx} = 0$, $u'_{By} = -u_0$, the y component of the velocity of B's ball in frame S is (Equation 1-25b) $u_{By} = -u_0/\gamma$. Thus, if the classical expression for momentum $\mathbf{p} = m\mathbf{u}$ is used, the vertical components of momentum of the two balls are not equal and opposite as seen by observer A. Since the balls are reversed by the collision, momentum is not conserved. Of course, the same result is observed by B. In the classical limit, when u is much less than c, γ is approximately 1, and the momentum of the system is conserved as seen by either observer.

The reason that the total momentum of a system is important in classical mechanics is that it is conserved when there are no external forces acting on the system, as is the case in collisions. We now see that the quantity $\Sigma m\mathbf{u}$ is conserved only in the approximation that $u \ll c$. We will define the relativistic momentum $\mathbf{p}$ of a particle to have the following properties:

1. In collisions, $\mathbf{p}$ is conserved.
2. As u/c approaches zero, $\mathbf{p}$ approaches $m\mathbf{u}$.

We will now show that the quantity

$$\mathbf{p} = \frac{m\mathbf{u}}{\sqrt{1 - u^2/c^2}} \qquad 1\text{-}27$$

is conserved in the elastic collision shown in Figure 1-15. Since this quantity also approaches $m\mathbf{u}$ as u/c approaches zero, we take this equation for the definition of the **relativistic momentum** of a particle.

We will compute the y component of the relativistic momentum of each particle in reference frame S and show that the y component of the total relativistic momentum is zero. The speed of ball A in S is u_0, so the y component of its relativistic momentum is

$$p_{Ay} = \frac{mu_0}{\sqrt{1 - u_0^2/c^2}}$$

The speed of ball B in S is more complicated. Its x component is V and its y component is $-u_0/\gamma$. Thus

$$u_B^2 = u_{Bx}^2 + u_{By}^2 = V^2 + (-u_0\sqrt{1 - V^2/c^2})^2 = V^2 + u_0^2 - \frac{u_0^2 V^2}{c^2}$$

Using this result to compute $\sqrt{1 - u_B^2/c^2}$, we obtain

$$1 - \frac{u_B^2}{c^2} = 1 - \frac{V^2}{c^2} - \frac{u_0^2}{c^2} + \frac{u_0^2 V^2}{c^4} = (1 - V^2/c^2)(1 - u_0^2/c^2)$$

and

$$\sqrt{1 - u_B^2/c^2} = \sqrt{1 - V^2/c^2}\sqrt{1 - u_0^2/c^2} = \frac{1}{\gamma}\sqrt{1 - u_0^2/c^2}$$

The y component of the relativistic momentum of ball B as seen in S is therefore

$$p_{By} = \frac{mu_{By}}{\sqrt{1 - u_B^2/c^2}} = \frac{-mu_0/\gamma}{(1/\gamma)\sqrt{1 - u_0^2/c^2}} = \frac{-mu_0}{\sqrt{1 - u_0^2/c^2}}$$

Since $p_{By} = -p_{Ay}$, the y component of the total momentum of the two balls is zero. If the speed of each ball is reversed by the collision, the total momentum will remain zero and momentum will be conserved.

One interpretation of Equation 1-27 is that the mass of an object increases with speed. The quantity $m/\sqrt{1 - u^2/c^2}$ is called the **relativistic mass** of a particle. The mass of a particle when it is at rest in some reference frame is called its **rest mass** m_0. The mass thus increases from m_0 at rest to $m_r = m_0/\sqrt{1 - u^2/c^2}$ when it is moving at speed u. To avoid confusion, we will label the rest mass m_0 and use $m_0/\sqrt{1 - u^2/c^2}$ for the relativistic mass in this chapter. The rest mass of a particle is the same in all reference frames. Using this notation, the relativistic momentum of a particle is then

$$\mathbf{p} = \frac{m_0\mathbf{u}}{\sqrt{1 - u^2/c^2}} \qquad 1\text{-}28 \quad \textit{Relativistic momentum}$$

The creation of elementary particles demonstrates the conversion of kinetic energy to rest energy. In this 1950 photograph of a cosmic ray shower, a high-energy sulfur nucleus (red) collides in a photographic emulsion and produces a spray of particles, including a fluorine nucleus (green), other nuclear fragments (blue), and about 16 pions (yellow).

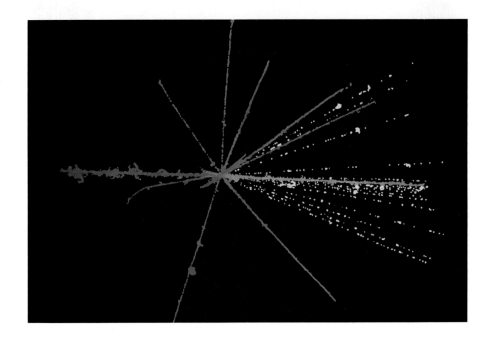

1-10 Relativistic Energy

In classical mechanics, the work done by an unbalanced force acting on a particle equals the change in the kinetic energy of the particle. In relativistic mechanics, we equate the unbalanced force to the rate of change of the relativistic momentum. The work done by such a force can then be calculated and set equal to the change in kinetic energy. As in classical mechanics, we will define kinetic energy as the work done by an unbalanced force in accelerating a particle from rest to some velocity. Considering one dimension only, we have

$$K = \int_{u=0}^{u} \sum F \, ds = \int_{0}^{u} \frac{dp}{dt} \, ds = \int_{0}^{u} u \, dp = \int_{0}^{u} u \, d\left(\frac{m_0 u}{\sqrt{1 - u^2/c^2}}\right) \quad \text{1-29}$$

where we have used $u = ds/dt$. It is left as a problem (Problem 70) for you to show that

$$d\left(\frac{m_0 u}{\sqrt{1 - u^2/c^2}}\right) = m_0 \left(1 - \frac{u^2}{c^2}\right)^{-3/2} du$$

If we substitute this expression into the integrand in Equation 1-29, we obtain

$$K = \int_{0}^{u} u \, d\left(\frac{m_0 u}{\sqrt{1 - u^2/c^2}}\right) = \int_{0}^{u} m_0 \left(1 - \frac{u^2}{c^2}\right)^{-3/2} u \, du$$

$$= m_0 c^2 \left(\frac{1}{\sqrt{1 - u^2/c^2}} - 1\right)$$

or

$$K = \frac{m_0 c^2}{\sqrt{1 - u^2/c^2}} - m_0 c^2 \quad \text{1-30}$$

The expression for kinetic energy consists of two terms. The first term depends on the speed of the particle. The second, $m_0 c^2$, is independent of the speed. The quantity m_0c^2 is called the **rest energy** of the particle E_0. The rest energy is the product of the rest mass and c^2:

$$E_0 = m_0c^2 \qquad \text{1-31} \qquad \textit{Rest energy}$$

The total **relativistic energy** E is then defined to be the sum of the kinetic energy and the rest energy:

$$E = K + m_0c^2 = \frac{m_0c^2}{\sqrt{1 - u^2/c^2}} \qquad \text{1-32} \qquad \textit{Relativistic energy}$$

Thus, the work done by an unbalanced force increases the energy from the rest energy m_0c^2 to the final energy $m_0c^2/\sqrt{1 - u^2/c^2} = m_rc^2$, where $m_r = m_0/\sqrt{1 - u^2/c^2}$ is the relativistic mass. We can obtain a useful expression for the velocity of a particle by multiplying Equation 1-28 for the relativistic momentum by c^2 and comparing the result with Equation 1-32 for the relativistic energy. We have

$$pc^2 = \frac{m_0c^2 u}{\sqrt{1 - u^2/c^2}} = Eu$$

or

$$\frac{u}{c} = \frac{pc}{E} \qquad \text{1-33}$$

Example 1-9

An electron with rest energy 0.511 MeV moves with speed $u = 0.8c$. Find its total energy, kinetic energy, and momentum.

We first calculate the factor $1/\sqrt{1 - u^2/c^2}$.

$$\frac{1}{\sqrt{1 - u^2/c^2}} = \frac{1}{\sqrt{1 - 0.64}} = \frac{5}{3} = 1.67$$

The total energy is then

$$E = \frac{m_0c^2}{\sqrt{1 - u^2/c^2}} = 1.67(0.511 \text{ MeV}) = 0.853 \text{ MeV}$$

The kinetic energy is the total energy minus the rest energy:

$$K = E - m_0c^2 = 0.853 \text{ MeV} - 0.511 \text{ MeV} = 0.342 \text{ MeV}$$

The magnitude of the momentum is

$$p = \frac{m_0 u}{\sqrt{1 - u^2/c^2}} = (1.67)m_0(0.8c) = \frac{1.33 m_0 c^2}{c}$$

$$= \frac{(1.33)(0.511 \text{ MeV})}{c} = 0.680 \text{ MeV}/c$$

The unit MeV/c is a convenient unit for momentum.

The expression for kinetic energy given by Equation 1-30 doesn't look much like the classical expression $\frac{1}{2}m_0u^2$. However, when u is much less than c, we can approximate $1/\sqrt{1-u^2/c^2}$ using the binomial expansion (Equation 1-2):

$$\frac{1}{\sqrt{1-u^2/c^2}} = \left(1 - \frac{u^2}{c^2}\right)^{-1/2}$$

$$\approx 1 + \frac{1}{2} \cdot \frac{u^2}{c^2}$$

From this result, when u is much less than c, the expression for relativistic kinetic energy becomes

$$K = m_0c^2\left(\frac{1}{\sqrt{1-u^2/c^2}} - 1\right)$$

$$\approx m_0c^2\left(1 + \frac{1}{2}\frac{u^2}{c^2} - 1\right)$$

$$= \frac{1}{2}m_0u^2$$

Thus at low speeds, the relativistic expression is the same as the classical expression.

We note from Equation 1-32 that as the speed u approaches the speed of light c, the energy of the particle becomes very large because $1/\sqrt{1-u^2/c^2}$ becomes very large. At $u = c$, the energy becomes infinite. For u greater than c, $\sqrt{1-u^2/c^2}$ is the square root of a negative number and is therefore imaginary. A simple interpretation of the result that it takes an infinite amount of energy to accelerate a particle to the speed of light is that no particle that is ever at rest in any inertial reference frame can travel as fast or faster than the speed of light c. As we noted in Example 1-7, if the speed of a particle is less than c in one reference frame, it is less than c in all other reference frames moving relative to that frame at speeds less than c.

In practical applications, the momentum or energy of a particle is often known rather than the speed. Equation 1-28 for the relativistic momentum and Equation 1-32 for the relativistic energy can be combined to eliminate the speed u. (See Problem 48.) The result is

$$E^2 = p^2c^2 + (m_0c^2)^2 \qquad 1\text{-}34$$

Relation for total energy, momentum, and rest energy

This useful equation can be conveniently remembered from the right triangle shown in Figure 1-16. If the energy of a particle is much greater than its rest energy mc^2, the second term on the right of Equation 1-34 can be neglected, giving the useful approximation

$$E \approx pc \qquad \text{for } E \gg m_0c^2 \qquad 1\text{-}35$$

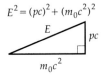

Figure 1-16 Right triangle for remembering Equation 1-34.

Equation 1-35 is an exact relation between energy and momentum for particles with no rest mass, such as photons and neutrinos.

Exercise

A proton with a rest mass of 938 MeV/c^2 has a total energy of 1400 MeV. Find (a) $1/\sqrt{1-u^2/c^2}$, (b) its momentum, and (c) its speed u. [Answers: (a) 1.49, (b) $p = 1040$ MeV/c, (c) $u = 0.74c$]

The identification of the term m_0c^2 as rest energy is not merely a convenience. The conversion of rest energy to kinetic energy with a corresponding loss in rest mass is a common occurrence in radioactive decay and nuclear reactions, including nuclear fission and nuclear fusion. We will give some examples of this in this section. Einstein considered Equation 1-31 relating the energy of a particle to its mass to be the most significant result of the theory of relativity. Energy and inertia, which were formerly two distinct concepts, are related through this famous equation.

To illustrate the interrelation of rest mass and energy, we consider a perfectly inelastic collision of two particles. Classically, kinetic energy is lost in such a collision. For example, in the zero-momentum reference frame, the particles are moving toward one another with equal and opposite momenta and are at rest after the collision. In this frame, the total kinetic energy of the system before the collision is lost. In any other reference frame, the particles move with the velocity of the center of mass, but the amount of kinetic energy lost is the same. We will now see that if we assume that the total relativistic energy is conserved, the loss in kinetic energy equals the gain in rest energy of the system. Consider a particle of rest mass m_{10} moving with initial speed u_1 that collides with a particle of rest mass m_{20} moving with initial speed u_2. The particles collide and stick together, forming a particle of rest mass M_0 that moves with speed u_f, as shown in Figure 1-17. Let E_1 be the initial total energy and K_1 be the initial kinetic energy of particle 1 and E_2 be the initial total energy and K_2 be the initial kinetic energy of particle 2. The initial total energy of the system is

$$E_i = E_1 + E_2$$

and the initial kinetic energy of the system is

$$K_i = K_1 + K_2 = (E_1 - m_{10}c^2) + (E_2 - m_{20}c^2)$$

After the collision, the composite particle has a rest mass M_0, total energy E_f, and kinetic energy $K_f = E_f - M_0c^2$. The loss in kinetic energy of the system is thus

$$K_i - K_f = (E_1 + E_2 - m_{10}c^2 - m_{20}c^2) - (E_f - M_0c^2) \quad \text{1-36}$$

If we assume the conservation of total energy, we have $E_f = E_i = E_1 + E_2$. Substituting $E_1 + E_2 - E_f = 0$ in Equation 1-36 and rearranging, we obtain

$$K_i - K_f = [M_0 - (m_{10} + m_{20})]c^2 = (\Delta m_0)c^2 \quad \text{1-37}$$

where $\Delta m_0 = M_0 - (m_{10} + m_{20})$ is the increase in rest mass of the system.

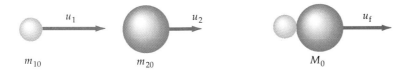

Figure 1-17 A perfectly inelastic collision between two particles. One particle of rest mass m_{10} collides with another particle of rest mass m_{20}. After the collision, the particles stick together, forming a composite particle of rest mass M_0 that moves with speed u_f such that relativistic momentum is conserved. Kinetic energy is lost in this process. If we assume that the total energy is conserved, the loss in kinetic energy must equal c^2 times the increase in the rest mass of the system.

Some numerical examples from atomic and nuclear physics will illustrate changes in rest mass and rest energy. Energies in atomic and nuclear physics are usually expressed in units of electron volts (eV) or mega electron volts (MeV):

$$1 \text{ eV} = 1.6 \times 10^{-19} \text{ J}$$

A convenient unit for the masses of atomic particles is eV/c^2 or MeV/c^2, which is just the rest energy of the particle divided by c^2. The rest masses and rest energies of some elementary particles and light nuclei are given in Table 1-1, from which we can see that the mass of a nucleus is not the same as the sum of the masses of its parts.

Table 1-1 **Rest Energies of Some Elementary Particles and Light Nuclei**

Particle	Symbol	Rest energy, MeV
Photon	γ	0
Electron (positron)	e or e^- (e^+)	0.5110
Muon	$\mu^\pm$	105.7
Pion	π^0	135
	$\pi^\pm$	139.6
Proton	p	938.280
Neutron	n	939.573
Deuteron	^2H or d	1875.628
Triton	^3H or t	2808.944
Alpha particle	^4He or α	3727.409

Example 1-10

A deuteron consists of a proton and neutron bound together. It is the nucleus of the deuterium atom, which is an isotope of hydrogen called heavy hydrogen and written ^2H. How much energy is required to separate the proton from the neutron in the deuteron?

From Table 1-1, we can see that the rest energy of the deuteron is 1875.63 MeV. The rest energy of the proton is 938.28 Mev, and that of the neutron is 939.57 MeV. The sum of the rest energies of the proton and neutron is 938.28 MeV + 939.57 MeV = 1877.85 MeV. This is greater than the rest energy of the deuteron by 1877.85 − 1875.63 = 2.22 MeV. The energy needed to break up a nucleus into its constituent parts is called the **binding energy** of the nucleus. The binding energy of the deuteron is 2.22 MeV. This is the energy that must be added to the deuteron to break it up into a proton plus a neutron. This can be done by bombarding deuterons with energetic particles or with electromagnetic radiation with energy of at least 2.22 MeV.

When a deuteron is formed by the combination of a neutron and proton, energy must be released. When neutrons from a reactor collide with protons, some neutrons are captured to form deuterons. In the capture process, 2.22 MeV of energy is released, usually in the form of electromagnetic radiation.

Example 1-10 illustrates an important property of atoms and nuclei. Any stable composite particle, such as a deuteron or a helium nucleus (2 neutrons plus 2 protons), that is made up of other particles has a rest energy that is less than the sum of the rest energies of its parts. The difference is the binding energy of the composite particle. The binding energies of atoms and molecules are of the order of a few electron volts, which leads to a negligible difference in mass between the composite particle and its parts. The binding energies of nuclei are of the order of several MeV, which leads to a noticeable difference in mass. Some very heavy nuclei, such as radium, are radioactive and decay into a lighter nucleus plus an alpha particle. In this case, the original nucleus has a rest energy greater than that of the decay particles. The excess energy appears as the kinetic energy of the decay products.

Example 1-11

In a typical nuclear fusion reaction, a tritium nucleus (^{3}H) and a deuterium nucleus (^{2}H) fuse together to form a helium nucleus (^{4}He) plus a neutron. How much energy is released in this fusion reaction?

The reaction is written

$$^2H + {}^3H \longrightarrow {}^4He + n$$

From Table 1-1, we see that the rest energy of the deuterium plus tritium nuclei is 1875.628 MeV + 2808.944 MeV = 4684.572 MeV. The rest energy of the helium nucleus plus the neutron is 3727.409 + 939.573 = 4666.982 MeV. This is less than that of the deuterium plus tritium by 4684.572 − 4666.982 = 17.59 MeV. The energy released in this reaction is thus 17.59 MeV. This and other fusion reactions occur in the sun and are responsible for the energy supplied to the earth. As the sun gives off energy, its rest mass continually decreases.

Example 1-12

A hydrogen atom consisting of a proton and an electron has a binding energy of 13.6 eV. By what percentage is the mass of the proton plus the electron greater than that of the hydrogen atom?

The rest energy of a proton plus that of an electron is 938.28 MeV + 0.511 eV = 938.791 MeV. The sum of the masses of these two particles is 938.791 MeV/c^2. The mass of the hydrogen atom is less than this by 13.6 eV/c^2. The percentage difference is

$$\frac{13.6 \text{ eV}/c^2}{938.791 \times 10^6 \text{ eV}/c^2} = 1.45 \times 10^{-8} = 1.45 \times 10^{-6}\%$$

This mass difference is so small as to be hardly measurable.

Example 1-13

A particle of rest mass 2 MeV/c^2 and kinetic energy 3 MeV collides with a stationary particle of rest mass 4 MeV/c^2. After the collision, the two particles stick together. Find (a) the initial momentum of the system, (b) the final velocity of the two-particle system, and (c) the rest mass of the two-particle system.

(a) Since the moving particle has kinetic energy of 3 MeV and rest energy of 2 MeV, its total energy is $E_1 = 5$ MeV. We obtain its momentum from Equation 1-34,

$$pc = \sqrt{E_1^2 - (m_0^2 c^2)^2} = \sqrt{(5 \text{ MeV})^2 - (2 \text{ MeV})^2} = \sqrt{21} \text{ MeV}$$

or

$$p = 4.58 \text{ MeV}/c$$

Since the other particle is rest, this is the total momentum of the system.

(b) We can find the final velocity of the two-particle system from its total energy E and its momentum p using Equation 1-33. By the conservation of total energy, the final energy of the system equals the initial total energy of the two particles:

$$E_f = E_i = E_1 + E_2 = 5 \text{ MeV} + 4 \text{ MeV} = 9 \text{ MeV}$$

By the conservation of momentum, the final momentum of the two-particle system equals the initial momentum, $p = 4.58$ MeV/c. The velocity of the two-particle system is thus given by

$$\frac{u}{c} = \frac{pc}{E} = \frac{4.58 \text{ MeV}}{9 \text{ MeV}} = 0.509$$

(c) We can find the rest mass of the final two-particle system from Equation 1-34 using $pc = 4.58$ MeV and $E = 9$ MeV. We have

$$E^2 = (pc)^2 + (M_0 c^2)^2$$
$$(9 \text{ MeV})^2 = (4.58 \text{ MeV})^2 + (M_0 c^2)^2$$
$$M_0 c^2 = \sqrt{81 - 21} \text{ MeV} = 7.75 \text{ MeV}$$
$$M_0 = 7.75 \text{ MeV}/c^2$$

It is instructive to check our answers by computing the initial and final kinetic energies. The initial kinetic energy is $K_i = 3$ MeV. The final kinetic energy is

$$K_f = E - M_0 c^2 = 9 \text{ MeV} - 7.75 \text{ MeV} = 1.25 \text{ MeV}$$

The loss in kinetic energy is

$$K_i - K_f = 3 \text{ MeV} - 1.25 \text{ MeV} = 1.75 \text{ MeV}$$

Since the initial rest energy is 2 MeV + 4 MeV = 6 MeV and the final rest energy is $M_0 c^2 = 7.75$ MeV, the gain in rest energy is 7.75 MeV − 6 MeV = 1.75 MeV.

1-11 General Relativity

The generalization of the theory of relativity to noninertial reference frames by Einstein in 1916 is known as the general theory of relativity. It is much more difficult mathematically than the special theory of relativity, and there are fewer situations in which it can be tested. Nevertheless, its importance calls for a brief qualitative discussion.

The basis of the general theory of relativity is the **principle of equivalence**:

> A homogeneous gravitational field is completely equivalent to a uniformly accelerated reference frame.

This principle arises in newtonian mechanics because of the apparent identity of gravitational mass and inertial mass. In a uniform gravitational field, all objects fall with the same acceleration **g** independent of their mass because the gravitational force is proportional to the (gravitational) mass, whereas the acceleration varies inversely with the (inertial) mass. Consider a compartment in space far from any matter and undergoing a uniform acceleration **a**, as shown in Figure 1-18a. No mechanics experiment can be performed *inside* the compartment that will distinguish whether the compartment is actually accelerating in space or is at rest (or is moving with uniform

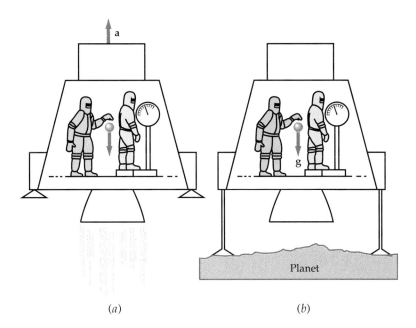

Figure 1-18 The results of experiments in a uniformly accelerated reference frame (a) cannot be distinguished from those in a uniform gravitational field (b) if the acceleration **a** and the gravitational field **g** have the same magnitude.

velocity) in the presence of a uniform gravitational field **g** = −**a**, as shown in Figure 1-18b. If objects are dropped in the compartment, they will fall to the "floor" with an acceleration **g** = −**a**. If people stand on a spring scale, it will read their "weight" of magnitude ma.

Einstein assumed that the principle of equivalence applies to all physics and not just to mechanics. In effect, he assumed that there is no experiment of any kind that can distinguish uniformly accelerated motion from the presence of a gravitational field. We will look qualitatively at a few of the consequences of this assumption.

The first consequence of the principle of equivalence we will discuss, the deflection of a light beam in a gravitational field, was one of the first to be tested experimentally. Figure 1-19 shows a beam of light entering a compartment that is accelerating. Successive positions of the compartment at equal time intervals are shown in Figure 1-19a. Because the compartment is accelerating, the distance it moves in each time interval increases with time. The path of the beam of light as observed from inside the compartment is therefore a parabola, as shown in Figure 1-19b. But according to the principle of equivalence, there is no way to distinguish between an accelerating compartment and one moving with uniform velocity in a uniform gravitational field. We conclude, therefore, that a beam of light, like objects having mass, will accelerate in a gravitational field. For example, near the surface of the earth, light will fall with an acceleration of 9.81 m/s^2. This is difficult to observe because of the enormous speed of light. For example, in a distance of 3000 km, which takes about 0.01 s to traverse, a beam of light should fall

Figure 1-19 (a) A light beam moving in a straight line through a compartment that is undergoing uniform acceleration. The position of the beam is shown at equally spaced times t_1, t_2, t_3, and t_4. (b) In the reference frame of the compartment, the light travels in a parabolic path as a ball would if it were projected horizontally. The vertical displacements are greatly exaggerated in both (a) and (b) for emphasis.

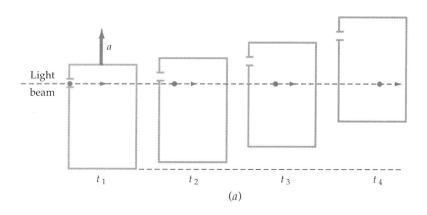

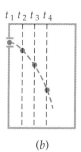

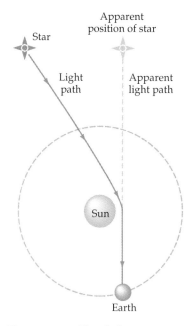

Figure 1-20 The deflection (greatly exaggerated) of a beam of light due to the gravitational attraction of the sun.

about 0.5 mm. Einstein pointed out that the deflection of a light beam in a gravitational field might be observed when light from a distant star passes close to the sun, as illustrated in Figure 1-20. Because of the brightness of the sun, such a star cannot ordinarily be seen. Such a deflection was first observed in 1919 during an eclipse of the sun. This well-publicized observation brought instant worldwide fame to Einstein.

A second prediction from Einstein's theory of general relativity, which we will not discuss in detail, is the excess precession of the perihelion of the orbit of Mercury of about 0.01° per century. This effect had been known and unexplained for some time, so, in a sense, explaining it constituted an immediate success of the theory.

A third prediction of general relativity concerns the change in time intervals and frequencies of light in a gravitational field. The gravitational potential energy between two masses M and m a distance r apart is

$$U = -\frac{GMm}{r}$$

where G is the universal gravitational constant, and the point of zero potential energy has been chosen to be when the separation of the masses is infinite. The potential energy per unit mass near a mass M is called the *gravitational potential* ϕ:

$$\phi = -\frac{GM}{r} \qquad 1\text{-}38$$

According to the general theory of relativity, clocks run more slowly in regions of low gravitational potential. (Since the gravitational potential is negative, as can be seen from Equation 1-38, low gravitational potential occurs near the mass where the *magnitude* of the potential is large.) If Δt_1 is a time interval between two events measured by a clock where the gravitational potential is ϕ_1 and Δt_2 is the interval between the same events as measured by a clock where the gravitational potential is ϕ_2, general relativity predicts

(*a*) This quartz sphere in the top part of the container is probably the world's most perfectly round object. It is designed to spin as a gyroscope in a satellite orbiting the earth. General relativity predicts that the rotation of the earth will cause the axis of rotation of the gyroscope to precess in a circle at a rate of about 1 revolution in 100,000 years. (*b*) This extremely accurate hydrogen maser clock was launched in a satellite in 1976, and its time was compared to that of an identical clock on earth. In accordance with the prediction of general relativity, the clock on earth, where the gravitational potential was lower, "lost" about 4.3×10^{-10} s each second compared with the clock orbiting the earth at an altitude of about 10,000 km.

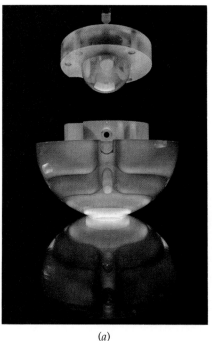
(*a*)

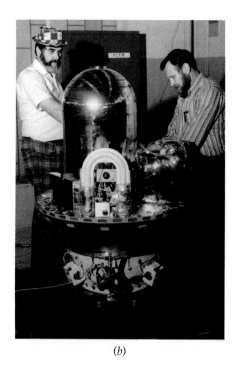
(*b*)

that the fractional difference between these times will be approximately

$$\frac{\Delta t_2 - \Delta t_1}{\Delta t} = \frac{1}{c^2}(\phi_2 - \phi_1) \qquad 1\text{-}39$$

(Since this shift is usually very small, it does not matter by which interval we divide on the left side of the equation.) A clock in a region of low gravitational potential will therefore run slower than one in a region of high potential. Since a vibrating atom can be considered to be a clock, the frequency of vibration of an atom in a region of low potential, such as near the sun, will be lower than that of the same atom on earth. This shift toward a lower frequency and therefore a longer wavelength is called the **gravitational redshift.**

As our final example of the predictions of general relativity, we mention **black holes,** which were first predicted by Oppenheimer and Snyder in 1939. According to the general theory of relativity, if the density of an object such as a star is great enough, its gravitational attraction will be so great that once inside a critical radius, nothing can escape, not even light or other electromagnetic radiation. (The effect of a black hole on objects outside the critical radius is the same as that of any other mass.) A remarkable property of such an object is that nothing that happens inside it can be communicated to the outside. As sometimes occurs in physics, a simple but incorrect calculation gives the correct results for the relation between the mass and the critical radius of a black hole. In newtonian mechanics, the speed needed for a particle to escape from the surface of a planet or star of mass M and radius R is given by

$$v_e = \sqrt{\frac{2GM}{R}}$$

If we set the escape speed equal to the speed of light and solve for the radius, we obtain the critical radius R_S, called the **Schwarzschild radius:**

$$R_S = \frac{2GM}{c^2} \qquad 1\text{-}40$$

For an object with a mass equal to that of our sun to be a black hole, its radius would have to be about 3 km. Since no radiation is emitted from a black hole and its radius is expected to be small, the detection of a black hole is not easy. The best chance of detection would occur if a black hole were a companion to a normal star in a binary star system. The black hole would affect a number of properties of its visible companion. Measurements of the doppler shift of the light from the normal star, for example, might allow a computation of the mass of the unseen companion to determine whether it is great enough to be a black hole. At present there are several excellent candidates—one in the constellation Cygnus, one in the Small Magellanic Cloud, and perhaps one in our own galaxy—but the evidence is not conclusive.

This antenna, consisting of a 1400-kg aluminum cylinder freely suspended by a steel cable, was built by Joseph Weber, David Zippy, and Robert Foward at the University of Maryland to detect gravitational waves. In theory, the antenna should vibrate as the gravity waves pass through it.

Summary

1. The special theory of relativity is based on two postulates of Albert Einstein:
   Postulate 1. Absolute, uniform motion cannot be detected.

Postulate 2. The speed of light is independent of the motion of the source.

An important implication of these postulates is

Postulate 2 (Alternate). Every observer measures the same value for the speed of light independent of the relative motion of the sources and observer.

All of the results of special relativity can be derived from these postulates.

2. The Michelson–Morley experiment was an attempt to measure the absolute velocity of the earth by comparing the speed of light in the direction of motion of the earth with that in a direction perpendicular to that motion. Their null result for the difference in these speeds is consistent with Einstein's postulates.

3. The Lorentz transformation relates the coordinates x, y, and z and the time t of an event seen in frame S to the coordinates x', y', and z' and the time t' of the same event as seen in frame S', which is moving with speed V relative to S:

$$x = \gamma(x' + Vt') \qquad y = y' \qquad z = z'$$

$$t = \gamma\left(t' + \frac{Vx'}{c^2}\right)$$

where

$$\gamma = \frac{1}{\sqrt{1 - V^2/c^2}}$$

The inverse transformation is

$$x' = \gamma(x - Vt) \qquad y' = y \qquad z' = z$$

$$t' = \gamma\left(t - \frac{Vx}{c^2}\right)$$

The transformation equations for velocities are

$$u_x = \frac{u'_x + V}{1 + Vu'_x/c^2}$$

$$u_y = \frac{u'_y}{\gamma(1 + Vu'_x/c^2)}$$

$$u_z = \frac{u'_z}{\gamma(1 + Vu'_x/c^2)}$$

The inverse velocity transformation equations are

$$u'_x = \frac{u_x - V}{1 - Vu_x/c^2}$$

$$u'_y = \frac{u_y}{\gamma(1 - Vu_x/c^2)}$$

$$u'_z = \frac{u_z}{\gamma(1 - Vu_x/c^2)}$$

4. The time interval measured between two events that occur at the same point in space in some reference frame is called the proper time. In another reference frame in which the events occur at different places,

the time interval between the events is longer by the factor γ. This result is known as time dilation. A related phenomenon is length contraction. The length of an object measured in a frame in which it is at rest is called its proper length L_p. When measured in another reference frame, the length of the object is L_p/γ.

5. Two events that are simultaneous in one reference frame are not simultaneous in another frame that is moving relative to the first. If two clocks are synchronized in the frame in which they are at rest, they will be out of synchronization in another frame. In the frame in which they are moving, the "chasing" clock leads by an amount $\Delta t_s = L_p V/c^2$, where L_p is the proper distance between the clocks.

6. The relativistic momentum of a particle is related to its mass and velocity by

$$\mathbf{p} = \frac{m_0 \mathbf{u}}{\sqrt{1 - u^2/c^2}}$$

where m_0 is the rest mass of the particle.

7. The kinetic energy of a particle is given by

$$K = \frac{m_0 c^2}{\sqrt{1 - u^2/c^2}} - m_0 c^2 = \frac{m_0 c^2}{\sqrt{1 - u^2/c^2}} - E_0$$

where

$$E_0 = m_0 c^2$$

is the rest energy. The total energy is

$$E = K + E_0 = \frac{m_0 c^2}{\sqrt{1 - u^2/c^2}}$$

The speed of a particle is related to its momentum and its total energy by

$$\frac{u}{c} = \frac{pc}{E}$$

The total energy is related to the momentum and rest energy by

$$E^2 = p^2 c^2 + (m_0 c^2)^2$$

For particles with energies much greater than their rest energies, a useful approximation is

$$E \approx pc \quad \text{for } E \gg m_0 c^2$$

This is an exact equation for particles of zero rest mass such as photons.

8. The total rest mass of bound systems of particles, such as nuclei or atoms, is less than the sum of the rest masses of the particles making up the system. The difference in mass times c^2 equals the binding energy of the system. The binding energy is the energy that must be added to break up the system into its parts. The binding energies of electrons in atoms are of the order of eV or keV, leading to a negligible difference in rest mass. The binding energies of nuclei are of the order of several MeV, and the difference in rest mass is noticeable.

9. The basis of the general theory of relativity is the principle of equivalence: A homogeneous gravitational field is completely equivalent to a uniformly accelerated reference frame. Important consequences of general relativity include the bending of light in a gravitational field, the prediction of the precession of the perihelion of the orbit of Mercury, the gravitational redshift, and probably the existence of black holes.

Suggestions for Further Reading

Bondi, Hermann: *Relativity and Common Sense: A New Approach to Einstein*, Doubleday, Garden City, New York, 1964.

This book uses familiar phenomena to help show how logical and easy it is to understand the ideas of special relativity.

Chaffee, Frederick H., Jr.: "The Discovery of a Gravitational Lens," *Scientific American*, November 1980, p. 70.

General relativity predicts that light should be deflected by concentrations of matter. This article describes how an elliptical galaxy can act as a giant lens in space.

Gamow, George: "Gravity," *Scientific American*, March 1961, p. 94.

Einstein's general theory of relativity is explained in an entertaining and nonmathematical fashion.

Goldberg, Stanley: *Understanding Relativity: Origin and Impact of a Scientific Revolution*, Birkhaeuser, Boston, 1984.

This book examines the intellectual and social context from which Einstein's special theory grew and the theory's early reception by communities of scientists in four countries.

MacKeown, P. K.: "Gravity is Geometry," *The Physics Teacher*, vol. 22, 1984, p. 557.

This article is an excellent, brief exposition of the ideas of general relativity.

Marder, L.: *Time and the Space Traveller*, George Allen & Unwin, Ltd., London, 1971.

This book presents some of the arguments which have been made in the long and colorful debate over the twin paradox. It also examines some practical limitations of space travel, the implications of time dilation for the long-distance space traveller, and the nature of living clocks.

Mook, Delo E., and Thomas Vargish: *Inside Relativity*, Princeton University Press, Princeton, 1987.

This is a book for nonscientists written by two scholars, one working in the physical sciences and the other in the humanities. The book provides a historical and scientific context for Einstein's work, and explains the special and general theories with the aid of drawings and graphs but no mathematics.

Schwinger, Julian: *Einstein's Legacy: The Unity of Space and Time*, Scientific American Books, Inc., New York, 1986.

A modern and well-illustrated exposition of the special and general theories of relativity and some of their consequences.

Shankland, R. S.: "The Michelson–Morley Experiment," *Scientific American*, November 1964, p. 107.

This article sets the experiment in its historical context and considers its influence on the development of the theory of relativity.

Will, Clifford M.: *Was Einstein Right?: Putting General Relativity to the Test*, Basic Books, Inc., New York, 1986.

Starting around 1960, new discoveries in astronomy motivated a renewed interest in experimentally testing predictions of general relativity. This book, written by a physicist who began his career during this relativity "renaissance," describes the tests with great enthusiasm.

Review

A. Objectives: After studying this chapter, you should:

1. Be able to discuss the results and significance of the Michelson–Morley experiment.

2. Be able to state the Einstein postulates of special relativity.

3. Be able to use the Lorentz transformation to derive expressions for time dilation and length contraction and to solve problems in which time and space intervals in different reference frames are compared.

4. Be able to discuss the lack of synchronization of clocks in moving reference frames.

5. Be able to discuss the twin paradox.

6. Be able to state the definition of relativistic momentum and the equations relating to kinetic energy and total energy of a particle to its speed.

7. Be able to discuss the relation between mass and energy in special relativity and compute the binding energy of various systems from the known rest masses of their constituents.

8. Be able to state the principle of equivalence and discuss three predictions derived from it.

B. Define, explain, or otherwise identify:

Reference frame
Inertial reference frame
Newtonian relativity
Ether
Michelson–Morley experiment
Einstein's postulates
Galilean transformation
Lorentz transformation
Proper time

Time dilation
Length contraction
Proper length
Lorentz–FitzGerald contraction
Synchronized clocks
Simultaneity
Redshift
Twin paradox
Relativistic momentum
Relativistic mass
Rest mass
Rest energy
Relativistic energy
Binding energy
Principle of equivalence
Gravitational redshift
Black hole
Schwarzschild radius

C. True or false: If the statement is true, explain why it is true. If it is false, give a counterexample.

1. The speed of light is the same in all reference frames.

2. Proper time is the shortest time interval between two events.

3. Absolute motion can be determined by means of length contraction.

4. The light-year is a unit of distance.

5. Simultaneous events must occur at the same place.

6. If two events are not simultaneous in one frame, they cannot be simultaneous in any other frame.

7. If two particles are tightly bound together by strong attractive forces, the rest mass of the system is less than the sum of the masses of the individual particles when separated.

Problems

Level I

1-1 Newtonian Relativity

There are no problems for this section.

1-2 The Michelson–Morley Experiment

1. In one series of measurements of the speed of light, Michelson used a path length L of 27.4 km (17 mi). (*a*) What is the time needed for light to make the round-trip distance of $2L$? (*b*) What is the classical correction term in seconds in Equation 1-1, assuming earth's speed is $v = 10^{-4}c$? (*c*) From about 1600 measurements, Michelson quoted the result for the speed of light as 299,796 ± 4 km/s. Is this experiment accurate enough to be sensitive to the correction term in Equation 1-1?

2. An airplane flies with speed u relative to still air from point A to point B and returns. Compare the time required for the round trip when the wind blows from A to B with speed v with that when the wind blows perpendicularly to the line AB with speed v.

1-3 Einstein's Postulates

There are no problems for this section.

1-4 The Lorentz Transformation

3. The proper mean lifetime of pions is 2.6×10^{-8} s. If a beam of pions has a speed of $0.85c$, (*a*) what would their mean lifetime be as measured in the laboratory? (*b*) How far would they travel, on the average, before they decay? (*c*) What would your answer be to part (*b*) if you neglect time dilation?

4. (*a*) In the reference frame of the pion in Problem 3, how far does the laboratory travel in a typical lifetime of 2.6×10^{-8} s? (*b*) What is this distance in the laboratory's frame?

5. The proper mean lifetime of a muon is 2 μs. Muons in a beam are traveling at $0.999c$. (*a*) What is their mean lifetime as measured in the laboratory? (*b*) How far do they travel, on the average, before they decay?

6. (*a*) In the reference frame of the muon in Problem 5, how far does the laboratory travel in a typical lifetime of 2 μs? (*b*) What is this distance in the laboratory's frame?

7. A spaceship of proper length 100 m passes you at a high speed. You measure the length of the spaceship to be 85 m. What was the speed of the spaceship?

8. A spaceship departs from earth for the star Alpha Centauri, which is 4 light-years away. The spaceship travels at $0.75c$. How long does it take to get there (*a*) as measured on earth and (*b*) as measured by a passenger on the spaceship?

9. A spaceship travels to a star 95 light-years away at a speed of 2.2×10^8 m/s. How long does it take to get there (*a*) as measured on earth and (*b*) as measured by a passenger on the spaceship?

10. The mean lifetime of a pion traveling at high speed is measured to be 7.5×10^{-8} s. Its lifetime when measured at rest is 2.6×10^{-8} s. How fast is the pion traveling?

11. How fast must a muon travel so that its mean lifetime is 46 μs if its mean lifetime at rest is 2 μs?

12. A meterstick moves with speed $V = 0.8c$ relative to you in the direction parallel to the stick. (*a*) Find the length of the stick as measured by you. (*b*) How long does it take for the stick to pass you?

13. How fast must a meterstick travel relative to you in the direction parallel to the stick so that its length as measured by you is 50 cm?

14. Use the binomial expansion (Equation 1-2) to derive the following results for the case when V is much less than c, and use the results when applicable in the following problems:

(*a*) $\gamma \approx 1 + \dfrac{1}{2}\dfrac{V^2}{c^2}$ (*b*) $\dfrac{1}{\gamma} \approx 1 - \dfrac{1}{2}\dfrac{V^2}{c^2}$

(*c*) $\gamma - 1 \approx 1 - \dfrac{1}{\gamma} \approx \dfrac{1}{2}\dfrac{V^2}{c^2}$

15. Supersonic jets achieve maximum speeds of about $(3 \times 10^{-6})c$. (a) By what percentage would you see a jet traveling at this speed contracted in length? (b) During a time of 1 y = 3.15×10^7 s on your clock, how much time would elapse on the pilot's clock? How many minutes are lost by the pilot's clock in 1 y of your time?

16. How great must the relative speed of two observers be for the time-interval measurements to differ by 1 percent? (See Problem 14.)

1-5 Clock Synchronization and Simultaneity

Problems 17 through 21 refer to the following situation: An observer in S' lays out a distance L' = 100c·min between points A' and B' and places a flashbulb at the midpoint C'. She arranges for the bulb to flash and for clocks at A' and B' to be started at zero when the light from the flash reaches them (see Figure 1-21). Frame S' is moving to the right with speed 0.6c relative to an observer C in S who is at the midpoint between A' and B' when the bulb flashes. At the instant he sees the flash, observer C sets his clock to zero.

Figure 1-21 Problems 17 through 21.

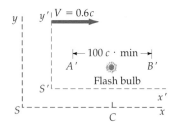

17. What is the separation distance between clocks A' and B' according to the observer in S?

18. As the light pulse from the flashbulb travels toward A' with speed c, A' travels toward C with speed 0.6c. Show that the clock in S reads 25 min when the flash reaches A'. (*Hint*: In time t, the light travels a distance ct and A' travels 0.6ct. The sum of these distances must equal the distance between A' and the flashbulb as seen in S.)

19. Show that the clock in S reads 100 min when the light flash reaches B', which is traveling away from C with speed 0.6c. (See the hint for Problem 18.)

20. The time interval between the reception of the flashes at A' and B' in Problems 18 and 19 is 75 min according to the observer in S. How much time does he expect to have elapsed on the clock at A' during this 75-min interval?

21. The time interval calculated in Problem 20 is the amount that the clock at A' leads that at B' according to the observer in S. Compare this result with $L_p V/c^2$.

1-6 The Doppler Effect

22. How fast must you be moving toward a red light (λ = 650 nm) for it to appear green (λ = 525 nm)?

23. A distant galaxy is moving away from us at a speed of 1.85×10^7 m/s. Calculate the fractional redshift $(\lambda' - \lambda_0)/\lambda_0$ in the light from this galaxy.

24. Show that if V is much less than c, the doppler shift is given approximately by $\Delta f/f_0 \approx \pm V/c$.

25. A distant galaxy is moving away from the earth with a speed that results in each wavelength received on earth being shifted such that $\lambda' = 2\lambda_0$. Find the speed of the galaxy relative to the earth.

26. Sodium light of wavelength 589 nm is emitted by a source that is receding from the earth with speed V. The wavelength measured in the frame of the earth is 620 nm. Find V.

27. A student on earth hears a tune on her radio that seems to be coming from a record that is being played too fast. She has a 33-rev/min record of that tune and determines that the tune sounds the same as when her record is played at 78 rev/min, that is, the frequencies are all too high by a factor of 78/33. If the tune is being played correctly, but is being broadcast by a spaceship that is approaching the earth at speed V, determine V.

1-7 The Twin Paradox

28. A friend of yours who is the same age as you travels at 0.999c to a star 15 light-years away. She spends 10 y on one of the star's planets and then returns at 0.999c. How long has she been away (a) as measured by you and (b) as measured by her?

1-8 The Velocity Transformation

29. Two spaceships are approaching each other. (a) If the speed of each is 0.6c relative to the earth, what is the speed of one relative to the other? (b) If the speed of each relative to the earth is 30,000 m/s (about 100 times the speed of sound), what is the speed of one relative to the other?

30. A light beam moves along the y' axis with speed c in frame S', which is moving to the right with speed V relative to frame S. (a) Find the x and y components of the velocity of the light beam in frame S. (b) Show that the magnitude of the velocity of the light beam in S is c.

31. A spaceship is moving east at speed 0.90c relative to the earth. A second spaceship is moving west at speed 0.90c relative to the earth. What is the speed of one spaceship relative to the other?

32. A particle moves with speed 0.8c along the x" axis of frame S", which moves with speed 0.8c along the x' axis relative to frame S'. Frame S' moves with speed 0.8c along the x axis relative to frame S. (a) Find the speed of the particle relative to frame S'. (b) Find the speed of the particle relative to frame S.

1-9 Relativistic Momentum;
1-10 Relativistic Energy

33. How much rest mass must be converted into energy (a) to produce 1 J and (b) to keep a 100-W light bulb burning for 10 years?

34. Sketch a graph of the momentum p of a particle versus its speed u.

35. (a) Calculate the rest energy in 1 g of dirt. (b) If you could convert this energy into electrical energy and sell it for 10 cents per kilowatt-hour, how much money would you get? (c) If you could power a 100-W light bulb with this energy, for how long could you keep the bulb lit?

36. Find the ratio of the total energy to the rest energy of a particle of rest mass m_0 moving with speed (a) $0.1c$, (b) $0.5c$, (c) $0.8c$, and (d) $0.99c$.

37. An electron with rest energy of 0.511 MeV moves with speed $u = 0.2c$. Find its total energy, kinetic energy, and momentum.

38. A muon has a rest energy of 105.7 MeV. Calculate its rest mass in kilograms.

39. A proton with rest energy of 938 MeV has a total energy of 1400 MeV. (a) What is its speed? (b) What is its momentum?

40. The total energy of a particle is twice its rest energy. (a) Find u/c for the particle. (b) Show that its momentum is given by $p = \sqrt{3}m_0 c$.

41. For the fusion reaction of Example 1-11, calculate the number of reactions per second that are necessary to generate 1 kW of power.

42. Use Table 1-1 to find how much energy is needed to remove one proton from ^{4}He, leaving ^{3}He plus a proton.

43. A free neutron decays into a proton plus an electron:

$$n \rightarrow p + e$$

Use Table 1-1 to calculate the energy released in this reaction.

44. How much energy would be required to accelerate a particle of mass m_0 from rest to (a) $0.5c$, (b) $0.9c$, and (c) $0.99c$? Express your answers as multiples of the rest energy.

45. If the kinetic energy of a particle equals its rest energy, what error is made by using $p = m_0 u$ for its momentum?

46. In another nuclear fusion reaction, ^{2}H nuclei are combined to produce ^{4}He. (a) How much energy is released in this reaction? (b) How many such reactions must take place per second to produce 1 kW of power?

1-11 General Relativity

There are no problems for this section.

Level II

47. A friend of yours who is the same age as you travels to the star Alpha Centauri, which is 4 light-years away and returns immediately. He claims that the entire trip took just 6 y. How fast did he travel?

48. Use Equations 1-28 and 1-32 to derive the equation $E^2 = p^2 c^2 + m_0^2 c^4$.

49. If a plane flies at a speed of 2000 km/h, for how long must it fly before its clock loses 1 s because of time dilation?

50. Use the binomial expansion (Equation 1-2) and Equation 1-34 to show that when $pc \ll m_0 c^2$, the total energy is given approximately by

$$E \approx m_0 c^2 + \frac{p^2}{2m_0}$$

51. A clock is placed in a satellite that orbits the earth with a period of 90 min. By what time interval will this clock differ from an identical clock on earth after 1 y? (Assume that special relativity applies.)

52. A and B are twins. A travels at $0.6c$ to Alpha Centauri (which is $4c$·years from earth as measured in the reference frame of the earth) and returns immediately. Each twin sends the other a light signal every 0.01 year as measured in her own reference frame. (a) At what rate does B receive signals as A is moving away from her? (b) How many signals does B receive at this rate? (c) How many total signals are received by B before A has returned? (d) At what rate does A receive signals as B is receding from her? (e) How many signals does A receive at this rate? (f) What is the total number of signals received by A? (g) Which twin is younger at the end of the trip, and by how many years?

53. In frame S, event B occurs 2 μs after event A, which occurs at $\Delta x = 1.5$ km from event A. How fast must an observer be moving along the $+x$ axis so that events A and B occur simultaneously? Is it possible for event B to precede event A for some observer?

54. Observers in reference frame S see an explosion located at $x_1 = 480$ m. A second explosion occurs 5 μs later at $x_2 = 1200$ m. In reference frame S', which is moving along the $+x$ axis at speed V, the explosions occur at the same point in space. What is the separation in time between the two explosions as measured in S'?

55. An interstellar spaceship travels from the earth to a distant star system $12c$·years away (as measured in the earth's frame). The trip takes 15 years as measured on the ship. (a) What is the speed of the ship relative to the earth? (b) When the ship arrives, it sends a signal to the earth. How long after the ship leaves the earth will it be before the earth receives the signal?

56. Show that the speed u of a particle of mass m_0 and total energy E is given by

$$\frac{u}{c} = \left[1 - \frac{(m_0 c^2)^2}{E^2}\right]^{1/2}$$

and that when E is much greater than $m_0 c^2$, this can be approximated by

$$\frac{u}{c} \approx 1 - \frac{(m_0 c^2)^2}{2E^2}$$

Find the speed of an electron with kinetic energy of (b) 0.51 MeV and (c) 10 MeV.

57. Two spaceships, each 100 m long when measured at rest, travel toward each other with speeds of $0.85c$ relative to the earth. (a) How long is each ship as measured by someone on earth? (b) How fast is each ship traveling as measured by an observer on the other? (c) How long is one ship when measured by an observer on the other? (d) At time $t = 0$ on earth, the fronts of the ships are together as they just begin to pass each other. At what time on earth are their ends together? (e) Sketch diagrams in the frame of one of the ships showing the passing of the other ship.

58. In the Stanford linear collider, small bundles of electrons and positrons are fired at each other. In the laboratory's frame of reference, each bundle is about 1 cm long and 10 μm in diameter. In the collision region, each particle has an energy of 50 GeV, and the electrons and positrons are moving in opposite directions. (a) How long and how wide is each bundle in its own reference frame? (b) What must be the minimum proper length of the accelerator for a bundle to have both its ends simultaneously in the accelerator in its own reference frame? (The actual length of the accelerator is less than 1000 m.) (c) What is the length of a positron bundle in the reference frame of the electron bundle?

59. An electron with rest energy of 0.511 MeV has a total energy of 5 MeV. (a) Find its momentum in units of MeV/c from Equation 1-34. (b) Find the ratio of its speed u to the speed of light.

60. The rest energy of a proton is about 938 MeV. If its kinetic energy is also 938 MeV, find (a) its momentum and (b) its speed.

61. What percent error is made in using $\frac{1}{2}m_0u^2$ for the kinetic energy of a particle if its speed is (a) $0.1c$ and (b) $0.9c$?

62. A rocket with a proper length of 1000 m moves in the $+x$ direction at $0.6c$ with respect to an observer on the ground. An astronaut stands at the rear of the rocket and fires a bullet toward the front of the rocket at $0.8c$ relative to the rocket. How long does it take the bullet to reach the front of the rocket (a) as measured in the frame of the rocket, (b) as measured in the frame of the ground, and (c) as measured in the frame of the bullet?

63. A rocket with a proper length of 700 m is moving to the right at a speed of $0.9c$. It has two clocks, one in the nose and one in the tail, that have been synchronized in the frame of the rocket. A clock on the ground and the nose clock on the rocket both read $t = 0$ as they pass. (a) At $t = 0$, what does the tail clock on the rocket read as seen by an observer on the ground? When the tail clock on the rocket passes the ground clock, (b) what does the tail clock read as seen by an observer on the ground, (c) what does the nose clock read as seen by an observer on the ground, and (d) what does the nose clock read as seen by an observer on the rocket? (e) At $t = 1$ h, as measured on the rocket, a light signal is sent from the nose of the rocket to an observer standing by the ground clock. What does the ground clock read when the observer receives this signal? (f) When the observer on the ground receives the signal, he sends a return signal to the nose of the rocket. When is this signal received at the nose of the rocket as seen on the rocket?

64. Derive Equation 1-24a for the frequency received by an observer moving with speed V toward a stationary source of electromagnetic waves.

65. Frames S and S' are moving relative to each other along the x and x' axis. They set their clocks to $t = 0$ when their origins coincide. In frame S, event 1 occurs at $x_1 = 1.0c$·year and $t_1 = 1$ y and event 2 occurs at $x_2 = 2.0c$·year and $t_2 = 0.5$ y. These events occur simultaneously in frame S'. (a) Find the magnitude and direction of the velocity of S' relative to S. (b) At what time do both these events occur as measured in S'?

66. An observer in frame S standing at the origin observes two flashes of colored light separated spatially by $\Delta x = 2400$ m. A blue flash occurs first, followed by a red flash 5 μs later. An observer in S' moving along the x axis at speed V relative to S also observes the flashes 5 μs apart and with a separation of 2400 m, but the red flash is observed first. Find the magnitude and direction of V.

67. The sun radiates energy at the rate of about 4×10^{26} W. Assume that this energy is produced by a reaction whose net result is the fusion of 4 H nuclei to form 1 He nucleus, with the release of 25 MeV for each He nucleus formed. Calculate the sun's loss of rest mass per day.

68. A spaceship of mass 10^6 kg is coasting through space when it suddenly becomes necessary to accelerate. The ship ejects 10^3 kg of fuel in a very short time at a speed of $c/2$ relative to the ship. (a) Neglecting any change in the rest mass of the system, calculate the speed of the ship in the frame in which it was initially at rest. (b) Calculate the speed of the ship using classical, newtonian mechanics. (c) Use your results from (a) to estimate the change in the rest mass of the system.

69. Reference frame S' is moving along the x' axis at $0.6c$ relative to frame S. A particle that is originally at $x' = 10$ m at $t'_1 = 0$ is suddenly accelerated and then moves at a constant speed of $c/3$ in the $-x'$ direction until time $t'_2 = 60$ m/c, when it is suddenly brought to rest. As observed in frame S, find (a) the speed of the particle, (b) the distance and direction the particle traveled from t'_1 to t'_2, and (c) the time the particle traveled.

70. Show that

$$d\left(\frac{m_0 u}{\sqrt{1 - u^2/c^2}}\right) = m_0\left(1 - \frac{u^2}{c^2}\right)^{-3/2} du$$

71. Two protons approach each other head on at $0.5c$ relative to reference frame S'. (a) Calculate the total kinetic energy of the two protons as seen in frame S'. (b) Calculate the total kinetic energy of the protons as seen in reference frame S, which is moving with speed $0.5c$ relative to S' such that one of the protons is at rest.

72. A particle of rest mass 1 MeV/c^2 and kinetic energy 2 MeV collides with a stationary particle of rest mass 2 MeV/c^2. After the collision, the particles stick together. Find (a) the speed of the first particle before the collision, (b) the total energy of the first particle before the collision, (c) the initial total momentum of the system, (d) the total kinetic energy after the collision, and (e) the rest mass of the system after the collision.

73. The radius of the orbit of a charged particle in a magnetic field is related to the momentum of the particle by

$$p = BqR \qquad 1\text{-}41$$

This equation holds classically for $p = mu$ and relativistically for $p = m_0 u / \sqrt{1 - u^2/c^2}$. An electron with kinetic energy of 1.50 MeV moves in a circular orbit perpendicular to a uniform magnetic field $B = 5 \times 10^{-3}$ T. (a) Find the radius of the orbit. (b) What result would you obtain if you used the classical relations $p = mu$ and $K = p^2/2m$?

74. Oblivious to economics and politics, physicists propose building a circular accelerator around the earth's circumference using bending magnets that provide a magnetic field of magnitude 1.5 T. (a) What would be the kinetic energy of protons orbiting in this field in a circle of radius R_E? (See Problem 73.) (b) What would be the period of rotation of these protons?

75. In a simple thought experiment, Einstein showed that there is mass associated with electromagnetic radiation. Consider a box of length L and mass M resting on a frictionless surface. At the left wall of the box is a light source that emits radiation of energy E, which is absorbed at the right wall of the box. According to classical electromagnetic theory, this radiation carries momentum of magnitude $p = E/c$. (a) Find the recoil velocity of the box such that momentum is conserved when the light is emitted. (Since p is small and M is large, you may use classical mechanics.) (b) When the light is absorbed at the right wall of the box, the box stops, so the total momentum remains zero. If we neglect the very small velocity of the box, the time it takes for the radiation to travel across the box is $\Delta t = L/c$. Find the distance moved by the box in this time. (c) Show that if the center of mass of the system is to remain at the same place, the radiation must carry mass $m = E/c^2$.

76. An antiproton $\bar{p}$ has the same rest energy at a proton. It is created in the reaction $p + p \to p + p + p + \bar{p}$. In an experiment, protons at rest in the laboratory are bombarded with protons of kinetic energy K_L, which must be great enough so that kinetic energy equal to $2m_0c^2$ can be converted into the rest energy of the two particles. In the frame of the laboratory, the total kinetic energy cannot be converted into rest energy because of conservation of momentum. However, in the zero-momentum reference frame in which the two initial protons are moving toward each other with equal speed u, the total kinetic energy can be converted into rest energy. (a) Find the speed of each proton u such that the total kinetic energy in the zero-momentum frame is $2m_0c^2$. (b) Transform to the laboratory's frame in which one proton is at rest, and find the speed u' of the other proton. (c) Show that the kinetic energy of the moving proton in the laboratory's frame is $K_L = 6m_0c^2$.

Level III

77. A stick of proper length L_P makes an angle θ with the x axis in frame S. Show that the angle θ' made with the x' axis in frame S', which is moving along the $+x$ axis with speed V, is given by $\tan \theta' = \gamma \tan \theta$ and that the length of the stick in S' is

$$L' = L_P \left[\frac{1}{\gamma^2} \cos^2 \theta + \sin^2 \theta \right]^{1/2}$$

78. Show that if a particle moves at an angle θ with the x axis with speed u in frame S, it moves at an angle θ' with the x' axis in S' given by

$$\tan \theta' = \frac{\sin \theta}{\gamma(\cos \theta - V/u)}$$

where the frame S' is moving with speed V relative to S.

79. For the special case of a particle moving with speed u along the y axis in frame S, show that its momentum and energy in frame S' are related to its momentum and energy in S by the transformation equations

$$p_x' = \gamma\left(p_x - \frac{VE}{c^2}\right) \qquad p_y' = p_y \qquad p_z' = p_z$$

$$\frac{E'}{c} = \gamma\left(\frac{E}{c} - \frac{Vp_x}{c}\right)$$

Compare these equations with the Lorentz transformation for x', y', z', and t'. These equations show that the quantities p_x, p_y, p_z, and E/c transform in the same way as do x, y, z, and ct.

80. The equation for the spherical wavefront of a light pulse that begins at the origin at time $t = 0$ is $x^2 + y^2 + z^2 - (ct)^2 = 0$. Using the Lorentz transformation, show that such a light pulse also has a spherical wavefront in frame S' by showing that $x'^2 + y'^2 + z'^2 - (ct')^2 = 0$ in S'.

81. In Problem 80, you showed that the quantity $x^2 + y^2 + z^2 - (ct)^2$ has the same value (0) in both S and S'. Such a quantity is called an *invariant*. From the results of Problem 79, the quantity $p_x^2 + p_y^2 + p_z^2 - (E/c)^2$ must also be an invariant. Show that this quantity has the value $-m_0c^2$ in both the S and S' reference frames.

82. Two events in S are separated by a distance $D = x_2 - x_1$ and a time $T = t_2 - t_1$. (a) Use the Lorentz transformation to show that in frame S', which is moving with speed V relative to S, the time separation is $t_2' - t_1' = \gamma(T - VD/c^2)$. (b) Show that the events can be simultaneous in frame S' only if D is greater than cT. (c) If one of the events is the *cause* of the other, the separation D must be less than cT since D/c is the smallest time that a signal can take to travel from x_1 to x_2 in frame S. Show that if D is less than cT, t_2' is greater than t_1' in all reference frames. This shows that if the cause precedes the effect in one frame, it must precede it in all reference frames. (d) Suppose that a signal could be sent with speed $c' > c$ so that in frame S the cause precedes the effect by the time $T = D/c'$. Show that there is then a reference frame moving with speed V less than c in which the effect precedes the cause.

83. Two identical particles of rest mass m_0 are each moving toward the other with speed u in frame S. The particles collide inelastically with a spring that locks shut (Figure 1-22) and come to rest in S, and their initial kinetic energy is transformed into potential energy. In this problem you are going to show that the conservation of momentum in reference frame S', in which one of the particles is initially at rest, requires that the total rest mass of the system after the collision be $2m_0/\sqrt{1-u^2/c^2}$. (a) Show that the speed of the particle not at rest in frame S' is

$$u' = \frac{2u}{1+u^2/c^2}$$

and use this result to show that

$$\sqrt{1-\frac{u'^2}{c^2}} = \frac{1-u^2/c^2}{1+u^2/c^2}$$

(b) Show that the initial momentum in frame S' is $p' = 2m_0 u/(1-u^2/c^2)$. (c) After the collision, the composite particle moves with speed u in S' (since it is at rest in S). Write the total momentum after the collision in terms of the final rest M_0, and show that the conservation of momentum implies that $M_0 = 2m_0/\sqrt{1-u^2/c^2}$. (d) Show that the total energy is conserved in each reference frame.

84. A horizontal turntable rotates with angular speed ω. There is a clock at the center of the turntable and one at a distance r from the center. In an inertial reference frame, the clock at distance r is moving with speed $u = r\omega$. (a) Show that from time dilation in special relativity, time intervals Δt_0 for the clock at rest and Δt_r for the moving clock are related by

$$\frac{\Delta t_r - \Delta t_0}{\Delta t_0} \approx -\frac{r^2\omega^2}{2c^2} \quad \text{if } r\omega \ll c$$

(b) In a reference frame rotating with the table, both clocks are at rest. Show that the clock at distance r experiences a pseudoforce $F_r = mr\omega^2$ in this accelerated frame and that this is equivalent to a difference in gravitational potential between r and the origin of $\phi_r - \phi_0 = -\frac{1}{2}r^2\omega^2$. Use this potential difference in Equation 1-39 to show that in this frame the difference in time intervals is the same as in the inertial frame.

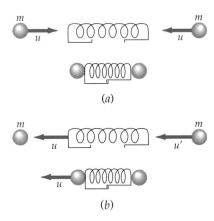

Figure 1-22 Problem 83. An inelastic collision between two identical objects (a) in the zero-momentum reference frame S and (b) in frame S', which is moving to the right with speed $V = u$ relative to frame S such that one of the particles is initially at rest. The spring, which is assumed to be massless, is merely a device for visualizing the storage of potential energy.

Chapter 2

The Origins of Quantum Theory

The continuous visible spectrum (*top*) along with the characteristic optical line spectrum (*from top to bottom*) emitted by hydrogen, helium, sodium, mercury, and barium. Niels Bohr's model of the hydrogen atom, which could be used to predict the wavelengths of the observed spectrum of hydrogen, was one of the great triumphs of twentieth century physics and was an important step in the development of our understanding of the microscopic world.

In Chapter 1, we saw that Newton's laws must be modified when they are applied to objects that move at speeds comparable to the speed of light. In the last 20 years of the nineteenth century and the first 30 years of the twentieth century, many startling discoveries, both experimental and theoretical, demonstrated that the laws of classical physics also break down when they are applied to microscopic systems, such as the particles within an atom. This failure is as drastic as the failure of newtonian mechanics at high speeds. The interior of the atom can be described only in terms of *quantum theory* (sometimes called *quantum mechanics* or *wave mechanics*), which requires the modification of some of our fundamental ideas about the relationships between physical theory and the physical world. Table 2-1 lists the approximate dates of some of the important experiments performed and theories proposed between 1881 and 1932.

The development of quantum theory was very different from that of the theory of relativity. In a sense, special relativity was presented as a complete theory in 1905 (and the general theory, in 1916) by a single scientist, Albert Einstein. Quantum theory, on the other hand, was developed over a long period by many different people. Many of the discoveries initially seemed unrelated, and it wasn't until the late 1920s that any consistent theory emerged. This theory is now the basis of our understanding of the micro-

Table 2-1 **Approximate Dates of Some Important Experiments and Theories, 1881–1932**

1881	Michelson obtains null result for absolute velocity of earth
1884	Balmer finds empirical formula for spectral lines of hydrogen
1887	Hertz produces electromagnetic waves, verifying Maxwell's theory and accidently discovering photoelectric effect
1887	Michelson repeats his experiment with Morley, again obtaining null result
1895	Röntgen discovers x rays
1896	Becquerel discovers nuclear radioactivity
1897	J. J. Thomson measures e/m for cathode rays, showing that electrons are fundamental constituents of atoms
1900	Planck explains blackbody radiation using energy quantization involving new constant h
1900	Lenard investigates photoelectric effect and finds energy of electrons independent of light intensity
1905	Einstein proposes special theory of relativity
1905	Einstein explains photoelectric effect by suggesting quantization of radiation
1907	Einstein applies energy quantization to explain temperature dependence of heat capacities of solids
1908	Rydberg and Ritz generalize Balmer's formula to fit spectra of many elements
1909	Millikan's oil-drop experiment shows quantization of electric charge
1911	Rutherford proposes nuclear model of atom based on alpha-particle scattering experiments of Geiger and Marsden
1912	Friedrich and Knipping and von Laue demonstrate diffraction of x rays by crystals showing that x rays are waves and that crystals are regular arrays
1913	Bohr proposes model of hydrogen atom
1914	Moseley analyzes x-ray spectra using Bohr model to explain periodic table in terms of atomic number
1914	Franck and Hertz demonstrate atomic energy quantization
1915	Duane and Hunt show that the short-wavelength limit of x rays is determined from quantum theory
1916	Wilson and Sommerfeld propose rules for quantization of periodic systems
1916	Millikan verifies Einstein's photoelectric equation
1923	Compton explains x-ray scattering by electrons as collision of photon and electron and verifies results experimentally
1924	De Broglie proposes electron waves of wavelength h/p
1925	Schrödinger develops mathematics of electron wave mechanics
1925	Heisenberg invents matrix mechanics
1925	Pauli states exclusion principle
1927	Heisenberg formulates uncertainty principle
1927	Davisson and Germer observe electron wave diffraction by single crystal
1927	G. P. Thomson observes electron wave diffraction in metal foil
1928	Gamow and Condon and Gurney apply quantum mechanics to explain alpha-decay lifetimes
1928	Dirac develops relativistic quantum mechanics and predicts existence of positron
1932	Chadwick discovers neutron
1932	Anderson discovers positron

scopic world. It is extremely successful, yet there is still debate about its philosophical interpretations. As with special relativity, quantum theory reduces to classical physics when it is applied to macroscopic (large-scale) systems, that is, to objects in our familiar, everyday world.

The origins of quantum theory were not, strangely enough, in the discoveries of radioactivity or x rays or atomic spectra but in thermodynamics. In his study of the radiation spectrum of a blackbody, Max Planck found that he could reconcile theory and experiment if he assumed that radiant energy was emitted and absorbed not continuously but in discrete lumps or quanta. It was Einstein who first recognized that this quantization of radiant energy was not just a calculational device but a general property of radiation. Niels Bohr then applied Einstein's ideas of energy quantization to the energy of an atom and proposed a model of the hydrogen atom that was spectacularly successful in calculations of the wavelengths of the radiation emitted by hydrogen. In this chapter, we will look qualitatively at the origins of the idea of energy quantization.

2-1 The Origin of the Quantum Constant: Blackbody Radiation

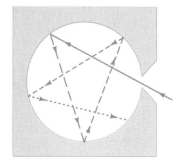

Figure 2-1 Cavity approximating an ideal blackbody. Radiation entering the cavity has little chance of leaving. It is usually completely absorbed.

One of the most puzzling phenomena studied near the end of the nineteenth century was the spectral distribution of blackbody radiation. A blackbody is an ideal system that absorbs all the radiation incident on it. It can be approximated by a cavity with a very small opening, as illustrated in Figure 2-1. The characteristics of the radiation in such a cavity depend only on the temperature of the walls. At ordinary temperatures (below about 600°C), the thermal radiation emitted by a blackbody is not visible because the energy is concentrated in the infrared region of the electromagnetic spectrum. As the body is heated, the amount of energy radiated increases according to the Stefan–Boltzmann law, and the concentration of energy moves to shorter wavelengths. Between about 600 and 700°C, there is enough energy in the visible spectrum for the body to glow a dull red. At higher temperatures, it becomes bright red or even "white hot."

Figure 2-2 shows the power radiated by a blackbody as a function of wavelength for three different temperatures. These curves are known as spectral distribution curves. The quantity P in this figure is the power radiated per unit wavelength. It is a function of both the wavelength λ and the temperature T and is called the spectral distribution function. This function $P(\lambda,T)$ has a maximum at a wavelength λ_{max} that varies inversely with temperature according to Wien's displacement law:

$$\lambda_{max} = \frac{2.898 \text{ mm} \cdot \text{K}}{T}$$

The spectral distribution function $P(\lambda,T)$ can be calculated from classical thermodynamics in a straightforward way, and the result can be compared with the experimentally obtained curves of Figure 2-2. The result of this classical calculation, known as the **Rayleigh–Jeans law,** is

$$P(\lambda,T) = 8\pi kT\lambda^{-4} \qquad 2\text{-}1$$

where k is Boltzmann's constant. This result agrees with experimental results in the region of long wavelengths, but it disagrees violently at short wavelengths. As λ approaches zero, the experimentally determined $P(\lambda,T)$ also approaches zero, but the calculated function approaches infinity be-

Portrait of Max Planck (1858–1947).

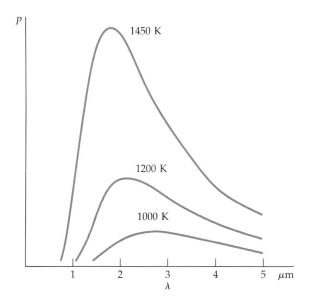

Figure 2-2 Spectral distribution of radiation from a blackbody for three different temperatures.

cause it is proportional to λ^{-4}. Thus, according to the classical calculation, blackbodies radiate an infinite amount of energy concentrated in the very short wavelengths. This result was known as the **ultraviolet catastrophe.**

In 1900, the German physicist Max Planck announced that by making a strange modification in the classical calculation he could derive a function $P(\lambda, T)$ that agreed with the experimental data at all wavelengths. Planck's result is shown in Figure 2-3 along with experimental data and the Rayleigh–Jeans law. Planck first found an empirical function that fit the data and then searched for a way to modify the usual calculation. He found that he could "derive" this function if he made the unusual assumption that the energy emitted and absorbed by the blackbody was not continuous but was instead emitted or absorbed in discrete packets or **quanta.** Planck found that the size of an energy quantum is proportional to the frequency of the radiation:

Quantization of energy of radiation

$$E = hf \qquad 2\text{-}2$$

where h is the proportionality constant now known as **Planck's constant.** The value of h was determined by Planck by fitting his function to the experimentally obtained data. The accepted value of this constant is now

$$h = 6.626 \times 10^{-34} \text{ J·s} = 4.136 \times 10^{-15} \text{ eV·s} \qquad 2\text{-}3$$

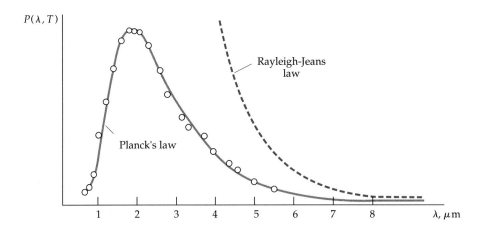

Figure 2-3 Spectral distribution of blackbody radiation versus wavelength at $T = 1600$ K. The classical theoretical calculation gives the Rayleigh–Jeans law, which agrees with experimental results at very long wavelengths but disagrees with them violently at short wavelengths.

Planck was unable to fit the constant h into the framework of classical physics. The fundamental importance of his assumption of energy quantization, implied by Equation 2-2, was not generally appreciated until Einstein applied similar ideas to explain the photoelectric effect and suggested that quantization is a fundamental property of electromagnetic radiation.

2-2 The Photoelectric Effect

In 1905, Einstein used Planck's idea of energy quantization to explain the photoelectric effect. (His paper on the photoelectric effect appeared in the same journal that contained his special theory of relativity.) Einstein's work marked the beginning of quantum theory, and for it he received the Nobel prize for physics. Whereas Planck looked at energy quantization in his blackbody-radiation theory as a calculational device, Einstein made the bold suggestion that energy quantization is a fundamental property of electromagnetic energy. Three years later, he applied the idea of energy quantization to molecular energies to clear up another puzzle in physics—the discrepancy between the specific heats calculated from the equipartition theorem and those observed experimentally at low temperatures. Later, the ideas of energy quantization were applied to atomic energies by Niels Bohr in the first explanation of atomic spectra.

The photoelectric effect was discovered by Hertz in 1887 and was studied by Lenard in 1900. Figure 2-4 shows a schematic diagram of the basic apparatus. When light is incident on the clean metal surface of the cathode C, electrons are emitted. If some of these electrons strike the anode A, there is a current in the external circuit. The number of emitted electrons that reach the anode can be increased or decreased by making the anode positive or negative with respect to the cathode. Let V be the difference in potential between the cathode and the anode. Figure 2-5 shows the current versus V for two values of the intensity of the light incident on the cathode. When V is positive, the electrons are attracted to the anode. At sufficiently large values of V, all the emitted electrons reach the anode, and the current is at its maximum value. A further increase in V does not affect the current. Lenard observed that the maximum current is proportional to the light intensity. When V is negative, the electrons are repelled from the anode. Only electrons with initial kinetic energies $\frac{1}{2}mv^2$ that are greater than $|eV|$ can then reach the anode. From Figure 2-5, we can see that if V is less than $-V_0$, no electrons reach the anode. The potential V_0 is called the **stopping potential**. It is related to the maximum kinetic energy of the emitted electrons by

$$(\tfrac{1}{2}mv^2)_{\max} = eV_0$$

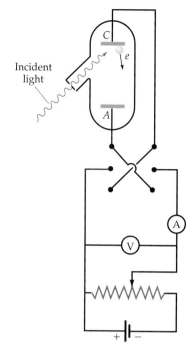

Figure 2-4 Schematic drawing of the apparatus for studying the photoelectric effect. Light strikes the cathode C and ejects electrons. The number of electrons that reach the anode A is measured by the current in the ammeter. The anode can be made positive or negative with respect to the cathode to attract or repel the electrons.

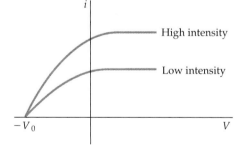

Figure 2-5 Photoelectric current i versus voltage V for two values of light intensity. There is no current when V is less than $-V_0$. The saturation current observed for large values of V is proportional to the intensity of the incident light.

The experimental result that V_0 is independent of the intensity of the incident light was surprising. Classically, increasing the rate of light energy falling on the cathode should increase the energy absorbed by an electron and should therefore increase the maximum kinetic energy of the electrons emitted. Apparently, this is not what happens. In 1905, Einstein demonstrated that this experimental result can be explained if light energy is not distributed continuously in space but rather is quantized in small bundles called **photons**. The energy of each photon is hf, where f is the frequency and h is Planck's constant. An electron emitted from a metal surface exposed to light receives its energy from a single photon. When the intensity of light of a given frequency is increased, more photons fall on the surface in unit time, but the energy absorbed by each electron is unchanged. If ϕ is the energy necessary to remove an electron from a metal surface, the maximum kinetic energy of the electrons emitted will be

Einstein's photoelectric equation

$$(\tfrac{1}{2}mv^2)_{max} = eV_0 = hf - \phi \qquad 2\text{-}4$$

The quantity ϕ, called the **work function**, is a characteristic of the particular metal. Some electrons will have kinetic energies less than $hf - \phi$ because of the loss of energy from traveling through the metal. Equation 2-4 is known as **Einstein's photoelectric equation.** From it, we can see that the slope of V_0 versus f should equal h/e.

Einstein's photoelectric equation was a bold prediction, for at the time it was made there was no evidence that Planck's constant had any applicability outside of blackbody radiation, and there were no experimental data on the stopping potential V_0 as a function of frequency. Experimental verification of Einstein's theory was quite difficult. Careful experiments by R. C. Millikan, reported first in 1914 and then in more detail in 1916, showed that Einstein's equation was correct and that measurements of h agreed with the value found by Planck. Figure 2-6 shows a plot of Millikan's data.

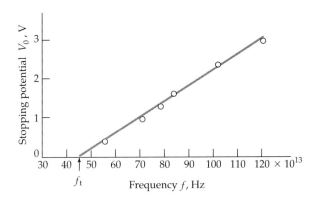

Figure 2-6 Millikan's data for the stopping potential V_0 versus frequency f for the photoelectric effect. The data fall on a straight line that has a slope h/e, as predicted by Einstein a decade before the experiment was performed.

Photons with frequencies less than a **threshold frequency** f_t, and therefore with wavelengths greater than a **threshold wavelength** λ_t, do not have enough energy to eject an electron from a particular metal. The threshold frequency and the corresponding threshold wavelength can be related to the work function ϕ by setting the maximum kinetic energy of the electrons equal to zero in Equation 2-4. Then

$$\phi = hf_t = \frac{hc}{\lambda_t} \qquad 2\text{-}5$$

Work functions for metals are typically a few electron volts. Since wavelengths are usually given in nanometers and energies in electron volts, it is useful to have the value of hc in electron volt–nanometers:

$$hc = (4.14 \times 10^{-15} \text{ eV·s})(3 \times 10^8 \text{ m/s}) = 1.24 \times 10^{-6} \text{ eV·m}$$

or

$$hc = 1240 \text{ eV·nm} \qquad 2\text{-}6$$

Example 2-1

Calculate the photon energy for light of wavelengths 400 nm (violet) and 700 nm (red). These are approximately the extreme wavelengths in the visible spectrum.

A collection of photomultiplier tubes used to detect very weak light. The face of each tube is a photosensitive area that emits electrons, via the photoelectric effect, when struck by photons. Each electron is accelerated and strikes a metal electrode, resulting in the emission of several more electrons—which are in turn accelerated and strike other electrodes. The electron beam cascades down the tube until it strikes the anode and produces a measurable electric current.

Using Equation 2-2, we have

$$E = hf = \frac{hc}{\lambda} = \frac{1240 \text{ eV} \cdot \text{nm}}{400 \text{ nm}} = 3.1 \text{ eV}$$

for $\lambda = 400$ nm. For $\lambda = 700$ nm, the photon energy is 4/7 that for $\lambda = 400$ nm or 1.77 eV. We can see from these calculations that visible light contains photons with energies that range from about 1.8 to 3.0 eV.

Example 2-2

The intensity of sunlight at the earth's surface is approximately 1400 W/m². Assuming the average photon energy is 2 eV (corresponding to a wavelength of about 600 nm), calculate the number of photons that strike an area of 1 cm² in one second.

Since 1 watt is 1 joule per second, the energy striking the earth's surface in one second is 1400 J/m². The energy per second per square centimeter is then

$$\frac{1400 \text{ J}}{\text{m}^2} \times \frac{1 \text{ m}^2}{(100 \text{ cm})^2} = 0.14 \text{ J/cm}^2$$

If N is the number of 2-eV photons that together have a total energy of 0.14 J, we have

$$N(2 \text{ eV}) = 0.14 \text{ J}$$

$$N = \frac{0.14 \text{ J}}{2 \text{ eV}} \times \frac{1 \text{ eV}}{1.6 \times 10^{-19} \text{ J}}$$

$$= 4.38 \times 10^{17} \text{ photons}$$

This is an enormous number. In most everyday situations, the number of photons is so great that a few more or less make no difference. That is, quantization is not noticed.

Example 2-3

The threshold wavelength for potassium is 564 nm. (a) What is the work function for potassium? (b) What is the stopping potential when light of wavelength 400 nm is incident on potassium?

(a) From Equation 2-5, we have for the work function

$$\phi = hf_t = \frac{hc}{\lambda_t} = \frac{1240 \text{ eV·nm}}{564 \text{ nm}} = 2.20 \text{ eV}$$

(b) The energy of a photon with a wavelength of 400 nm was calculated in Example 2-1 to be 3.1 eV. The maximum kinetic energy of the emitted electrons is then

$$(\tfrac{1}{2}mv^2)_{\max} = eV_0 = hf - \phi = 3.10 \text{ eV} - 2.20 \text{ eV} = 0.90 \text{ eV}$$

The stopping potential is therefore 0.90 V.

Exercise

Find the energy of a photon corresponding to electromagnetic radiation in the FM radio band of wavelength 3 m. (Answer: 4.13×10^{-7} eV)

Exercise

Find the wavelength of a photon whose energy is (a) 0.1 eV, (b) 1 keV, and (c) 1 MeV. [Answers: (a) 12.4 μm, (b) 1.24 nm, (c) 1.24 pm]

Another interesting feature of the photoelectric effect is the absence of any lag between the time the light first strikes the metal and the time the electrons appear. In the classical theory, given the intensity (the power per unit area), the time it takes for enough energy to fall on the area of an atom to eject an electron can be calculated. However, even when the intensity is so small that such a calculation gives a time lag of hours, essentially no time lag is observed. The explanation of this result is simple. When the intensity is low, the number of photons hitting the metal per unit time is very small, but each photon has enough energy to eject an electron. There is therefore a good chance that one photon will be absorbed immediately. The classical calculation gives the correct *average* number of electrons ejected per unit time.

2-3 X Rays

While working with a cathode-ray tube in 1895, W. Röntgen discovered that "rays" from the tube could pass through materials that were opaque to light and activate a fluorescent screen or photographic film. These rays originated from a point where the electrons in the tube hit a target within the tube or the glass tube itself. Röntgen was not able to deflect these rays in a magnetic field, as would be expected if they were charged particles, nor was he able to observe diffraction or interference, as would be expected if they were waves. He therefore gave the rays the somewhat mysterious name **x rays**. Röntgen investigated these rays extensively and found that all materials were transparent to them to some degree and that the degree of transparency decreased with increasing density of the material. This fact led to the medical use of x rays within months after Röntgen's first paper. Röntgen was the first recipient of the Nobel Prize for physics in 1901.

Since classical electromagnetic theory predicts that electric charges will radiate electromagnetic waves when they are accelerated (or decelerated), it

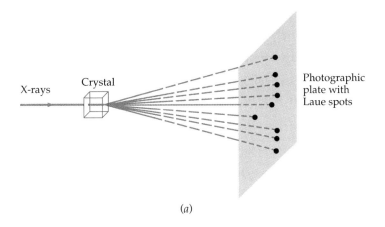
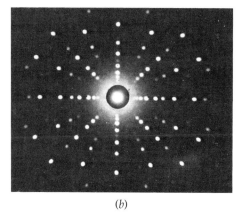

(a) (b)

Figure 2-7 (*a*) Schematic diagram of the Laue experiment. The crystal acts as a three-dimensional grating that diffracts the x-ray beam and produces a regular array of spots called a *Laue pattern* on a photographic plate. (*b*) A modern Laue x-ray diffraction pattern from a niobium diboride crystal using 20-kV x rays from a molybdenum target.

was natural to assume that x rays are electromagnetic waves produced when electrons decelerate as they are stopped by a target. A few years later, a slight broadening of an x-ray beam after it passed through slits a few thousandths of a millimeter wide was observed. This was assumed to be due to diffraction, and the wavelength of x rays was estimated to be about 0.1 nm. In 1912, M. Laue suggested that, since the wavelengths of x rays were of the same order of magnitude as the spacing of the atoms in a crystal, the regular array of atoms in a crystal might act as a three-dimensional grating for the diffraction of x rays. Acting on this suggestion, W. Friedrich and P. Knipping allowed a collimated beam of x rays to pass through a crystal and strike a photographic plate (Figure 2-7a). In addition to the central beam, they observed a regular array of spots like those shown in Figure 2-7b. From an analysis of the positions of the spots, they were able to calculate that their x-ray beam had wavelengths ranging from about 0.01 to 0.05 nm. This important experiment confirmed two important assumptions: (1) x rays are electromagnetic radiation and (2) the atoms in crystals are arranged in a regular array.

Figure 2-8 shows a plot of intensity versus wavelength for the spectrum emitted from a typical x-ray tube, in which a target (molybdenum in this case) is bombarded with electrons. The spectrum consists of a series of sharp lines called the **characteristic spectrum** superimposed on a continuous spectrum called a **bremsstrahlung spectrum** (from the German for "braking radiation"). The line spectrum is characteristic of the target material and varies from element to element. It is similar to the optical spectrum of the elements except the x-ray spectrum involves transitions of the inner atomic electrons whereas the optical spectrum involves transitions of the outer atomic electrons. We will discuss both the optical spectrum and the characteristic x-ray spectrum in Chapter 4. The continuous bremsstrahlung spectrum is produced by the rapid deceleration of the bombarding electrons when they crash into the target. If the voltage across the x-ray tube is V, the maximum kinetic energy of the electrons is eV when they hit the target. Often several photons are emitted as an electron slows down. However, sometimes just one photon with the maximum energy eV is emitted. Since the wavelength of a photon varies inversely with its energy ($\lambda = hc/hf = hc/E$), the minimum wavelength in the bremsstrahlung spectrum corresponds to a photon with the maximum energy eV. The minimum wavelength is called the **cutoff wavelength** and is labeled λ_m in the figure. The cutoff wavelength is related to the voltage of the x-ray tube by

$$\lambda_m = \frac{hc}{E} = \frac{hc}{eV} \qquad 2\text{-}7$$

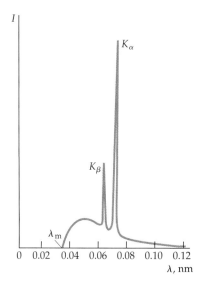

Figure 2-8 X-ray spectrum of molybdenum. The sharp peaks labeled K_α and K_β are characteristic of the target element. The cutoff wavelength λ_m is independent of the target element and is related to the voltage of the x-ray tube by $\lambda_m = hc/eV$.

Example 2-4

What is the minimum wavelength of the x rays emitted by a television picture tube with a voltage of 2000 V?

The maximum kinetic energy of the electrons is 2000 eV, so this will be the maximum energy of the photons in the x-ray spectrum. The wavelength of a photon of this energy is the cutoff wavelength, which from Equation 2-7 is

$$\lambda_m = \frac{hc}{E} = \frac{1240 \text{ eV}\cdot\text{nm}}{2000 \text{ eV}} = 0.62 \text{ nm}$$

Exercise

An x-ray tube operates at a potential of 30 kV. What is the minimum wavelength of the continuous x-ray spectrum from this tube? (Answer: 0.041 nm)

2-4 Compton Scattering

Further evidence of the correctness of the photon concept was furnished by Arthur H. Compton, who measured the scattering of x rays by free electrons. According to classical theory, when an electromagnetic wave of frequency f_1 is incident on material containing charges, the charges will oscillate with this frequency and will reradiate electromagnetic waves of the same frequency. Compton pointed out that if this interaction were described as a scattering process involving a collision between a photon and an electron, the electron would recoil and thus absorb energy. The scattered photon would then have less energy and therefore a lower frequency than the incident photon.

According to classical theory, the energy and momentum of an electromagnetic wave are related by

$$E = pc \qquad 2\text{-}8$$

This result is consistent with the relativistic expression relating the energy and momentum of a particle (Equation 1-34),

$$E^2 = p^2c^2 + (mc^2)^2$$

if the mass m of the photon is assumed to be zero. Figure 2-9 shows the geometry of a collision between a photon of wavelength λ_1 and an electron initially at rest. Compton related the scattering angle θ to the incident and scattered wavelengths λ_1 and λ_2 by treating the scattering as a relativistic-mechanics problem and using the conservation of energy and momentum. Let $\mathbf{p}_1$ be the momentum of the incident photon, $\mathbf{p}_2$ be that of the scattered photon, and $\mathbf{p}_e$ be that of the recoiling electron. The conservation of momentum gives

$$\mathbf{p}_1 = \mathbf{p}_2 + \mathbf{p}_e \qquad 2\text{-}9$$

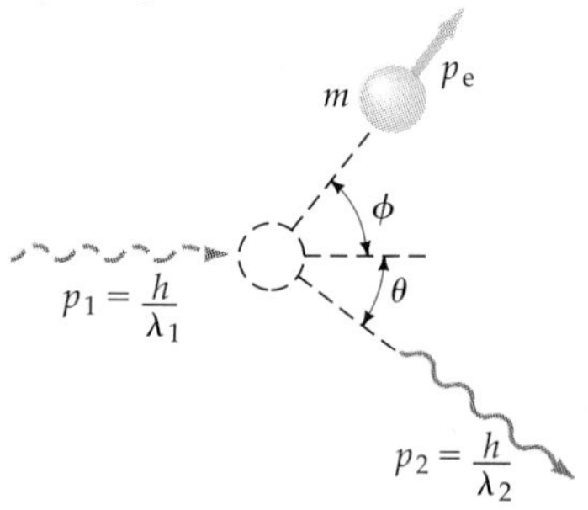

Figure 2-9 Compton scattering of an x ray by an electron. The scattered photon has less energy and therefore a greater wavelength than the incident photon because of the recoil energy of the electron. The change in wavelength is found from conservation of energy and momentum.

or
$$\mathbf{p}_e = \mathbf{p}_1 - \mathbf{p}_2$$
Taking the dot product of each side with itself, we obtain
$$p_e^2 = p_1^2 + p_2^2 - 2\mathbf{p}_1 \cdot \mathbf{p}_2$$
or
$$p_e^2 = p_1^2 + p_2^2 - 2p_1 p_2 \cos\theta \qquad 2\text{-}10$$

The energy before the collision is $p_1 c + mc^2$, where mc^2 is the rest energy of the electron. After the collision, the electron has energy $\sqrt{(mc^2)^2 + p_e^2 c^2}$. Conservation of energy then gives
$$p_1 c + mc^2 = p_2 c + \sqrt{(mc^2)^2 + p_e^2 c^2} \qquad 2\text{-}11$$

Compton eliminated the electron momentum p_e from Equations 2-10 and 2-11 and expressed the photon momenta in terms of the wavelengths to obtain an equation relating the incident and scattered wavelengths λ_1 and λ_2 and the angle θ. The algebraic details are left as a problem (see Problem 60). Compton's result is

$$\lambda_2 - \lambda_1 = \frac{h}{mc}(1 - \cos\theta) \qquad 2\text{-}12$$

The change in wavelengths is independent of the original wavelength. The quantity h/mc depends only on the mass of the electron. It has the dimension of length and is called the **Compton wavelength.** Its value is

$$\lambda_C = \frac{h}{mc} = \frac{hc}{mc^2} = \frac{1240 \text{ eV}\cdot\text{nm}}{5.11 \times 10^5 \text{ eV}} = 2.43 \times 10^{-12} \text{ m} = 2.43 \text{ pm} \qquad 2\text{-}13$$

Because $\lambda_2 - \lambda_1$ is small, it is difficult to observe unless λ_1 is so small that the fractional change $(\lambda_2 - \lambda_1)/\lambda_1$ is appreciable. Compton used x rays of wavelength 71.1 pm. The energy of a photon of this wavelength is $E = hc/\lambda = (1240 \text{ eV}\cdot\text{nm})/(0.0711 \text{ nm}) = 17.4$ keV. Since this is much greater than the binding energy of the valence electrons in atoms (which is of the order of a few electron volts), these electrons can be considered to be essentially free. Compton's experimental results for $\lambda_2 - \lambda_1$ as a function of the scattering angle θ agreed with Equation 2-12, thereby confirming the correctness of the photon concept.

Example 2-5

Calculate the percentage change in wavelength observed in the Compton scattering of 20-keV photons at $\theta = 60°$.

The change in wavelength at $\theta = 60°$ is given by Equation 2-12:
$$\lambda_2 - \lambda_1 = \lambda_C(1 - \cos\theta) = (2.43 \text{ pm})(1 - \cos 60°) = 1.22 \text{ pm}$$

The wavelength of the incident 20-keV photons is
$$\lambda_1 = \frac{1240 \text{ eV}\cdot\text{nm}}{20,000 \text{ eV}} = 0.062 \text{ nm} = 62 \text{ pm}$$

The percentage change in wavelength is thus
$$\frac{\lambda_2 - \lambda_1}{\lambda_1} = \frac{1.22 \text{ pm}}{62 \text{ pm}} \times 100\% = 1.97\%$$

2-5 Quantization of Atomic Energies: The Bohr Model

The most famous application of energy quantization to microscopic systems was that of Niels Bohr. In 1913, Bohr proposed a model of the hydrogen atom that had spectacular success in calculations of the wavelengths of the lines in the known hydrogen spectrum and in predicting new lines (later found experimentally) in the infrared and ultraviolet spectra.

Near the turn of the century, much data was collected on the emission of light by atoms in a gas when they are excited by an electric discharge. Viewed through a spectroscope with a narrow-slit aperture, this light appears as a discrete set of lines of different colors or wavelengths; the spacing and intensities of the lines are characteristic of the element. It was possible to determine the wavelengths of these lines accurately, and much effort went into finding regularities in the spectra. In 1884, a Swiss schoolteacher, Johann Balmer, found that the wavelengths of some of the lines in the spectrum of hydrogen can be represented by the formula

$$\lambda = (364.6 \text{ nm}) \frac{m^2}{m^2 - 4} \qquad 2\text{-}14$$

where m is an integer that takes on the values $m = 3, 4, 5, \ldots$. Figure 2-10 shows the set of spectral lines of hydrogen, now known as the **Balmer series,** whose wavelengths are given by Equation 2-14.

Figure 2-10 The Balmer series for light emitted from the hydrogen atom. The wavelengths of these lines are given by Equation 2-14 for different values of the integer m.

Balmer suggested that his formula might be a special case of a more general expression that would be applicable to the spectra of other elements. Such an expression, found by Johannes R. Rydberg and Walter Ritz and known as the **Rydberg–Ritz formula,** gives the reciprocal wavelength as

$$\frac{1}{\lambda} = RZ^2 \left(\frac{1}{n_2^2} - \frac{1}{n_1^2} \right) \qquad n_1 > n_2 \qquad 2\text{-}15$$

This formula is valid not only for hydrogen, with atomic number $Z = 1$, but also for heavier atoms of nuclear charge Ze from which all electrons but one have been removed. R, called the **Rydberg constant** or simply the **Rydberg,** is the same for all spectral series of the same element and varies only slightly in a regular way from element to element. For very massive elements, R approaches the value

$$R_\infty = 10.97373 \ \mu\text{m}^{-1} \qquad 2\text{-}16$$

If we take the reciprocal of Equation 2-14 for the Balmer series, we obtain

$$\frac{1}{\lambda} = \frac{1}{364.6 \text{ nm}} \left(\frac{m^2 - 4}{m^2} \right) = \frac{1}{364.6 \text{ nm}} \left(\frac{1}{1} - \frac{4}{m^2} \right)$$

$$= \frac{4}{364.6 \text{ nm}} \left(\frac{1}{4} - \frac{1}{m^2} \right) = 10.97 \ \mu\text{m}^{-1} \left(\frac{1}{2^2} - \frac{1}{m^2} \right)$$

We can thus see that the Balmer formula is indeed a special case of the Rydberg–Ritz formula (Equation 2-15) for hydrogen with $n_2 = 2$ and $n_1 = $

m. The Rydberg–Ritz formula and various modifications of it were very successful in predicting other spectra. For example, hydrogen lines outside the visible spectrum were predicted and found. Setting $n_2 = 1$ in Equation 2-15 leads to a series in the ultraviolet region called the *Lyman series*, whereas setting $n_2 = 3$ leads to the *Paschen series* in the infrared region.

Many attempts were made to construct a model of the atom that would yield these formulas for its radiation spectrum. The most popular early model, developed by J. J. Thomson, considered electrons to be embedded in various arrangements in some kind of fluid that contained most of the mass of the atom and had enough positive charge to make the atom electrically neutral. Thomson's model, called the "plum pudding" model, is illustrated in Figure 2-11. Since classical electromagnetic theory predicted that a charge oscillating with a frequency f would radiate light of the same frequency, Thomson searched for configurations of electrons that were stable and had normal modes of vibration with frequencies equal to those of the spectrum of the atom. A difficulty with this model and all others was that electric forces alone cannot produce stable equilibrium. Thomson was unable to find a configuration of electrons that predicted the observed frequencies for any atom.

Figure 2-11 J. J. Thomson's "plum pudding" model of the atom. In this model, the negative electrons are embedded in a fluid of positive charge. For a given configuration of electrons in such a system, the resonance frequencies of the oscillations of the electrons can be calculated. According to classical theory, the atom should radiate light with a frequency equal to the frequency of oscillation of the electrons. Thomson could not find any configuration of electrons that would give frequencies that agreed with the measured frequencies of the spectrum of any atom.

The Thomson model was essentially ruled out by a set of experiments performed by H. W. Geiger and E. Marsden under the supervision of E. Rutherford around 1911 in which alpha particles from radioactive radium were scattered by atoms in a gold foil. Rutherford showed that the number of alpha particles scattered at large angles could not be accounted for by an atom in which the positive charge was distributed throughout the atomic volume (known to be about 0.1 nm in diameter). The findings required that the positive charge and most of the mass of the atom be concentrated in a very small region, now called the nucleus, with a diameter of the order of 10^{-6} nm = 1 fm. (Before the establishment of the SI, the femtometer, 1 fm = 10^{-15} m, was called a *fermi* after the Italian physicist Enrico Fermi.)

Niels Bohr, who was working in the Rutherford laboratory at the time, proposed a model of the hydrogen atom that combined the work of Planck, Einstein, and Rutherford and successfully predicted the observed spectra. Bohr assumed that the electron in the hydrogen atom moved under the influence of the Coulomb attraction between it and the positive nucleus according to classical mechanics, which predicts circular or elliptical orbits with the force center at one focus, as in the motion of the planets around the sun. For simplicity he chose a circular orbit as shown in Figure 2-12. Although mechanical stability is achieved because the Coulomb attractive force provides the centripetal force necessary for the electron to remain in orbit, such an atom would be unstable electrically according to classical theory because the electron must accelerate when moving in a circle and must therefore radiate electromagnetic energy of a frequency equal to that of its motion. According to classical electromagnetic theory, such an atom would quickly collapse because the electron would spiral into the nucleus as it radiates away its energy.

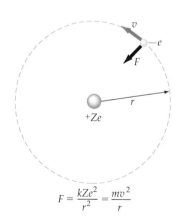

Figure 2-12 An electron of charge $-e$ traveling in a circular orbit of radius r around the nuclear charge $+Ze$. The attractive electrical force kZe^2/r^2 provides the centripetal force to hold the electron in its orbit.

60 Chapter 2 The Origins of Quantum Theory

Bohr's first postulate: nonradiating orbits

Bohr "solved" this difficulty by modifying the laws of electromagnetism and *postulating* that the electron could only move in certain nonradiating orbits. This idea is referred to as Bohr's first postulate. He called these stable orbits **stationary states.** The atom radiates only when the electron somehow makes a transition from one stationary state to another. The frequency of the radiation is not the frequency of the electron's motion in either stable orbit. Instead, it is related to the energies of the orbits by

Bohr's second postulate: photon frequency from energy conservation

$$f = \frac{E_i - E_f}{h} \qquad 2\text{-}17$$

where h is Planck's constant and E_i and E_f are the total energies in the initial and final orbits, respectively. This assumption, which is equivalent to the assumption of conservation of energy with the emission of a photon, is a key one in the Bohr theory because it deviates from classical theory, which requires the frequency of radiation to be that of the motion of the charged particle.

If the nuclear charge is $+Ze$ and there is only one electron of charge $-e$, the potential energy at a distance r is

$$U = -\frac{kZe^2}{r}$$

where k is the Coulomb constant. (For hydrogen, $Z = 1$, but it is convenient not to specify Z at this time so that the results can be applied to other hydrogen-like atoms.) The total energy of the electron moving in a circular orbit with a speed v is then

$$E = \frac{1}{2}mv^2 + U = \frac{1}{2}mv^2 - \frac{kZe^2}{r}$$

The kinetic energy can be obtained as a function of r by using Newton's second law $F = ma$. Setting the Coulomb attractive force equal to the mass times the centripetal acceleration, we obtain

$$\frac{kZe^2}{r^2} = m\frac{v^2}{r}$$

or

$$\frac{1}{2}mv^2 = \frac{1}{2}\frac{kZe^2}{r} \qquad 2\text{-}18$$

For circular orbits, the kinetic energy equals half the magnitude of the potential energy, a result that holds for circular motion in any inverse-square force field. The total energy is then

$$E = \frac{1}{2}\frac{kZe^2}{r} - \frac{kZe^2}{r} = -\frac{1}{2}\frac{kZe^2}{r} \qquad 2\text{-}19$$

Using Equation 2-17 for the frequency of radiation emitted when the electron changes from orbit 1 of radius r_1 to orbit 2 of radius r_2, we obtain

$$f = \frac{E_1 - E_2}{h} = \frac{1}{2}\frac{kZe^2}{h}\left(\frac{1}{r_2} - \frac{1}{r_1}\right) \qquad 2\text{-}20$$

To obtain the Rydberg–Ritz formula, $f = c/\lambda = cR(1/n_2^2 - 1/n_1^2)$, it is evident that the radii of stable orbits must be proportional to the squares of integers. Bohr searched for a quantum condition for the radii of the stable orbits that would yield this result. After much trial and error, he found that he could obtain it if he postulated that the angular momentum of the electron in a stable orbit equals an integer times Planck's constant divided by 2π. Since the angular momentum of a circular orbit is just mvr, this postulate is

$$mvr = \frac{nh}{2\pi} = n\hbar \qquad 2\text{-}21$$

Bohr's third postulate: quantized angular momentum

where

$$\hbar = \frac{h}{2\pi} = 1.05 \times 10^{-34} \text{ J·s}$$

(The constant $\hbar = h/2\pi$, read "h bar", is often more convenient to use than h itself, just as the angular frequency $\omega = 2\pi f$ is often more convenient to use than the frequency f.) We can determine r by eliminating v between Equations 2-18 and 2-21. Solving Equation 2-21 for v and squaring, and using Equation 2-18, we obtain

$$v^2 = n^2 \frac{\hbar^2}{m^2 r^2} = \frac{kZe^2}{mr}$$

Solving for r we obtain

$$r = n^2 \frac{\hbar^2}{mkZe^2} = n^2 \frac{a_0}{Z} \qquad 2\text{-}22$$

where

$$a_0 = \frac{\hbar^2}{mke^2} \approx 0.0529 \text{ nm} \qquad 2\text{-}23$$

Bohr radius

is called the first **Bohr radius.** Combining Equations 2-22 and 2-20, we get

$$f = Z^2 \frac{mk^2e^4}{4\pi\hbar^3}\left(\frac{1}{n_2^2} - \frac{1}{n_1^2}\right) \qquad 2\text{-}24$$

If we compare this expression for $f = c/\lambda$ with the empirical Rydberg–Ritz formula (Equation 2-15), we obtain for the Rydberg constant

$$R = \frac{mk^2e^4}{4\pi c \hbar^3} \qquad 2\text{-}25$$

Using the values of m, e, and $\hbar$ known in 1913, Bohr calculated R and found his result to agree (within the limits of the uncertainties of the constants) with the value obtained from spectroscopy. Figure 2-13 illustrates the Bohr model of the hydrogen atom.

The possible values of the energy of the hydrogen atom predicted by the Bohr model and given by Equation 2-19, when r is given by Equation 2-22, are

$$E_n = -\frac{k^2e^4m}{2\hbar^2}\frac{Z^2}{n^2} = -Z^2\frac{E_0}{n^2} \qquad 2\text{-}26$$

Energy levels

where

$$E_0 = \frac{k^2e^4m}{2\hbar^2} \approx 13.6 \text{ eV} \qquad 2\text{-}27$$

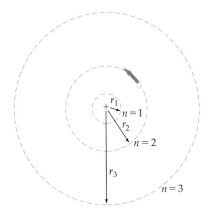

Figure 2-13 Stable orbits in the Bohr model of the hydrogen atom. The radii of the stable orbits are given by $r_n = n^2 a_0$, where n is an integer and a_0 is the smallest radius.

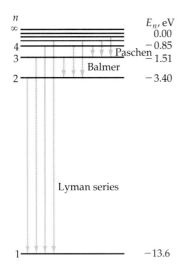

Figure 2-14 An energy-level diagram for hydrogen showing a few of the transitions in the Lyman, Balmer, and Paschen series. The energies of the levels are given by Equation 2-26.

It is sometimes convenient to represent these energies in an **energy-level diagram**, as in Figure 2-14. The lowest energy level is called the ground state. The energy of the hydrogen atom in the ground state is -13.6 eV. The highest energy state E_n is reached as $n \to \infty$ or $r \to \infty$, as can be seen from Equations 2-26 and 2-22. This process of removing the electron from an atom is termed **ionization**. The ionization energy of the hydrogen atom has been measured as 13.6 eV. This is thus the energy with which the electron is bound to the atom.

Various series of radiative transitions are indicated in Figure 2-14 by vertical arrows between the energy levels. The frequency of the light emitted in one of these transitions is the energy difference divided by h (Equation 2-17). At the time of Bohr's paper (1913), the Balmer series, corresponding to $n_2 = 2$ and $n_1 = 3, 4, 5, \ldots$, and the Paschen series, corresponding to $n_2 = 3$ and $n_1 = 4, 5, 6, \ldots$, were known. In 1916, T. Lyman found the series corresponding to $n_2 = 1$, and in 1922 and 1924, F. Brackett and H. A. Pfund, respectively, found series corresponding to $n_2 = 4$ and $n_2 = 5$. As can be determined by computing the wavelengths of these series, only the Balmer series lies in the visible portion of the electromagnetic spectrum.

In our derivations, we have assumed that the electron revolves around a stationary nucleus. This is equivalent to assuming that the nucleus has infinite mass. Since the mass of the hydrogen nucleus is not infinite but about 2000 times that of the electron, a correction must be made for the motion of the nucleus. This correction leads to a very slight dependence of the Rydberg constant, as given in Equation 2-25, on the nuclear mass, in precise agreement with the observed variation.

Example 2-6

Find the energy and wavelength of the line with the longest wavelength in the Lyman series.

From Figure 2-14, we can see that the Lyman series corresponds to transitions ending at the ground-state energy, $E_f = E_1 = -13.6$ eV. Since λ varies inversely with energy, the transition with the longest wavelength is the transition with the lowest energy, which is that from the first excited state $n = 2$ to the ground state $n = 1$. The energy of the first excited state is $E_2 = (-13.6 \text{ eV})/4 = -3.40$ eV. Since this is 10.2 eV above the ground-state energy, the energy of the photon emitted is 10.2 eV. The wavelength of this photon is

$$\lambda = \frac{hc}{\Delta E} = \frac{1240 \text{ eV} \cdot \text{nm}}{10.2 \text{ eV}} = 121.6 \text{ nm}$$

This photon is outside the visible spectrum and in the ultraviolet region. Since all the other lines in the Lyman series have even greater energies and shorter wavelengths, the Lyman series is completely in the ultraviolet region.

Exercise

Find the shortest wavelength for a line in the Lyman series. (Answer: 91.2 nm)

Questions

1. If an electron moves to a larger orbit, does its total energy increase or decrease? Does its kinetic energy increase or decrease?

2. How does the spacing of adjacent energy levels change as n increases?

3. What is the energy of the photon with the shortest wavelength emitted by the hydrogen atom?

2-6 Electron Waves and Quantum Theory

In 1924, a French student, L. de Broglie, suggested in his dissertation that electrons may have wave properties. His reasoning was based on the symmetry of nature. Since light was known to have both wave and particle properties, perhaps matter—especially electrons—might also have both wave and particle characteristics. This suggestion was highly speculative since there was no evidence at that time of any wave aspects of electrons. For the frequency and wavelength of electron waves, de Broglie chose the equations

$$f = \frac{E}{h} \qquad \text{2-28}$$

$$\lambda = \frac{h}{p} \qquad \text{2-29}$$

where p is the momentum and E is the energy of the electron. Equation 2-28 is the same as the Planck–Einstein equation for the energy of a photon. Equation 2-29 also holds for photons, as can be seen from

$$\lambda = \frac{c}{f} = \frac{hc}{hf} = \frac{hc}{E}$$

Since the momentum of a photon is related to its energy by $E = pc$, we have

$$\lambda = \frac{hc}{pc} = \frac{h}{p}$$

De Broglie's equations are thought to apply to all matter. However, for macroscopic objects, the wavelengths calculated from Equation 2-29 are so small that it is impossible to observe the usual wave properties of interference or diffraction. Even a particle as small as 1 μg is much too massive for any wave characteristics to be noticed, as we will see in the following example.

Example 2-7

Find the de Broglie wavelength of a particle of mass 10^{-6} g moving with a speed of 10^{-6} m/s.

From Equation 2-29, we have

$$\lambda = \frac{h}{p} = \frac{h}{mv} = \frac{6.63 \times 10^{-34} \text{ J·s}}{(10^{-9} \text{ kg})(10^{-6} \text{ m/s})}$$

$$= 6.63 \times 10^{-19} \text{ m}$$

Since the wavelength found in Example 2-7 is much smaller than any possible apertures or obstacles (the diameter of the nucleus of an atom is about 10^{-15} m, roughly 10,000 times this wavelength), diffraction or interference of such waves cannot be observed. As in classical physics, the propagation of waves of very small wavelength is indistinguishable from the propagation of particles. Note that the momentum in Example 2-7 is extremely small. A macroscopic particle with a greater momentum would have an even smaller de Broglie wavelength. We therefore do not observe the wave properties of such macroscopic objects as baseballs or billiard balls.

Exercise

Find the de Broglie wavelength of a baseball of mass 0.17 kg moving at 100 km/h. (Answer: 1.4×10^{-34} m)

The situation is different for low-energy electrons. Consider an electron with kinetic energy K. If the electron is nonrelativistic, its momentum is found from

$$K = \frac{p^2}{2m}$$

or

$$p = \sqrt{2mK}$$

Its wavelength is then

$$\lambda = \frac{h}{p} = \frac{h}{\sqrt{2mK}} = \frac{hc}{\sqrt{2mc^2 K}}$$

Using $hc = 1240$ eV·nm and $mc^2 = 0.511$ MeV, we obtain

$$\lambda = \frac{1240 \text{ eV·nm}}{\sqrt{2(0.511 \times 10^6 \text{ eV})K}}$$

or

$$\lambda = \frac{1.226}{\sqrt{K}} \text{ nm} \qquad K \text{ in electron volts} \qquad 2\text{-}30$$

From Equation 2-30, we can see that electrons with energies of the order of tens of electron volts have de Broglie wavelengths of the order of nanometers. This is the order of magnitude of the size of the atom and the spacing of atoms in a crystal. Thus, when electrons with energies of the order of 10 eV are incident on a crystal, they are scattered in much the same way as are x rays of the same wavelength.

Exercise

Find the wavelength of an electron whose kinetic energy is 10 eV. (Answer: 0.388 nm)

The crucial test for the existence of wave properties of electrons was the observation of diffraction and interference of electron waves. This was first accomplished accidentally in 1927 by C. J. Davisson and L. H. Germer as they were studying electron scattering from a nickel target at the Bell Telephone Laboratories. After heating the target to remove an oxide coating that had accumulated during an accidental break in the vacuum system, Davisson and Germer found that the intensity of the scattered electrons as a function of the scattering angle showed maxima and minima. Their target had crystallized, and by accident they had observed electron diffraction. They then prepared a target consisting of a single crystal of nickel and investigated this phenomenon extensively. Figure 2-15 illustrates their experiment. Electrons from an electron gun are directed at a crystal and are detected at some ϕ that can be varied. Figure 2-16 shows a typical pattern observed. There is a strong scattering maximum at an angle of 50°. The angle for maximum intensity of scattering of waves from a crystal depends on the wavelength of the waves and the spacing of the atoms in the crystal. Using the known spacing of atoms in their crystal, Davisson and Germer calculated the wavelength that could produce such a maximum and found that it agreed with the de Broglie equation (Equation 2-29) for the electron energy they were using. By varying the energy of the incident electrons, they could vary the electron wavelengths and produce maxima and minima at different locations in the diffraction patterns. In all cases, the measured wavelengths agreed with de Broglie's hypothesis.

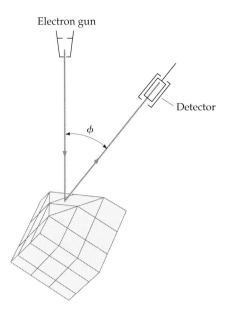
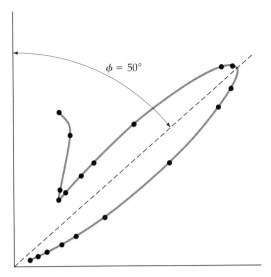

Figure 2-15 The Davisson–Germer experiment. Electrons from the electron gun incident on a crystal are scattered into a detector at some angle ϕ that can be varied.

Figure 2-16 Plot of intensity versus angle for the scattered electron in the Davisson–Germer experiment. If this pattern is assumed to be a diffraction–interference pattern, the wavelength of the electrons can be calculated from the known spacing of the atoms in the crystal and the position of the maximum. The result agrees with the de Broglie hypothesis for the wavelength of an electron.

In the same year G. P. Thomson (son of J. J. Thomson) also observed electron diffraction in the transmission of electrons through thin metal foils. A metal foil consists of tiny, randomly oriented crystals. The diffraction pattern resulting from such a foil is a set of concentric circles. Since Thomson performed his experiment, diffraction has been observed for neutrons, protons, and other particles. Figure 2-17a to c shows the diffraction patterns of x rays, electrons, and neutrons of similar wavelength transmitted through thin metal foils. Figure 2-17d shows a diffraction pattern produced by electrons incident on two narrow slits. This experiment is equivalent to Young's famous double-slit diffraction–interference experiment with light. The pattern is identical to that observed with photons of the same wavelength.

Shortly after the wave properties of the electron were demonstrated, it was suggested that electrons rather than light might be used to "see" small objects. Today, the electron microscope is an important research tool. Figure

(a)

(b)

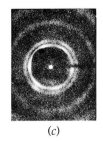
(c)

(d)

Figure 2-17 Diffraction pattern produced by (a) x rays and (b) electrons incident on an aluminum foil target and (c) neutrons incident on a target of polycrystalline copper. Note the similarity in the patterns produced. (d) A two-slit electron diffraction–interference pattern. This pattern is the same as that obtained with photons.

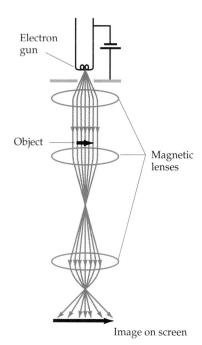

Figure 2-18 Electron microscope. Electrons from a heated filament (the electron gun) are accelerated by a large voltage difference. The electron beam is made parallel by a magnetic focusing lens. The electrons strike a thin target and are then focused by a second magnetic lens that is equivalent to the objective lens in an ordinary microscope. The third magnetic lens takes the place of the eyepiece in a microscope. It projects the electron beam onto a fluorescent screen for viewing the image.

2-18 illustrates the features of an electron microscope. The electron beam is made parallel and focused by specially designed magnets that serve as lenses. The energy of the electrons is typically 100 keV, resulting in a wavelength of about 0.004 nm. The target specimen must be very thin so that the transmitted beam will not be slowed down or scattered too much. The final image is projected onto a fluorescent screen or film. Various distortions resulting from focusing problems with the magnetic lenses limit the resolution to a few tenths of a nanometer, which is about a thousand times better than can be achieved with visible light.

Standing Waves and Energy Quantization

De Broglie pointed out that the Bohr quantum condition (Equation 2-21) for the angular momentum of the electron in a hydrogen atom is equivalent to a standing-wave condition. This condition states that

$$mvr = n\frac{h}{2\pi}$$

Substituting h/λ for the momentum mv gives

$$\frac{h}{\lambda}r = n\frac{h}{2\pi}$$

or

$$n\lambda = 2\pi r = C \qquad 2\text{-}31$$

where C is the circumference of the Bohr orbit. Thus, Bohr's quantum condition is equivalent to saying that an integral number of electron waves must fit into the circumference of the circular orbit as shown in Figure 2-19.

Example 2-8

The kinetic energy of the electron in the ground (lowest energy) state of the hydrogen atom is 13.6 eV. (Its potential energy is -27.2 eV and its total energy is -13.6 eV, leading to a binding energy of 13.6 eV.) Find the de Broglie wavelength for this electron.

Using $K = 13.6$ eV in Equation 2-30, we have

$$\lambda = \frac{1.226}{\sqrt{13.6}} \text{ nm} = 0.332 \text{ nm} = 2\pi(0.0529 \text{ nm})$$

This is the circumference of the first Bohr orbit in the hydrogen atom.

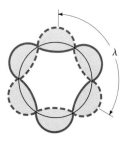

Figure 2-19 Standing waves around the circumference of a circle.

The fitting of an integral number of electron waves into the circumference of a Bohr orbit is similar to the fitting of an integral number of half wavelengths into the length of a string or organ pipe, as in standing waves on strings or standing sound waves. In classical wave theory, standing waves lead to a quantization of frequency. For example, for standing waves on a string of length L fixed at both ends (Figure 2-20), the standing wave condition is

$$n\frac{\lambda}{2} = L$$

For waves traveling with a speed v, the frequency of such standing waves on a string is then given by

$$f = \frac{v}{\lambda} = n\frac{v}{2L}$$

If energy is associated with the frequency of a standing wave, as in Equation 2-28, then standing waves imply quantized energies.

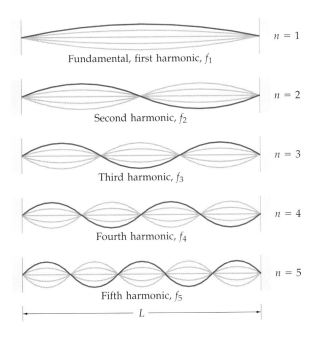

Figure 2-20 Standing waves on a string fixed at both ends. The frequencies of these waves are quantized; that is, they may have only certain values given by $f_n = nf_1$, where f_1 is the fundamental frequency.

The idea of explaining the discrete energy states of matter by standing waves led to the development of a detailed mathematical theory by Erwin Schrödinger and others in 1928. In this theory, known as **quantum theory, quantum mechanics,** or **wave mechanics,** the electron is described by a wave function ψ that obeys a wave equation that is somewhat similar to the classical wave equations for sound and light waves. The frequency and wavelength of electron waves are related to the energy and momentum of the electron just as the frequency and wavelength of light waves are related to the energy and momentum of photons. Schrödinger solved the standing-wave problem for the hydrogen atom, the simple harmonic oscillator, and other systems of interest. He found that the allowed frequencies combined with the de Broglie relation $E = hf$ led to the set of energy levels for the hydrogen atom found by Bohr (Equation 2-26), thereby demonstrating that quantum theory provides a general method of finding the quantized energy levels for a given system. Quantum theory is the basis for our understanding of the modern world, from the inner workings of the atomic nucleus to the radiation spectrum of distant galaxies in cosmology.

Summary

1. The energy in electromagnetic radiation is not continuous but comes in quanta with energies given by

$$E = hf = \frac{hc}{\lambda}$$

   where f is the frequency, λ is the wavelength, and h is Planck's constant, which has the value

$$h = 6.626 \times 10^{-34} \text{ J·s} = 4.136 \times 10^{-15} \text{ eV·s}$$

   The quantity hc occurs often in calculations and has the value

$$hc = 1240 \text{ eV·nm}$$

   The quantum nature of light is exhibited in the photoelectric effect, in which a photon is absorbed by an atom with the emission of an electron, and in Compton scattering, in which a photon collides with a free electron and emerges with reduced energy and therefore a greater wavelength.

2. X rays are emitted when electrons are decelerated by crashing into a target in an x-ray tube. An x-ray spectrum consists of a series of sharp lines called the characteristic spectrum superimposed on the continuous bremsstrahlung spectrum. The minimum wavelength in the bremsstrahlung spectrum λ_m corresponds to the maximum energy of the emitted photon, which equals the maximum kinetic energy of the electrons eV, where V is the voltage of the x-ray tube. The minimum wavelength is then given by

$$\lambda_m = \frac{hc}{eV}$$

3. The wavelengths of x rays are typically a few nanometers, which is also approximately equal to the spacing of atoms in a crystal. Diffraction maxima are observed when x rays are scattered from a crystal, indicating that x rays are electromagnetic waves and that the atoms in crystals are arranged in a regular array.

4. In order to derive the Balmer formula for the spectrum of the hydrogen atom, Bohr proposed the following postulates:

   Postulate 1: The electron in the hydrogen atom can move only in certain nonradiating circular orbits called stationary states.

   Postulate 2: The atom radiates a photon when the electron makes a transition from one stationary orbit to another. The frequency of the photon is given by

$$f = \frac{E_i - E_f}{h}$$

   where E_i and E_f are the initial and final energies of the atom.

   Postulate 3: The radius (and hence the energy) of a stationary state orbit is determined by classical physics together with the quantum condition that the angular momentum of the electron must equal an integer times Planck's constant divided by 2π:

$$mvr = \frac{nh}{2\pi} = n\hbar$$

   where $\hbar = h/2\pi = 1.05 \times 10^{-34}$ J·s.

These postulates lead to allowed energy levels in the hydrogen atom given by

$$E_n = -\frac{k^2 e^4 m}{2\hbar^2} \frac{Z^2}{n^2} = -Z^2 \frac{E_0}{n^2}$$

where n is an integer and

$$E_0 = \frac{k^2 e^4 m}{2\hbar^2} \approx 13.6 \text{ eV}$$

The radii of the stationary orbits are given by

$$r = n^2 \frac{\hbar^2}{mkZe^2} = n^2 \frac{a_0}{Z}$$

where

$$a_0 = \frac{\hbar^2}{mke^2} \approx 0.0529 \text{ nm}$$

is the first Bohr radius.

5. The wave nature of electrons was first suggested by de Broglie, who postulated the equations

$$f = \frac{E}{h} \quad \text{and} \quad \lambda = \frac{h}{p}$$

for the frequency and wavelength of electron waves. With these equations, the Bohr quantum condition can be understood as a standing-wave condition. The wave nature of electrons was observed experimentally first by Davisson and Germer and later by G. P. Thomson, who measured the diffraction and interference of electrons.

6. The mathematical theory of the wave nature of matter is known as quantum theory. In this theory, the electron is described by a wave function that obeys a wave equation. Energy quantization arises from standing-wave conditions applied to electrons in various systems. Quantum theory is the basis for our understanding of the physical nature of the modern world.

Suggestions for Further Reading

Feinberg, Gerald: "Light," *Scientific American*, September 1968, p. 50.

This article presents an introduction to our present understanding of light as a phenomenon of both wavelike and particlelike properties, as manifested by diffraction, two-slit interference, the photoelectric effect, and blackbody radiation.

Moran, Paul R., R. Jerome Nickles, and James A. Zagzebski: "The Physics of Medical Imaging," *Physics Today*, vol. 36, no. 7, 1983, p. 36.

This article briefly describes such new medical imaging techniques as digital subtraction angiography, computed tomography (CAT), nmr imaging (MRI), positron-emission tomography (PET), and ultrasound imaging.

Wheeler, John Archibald: "Niels Bohr, the Man," *Physics Today*, vol. 38, no. 10, 1985, p. 66.

Bohr's very personal approach to science is recounted by a former collaborator, who is himself a highly respected physicist. This article appears as part of a special issue commemorating the centennial of Bohr's birth.

Review

A. Objectives: After studying this chapter you should:

1. Be able to sketch the spectral distribution curve for blackbody radiation and the curve predicted by the Rayleigh–Jeans law.

2. Be able to discuss the photoelectric effect and state the Einstein equation describing it.

3. Be able to discuss how the photon concept explains all the features of the photoelectric effect and the Compton scattering of x rays.

4. Be able to sketch a typical x-ray spectrum and relate the minimum wavelength of the spectrum to the voltage of the x-ray tube.

5. Be able to state the Bohr postulates and describe the Bohr model of the hydrogen atom.

6. Be able to draw an energy-level diagram for hydrogen, indicate on it transitions involving the emission of a photon, and use it to calculate the wavelengths of the emitted photons.

7. Be able to state the de Broglie relations for the frequency and wavelength of electron waves and use them and the standing-wave condition to derive the Bohr condition for the quantization of angular momentum in the hydrogen atom.

8. Be able to discuss the experimental evidence for the existence of electron waves.

B. Define, explain, or otherwise identify:

Blackbody radiation
Rayleigh–Jeans law
Ultraviolet catastrophe
Quanta
Planck's constant
Photoelectric effect
Stopping potential
Photons
Work function
Einstein's photoelectric equation
Threshold frequency
Threshold wavelength
X rays
Characteristic spectrum
Bremsstrahlung spectrum
Cutoff wavelength
Compton wavelength
Balmer series
Rydberg–Ritz formula
Rydberg
Stationary states
Bohr radius
Energy-level diagram
Ionization
Quantum theory
Quantum mechanics
Wave mechanics

C. True or false: If the statement is true, explain why it is true. If it is false, give a counterexample.

1. The spectral distribution of radiation in a blackbody depends only on the temperature of the body.

2. In the photoelectric effect, the maximum current is proportional to the intensity of the incident light.

3. The work function of a metal depends on the frequency of the incident light.

4. The maximum kinetic energy of electrons emitted in the photoelectric effect varies linearly with the frequency of the incident light.

5. The energy of a photon is proportional to its frequency.

6. One of Bohr's assumptions is that atoms never radiate light.

7. In the Bohr model, the energy of a hydrogen atom is quantized.

8. In the ground state of the hydrogen atom, the potential energy is -27.2 eV.

9. The de Broglie wavelength of an electron varies inversely with its momentum.

10. Electrons can be diffracted.

11. Neutrons can be diffracted.

12. An electron microscope is used to look at electrons.

Problems

Level I

2-1 The Origin of the Quantum Constant: Blackbody Radiation

There are no problems for this section

2-2 The Photoelectric Effect

1. Find the photon energy in joules and in electron volts for an electromagnetic wave in the FM radio band of frequency 100 MHz.

2. Repeat Problem 1 for an electromagnetic wave in the AM radio band of frequency 900 kHz.

3. What is the frequency of a photon of energy (*a*) 1 eV, (*b*) 1 keV, and (*c*) 1 MeV?

4. Find the photon energy for light of wavelength (*a*) 450 nm, (*b*) 550 nm, and (*c*) 650 nm.

5. Find the range of photon energies in the visible spectrum, which ranges from wavelengths of 400 to 700 nm.

6. Find the photon energy if the wavelength is (*a*) 0.1 nm (about 1 atomic diameter) and (*b*) 1 fm (1 fm = 10^{-15} m, about 1 nuclear diameter).

7. The work function for tungsten is 4.58 eV. (*a*) Find the threshold frequency and wavelength for the photoelectric effect. Find the stopping potential if the wavelength of the incident light is (*b*) 200 nm and (*c*) 250 nm.

8. When light of wavelength 300 nm is incident on potassium, the emitted electrons have maximum kinetic energy of 2.03 eV. (*a*) What is the energy of the incident photon?

(b) What is the work function for potassium? (c) What would be the stopping potential if the incident light had a wavelength of 430 nm? (d) What is the threshold wavelength for the photoelectric effect with potassium?

9. The threshold wavelength for the photoelectric effect for silver is 262 nm. (a) Find the work function for silver. (b) Find the stopping potential if the incident radiation has a wavelength of 175 nm.

10. The work function for cesium is 1.9 eV. (a) Find the threshold frequency and wavelength for the photoelectric effect. Find the stopping potential if the wavelength of the incident light is (b) 250 nm and (c) 350 nm.

11. A light beam of wavelength 400 nm has an intensity of 100 W/m². (a) What is the energy of each photon in the beam? (b) How much energy strikes an area of 1 cm² perpendicular to the beam in 1 s? (c) How many photons strike this area in 1 s?

2-3 X Rays

12. An x-ray tube operates at a potential of 460 kV. What is the minimum wavelength of the continuous x-ray spectrum from this tube?

13. The minimum wavelength in the continuous x-ray spectrum from a television tube is 0.134 nm. What is the voltage of the tube?

14. What is the minimum wavelength of the continuous x-ray spectrum from a television tube operating at 2500 V?

2-4 Compton Scattering

15. Find the shift in wavelength of photons scattered at $\theta = 60°$.

16. When photons are scattered by electrons in carbon, the shift in wavelength is 0.33 pm. Find the scattering angle.

17. Find the momentum of a photon in eV/c and in kg·m/s if the wavelength is (a) 400 nm, (b) 2 nm, (c) 0.1 nm, and (d) 3 cm.

18. The wavelength of Compton-scattered photons is measured at $\theta = 90°$. If $\Delta\lambda/\lambda$ is to be 1.5 percent, what should the wavelength of the incident photons be?

19. Compton used photons of wavelength 0.0711 nm. (a) What is the energy of these photons? (b) What is the wavelength of the photon scattered at $\theta = 180°$? (c) What is the energy of the photon scattered at this angle?

20. For the photons used by Compton, find the momentum of the incident photon and that of the photon scattered at 180°, and use momentum conservation to find the momentum of the recoil electron in this experiment (see Problem 19).

2-5 Quantization of Atomic Energies: The Bohr Model

21. Use the known values of the constants in Equation 2-23 to show that a_0 is approximately 0.0529 nm.

22. The wavelength of the longest wavelength of the Lyman series was calculated in Example 2-6. Find the wavelengths for the transitions (a) $n_1 = 3$ to $n_2 = 1$ and (b) $n_1 = 4$ to $n_2 = 1$. (c) Find the shortest wavelength in the Lyman series.

23. Find the photon energy for the three longest wavelengths in the Balmer series and calculate the wavelengths.

24. (a) Find the photon energy and wavelength for the series limit (shortest wavelength) in the Paschen series ($n_2 = 3$). (b) Calculate the wavelengths for the three longest wavelengths in this series and indicate their positions on a horizontal linear scale.

25. Repeat Problem 24 for the Brackett series ($n_2 = 4$).

26. A hydrogen atom is in its tenth excited state according to the Bohr model ($n = 11$). (a) What is the radius of the Bohr orbit? (b) What is the angular momentum of the electron? (c) What is the electron's kinetic energy? (d) What is the electron's potential energy? (e) What is the electron's total energy?

2-6 Electron Waves and Quantum Theory

27. Use Equation 2-30 to calculate the de Broglie wavelength for an electron of kinetic energy (a) 2.5 eV, (b) 250 eV, (c) 2.5 keV, and (d) 25 keV.

28. An electron is moving at $v = 2.5 \times 10^5$ m/s. Find its de Broglie wavelength.

29. An electron has a wavelength of 200 nm. Find (a) its momentum and (b) its kinetic energy.

30. Through what potential must an electron be accelerated so that its de Broglie wavelength is (a) 5 nm and (b) 0.01 nm?

31. A thermal neutron in a reactor has kinetic energy of about 0.02 eV. Calculate the de Broglie wavelength of this neutron from

$$\lambda = \frac{hc}{\sqrt{2mc^2 K}}$$

where $mc^2 = 940$ MeV is the rest energy of the neutron.

32. Find the de Broglie wavelength of a proton (rest energy $mc^2 = 938$ MeV) that has a kinetic energy of 2 MeV. (See Problem 31.)

33. A proton is moving at $v = 0.003c$, where c is the speed of light. Find its de Broglie wavelength.

34. What is the kinetic energy of a proton whose de Broglie wavelength is (a) 1 nm and (b) 1 fm?

35. Find the de Broglie wavelength of a baseball of mass 0.145 kg moving at 30 m/s.

36. The energy of the electron beam in Davisson and Germer's experiment was 54 eV. Calculate the wavelength for these electrons.

37. The distance between Li⁺ and Cl⁻ ions in a LiCl crystal is 0.257 nm. Find the energy of electrons that have wavelengths equal to this spacing.

38. An electron microscope uses electrons of energy 70 keV. Find the wavelength of these electrons.

Level II

39. An x ray undergoes Compton scattering and emerges with a wavelength of 0.20 nm at a scattering angle of 100°. What was the initial energy of the x-ray photon?

40. When the kinetic energy of an electron is much greater than its rest energy, the relativistic approximation $E \approx pc$ is good. (a) Show that in this case photons and electrons of the same energy have the same wavelength. (b) Find the de Broglie wavelength of an electron of energy 200 MeV.

41. Suppose that a 100-W source radiates light of wavelength 600 nm uniformly in all directions and that the eye can detect this light if only 20 photons per second enter a dark-adapted eye having a 7-mm diameter pupil. How far from the source can the light be detected under these rather extreme conditions?

42. Data for stopping potential versus wavelength for the photoelectric effect using sodium are

λ, nm	200	300	400	500	600
V_0, V	4.20	2.06	1.05	0.41	0.03

Plot these data so as to obtain a straight line and from your plot find (a) the work function, (b) the threshold frequency, and (c) the ratio h/e.

43. The diameter of the pupil of the eye is about 5 mm. (It can vary from about 1 mm to 8 mm). Find the intensity of light of wavelength 600 nm such that 1 photon per second passes through the pupil.

44. Show that the speed of an electron in the nth Bohr orbit of hydrogen is given by $v_n = e^2/2\epsilon_0 hn$.

45. A light bulb radiates 90 W uniformly in all directions. (a) Find the intensity at a distance of 1.5 m. (b) If the wavelength is 650 nm, find the number of photons per second that strike a 1-cm^2 area oriented so that its normal is along the line to the bulb.

46. How many head-on Compton scattering events are necessary to double the wavelength of a photon having initial wavelength 200 pm?

47. An x-ray photon of wavelength 6 pm makes a head-on collision with an electron so that it is scattered by an angle of 180°. (a) What is the change in wavelength of the photon? (b) What is the energy lost by the photon? (c) What is the kinetic energy of the scattered electron?

48. A 0.200-pm photon scatters from a free electron that is initially at rest. For what photon scattering angle will the kinetic energy of the recoiling electron equal the energy of the scattered photon?

49. The binding energy of an electron is the minimum energy required to remove the electron from its ground state to a large distance from the nucleus. (a) What is the binding energy for the hydrogen atom? (b) What is the binding energy for He$^+$? (c) What is the binding energy for Li^{2+}?

50. A hydrogen atom has its electron in the $n = 2$ state. The electron makes a transition to the ground state. (a) What is the energy of the photon according to the Bohr model? (b) If angular momentum is conserved, what is the angular momentum of the photon? (c) The linear momentum of the emitted photon is E/c. If we assume conservation of linear momentum, what is the recoil velocity of the atom? (d) Find the recoil kinetic energy of the atom in electron volts. By what percent must the energy of the photon calculated in part (a) be corrected to account for this recoil energy?

51. A particle of mass m moves in a one-dimensional box of length L. (Take the potential energy of the particle in the box to be zero so that its total energy is its kinetic energy $p^2/2m$). Its energy is quantized by the condition $n(\lambda/2) = L$, where λ is the de Broglie wavelength of the particle and n is an integer. (a) Show that the allowed energies are given by

$$E_n = n^2 E_1 \quad \text{where } E_1 = h^2/8mL^2$$

(b) Evaluate E_n for an electron in a box of size $L = 0.1$ nm and make an energy-level diagram for the states from $n = 1$ to $n = 5$. Use Bohr's second postulate $f = \Delta E/h$ to calculate the wavelength of electromagnetic radiation emitted when the electron makes a transition from (c) $n = 2$ to $n = 1$, (d) $n = 3$ to $n = 2$, and (e) $n = 5$ to $n = 1$.

52. (a) Use the results of Problem 51 to find the energy of the ground state ($n = 1$) and the first two excited states of a proton in a one-dimensional box of length $L = 10^{-15}$ m = 1 fm. (These are of the order of magnitude of nuclear energies.) Calculate the wavelength of electromagnetic radiation emitted when the proton makes a transition from (b) $n = 2$ to $n = 1$, (c) $n = 3$ to $n = 2$, and (d) $n = 3$ to $n = 1$.

53. (a) Use the results of Problem 51 to find the energy of the ground state ($n = 1$) and the first two excited states of a proton in a one-dimensional box of length 0.2 nm (about the diameter of a H$_2$ molecule). Calculate the wavelength of electromagnetic radiation emitted when the proton makes a transition from (b) $n = 2$ to $n = 1$, (c) $n = 3$ to $n = 2$, and (d) $n = 3$ to $n = 1$.

54. (a) Use the results of Problem 51 to find the energy of the ground state ($n = 1$) and the first two excited states of a small particle of mass 1 μg confined to a one-dimensional box of length 1 cm. (b) If the particle moves with a speed of 1 mm/s, calculate its kinetic energy and find the approximate value of the quantum number n.

55. In the center-of-mass reference frame of the electron and the nucleus of an atom, the electron and nucleus have equal and opposite momenta of magnitude p. (a) Show that the total kinetic energy of the electron and nucleus can be written

$$K = \frac{p^2}{2\mu}$$

where

$$\mu = \frac{m_e M}{m_e + M} = \frac{m_e}{1 + m_e/M}$$

is called the reduced mass, m_e is the mass of the electron, and M is the mass of the nucleus. It can be shown that the motion of the nucleus can be accounted for by replacing the mass of the electron by the reduced mass. (b) Use Equation 2-25 with m replaced by μ to calculate the Rydberg for hydrogen ($M = m_p$) and for a very massive nucleus ($M = \infty$). (c) Find the percentage correction for the ground-state energy of the hydrogen atom due to the motion of the proton.

56. The kinetic energy of rotation of a diatomic molecule can be written $K = L^2/2I$, where L is its angular momentum and I is the moment of inertia. (a) Assuming that the angular momentum is quantized as in the Bohr model of the hydrogen atom, show that the energy is given by $K_n = n^2 K_1$, where $K_1 = \hbar^2/2I$. (b) Make an energy-level diagram for such a molecule. (c) Estimate K_1 for the hydrogen molecule assuming the separation of the atoms to be $r = 0.1$ nm and considering rotation about an axis through the center of mass and perpendicular to the line joining the atoms. Express your answer in electron volts. (d) When K_1 is greater than kT (where k is Boltzmann's constant), molecular collisions do not result in rotation and so rotational energy does not contribute to the internal energy of the gas. Use your result of part (c) to find the critical temperature $T_c = K_1/k$.

Level III

57. This problem is one of estimating the time lag (expected classically but not observed) in the photoelectric effect. Let the intensity of the incident radiation be 0.01 W/m². (a) If the area of the atom is 0.01 nm², find the energy per second falling on an atom. (b) If the work function is 2 eV, how long would it take classically for this much energy to fall on one atom?

58. A photon cannot transfer all of its energy to a single free electron. Prove this by considering the problem of conservation of energy and momentum.

59. An electron and positron are moving towards each other with equal speeds of 3×10^6 m/s. The two particles annihilate each other and produce two photons of equal energy. (a) What were the de Broglie wavelengths of the electron and positron? Find the (b) energy, (c) momentum, and (d) wavelength of each photon.

60. (a) Solve Equation 2-11 for p_e^2 to obtain $p_e^2 = p_1^2 + p_2^2 - 2p_1 p_2 + 2mc(p_1 - p_2)$. (b) Eliminate p_e^2 from your result in part (a) and Equation 2-10 to obtain $mc(p_1 - p_2) = p_1 p_2 (1 - \cos\theta)$. (c) Multiply both sides of your result in part (b) by $h/mcp_1 p_2$ and use the de Broglie relation $h/p = \lambda$ to obtain the Compton formula (Equation 2-12).

61. The total energy density of radiation in a blackbody is given by

$$\eta = \int f(\lambda, T)\, d\lambda$$

where $f(\lambda, T)$ is given by the Planck formula

$$f(\lambda, T) = \frac{8\pi hc \lambda^{-5}}{e^{hc/\lambda kT} - 1}$$

Change the variable to $x = hc/\lambda kT$ and show that the total energy density can be written

$$\eta = \left(\frac{kT}{hc}\right)^4 8\pi hc \int_0^\infty \frac{x^3}{e^x - 1} dx = \alpha T^4$$

where α is some constant independent of T. This shows that the energy density in a blackbody is proportional to T^4.

62. The frequency of revolution of an electron in a circular orbit of radius r is $f_{\text{rev}} = v/2\pi r$, where v is the speed. (a) Show that in the nth stationary state

$$f_{\text{rev}} = \frac{k^2 Z^2 e^4 m}{2\pi \hbar^3} \frac{1}{n^3}$$

(b) Show that when $n_1 = n$, $n_2 = n - 1$, and n is much greater than 1,

$$\frac{1}{n_2^2} - \frac{1}{n_1^2} \approx \frac{2}{n^3}$$

(c) Use your result in part (b) in Equation 2-24 to show that in this case the frequency of radiation emitted equals the frequency of motion. This result is an example of Bohr's correspondence principle: when n is large, so that the energy difference between adjacent states is a small fraction of the total energy, classical and quantum physics must give the same results.

Chapter 3

Quantum Mechanics

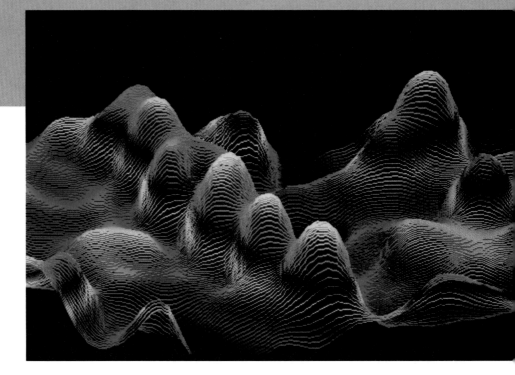

False color scanning tunneling micrograph of a segment of a DNA molecule magnified about one and a half million times. A sample of double-stranded DNA was dissolved in a salt solution and deposited on graphite prior to being imaged in air by the microscope. The row of orange-yellow peaks corresponds to the ridges of the DNA double helix.

In the previous chapter, we saw that light, which was thought to be a wave phenomenon, has particle properties and that electrons and other massive objects have wave properties. De Broglie's ideas about the wave nature of electrons were developed into a detailed mathematical theory by Erwin Schrödinger in 1926. In this theory, the electron is described by a wave function that obeys a wave equation called the Schrödinger equation, which is somewhat similar to the classical wave equations for sound and light waves. The frequency and wavelength of electron waves are related to the energy and momentum of the electron just as the frequency and wavelength of light waves are related to the energy and momentum of a photon. The diffraction and interference of electron waves observed by Davisson and Germer and others is a natural consequence of the propagation of these waves. The quantization of energy in atoms, molecules, and other microscopic systems results from standing-wave patterns of electron waves in these systems.

In this chapter, we will look at some of the properties of electron waves and see how the Schrödinger equation leads to the quantization of energy.

3-1 The Electron Wave Function

In classical waves, such as waves on a string, sound waves, or light waves, the energy density (the energy per unit volume in a wave) is proportional to the square of the wave function of the wave. The intensity, which equals the energy density times the wave speed, is also proportional to the square of the wave function. For waves on a string, the wave function is the displacement of the string $y(x, t)$. For sound waves in air, the wave function is the displacement of the air molecules from their equilibrium positions or, alternatively, the pressure variations due to the sound wave. For light and other electromagnetic waves, the wave function is the electric field $\mathcal{E}$ associated with the wave.*

The wave function for electron waves (or other matter waves) is designated by the Greek letter psi (Ψ). The wave function is a solution of a wave equation called the Schrödinger equation, which we will discuss later in this chapter, just as the wave function $\mathcal{E}$ for light waves is a solution of the classical wave equation for light. When Schrödinger first published his wave equation for electrons, it was not clear to him or to anyone else just what the wave function Ψ represents. We can get a hint as to how to interpret Ψ by considering the quantization of light waves.

Since the energy per unit volume in a light wave is proportional to $\mathcal{E}^2$ and the energy is quantized in units of hf for each photon, we expect that the number of photons in a unit volume is proportional to $\mathcal{E}^2$. Let us consider Young's famous double-slit experiment (Figure 3-1). The pattern observed on the screen is determined by the interference of the waves from the slits. At a point P_1 on the screen where the wave from one slit is 180° out of phase with that from the other, the resultant electric field $\mathcal{E}$ is zero. There is no light energy at that point and the point is dark. At points such as P_2, where the waves from the slits are in phase, $\mathcal{E}$ is maximum and the points are bright. If the intensity of the light incident on the slits is reduced, we can still observe the interference pattern if we replace the screen by a film and wait a sufficient time to expose the film.

The interaction of light with film is a quantum phenomenon. If we expose the film for a very short time with a low-intensity light source, we do not see merely a weaker version of the high-intensity pattern. Instead, we see "dots" on the film caused by the interactions of the individual photons

---

*We use a script $\mathcal{E}$ here for the electric field so as not to confuse it with the energy E.

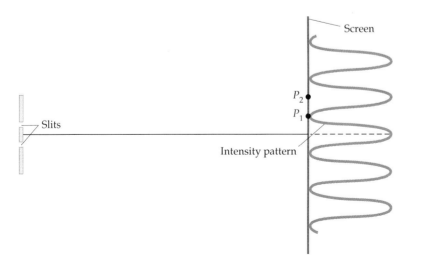

Figure 3-1 Young's double-slit experiment. Many photons go to point P_2 but no photons go to P_1. This experiment can be done with a light intensity that is so low that only one photon at a time arrives at the two slits and at the screen. The intensity pattern must then be interpreted as a measure of the probability of an individual photon arriving at a particular point.

Figure 3-2 Growth of the two-slit interference pattern. Drawing (a) shows the expected pattern after the film has been exposed to about 28 photons or electrons. Drawings (b) and (c) show the expected patterns for 1000 and 10,000 photons or electrons, respectively. Note that there are no dots in the regions of the interference minima. The photograph (d) is an actual two-slit electron interference pattern resulting from film being exposed to millions of electrons. The pattern is identical to that usually obtained with photons.

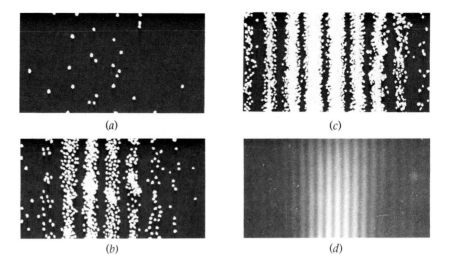

(Figure 3-2). At points where the waves from the slits interfere destructively, there are no dots; that is, no photons arrive at these points. At points where the waves interfere constructively, there are many dots, indicating that many photons arrive at these points. When the exposure time is very short and the light source is weak, random fluctuations from the average locations predicted by the wave theory occur, and the quantum nature of light is clearly evident. If the exposure time is long or the light source is strong so that many photons interact with the film, the fluctuations average out and the quantum nature of light is not noticed. The interference pattern depends on the total number of photons interacting with the film and not on the rate of interaction. *Even when the intensity is so low that only one photon at a time hits the film, the wave theory predicts the correct average pattern.* For low intensities, we therefore interpret $\mathcal{E}^2$ to be proportional to the probability of detecting a photon in a unit volume of space.

At points on the film or screen where $\mathcal{E}^2$ is zero, no photons are observed, whereas at points where $\mathcal{E}^2$ is large they are most likely to be observed. As we have seen, we obtain the same double-slit pattern if we use electrons instead of photons. Figure 3-3 shows the interference patterns with just a few electrons and with many electrons.

In the wave theory of electrons, the motion of a single electron is described by the wave function Ψ, which is a solution of the Schrödinger wave

Figure 3-3 Actual electron interference patterns at increasing electron-beam densities filmed from a television monitor.

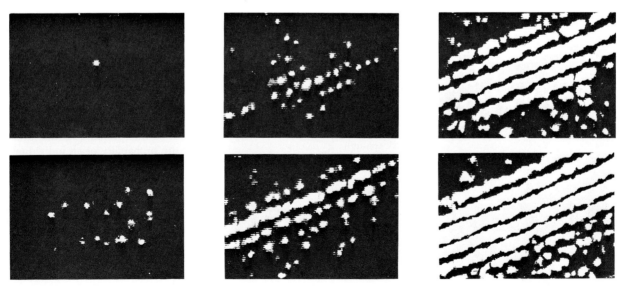

equation. Because the Schrödinger equation contains the imaginary number $i = \sqrt{-1}$ (Section 3-5), the wave functions that describe the motion of an electron are not necessarily real, that is, they may be complex. Since a probability must be a real number, the quantity analogous to $\mathcal{E}^2$ for photons that describes the probability of finding an electron in some region of space is $|\Psi|^2$. (Here $|\Psi|^2$ is the square modulus of Ψ. See appendix Equation A-8.)

> The probability of finding an electron in a unit volume of space is proportional to $|\Psi|^2$.

The probability of finding the electron in some volume element dV must also be proportional to the size of the volume element dV. In one dimension, the probability of finding an electron in some interval dx is $|\Psi|^2 \, dx$. If we call this probability $P(x) \, dx$, where $P(x)$ is the **probability distribution function** (also called the **probability density**), we have

$$P(x) = |\Psi|^2 \qquad \text{3-1} \quad \textit{Probability density}$$

Example 3-1

A classical point particle moves back and forth with constant speed between two walls at $x = 0$ and $x = 8$ cm. (a) What is the probability density $P(x)$? (b) What is the probability of finding the particle at $x = 2$ cm? (c) What is the probability of finding the particle between $x = 3.0$ cm and $x = 3.4$ cm?

(a) We do not know the initial position of the particle. Since the particle moves with constant speed, it is equally likely to be anywhere in the region $0 < x < 8$ cm. The probability density $P(x)$ is therefore constant, independent of x, for $0 < x < 8$ cm, and zero outside of this range:

$$P(x) = P_0 \qquad 0 < x < 8 \text{ cm}$$
$$= 0 \qquad x < 0 \text{ or } x > 8 \text{ cm}$$

The probability of finding the particle in the interval dx at point x_1 or at point x_2 is the sum of the separate probabilities $P(x_1) \, dx + P(x_2) \, dx$. Since the particle must certainly be somewhere, the sum of the probabilities over all possible values of x must equal 1:

$$\int_{-\infty}^{+\infty} P(x) \, dx = \int_0^{8 \text{ cm}} P_0 \, dx = P_0(8 \text{ cm}) = 1$$

Note that we need only integrate from 0 to 8 cm because $P(x)$ is zero outside this range. From this result we see that $P_0 = 1/(8 \text{ cm})$.

(b) The probability of finding the particle in some range dx is proportional to dx. Since $dx = 0$, the probability of finding the particle at the point $x = 2$ cm is 0. Alternatively, since there is an infinite number of points between $x = 0$ and $x = 8$ cm, and the particle is equally likely to be at any point, the chance that the particle will be at one particular point must be zero.

(c) Since the probability density is constant, the probability of a particle being in some range Δx in the region $0 < x < 8$ cm is $P_0 \Delta x$. The probability of the particle being in the region 3.0 cm $< x <$ 3.4 cm is thus

$$P_0 \Delta x = \left(\frac{1}{8 \text{ cm}}\right) 0.4 \text{ cm} = 0.05$$

3-2 Electron Wave Packets

A harmonic wave on a string is represented by the wave function

$$y(x, t) = A \sin (kx - \omega t) \qquad 3\text{-}2$$

where k is the wave number, which is related to the wavelength λ by

$$k = \frac{2\pi}{\lambda} \qquad 3\text{-}3$$

and ω is the angular frequency, which is related to the frequency by

$$\omega = 2\pi f \qquad 3\text{-}4$$

The velocity of the wave is related to the frequency and the wavelength by

$$v = f\lambda = \left(\frac{\omega}{2\pi}\right)\left(\frac{2\pi}{k}\right) = \frac{\omega}{k} \qquad 3\text{-}5$$

Equation 3-2 also describes a harmonic sound wave if the displacement of the string $y(x, t)$ is replaced by the displacement of air molecules $s(x, t)$ or by the pressure variation $p(x, t)$. (See Appendix B for a general review of classical waves.) We can use Equation 3-2 to describe a harmonic electron wave by replacing the displacement $y(x, t)$ with the electron wave function $\Psi(x, t)$.

An important property of a harmonic wave with a single frequency ω and wave number k is that it has no beginning or end in space or time. To represent a pulse that is localized in space, we need a wave packet, that is, a group of harmonic waves containing a continuous distribution of frequencies and wave numbers. An electron that is completely unlocalized, that is, one that can be anywhere in space, can be represented by a single harmonic wave. However, a wave packet is needed to describe an electron that is localized in space. We will illustrate some of the properties of wave packets by considering a very simple group consisting of just two waves of equal amplitude and nearly equal frequencies and wave numbers. Such a group is used to describe the phenomenon of beats. Let the wave numbers be k_1 and k_2 and the angular frequencies be ω_1 and ω_2. The sum of the two waves is

$$\Psi(x, t) = A_0 \sin (k_1 x - \omega_1 t) + A_0 \sin (k_2 x - \omega_2 t)$$

where A_0 is the amplitude of each wave. Using

$$\sin \theta_1 + \sin \theta_2 = 2 \cos \tfrac{1}{2}(\theta_1 - \theta_2) \sin \tfrac{1}{2}(\theta_1 + \theta_2)$$

for the sum of two sine functions, we obtain for the resultant wave

$$\Psi(x, t) = 2A_0 \cos [\tfrac{1}{2}(k_1 - k_2)x - \tfrac{1}{2}(\omega_1 - \omega_2)t] \sin [\tfrac{1}{2}(k_1 + k_2)x - \tfrac{1}{2}(\omega_1 + \omega_2)t]$$

Using $k_{av} = \tfrac{1}{2}(k_1 + k_2)$ and $\omega_{av} = \tfrac{1}{2}(\omega_1 + \omega_2)$ for the average wave number and average angular frequency and $\Delta k = k_1 - k_2$ and $\Delta \omega = \omega_1 - \omega_2$ for the difference in wave numbers and difference in angular frequencies, we have

$$\Psi(x, t) = [2A_0 \cos (\tfrac{1}{2} \Delta k \, x - \tfrac{1}{2} \Delta \omega \, t)] \sin (k_{av} x - \omega_{av} t) \qquad 3\text{-}6$$

Figure 3-4 shows a sketch of $\Psi(x, t)$ at some particular time as a function of x. The dashed curve is the envelope of the group of two waves given by the factor in brackets in Equation 3-6. The individual waves within the envelope move with the speed $v = \omega_{av}/k_{av}$, which is the phase velocity. If we write the modulating factor in brackets as $\cos \{\tfrac{1}{2} \Delta k \, [x - (\Delta\omega/\Delta k)t]\}$, we can see that the envelope moves with speed $\Delta\omega/\Delta k$. The speed of the envelope is the **group velocity**

$$V_g = \frac{\Delta\omega}{\Delta k} \qquad 3\text{-}7$$

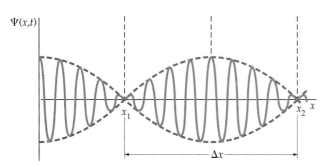

Figure 3-4 The spatial extent of the group Δx is inversely proportional to the difference in wave numbers Δk, where k is related to the wavelength by $k = 2\pi/\lambda$. Identical figures are obtained if $\Psi(x, t)$ is plotted versus time t at a fixed point x. In that case, the extent in time Δt is inversely proportional to the frequency difference $\Delta\omega$.

If x_1 and x_2 are two consecutive values of x for which the envelope is zero, we can take $\Delta x = x_2 - x_1$ to be a measure of the spatial extent of the group. Since the cosine function is zero when its argument is $\frac{1}{2}\pi, \frac{3}{2}\pi, \frac{5}{2}\pi$, and so on, the values x_1 and x_2 are related by

$$\tfrac{1}{2} \Delta k\, x_2 - \tfrac{1}{2} \Delta k\, x_1 = \pi$$

or

$$\Delta k\, \Delta x = 2\pi$$

For a particular value of x, the function $\Psi(x, t)$ versus t looks like Figure 3-4 with t replacing x. The extent in time Δt is thus related to $\Delta\omega$ by

$$\Delta\omega\, \Delta t = 2\pi$$

These results are consistent with those discussed in Appendix B (Equations B-16 and B-17):

$$\Delta k\, \Delta x \sim 1 \qquad 3\text{-}8$$

$$\Delta\omega\, \Delta t \sim 1 \qquad 3\text{-}9$$

However, the ranges Δx and Δt for our group of just two waves are artificial because the envelope does not remain small outside these ranges.

The wave function for a general wave packet made up of a discrete set of harmonic waves can be written

$$\Psi(x, t) = \Sigma\, A_i \sin(k_i x - \omega_i t) \qquad 3\text{-}10$$

where A_i is the amplitude of the wave of wave number k_i and angular frequency ω_i. The calculation of the amplitudes A_i needed to construct a wave packet of a given shape is a problem in Fourier series. If we use only a finite number of waves, it is not possible to obtain a wave packet that is small everywhere outside some region of space. To describe an electron that is localized in space, we must construct a wave packet from a continuous distribution of waves. We represent this mathematically by replacing A_i in Equation 3-10 with $A(k)\, dk$ and replacing the sum with an integral. The quantity $A(k)$ is called the distribution function for the wave number k. Either the shape of the wave packet at some fixed time $\Psi(x)$ or the distribution of wave numbers $A(k)$ can be found from the other by the methods of

Figure 3-5 A gaussian-shaped wave packet $\psi(x)$ and the corresponding gaussian distribution of wave numbers $A(k)$. The standard deviations of these packets are related by $\sigma_x \sigma_k = \tfrac{1}{2}$.

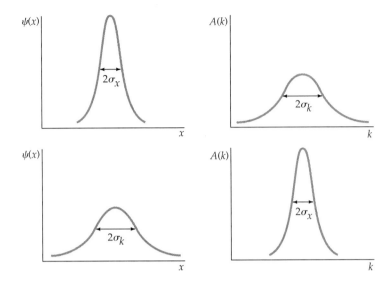

Fourier analysis. Figure 3-5 shows a gaussian-shaped wave packet and the corresponding wave-number distribution function for a narrow packet and a wide packet. (See also the section on the gaussian distribution in Appendix C.) For this special case, $A(k)$ is also a gaussian function. The standard deviations of these gaussian functions σ_x and σ_k are related by

$$\sigma_k \sigma_x = \tfrac{1}{2} \qquad 3\text{-}11$$

It can be shown that the product of the standard deviations is greater than $\tfrac{1}{2}$ for a wave packet of any other shape. For a continuous distribution of waves, Equation 3-7 for the group velocity of a wave packet becomes

$$V_g = \frac{d\omega}{dk} \qquad 3\text{-}12$$

The energy and momentum of an electron are related to the frequency and wavelength of the associated electron wave by the de Broglie equations. They are therefore also related to the angular frequency and wave number. Thus, we have

$$p = \frac{h}{\lambda} = \frac{h}{2\pi/k} = \frac{hk}{2\pi}$$

and

$$E = hf = h\frac{\omega}{2\pi}$$

In terms of $\hbar = h/2\pi$, these relations become

$$p = \hbar k \qquad 3\text{-}13$$

and

$$E = \hbar \omega \qquad 3\text{-}14$$

The kinetic energy of an electron moving in free space with no forces acting on it is given by

$$E = \tfrac{1}{2}mv^2 = \frac{p^2}{2m}$$

Substituting $\hbar\omega$ for E and $\hbar k$ for p, we obtain

$$\hbar\omega = \frac{\hbar^2 k^2}{2m} \qquad 3\text{-}15$$

Using Equation 3-12 for the group velocity, we obtain

$$V_g = \frac{d\omega}{dk} = \frac{d}{dk}\left(\frac{\hbar k^2}{2m}\right) = \frac{\hbar k}{m} = \frac{p}{m} = v \qquad 3\text{-}16$$

Thus, the group velocity equals the velocity of the electron, as we should expect. Note that the phase velocity of the individual waves in the wave packet is not equal to the velocity of the electron:

$$V_p = \frac{\omega}{k} = \frac{\hbar\omega}{\hbar k} = \frac{E}{p} = \frac{p}{2m} = \frac{v}{2}$$

Questions

1. What are Δx and Δk for a purely harmonic wave of a single frequency and wavelength?
2. Which is more important for communication, the phase velocity or the group velocity?

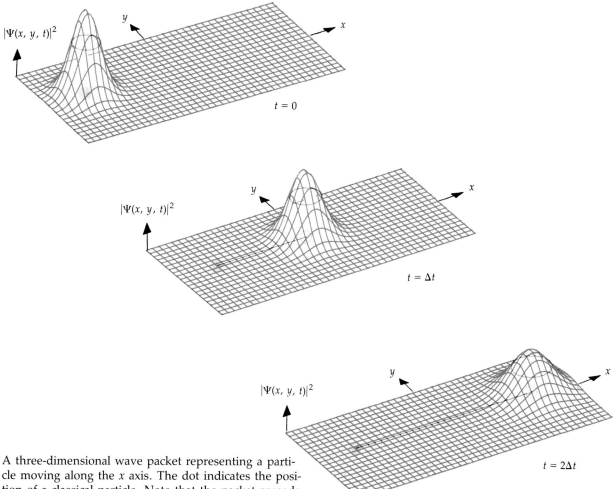

A three-dimensional wave packet representing a particle moving along the x axis. The dot indicates the position of a classical particle. Note that the packet spreads out in the x and y directions. This spreading is due to dispersion, resulting from the fact that the phase velocity of the individual waves making up the packet depends on the wavelength of the waves.

3-3 The Uncertainty Principle

The wave nature of electrons (and other particles) has important consequences. Consider a wave packet $\Psi(x, t)$ representing an electron. The most probable position of the electron is the value of x for which $|\Psi(x, t)|^2$ is a maximum. Since $|\Psi(x, t)|^2$ is proportional to the probability that the electron is at x and $|\Psi(x, t)|^2$ is nonzero for a *range* of values of x, there is an uncertainty in the position of the electron. If we make a number of position measurements on identical electrons, that is, electrons with the same wave function, we will not always obtain the same result. In fact, the distribution function for the results of the measurements will be given by $|\Psi(x, t)|^2$. If the wave packet is very narrow, the uncertainty in position will be small. However, a narrow wave packet must contain a wide range of wave numbers k. Since the momentum is related to the wave number by $p = \hbar k$, a wide range of k values means a wide range of momentum values. If we make a number of momentum measurements on identical electrons, we obtain a distribution of results corresponding to the distribution of wave numbers in the wave packet. Thus, a narrow wave packet that corresponds to a small uncertainty in position also corresponds to a wide distribution of momentum values and therefore to a large uncertainty in momentum. In general, the ranges Δx and Δk are related by Equation 3-8:

$$\Delta k \, \Delta x \sim 1$$

Similarly, a wave packet that is localized in time Δt must contain a range of frequencies $\Delta \omega$, where the ranges are related by Equation 3-9:

$$\Delta \omega \, \Delta t \sim 1$$

These results are inherent properties of waves. If we multiply these equations by $\hbar$ and use $p = \hbar k$ and $E = \hbar \omega$, we obtain

$$\Delta p \, \Delta x \sim \hbar \qquad \qquad 3\text{-}17$$

and

$$\Delta E \, \Delta t \sim \hbar \qquad \qquad 3\text{-}18$$

Equations 3-17 and 3-18 provide a statement of the **uncertainty principle** first enunciated by Werner Heisenberg in 1927. Equation 3-17 expresses the fact that the distribution functions for position and momentum cannot both be made arbitrarily narrow; thus, measurements of positions and momentum will have uncertainties whose product is at least as great as $\hbar$. Equation 3-18 expresses the fact that if Δt is the time available for the measurement of the energy of a system, the measurement of the energy will be uncertain by an amount ΔE that is at least as great as $\hbar/\Delta t$. Equation 3-18 has important applications in the determination of the excitation energy of atoms, molecules, and nuclei. For example, if an excited state of an atom has a lifetime τ, its energy can be known only to within about $\hbar/\tau$.

If we define precisely what we mean by the uncertainty in the measurements of position and momentum, we can give a precise statement of the uncertainty principle. If σ_x is the standard deviation for measurements of the wave number k, the product $\sigma_x \sigma_k$ has its minimum value of $\frac{1}{2}$ when the distribution functions are gaussian. If we define Δx and Δp to be the standard deviations, the minimum value of their product is $\frac{1}{2}\hbar$. Thus,

Uncertainty principle for position and momentum

$$\Delta p \, \Delta x \geq \tfrac{1}{2}\hbar \qquad \qquad 3\text{-}19$$

Similarly,

$$\Delta E \, \Delta t \geq \tfrac{1}{2}\hbar \qquad \qquad 3\text{-}20$$

Uncertainty principle for energy and time

Usually, the uncertainty product is much greater than $\hbar/2$. The equality holds only if the measurements of both x and p or both E and t are ideal.

We can get a qualitative understanding of the uncertainty principle by considering the measurement of the position and momentum of a particle. If we know the mass of the particle, we can determine its momentum by measuring its position at two nearby times and computing its velocity. A common way to measure the position of an object is to look at it with light. When we do this, we scatter light from the object and determine the position from the direction in which the light is scattered. If we use light of wavelength λ, we can measure the position only to an uncertainty of the order of λ because of diffraction effects. To reduce the uncertainty in position, we use light of very short wavelength, perhaps even x rays. In principle, there is no limit to the accuracy of such a position measurement because there is no limit on how small a wavelength λ we can use. However, since all electromagnetic radiation carries momentum, the scattering of the radiation by the particle will deflect the radiation and change its original momentum in an uncontrollable way. Since momentum is conserved, the momentum of the particle also changes in an uncontrollable way. According to classical wave theory, this effect on the momentum of the particle could be reduced by reducing the intensity of the radiation. However, the energy and momentum of the radiation are quantized; each photon has momentum h/λ. When the wavelength of the radiation is small, the momentum of each photon will be large and the momentum measurement will have a large uncertainty. This uncertainty cannot be eliminated by reducing the intensity of light; such a reduction merely reduces the number of photons in the beam. To "see" the particle, we must scatter at least one photon. Therefore, the uncertainty in the momentum measurement of the particle will be large if λ is small, and the uncertainty in the position measurement of the particle will be large if λ is large. A detailed analysis shows that the product of these uncertainties will always be at least of the order of Planck's constant h. Of course, we could always "look" at the particles by scattering electrons instead of photons, but we would still have the same difficulty. If we use low-momentum electrons to reduce the uncertainty in the momentum measurement, we have a large uncertainty in the position measurement because of the diffraction of the electrons. The relation between the wavelength and momentum, $\lambda = h/p$, is the same for electrons as for photons.

One consequence of the uncertainty principle is that when a particle is confined in some region of space, it cannot have zero kinetic energy. The minimum energy of a particle is called its **zero-point energy**. Suppose, for example, that a particle is confined to some region of space of length L. The uncertainty in its position is then no greater than L. Consequently, we see from Equation 3-19 that the uncertainty in its momentum Δp is

$$\Delta p \geq \frac{\hbar}{2L} \qquad \qquad 3\text{-}21$$

The kinetic energy of the particle is

$$K = \frac{1}{2}mv^2 = \frac{p^2}{2m} \qquad \qquad 3\text{-}22$$

The magnitude of the momentum p must be at least as large as its uncertainty Δp ($p = 0 \pm \Delta p$). Therefore,

$$K = \frac{p^2}{2m} \geq \frac{(\hbar^2/4L^2)}{2m} = \frac{\hbar^2}{8mL^2} \qquad 3\text{-}23$$

The smaller the region of space L, the greater the minimum kinetic energy.

Example 3-2

A marble of mass 25 g is in a box of length 10 cm. Find the minimum uncertainty in its momentum, its speed v, and its minimum kinetic energy, assuming that $p = \Delta p$.

From Equation 3-19, with $\Delta x = 10$ cm, we have

$$(\Delta p)_{min} = \frac{\hbar}{2\,\Delta x} = \frac{1.05 \times 10^{-34}\ \text{J·s}}{2(0.1\ \text{m})}$$

$$= 5.3 \times 10^{-34}\ \text{kg·m/s}$$

The speed corresponding to a momentum of this magnitude is

$$v = \frac{p}{m} = \frac{5.3 \times 10^{-34}\ \text{kg·m/s}}{0.025\ \text{kg}} = 2.1 \times 10^{-32}\ \text{m/s}$$

We would be quite safe in saying that the marble is at rest. The minimum kinetic energy is

$$K_{min} = \frac{(\Delta p_{min})^2}{2m} = \frac{(5.3 \times 10^{-34}\ \text{kg·m/s})^2}{0.050\ \text{kg}}$$

$$= 5.6 \times 10^{-66}\ \text{J}$$

Because Planck's constant is so small, the uncertainty relation of Equation 3-19 is not significant for macroscopic systems.

Example 3-3

Work Example 3-2 for an electron confined to a region of space of length $L = 0.1$ nm. This distance is of the order of the diameter of an atom.

In this case, the minimum uncertainty in the momentum is

$$(\Delta p)_{min} = \frac{\hbar}{2\,\Delta x} = \frac{1.05 \times 10^{-34}\ \text{J·s}}{2(10^{-10}\ \text{m})}$$

$$= 5.3 \times 10^{-25}\ \text{kg·m/s}$$

The speed of an electron with momentum of this magnitude is

$$v = \frac{p}{m} = \frac{5.3 \times 10^{-25}\ \text{kg·m/s}}{9.1 \times 10^{-31}\ \text{kg}}$$

$$= 5.8 \times 10^5\ \text{m/s}$$

Note that this is a significant speed. The minimum kinetic energy is

$$K_{min} = \frac{(\Delta p_{min})^2}{2m} = \frac{(5.3 \times 10^{-25}\ \text{kg·m/s})^2}{2(9.1 \times 10^{-31}\ \text{kg})}$$

$$= 1.5 \times 10^{-19}\ \text{J}$$

This is approximately 1 eV, which is about the order of magnitude of the kinetic energy of an electron in an atom.

Questions

3. Does the uncertainty principle say that the momentum of an electron can never be precisely known?
4. Why is the uncertainty principle not important for macroscopic objects?

3-4 Wave–Particle Duality

We have seen that light, which we ordinarily think of as a wave, exhibits particle properties when it interacts with matter, as in the photoelectric effect or in Compton scattering, and that electrons, which we usually think of as particles, exhibit the wave properties of interference and diffraction. All carriers of momentum and energy, such as electrons, atoms, light, or sound, have both particle and wave characteristics. It might be tempting to say that an electron, for example, is both a wave and a particle, but the meaning of such a statement is not clear. In classical physics, the concepts of waves and particles are mutually exclusive. A **classical particle** behaves like a piece of shot; it can be localized and scattered, it exchanges energy suddenly at a point in space, and it obeys the laws of conservation of energy and momentum in collisions. It does *not* exhibit interference or diffraction. A **classical wave,** on the other hand, behaves like a water wave; it exhibits diffraction and interference, and its energy is spread out continuously in space and time. Nothing can be both a classical particle and a classical wave at the same time.

After Thomas Young observed the two-slit interference pattern with light in 1801, light was thought to be a classical wave. Similarly, after J. J. Thomson's experiment in 1897, in which he deflected electrons in electric and magnetic fields, electrons were thought to be classical particles. We now know that these concepts of classical waves and particles do not adequately describe the complete behavior of any phenomenon.

> Everything propagates like a wave and exchanges energy like a particle.

Often the concepts of the classical particle and the classical wave give the same results. When the wavelength is very small, the propagation of a classical wave cannot be distinguished from that of a classical particle. For waves of very small wavelengths, diffraction effects are negligible, so the waves travel in straight lines. Similarly, interference is not seen for waves of very small wavelength because the interference fringes are too closely spaced to be observed. It then makes no difference which concept we use. When diffraction is negligible, we can think of light as a wave propagating along rays, as in geometrical optics, or as a beam of photon particles. Similarly, we can think of an electron as a wave propagating in straight lines along rays or, more commonly, as a particle.

We can also use either the wave or particle concept to describe exchanges of energy if we have a large number of particles and we are interested only in the average values of energy and momentum exchanges. For example, if we are interested only in the total current (above the threshold) in the photoelectric effect, the wave theory of light correctly predicts that this current is proportional to the intensity of the light.

Question

5. Which is the better model, the classical wave or the classical particle for the description of the propagation of electrons through a crystal? Which is better to describe the interaction of light with a photographic film?

3-5 The Schrödinger Equation

The wave equation governing the motion of electrons (and other particles with mass), which is analogous to the classical wave equation (Equation B-1 in Appendix B), was developed by Schrödinger in 1926 and is now known as the **Schrödinger equation.** Like the classical wave equation, the Schrödinger equation relates the time and space derivatives of the wave function. Schrödinger's reasoning is somewhat difficult to follow and is not important for our purposes. In any case, the Schrödinger equation cannot be derived, just as Newton's laws of motion cannot be derived. The validity of any fundamental equation lies in its agreement with experiment. Although it would be logical merely to postulate the Schrödinger equation, it is helpful to get some idea of what to expect by first considering the wave equation for photons, which is Equation B-1 with speed $v = c$ and with $y(x, t)$ replaced by the electric field $\mathcal{E}(x, t)$:

$$\frac{\partial^2 \mathcal{E}(x, t)}{\partial x^2} = \frac{1}{c^2} \frac{\partial^2 \mathcal{E}(x, t)}{\partial t^2} \qquad 3\text{-}24$$

A particularly important solution of this equation is the harmonic wave function $\mathcal{E}(x, t) = \mathcal{E}_0 \sin(kx - \omega t)$. Differentiating this function twice with respect to time yields $\partial^2 \mathcal{E}/\partial t^2 = -\omega^2 \mathcal{E}_0 \sin(kx - \omega t)$, whereas differentiating it twice with respect to x gives $\partial^2 \mathcal{E}/\partial x^2 = -k^2 \mathcal{E}_0 \sin(kx - \omega t)$. Substitution of $\mathcal{E} = \mathcal{E}_0 \sin(kx - \omega t)$ into Equation 3-24 then gives

$$-k^2 = -\frac{\omega^2}{c^2}$$

or

$$\omega = kc \qquad 3\text{-}25$$

Multiplying each side by $\hbar$, we obtain

$$\hbar \omega = \hbar k c$$

or

$$E = pc \qquad 3\text{-}26$$

which is the relation between the energy and momentum of a photon.

Let us now use the de Broglie equations for a particle with mass to find the relation between ω and k for an electron that is analogous to Equation 3-25. The energy of a particle of mass m is

$$E = \frac{p^2}{2m} + U \qquad 3\text{-}27$$

where U is the potential energy. The de Broglie equations are

$$E = hf = \hbar \omega$$

and

$$p = h/\lambda = \hbar k$$

Substituting these values for E and p into Equation 3-27, we obtain

$$\hbar \omega = \frac{\hbar^2 k^2}{2m} + U \qquad 3\text{-}28$$

This differs from Equation 3-25 for photons in that it contains the potential energy U and the angular frequency ω does not vary linearly with k. Note that we get a factor of ω when we differentiate a harmonic wave function with respect to time and a factor of k when we differentiate it with respect to position. We expect, therefore, that the wave equation that applies to electrons will relate the *first* time derivative to the second space derivative and will also involve the potential energy of the electron.

We are now ready to postulate the Schrödinger equation. In one dimension, it has the form

$$-\frac{\hbar^2}{2m}\frac{\partial^2 \Psi(x,t)}{\partial x^2} + U(x)\Psi(x,t) = i\hbar \frac{\partial \Psi(x,t)}{\partial t} \qquad 3\text{-}29$$

Time-dependent Schrödinger equation

Equation 3-29 is called the **time-dependent Schrödinger equation.** An important difference between the time-dependent Schrödinger equation and the classical wave equation is the explicit appearance of the imaginary number $i = \sqrt{-1}$. The wave functions that satisfy the Schrödinger equation may therefore be complex functions.

Schrödinger's first application of his wave equation was to systems such as the hydrogen atom and the simple harmonic oscillator. He showed that energy quantization for these systems can be explained naturally in terms of standing waves. For such problems we need not consider the time dependence of the wave function. For any wave function that describes a particle in a state of definite energy, the general Schrödinger equation can be simplified by writing the wave function in the form

$$\Psi(x,t) = \psi(x)e^{-i\omega t}$$

where $\psi(x)$ is a function of x only.* The right side of Equation 3-29 is then

$$i\hbar \frac{\partial \Psi(x,t)}{\partial t} = i\hbar(-i\omega)\psi(x)e^{-i\omega t} = \hbar\omega\psi(x)e^{-i\omega t} = E\psi(x)e^{-i\omega t}$$

where we have used $E = \hbar\omega$. Substituting $\Psi(x,t) = \psi(x)e^{-i\omega t}$ into Equation 3-29, we obtain

$$-\frac{\hbar^2}{2m}\frac{\partial^2 \psi(x)}{\partial x^2}e^{-i\omega t} + U(x)\psi(x)e^{-i\omega t} = E\psi(x)e^{-i\omega t}$$

Canceling the common factor $e^{-i\omega t}$, we obtain an equation for $\psi(x)$ called the **time-independent Schrödinger equation:**

$$-\frac{\hbar^2}{2m}\frac{d^2\psi(x)}{dx^2} + U(x)\psi(x) = E\psi(x) \qquad 3\text{-}30$$

Time-independent Schrödinger equation

The time-independent Schrödinger equation in one dimension is an ordinary differential equation in one variable x and is therefore much easier to handle than Equation 3-29, which is a partial differential equation containing the two variables x and t.

There are important conditions that the wave function $\psi(x)$ must satisfy in addition to being the solution of Equation 3-30. The probability of finding the electron in a region dx at x is $|\psi(x)|^2\,dx$. Since the probability of finding an electron cannot jump discontinuously as we move from one point to a nearby point, the wave function $\psi(x)$ must be continuous in x. Also, for potential-energy functions $U(x)$ that are not infinite, the first derivative $d\psi/dx$ must be continuous. We can see this from Equation 3-30 by multiply-

---

*It is customary to denote the spatial part of the wave function by a lower case psi (ψ) and the complete wave function with time dependence by an upper case psi (Ψ).

ing each term by $2m/\hbar^2$ and writing $d^2\psi/dx^2$ as $\dfrac{d}{dx}(d\psi/dx)$. We then have

$$\frac{d}{dx}\left(\frac{d\psi}{dx}\right) = \frac{2m}{\hbar^2}[U(x) - E]\psi(x)$$

or

$$d\left(\frac{d\psi}{dx}\right) = \frac{2m}{\hbar^2}[U(x) - E]\psi(x)\,dx$$

If we let dx approach zero, then $d(d\psi/dx)$, the change in the first derivative, also approaches zero as long as $U(x)$ is not infinite. This is equivalent to saying that the first derivative is continuous in x if $U(x)$ is not infinite.

The probability of finding the electron in dx at point x_1 or at point x_2 is the sum of the separate probabilities $P(x_1)\,dx + P(x_2)\,dx$. Since the electron must certainly be somewhere, the sum of the probabilities over all the possible values of x must equal 1. That is,

Normalization condition

$$\int_{-\infty}^{\infty} |\psi|^2 \, dx = 1 \qquad 3\text{-}31$$

Equation 3-31 is called the **normalization condition.** This condition plays an important role in quantum mechanics. If $\psi(x)$ is to satisfy the normalization condition of Equation 3-31, it must approach zero as x approaches infinity. This places a restriction on the possible solutions of the Schrödinger equation and leads to energy quantization.

3-6 A Particle in a Box

We will now apply the time-independent Schrödinger equation to a traditional though somewhat artificial problem of a particle, such as an electron, confined to a one-dimensional box of length L. Classically, the particle bounces back and forth between the walls of the box, which we assume are at $x = 0$ and $x = L$. The particle is equally likely to be found anywhere in the box, and its energy and momentum can take on any values.

According to quantum theory, the particle is described by a wave function ψ that obeys Equation 3-30. The potential energy for this problem is shown in Figure 3-6. It is called an **infinite square-well potential** and is described mathematically by

$$\begin{aligned} U(x) &= 0 & 0 < x < L \\ U(x) &= \infty & x < 0 \text{ or } x > L \end{aligned} \qquad 3\text{-}32$$

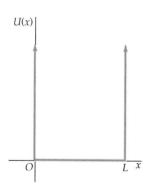

Figure 3-6 The infinite square-well potential energy. For $x < 0$ and $x > L$, the potential energy $U(x)$ is infinite. The particle is confined to the region in the well $0 < x < L$.

That is, inside the box, the potential energy is zero, whereas outside the box it is infinite. Since we require the particle to be in the box, we have $\psi(x) = 0$ everywhere outside the box. We then need to solve Equation 3-30 for inside the box subject to the condition that, since the wave function must be continuous, $\psi(x)$ must be zero at $x = 0$ and at $x = L$. Such a condition on the wave function is called a **boundary condition.** (The derivative of the wave function $d\psi/dx$ need not be continuous because the potential energy is infinite for $x < 0$ and $x > L$ in this problem.) Inside the box, Equation 3-30 is

$$-\frac{\hbar^2}{2m}\frac{d^2\psi(x)}{dx^2} = E\psi(x)$$

or

$$\frac{d^2\psi(x)}{dx^2} = -\frac{2mE}{\hbar^2}\psi(x) = -k^2\psi(x) \qquad 3\text{-}33$$

where

$$k^2 = \frac{2mE}{\hbar^2} \qquad 3\text{-}34$$

The general solution of Equation 3-33 can be written as

$$\psi(x) = A \sin kx + B \cos kx \qquad 3\text{-}35$$

where A and B are constants. At $x = 0$, we have

$$\psi(0) = A \sin (k0) + B \cos (k0) = 0 + B$$

The boundary condition $\psi(x) = 0$ at $x = 0$ thus gives $B = 0$, and Equation 3-35 becomes

$$\psi(x) = A \sin kx \qquad 3\text{-}36$$

The boundary condition $\psi(x) = 0$ at $x = L$ gives

$$\psi(L) = A \sin kL = 0 \qquad 3\text{-}37$$

This condition is satisfied if kL is π or any integer times π, that is, if k is restricted to the values k_n given by

$$k_n = n\frac{\pi}{L} \qquad n = 1, 2, 3, \ldots \qquad 3\text{-}38$$

For each value of n there is wave function ψ_n given by

$$\psi_n = A_n \sin \frac{n\pi x}{L} \qquad 3\text{-}39$$

If we write k in terms of the wavelength, $k = 2\pi/\lambda$, Equation 3-38 becomes

$$\frac{2\pi}{\lambda_n} = n\frac{\pi}{L}$$

or

$$L = n\frac{\lambda_n}{2}$$

which is the same as the standing-wave condition for waves on a string fixed at $x = 0$ and at $x = L$. The energy E is related to the wave number k by Equation 3-34:

$$E_n = \frac{\hbar^2 k_n^2}{2m} = n^2 \frac{\hbar^2 \pi^2}{2mL^2} = n^2 \frac{h^2}{8mL^2}$$

or

$$E_n = n^2 E_1 \qquad 3\text{-}40 \qquad \textit{Allowed energies for an infinite square-well potential}$$

where

$$E_1 = \frac{h^2}{8mL^2} \qquad 3\text{-}41 \qquad \textit{Ground-state energy for an infinite square-well potential}$$

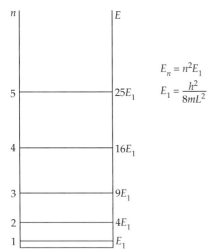

Figure 3-7 Energy-level diagram for the infinite square-well potential. Classically, a particle can have any value of energy. Quantum mechanically, only certain values of energy given by $E_n = n^2 E_1 = n^2(h^2/8mL^2)$ are allowed.

is the energy of the lowest state, the ground state. The energy is thus quantized. Figure 3-7 shows the energy-level diagram for the infinite square-well potential. Note that the lowest energy is not zero. According to quantum theory, the particle cannot remain at rest in the box. This result, which is a consequence of the uncertainty principle as discussed in Section 3-3, is a general feature of quantum theory. When a particle is confined to some region of space, it must have a minimum kinetic energy, the zero-point energy. The smaller the region of space, the greater the zero-point energy, as indicated by the fact that E_1 varies as $1/L^2$ in Equation 3-41.

The constant A_n is determined by the normalization condition (Equation 3-31):

$$\int_{-\infty}^{\infty} \psi^2 \, dx = \int_0^L A_n^2 \sin^2 \frac{n\pi x}{L} \, dx = 1$$

Note that we need integrate only from $x = 0$ to $x = L$ because $\psi(x)$ is zero everywhere else. Substituting $\theta = n\pi x/L$, we have

$$\int_0^L A_n^2 \sin^2 \frac{n\pi x}{L} \, dx = A_n^2 \frac{L}{n\pi} \int_0^{n\pi} \sin^2 \theta \, d\theta = 1$$

The integral of $\sin^2 \theta$ can be found in tables:

$$\int_0^{n\pi} \sin^2 \theta \, d\theta = \left.\frac{\theta}{2}\right|_0^{n\pi} - \left.\frac{\sin 2\theta}{4}\right|_0^{n\pi} = \frac{n\pi}{2}$$

The normalization condition thus gives

$$A_n^2 \frac{L}{n\pi} \frac{n\pi}{2} = 1$$

or

$$A_n = \sqrt{\frac{2}{L}}$$

independent of n. The normalized wave functions for the infinite square-well potential are thus

Wave functions for the infinite square-well potential

$$\psi_n = \sqrt{\frac{2}{L}} \sin \frac{n\pi x}{L} \qquad \text{3-42}$$

The number n is called a **quantum number.** It characterizes the wave function for a particular state and the energy of that state. In our one-dimensional problem, it arises from the boundary condition on the wave function that it be zero at $x = 0$ and $x = L$. In three-dimensional problems, three quantum numbers arise, each associated with a boundary condition in each dimension.

Figure 3-8 shows plots of ψ^2 for the ground state $n = 1$, the first excited state $n = 2$, the second excited state $n = 3$, and the state $n = 10$.* In the ground state, the particle is most likely to be found near the center of the box, as indicated by the maximum value of ψ^2 at $x = L/2$. In the first excited state, the particle is never found exactly in the center of the box because ψ^2 is zero at $x = L/2$. For very large values of n, the maxima and minima of ψ^2 are very close together, as illustrated for $n = 10$. The average value of ψ^2 is

---

*Since $\psi(x)$ is real in this case, $|\psi|^2 = \psi^2$.

indicated in this figure by the dashed line. For very large values of n, the maxima are so closely spaced that ψ^2 cannot be distinguished from its average value. The fact that $(\psi^2)_{av}$ is constant across the whole box means that the particle is equally likely to be found anywhere in the box—the same as the classical result. This is an example of **Bohr's correspondence principle:**

> In the limit of very large quantum numbers, the classical calculation and the quantum calculation must yield the same results.

Bohr's correspondence principle

The region of very large quantum numbers is also the region of very large energies. It can be shown that for large energies, the percentage change in energy between adjacent quantum states is very small, so energy quantization is not important (see Problem 48).

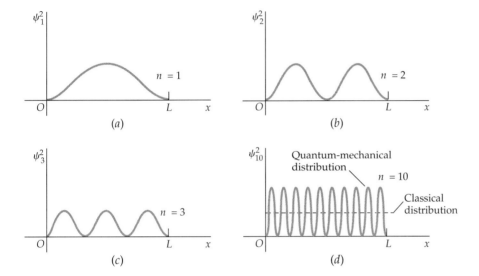

Figure 3-8 ψ^2 versus x for a particle in a box of length L for (*a*) the ground state, $n = 1$; (*b*) the first excited state, $n = 2$; (*c*) the second excited state, $n = 3$; and (*d*) the state $n = 10$. For $n = 10$, the maxima and minima of ψ^2 are so close together that individual maxima may be hard to distinguish. The average value of ψ^2 is indicated by the dashed line. It gives the classical prediction that the particle is equally likely to be found at any point in the box.

We are so accustomed to thinking of the electron as a classical particle that we tend to think of an electron in a box as a particle bounding back and forth between the walls. But the probability distributions shown in Figure 3-8 are stationary; that is, they do not depend on time. (The analogous patterns for the electron in the hydrogen atom discussed in the next chapter are the quantum theoretical counterparts of the stationary orbits of the Bohr model.) An alternative picture of the electron is a cloud of charge with the charge density proportional to ψ^2. Figure 3-8 can then be thought of as plots of the charge density versus x for the various states. In the ground state, $n = 1$, the electron cloud is centered in the middle of the box and is spread out over most of the box, as indicated in Figure 3-8*a*. In the first excited state, $n = 2$, the charge density of the electron cloud has two maxima, as indicated in Figure 3-8*b*. For very large values of n, there are many closely spaced maxima and minima in the charge density resulting in an average charge density that is approximately uniform throughout the box. This electron-cloud picture is very useful in understanding the structure of atoms and molecules, which will be discussed in the following chapters. However, it should be noted that whenever an electron is observed to interact with matter or radiation, it is always observed as a single charge.

Example 3-4

(a) Find the energy in the ground state of an electron confined to a one-dimensional box of length $L = 0.1$ nm. (This box is roughly the size of an atom.) (b) Make an energy-level diagram and find the wavelengths of the photons emitted for all transitions beginning at state $n = 3$ or less and ending at a lower energy state.

(a) The energy in the ground state is given by Equation 3-41. Multiplying the numerator and denominator by c^2, we obtain an expression in terms of hc and the rest energy mc^2:

$$E_1 = \frac{(hc)^2}{8mc^2L^2} \qquad \text{3-43}$$

Substituting $hc = 1240$ eV·nm and $mc^2 = 0.511$ MeV, we obtain

$$E_1 = \frac{(1240 \text{ eV·nm})^2}{8(5.11 \times 10^5 \text{ eV})(0.1 \text{ nm})^2} = 37.6 \text{ eV}$$

This is greater than the minimum energy of about 1 eV that we found from the uncertainty principle in Example 3-3. It is of the same order of magnitude as the kinetic energy of the electron in the ground state of the hydrogen atom, which is 13.6 eV. In that case, the wavelength of the electron equals the circumference of a circle of radius 0.0529 nm, or about 0.33 nm, whereas for the electron in a one-dimensional box of length 0.1 nm, the wavelength in the ground state is $2L = 0.2$ nm.

(b) The energies of this system are given by

$$E_n = n^2 E_1 = n^2(37.6 \text{ eV})$$

Figure 3-9 shows these energies in an energy-level diagram. The energy of the first excited state is $E_2 = 4(37.6 \text{ eV}) = 150.4$ eV, and that of the second excited state is $E_3 = 9(37.6 \text{ eV}) = 338.4$ eV. The possible transitions from level 3 to level 2, from level 3 to level 1, and from level 2 to level 1 are indicated by the vertical arrows on the diagram. The energies of these transitions are

$$\Delta E_{3\rightarrow 2} = 338.4 \text{ eV} - 150.4 \text{ eV} = 188 \text{ eV}$$

$$\Delta E_{3\rightarrow 1} = 338.4 \text{ eV} - 37.6 \text{ eV} = 300.8 \text{ eV}$$

and

$$\Delta E_{2\rightarrow 1} = 150.4 \text{ eV} - 37.6 \text{ eV} = 112.8 \text{ eV}$$

The photon wavelengths for these transitions are

$$\lambda_{3\rightarrow 2} = \frac{hc}{\Delta E_{3\rightarrow 2}} = \frac{1240 \text{ eV·nm}}{188 \text{ eV}} = 6.60 \text{ nm}$$

$$\lambda_{3\rightarrow 1} = \frac{hc}{\Delta E_{3\rightarrow 1}} = \frac{1240 \text{ eV·nm}}{300.8 \text{ eV}} = 4.12 \text{ nm}$$

$$\lambda_{2\rightarrow 1} = \frac{hc}{\Delta E_{2\rightarrow 1}} = \frac{1240 \text{ eV·nm}}{112.8 \text{ eV}} = 11.0 \text{ nm}$$

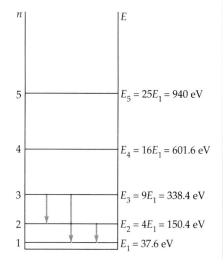

Figure 3-9 Energy-level diagram for Example 3-4. Transitions from the state $n = 3$ to the states $n = 2$ and $n = 1$, and from the state $n = 2$ to $n = 1$, are indicated by the vertical arrows.

Exercise

Calculate the wavelength of the photon emitted if the electron in Example 3-4 makes a transition from $n = 4$ to $n = 3$. (Answer: 4.71 nm)

Example 3-5

A particle is in the ground state of an infinite square-well potential. Find the probability of finding the particle (a) in the region $0 < x < \frac{1}{4}L$, and (b) in $\Delta x = 0.01L$ at $x = \frac{1}{2}L$.

(a) The probability of finding the particle in some range dx is

$$P(x)\,dx = \psi^2(x)\,dx = \frac{2}{L}\sin^2\frac{\pi x}{L}\,dx$$

We find the probability of finding the particle in the region $0 < x < \frac{1}{4}L$ by integrating this expression over this range:

$$P = \int_0^{L/4} \frac{2}{L}\sin^2\frac{\pi x}{L}\,dx$$

Substituting $\theta = \pi x/L$, we obtain

$$P = \frac{2}{L}\frac{L}{\pi}\int_0^{\pi/4}\sin^2\theta\,d\theta$$

$$= \frac{2}{\pi}\left[\frac{\theta}{2}\bigg|_0^{\pi/4} - \frac{\sin 2\theta}{4}\bigg|_0^{\pi/4}\right]$$

$$= \frac{2}{\pi}\left(\frac{\pi}{8} - \frac{1}{4}\right) = 0.091$$

The chance of finding the particle in the region $0 < x < \frac{1}{4}L$ is thus about 9.1 percent. This probability is indicated by the shaded region in Figure 3-10.

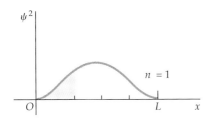

Figure 3-10 The probability density $\psi^2(x)$ versus x for a particle in the ground state of an infinite square-well potential. The probability of finding the particle in the region $0 < x < \frac{1}{4}L$ equals the shaded area.

(b) Since the region $\Delta x = 0.01L$ is very small compared with L, we do not need to integrate. The approximate probability is

$$P = \psi^2(x)\,\Delta x = \frac{2}{L}\sin^2\frac{\pi x}{L}\,\Delta x$$

Substituting $x = \frac{1}{2}L$ and $\Delta x = 0.01L$, we obtain

$$P = \frac{2}{L}\left(\sin^2\frac{\pi}{2}\right)(0.01L)$$

$$= \frac{2}{L}(1.0)(0.01L) = 0.02$$

There is thus a 2 percent chance of finding the particle in the region $\Delta x = 0.01L$ at $x = \frac{1}{2}L$.

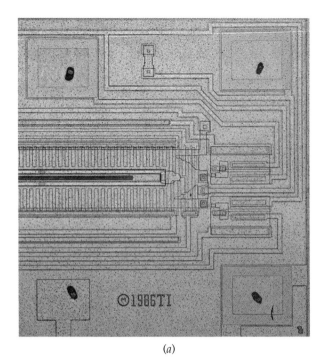

(a) (b)

Charge-coupled devices (CCDs) are on the forefront of imaging technology. They are efficient and fast, and their output is stored electronically so that it is easily processed and manipulated by computer. Typically 40 to 80 percent of the photons incident on a CCD surface are converted into a stored electrical signal, allowing for short exposure times and a very low detection threshold. This compares to the 2 or 3 percent of incoming photons that react with a film's light-sensitive atoms to produce exposed film grains. Also, unlike a photographic film, the response of a CCD is directly proportional to the amount of incoming light, making possible a much more precise measurement of data.

A CCD is a three-layer semiconductor—the top layer is a series of metallic electrodes, the bottom layer is a silicon crystal, and the middle layer is an insulator separating the two. Light striking silicon in the semiconductor frees electrons, which accumulate in potential wells at the surface of the silicon. Each well in the two-dimensional array on the silicon surface stores an amount of charge that is proportional to the number of photons that strike

Continued

3-7 A Particle in a Finite Square Well

The quantization of energy that we found for a particle in an infinite square well is a general result that follows from the solution of the Schrödinger equation for any particle confined to some region of space. We will illustrate this by considering the qualitative behavior of the wave function for a slightly more general potential-energy function, the finite square well shown in Figure 3-11. This potential-energy function is described mathematically by

$$U(x) = U_0 \quad x < 0$$
$$U(x) = 0 \quad 0 < x < L$$
$$U(x) = U_0 \quad x > L$$

This potential-energy function is discontinuous at $x = 0$ and $x = L$, but it is finite everywhere. The solutions of the Schrödinger equation for this type of potential-energy function depend on whether the total energy E is greater or less than U_0. We will not discuss the case of $E > U_0$, except to remark that in that case the particle is not confined and any value of the energy is allowed; that is, there is no energy quantization. Here we assume that $E < U_0$.

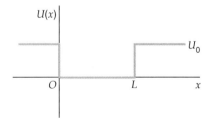

Figure 3-11 The finite square-well potential energy.

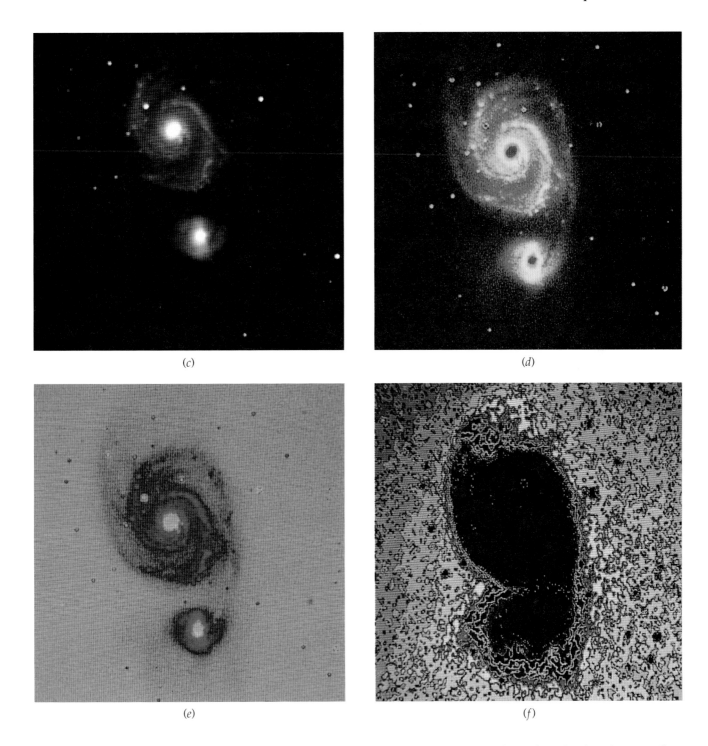

(c)

(d)

(e)

(f)

the surface in the region of the well. The charge is dumped electronically into a computer that records the location and amount of charge in each well. A conventional TV monitor can then be used to reconstruct a digitized version of the original image.

(a) A close-up of part of a CCD. The horizontal bar emerging from the left is the photosensitive area (called a "pixel"). The vertical segments above and below it (called a "transfer register") contain the succession of electrodes that transfer accumulated charge packets along a line of potential wells, from left to right, eventually depositing them in an amplifier located in the central right portion of the chip. (b) This platinum-silicide CCD chip responds to infrared wavelengths. It contains pixels in a 320 by 244 array. (c) An unprocessed CCD image of spiral galaxy Messier 51 and companion galaxy. (d) An image generated from the data contained in c in which false colors have been assigned, corresponding to different intensity ranges. (e) An image that again, like d, has been generated from the data in c and enhanced and colorized by computer. (f) This time the image has been processed for maximum contrast and contoured to show detail in the outer rims of the galaxies.

Inside the well, $U(x) = 0$, and the time-independent Schrödinger equation is the same as for the infinite well (Equation 3-33):

$$\frac{d^2\psi(x)}{dx^2} = -k^2\psi(x)$$

with

$$k^2 = \frac{2mE}{\hbar^2}$$

The general solution is of the form

$$\psi(x) = A \sin kx + B \cos kx$$

In this case, $\psi(x)$ is not zero at $x = 0$, so B is not zero. Outside the well, the time-independent Schrödinger equation is

$$\frac{d^2\psi(x)}{dx^2} = \frac{2m}{\hbar^2}(U_0 - E)\psi(x) = \alpha^2\psi(x) \qquad 3\text{-}44$$

where

$$\alpha^2 = \frac{2m}{\hbar^2}(U_0 - E) > 0 \qquad 3\text{-}45$$

The wave functions and allowed energies can be found by solving Equation 3-44 for $\psi(x)$ outside the well and then requiring that $\psi(x)$ and $d\psi(x)/dx$ be continuous at the boundaries $x = 0$ and $x = L$. The solution of Equation 3-44 is not difficult [for positive values of x, it is of the form $\psi(x) = Ce^{-\alpha x}$], but applying the boundary conditions involves much tedious algebra and is not important for our purpose. The important feature of Equation 3-44 is that the second derivative of $\psi(x)$, which is related to the curvature of the wave function, has the same sign as the wave function ψ. If ψ is positive, $d^2\psi/dx^2$ is also positive and the wave function curves away from the axis as shown in Figure 3-12a. Similarly, if ψ is negative, $d^2\psi/dx^2$ is negative and ψ again curves away from the axis as shown in Figure 3-12b. This behavior is very different from that inside the well, where ψ and $d^2\psi/dx^2$ have opposite signs so that ψ always curves toward the axis like a sine or cosine function. Because of this behavior outside the well, $\psi(x)$ becomes infinite as x approaches $\pm\infty$ for most values of the energy. That is, $\psi(x)$ is not well behaved outside the well. Though they satisfy the Schrödinger equation, such functions are not proper wave functions because they cannot be normalized. Only for certain values of the energy do the wave functions approach 0 as $|x|$ becomes very large. These energy values are the allowed energies for the finite square well.

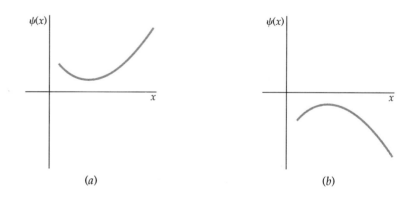

Figure 3-12 (a) A positive function with positive curvature. (b) A negative function with negative curvature.

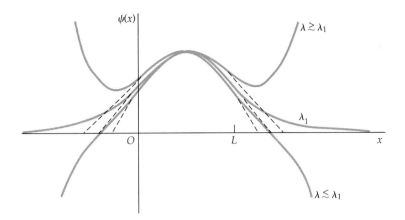

Figure 3-13 shows a well-behaved wave function with a wavelength λ_1 inside the well corresponding to the ground-state energy. The behavior of the wave functions corresponding to nearby wavelengths and energies are also shown. Figure 3-14 shows the wave functions and probability distributions for the ground state and first two excited states. From this figure, we can see that the wavelengths inside the well are slightly longer than the corresponding wavelengths for the infinite well (Figure 3-8), so the corresponding energies are slightly less than those for the infinite well. Another feature of the finite-well problem is that there are only a finite number of allowed energies depending on the size of U_0. For very small values of U_0, there is only one allowed energy.

Note that the wave function penetrates beyond the edges of the well at $x = L$ and $x = 0$, indicating that there is some small probability of finding the particle in the region in which its total energy E is less than its potential energy U_0. This region is called the *classically forbidden region* because the kinetic energy, which is $E - U_0$, would be negative when $U_0 > E$. Since negative kinetic energy has no meaning in classical physics, it is interesting

Figure 3-13 Functions satisfying the Schrödinger equation with wavelengths near the wavelength λ_1 corresponding to the ground-state energy $E_1 = h^2/2m\lambda_1^2$ in the finite well. If λ is slightly greater than λ_1, the function approaches infinity as shown in Figure 3-12a. At the critical wavelength λ_1, the function and its slope approach zero together. If λ is slightly less than λ_1, the function crosses the x axis while the slope is still negative. The slope then becomes more negative because its rate of change $d^2\psi/dx^2$ is now negative. This function approaches negative infinity as x approaches infinity.

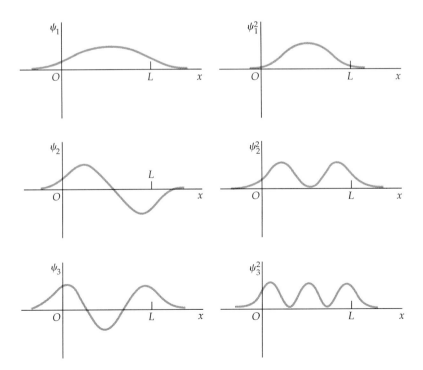

Figure 3-14 Graphs of the wave functions $\psi_n(x)$ and probability distributions $\psi_n^2(x)$ for $n = 1, 2,$ and 3 for the finite square well. Compare these graphs with those of Figure 3-8 for the infinite square well, where the wave functions are zero at $x = 0$ and $x = L$. The wavelengths here are slightly longer than the corresponding ones for the infinite well, so the allowed energies are somewhat smaller.

to speculate on the result of an attempt to observe the particle in the classically forbidden region. It can be shown from the uncertainty principle that if an attempt is made to localize the particle in the classically forbidden region, such a measurement introduces an uncertainty in the momentum of the particle corresponding to a minimum kinetic energy that is greater than $U_0 - E$, which is just great enough to prevent us from measuring a negative kinetic energy. The penetration of the wave function into a classically forbidden region does have important consequences in barrier penetration, which will be discussed in Section 3-9.

Much of our discussion of the finite-well problem applies to any problem in which $E > U(x)$ in some region and $E < U(x)$ outside that region. Consider the potential energy $U(x)$ shown in Figure 3-15. Inside the well, the Schrödinger equation is

$$\frac{d^2\psi(x)}{dx^2} = -k^2\psi(x)$$

where $k^2 = 2m[E - U(x)]/\hbar^2$ now depends on x. The solutions of this equation are no longer simple sine or cosine functions because the wave number $k = 2\pi/\lambda$ now varies with x, but since $d^2\psi/dx^2$ and ψ have opposite signs, ψ will always curve toward the axis and the solutions will oscillate. Outside the well, $d^2\psi/dx^2$ and ψ have the same sign, so ψ will curve away from the axis and there will be only certain values of E for which solutions exist that approach zero as x approaches infinity.

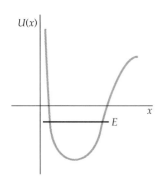

Figure 3-15 Arbitrary well-type potential with possible energy E. Inside the well where E is greater than $U(x)$, $\psi(x)$ and $d^2\psi/dx^2$ have opposite signs, and the wave function will oscillate. Outside the well, $\psi(x)$ and $d^2\psi/dx^2$ have the same sign and, except for certain values of E, the wave functions will not be well behaved.

3-8 Expectation Values

The solution of a classical mechanics problem is typically specified by giving the position of a particle as a function of time. As we have discussed, the wave nature of matter prevents us from doing this for microscopic systems. Instead, we find the wave function $\Psi(x, t)$ and the probability distribution $|\Psi(x, t)|^2$. The most that we can know is the probability of measuring a certain value of position x. If we measure the position for a large number of identical systems, we get a range of values corresponding to the probability distribution. The average value of x obtained from such measurements is called the **expectation value** and written $\langle x \rangle$.

Expectation value

The expectation value of x is the same as the average value of x that we would expect to obtain from a measurement of the positions of a large number of particles with the same wave function $\Psi(x, t)$.

Suppose we have a large number N of identical systems. The number of systems for which the particle is in a particular region dx at some value x is given by $NP(x)\, dx = N|\Psi(x, t)|^2\, dx$. To find the average value of x, we sum all the values of x and then divide by the number of terms in the sum. The values of x in this sum are given by $xNP(x)\, dx = xN|\Psi(x, t)|^2\, dx$, which is the value of x times the number of times it occurs. Since x is continuous, a sum over all the possible values of x is performed by integrating over dx. When we divide by the total number of measurements N, we obtain for the average or expectation value

Expectation value of x defined

$$\langle x \rangle = \int x|\Psi(x, t)|^2\, dx \qquad 3\text{-}46$$

As we have seen, for a particle in a state of definite energy, the probability distribution is independent of time. The expectation value is then given by

$$\langle x \rangle = \int x|\psi(x)|^2 \, dx \qquad \text{3-47}$$

Expectation value of x for time-independent state

The expectation value of any function $f(x)$ is given by

$$\langle f(x) \rangle = \int f(x)|\psi(x)|^2 \, dx \qquad \text{3-48}$$

Expectation value of f(x) defined

Example 3-6

Find (a) $\langle x \rangle$ and (b) $\langle x^2 \rangle$ for a particle in the ground state of an infinite square-well potential.

(a) The wave function for the ground state in an infinite square-well potential is given by Equation 3-42 with $n = 1$:

$$\psi = \sqrt{\frac{2}{L}} \sin \frac{\pi x}{L}$$

The expectation value of x is then

$$\langle x \rangle = \int_{-\infty}^{+\infty} x\psi^2(x) \, dx = \int_0^L x \frac{2}{L} \sin^2 \frac{\pi x}{L} \, dx$$

$$= \frac{2}{L}\left(\frac{L}{\pi}\right)^2 \int_0^\pi \theta \sin^2 \theta \, d\theta$$

where we have substituted $\theta = \pi x/L$. This integral can be found in tables. Its value is

$$\int_0^\pi \theta \sin^2 \theta \, d\theta = \left[\frac{\theta^2}{4} - \frac{\theta \sin 2\theta}{4} - \frac{\cos 2\theta}{8}\right]_0^\pi = \frac{\pi^2}{4}$$

The expectation value of x is thus

$$\langle x \rangle = \frac{2L}{\pi^2}\left(\frac{\pi^2}{4}\right) = \frac{L}{2}$$

which is what we would expect because the probability distribution is symmetric about the midpoint of the well.

(b) The expectation value of x^2 is

$$\langle x^2 \rangle = \int_{-\infty}^{+\infty} x^2 \psi^2(x) \, dx = \int_0^L x^2 \frac{2}{L} \sin^2 \frac{\pi x}{L} \, dx$$

$$= \frac{2}{L}\left(\frac{L}{\pi}\right)^3 \int_0^\pi \theta^2 \sin^2 \theta \, d\theta$$

Again, we obtain the integral from tables:

$$\int_0^\pi \theta^2 \sin^2 \theta \, d\theta = \left[\frac{\theta^3}{6} - \left(\frac{\theta^2}{4} - \frac{1}{8}\right)\sin 2\theta - \frac{\theta \cos 2\theta}{4}\right]_0^\pi = \frac{\pi^3}{6} - \frac{\pi}{4}$$

Then,

$$\langle x^2 \rangle = \frac{2L^2}{\pi^3}\left(\frac{\pi^3}{6} - \frac{\pi}{4}\right) = L^2\left(\frac{1}{3} - \frac{1}{2\pi^2}\right) = 0.283L^2$$

Exercise

A six-sided die has the number 1 painted on three sides and the number 2 painted on the other three sides. (*a*) What is the probability of a 1 coming up when the die is thrown? (*b*) What is the expectation value of the number that comes up when the die is thrown? [Answers: (*a*) 0.50, (*b*) 1.5]

Question

6. Can the expectation value of x ever equal a value that has zero probability of being measured?

3-9 Reflection and Transmission of Electron Waves: Barrier Penetration

In Sections 3-6 and 3-7, we were concerned with bound-state problems in which the potential energy is larger than the total energy for large values of $|x|$. In this section, we consider some simple examples of unbound states for which E is greater than $U(x)$. For these problems, $d^2\psi/dx^2$ and ψ have opposite signs, so $\psi(x)$ curves toward the axis and does not become infinite at large values of $|x|$.

Step Potential

Consider a particle of energy E moving in a region in which the potential energy is the step function

$$U(x) = 0 \qquad x < 0$$
$$U(x) = U_0 \qquad x > 0$$

as shown in Figure 3-16. We are interested in what happens when a particle moving from left to right encounters the step.

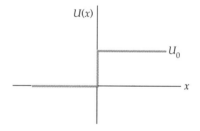

Figure 3-16 Step potential. A classical particle incident from the left, with total energy E greater than U_0, is always transmitted. The change in potential energy at $x = 0$ merely provides an impulsive force that reduces the speed of the particle. A wave incident from the left is partially transmitted and partially reflected because the wavelength changes abruptly at $x = 0$.

The classical answer is simple. To the left of the step, the particle moves with a speed $v = \sqrt{2E/m}$. At $x = 0$, an impulsive force acts on the particle. If the initial energy E is less than U_0, the particle will be turned around and will then move to the left at its original speed; that is, the particle will be reflected by the step. If E is greater than U_0, the particle will continue to move to the right but with reduced speed given by $v = \sqrt{2(E - U_0)/m}$. We can picture this classical problem as a ball rolling along a level surface and coming to a steep hill of height h given by $mgh = U_0$. If the initial kinetic energy of the ball is less than mgh, the ball will roll part way up the hill and then back down and to the left along the lower surface at its original speed.

If E is greater than mgh, the ball will roll up the hill and proceed to the right at a lesser speed.

The quantum-mechanical result is similar when E is less than U_0. Figure 3-17 shows the wave function for the case $E < U_0$. The wave function does not go to zero at $x = 0$ but rather decays exponentially, like the wave function for the bound state in a finite square-well problem. The wave penetrates slightly into the classically forbidden region $x > 0$, but it is eventually completely reflected. This problem is somewhat similar to that of total internal reflection in optics.

For $E > U_0$, the quantum-mechanical result differs markedly from the classical result. At $x = 0$, the wavelength changes abruptly from $\lambda_1 = h/p_1 = h/\sqrt{2mE}$ to $\lambda_2 = h/p_2 = h/\sqrt{2m(E - U_0)}$. We know from optics that when the wavelength changes suddenly part of the wave is reflected and part is transmitted. Since the motion of an electron (or other particle) is governed by a wave equation, the electron will likewise be sometimes transmitted and sometimes reflected. The probabilities of reflection and transmission can be calculated by solving the Schrödinger equation in each region of space and comparing the amplitudes of the transmitted and reflected waves with that of the incident wave. This calculation and its result are similar to finding the fraction of light reflected from an air–glass surface. If R is the probability of reflection, called the **reflection coefficient,** this calculation gives

$$R = \frac{(k_1 - k_2)^2}{(k_1 + k_2)^2} \quad\quad 3\text{-}49$$

where k_1 is the wave number for the incident wave and k_2 is that for the transmitted wave. This result is the same as that in optics for the reflection of light at normal incidence from the boundary between two media having different indexes of refraction n. The probability of transmission T, called the **transmission coefficient,** can be calculated from the reflection coefficient since the probability of transmission plus the probability of reflection must equal 1:

$$T + R = 1 \quad\quad 3\text{-}50$$

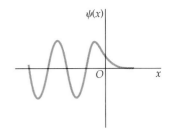

Figure 3-17 When the total energy E is less than U_0, the wave function penetrates slightly into the region $x > 0$. However, the probability of reflection for this case is 1, so no energy is transmitted.

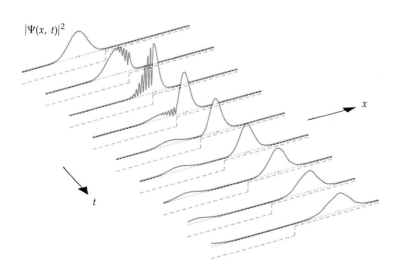

Time development of a one-dimensional wave packet representing a particle incident on a step potential for $E > U_0$. The position of a classical particle is indicated by the dot. Note that part of the packet is transmitted and part is reflected.

Barrier Penetration

Figure 3-18a shows a rectangular barrier potential of height U_0 and width a given by

$$U(x) = 0 \quad x < 0$$
$$U(x) = U_0 \quad 0 < x < a$$
$$U(x) = 0 \quad x > a$$

We consider a particle of energy E, which is slightly less than U_0, that is incident on the barrier from the left. Classically, the particle would always be reflected. However, a wave incident from the left does not decrease immediately to zero at the barrier but will instead decay exponentially in the classically forbidden region $0 < x < a$. Upon reaching the far wall of the barrier ($x = a$), the wave function must join smoothly to a sinusoidal wave function to the right of the barrier as shown in Figure 3-18b. This implies that there is some probability of the particle (which is represented by the wave function) being found on the far side of the barrier even though, classically, it should never pass through the barrier. For the case in which the quantity $\alpha a = \sqrt{2ma^2(U_0 - E)/\hbar^2}$ is much greater than 1, the transmission coefficient is proportional to $e^{-2\alpha a}$:

Transmission through a barrier

$$T \propto e^{-2\alpha a} \qquad 3\text{-}51$$

with $\alpha = \sqrt{2m(U_0 - E)/\hbar^2}$.

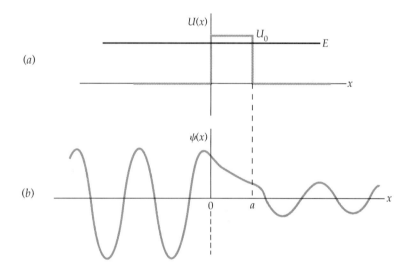

Figure 3-18 (a) Square-barrier potential. (b) Penetration of the barrier by a wave with total energy less than the barrier energy. Part of the wave is transmitted by the barrier even though, classically, the particle cannot enter the region $0 < x < a$ in which the potential energy is greater than the total energy.

The probability of penetration of the barrier thus decreases exponentially with the barrier thickness a and with the square root of the relative barrier height $(U_0 - E)$.

The penetration of a barrier is not unique to quantum mechanics. In Figure 3-19, a light ray is incident on a glass–air surface. At an angle greater than the critical angle, total reflection occurs. But because of the wave nature of light, the electric field $\mathcal{E}$ does not immediately drop to zero at the surface but rather decreases exponentially and becomes negligible within a few wavelengths of the surface. If another piece of glass is brought near the surface—as shown in Figure 3-19—some of the light is transmitted across the barrier. (This effect can be demonstrated with a laser beam and two 45° prisms.) Similarly, Figure 3-20 shows barrier penetration by water waves in a ripple tank.

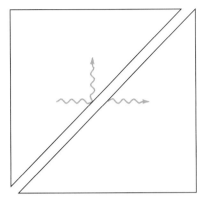

Figure 3-19 Penetration of an optical barrier. If the second prism is close enough to the first, part of the wave penetrates the air barrier even when the angle of incidence in the first prism is greater than the critical angle.

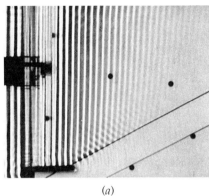

Figure 3-20 Penetration of a barrier by water waves in a ripple tank. In (a) the waves are totally reflected from a gap of deeper water. When the gap is very narrow, as in (b), a transmitted wave appears.

The theory of barrier penetration was used by George Gamow in 1928 to explain the enormous variation in the half-lives for α decay of radioactive nuclei. (Alpha particles are helium nuclei, which consist of two protons and two neutrons tightly bound together.) In general, the smaller the energy of the emitted α particle, the longer the half-life. The energies of α particles from natural radioactive sources range from about 4 to 7 MeV, whereas the half-lives range from about 10^{-5} second to 10^{10} years. Gamow represented a radioactive nucleus by a potential well containing an α particle as shown in Figure 3-21. Without knowing very much about the nuclear force that is exerted on the α particle within the nucleus, he represented it by a square well. Just outside the well, the α particle with its charge of $+2e$ is repelled by the nucleus with its charge $+Ze$, where Ze is the remaining nuclear charge. This force is represented by the Coulomb potential energy $+k(2e)(Ze)/r$. The energy E is the measured kinetic energy of the emitted α particle, because when it is far from the nucleus its potential energy is zero. After the α particle is formed inside the radioactive nucleus, it bounces back and forth inside the nucleus, hitting the barrier at the nuclear radius R. Each time it strikes the barrier, there is some small probability of its penetrating and appearing outside the nucleus. We can see from Figure 3-21 that a small increase in E reduces the relative height of the barrier $U - E$ and also its thickness. Because the probability of penetration is so sensitive to the barrier thickness and relative height, a small increase in E leads to a large increase in the probability of transmission and therefore to a shorter lifetime. Gamow was able to derive an expression for the half-life as a function of E that is in excellent agreement with experimental results.

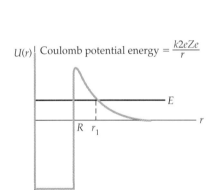

Figure 3-21 Model of a potential-energy function for an α particle in a radioactive nucleus. The strong attractive nuclear force when r is less than the nuclear radius R can be approximately described by the potential well shown. Outside the nucleus the nuclear force is negligible, and the potential is given by Coulomb's law, $U(r) = +k(2e)(Ze)/r$, where Ze is the nuclear charge and $2e$ is the charge of the α particle.

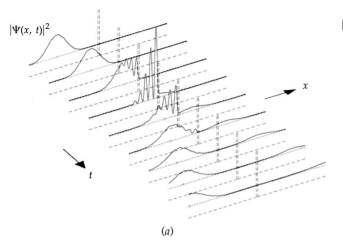
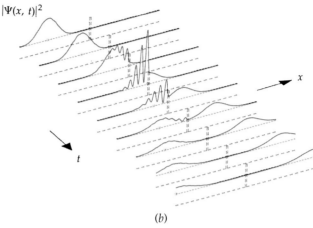

(a)   (b)

Barrier penetration. (a) A wave packet representing a particle incident on a barrier of height greater than the energy of the particle. A small part of the packet tunnels through the barrier. (b) The same particle incident on a barrier of half the height of the one in (a), but still greater than the energy of the particle. The probability of transmission is much greater as indicated by the size of the transmitted packet. In both drawings, the position of a classical particle is indicated by a dot.

In the **scanning tunneling electron microscope** developed in the 1980s, a thin space between a material specimen and a tiny probe acts as a barrier to electrons bound in the specimen. A small voltage applied between the probe and specimen causes the electrons to tunnel through the vacuum separating the two surfaces if the surfaces are close enough together. The tunneling current is extremely sensitive to the size of the gap between the probe and specimen. If a constant tunneling current is maintained as the probe scans the specimen, the surface of the specimen can be mapped out by the motions of the probe. In this way, the surface features of a specimen can be measured with a resolution of the order of the size of an atom.

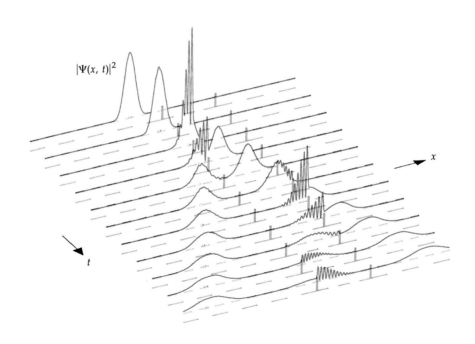

Wave packet representing a particle incident on two barriers. At each encounter, part of the packet is transmitted and part reflected, resulting in part of the packet being trapped between the barriers.

3-10 The Schrödinger Equation in Three Dimensions

The one-dimensional time-independent Schrödinger equation (Equation 3-30) is easily extended to three dimensions. In rectangular coordinates, it is

$$-\frac{\hbar^2}{2m}\left(\frac{\partial^2 \psi}{\partial x^2} + \frac{\partial^2 \psi}{\partial y^2} + \frac{\partial^2 \psi}{\partial z^2}\right) + U\psi = E\psi \qquad 3\text{-}52$$

where the wave function ψ and the potential energy U are generally functions of all three coordinates, x, y, and z. To illustrate some of the features of problems in three dimensions, we consider a particle in a three-dimensional infinite square well given by $U(x, y, z) = 0$ for $0 < x < L$, $0 < y < L$, and $0 < z < L$. Outside this cubical region $U(x, y, z) = \infty$. For this problem, the wave function must be zero at the edges of the well. The solution of Equation 3-52 can be written as

$$\psi(x, y, z) = A \sin k_1 x \sin k_2 y \sin k_3 z \qquad 3\text{-}53$$

where the constant A is determined by normalization. Inserting this solution into Equation 3-52, we obtain for the energy

$$E = \frac{\hbar^2}{2m}(k_1^2 + k_2^2 + k_3^2)$$

which is equivalent to $E = (p_x^2 + p_y^2 + p_z^2)/2m$, with $p_x = \hbar k_1$ and so on. The wave function will be zero at $x = L$ if $k_1 = n_1\pi/L$, where n_1 is an integer. Similarly, the wave function will be zero at $y = L$ if $k_2 = n_2\pi/L$, and it will be zero at $z = L$ if $k_3 = n_3\pi/L$. The energy is thus quantized to the values

$$E_{n_1,n_2,n_3} = \frac{\hbar^2 \pi^2}{2mL^2}(n_1^2 + n_2^2 + n_3^2) \qquad 3\text{-}54$$

where n_1, n_2, and n_3 are integers. Note that the energy and wave function are characterized by three quantum numbers, each arising from a boundary condition for one of the coordinates.

The lowest energy state (the ground state) for the cubical well occurs when $n_1 = n_2 = n_3 = 1$ and has the value

$$E_{1,1,1} = \frac{3\hbar^2 \pi^2}{2mL^2}$$

The first excited energy level can be obtained in three different ways: $n_1 = 2$, $n_2 = n_3 = 1$; $n_2 = 2$, $n_1 = n_3 = 1$; or $n_3 = 2$, $n_1 = n_2 = 1$. Each has a different wave function. For example, the wave function for $n_1 = 2$ and $n_2 = n_3 = 1$ is

$$\psi_{2,1,1} = A \sin \frac{2\pi x}{L} \sin \frac{\pi y}{L} \sin \frac{\pi z}{L} \qquad 3\text{-}55$$

An energy level with which more than one wave function is associated is said to be **degenerate**. In this case, there is threefold degeneracy. Degeneracy is related to the spatial symmetry of the problem. If, for example, we consider a noncubic well where $U = 0$ for $0 < x < L_1$, $0 < y < L_2$, and $0 < z < L_3$, the boundary conditions at the edges would lead to the quantum conditions $k_1 L_1 = n_1 \pi$, $k_2 L_2 = n_2 \pi$, and $k_3 L_3 = n_3 \pi$, and the total energy would be

$$E_{n_1,n_2,n_3} = \frac{\hbar^2 \pi^2}{2m}\left(\frac{n_1^2}{L_1^2} + \frac{n_2^2}{L_2^2} + \frac{n_3^2}{L_3^2}\right) \qquad 3\text{-}56$$

Figure 3-22 Energy-level diagrams for (a) a cubic infinite well and (b) a noncubic infinite well. In (a) the energy levels are degenerate; that is, there are two or more wave functions having the same energy. The degeneracy is removed when the symmetry of the potential is removed, as in (b).

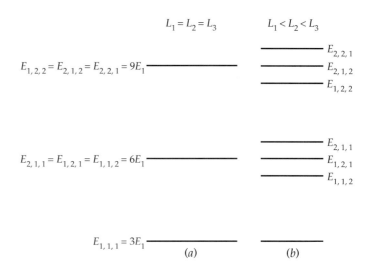

These energy levels are not degenerate if L_1, L_2, and L_3 are not equal. Figure 3-22 shows the energy levels for the ground state and first two excited states for an infinite cubic well in which the excited states are degenerate and for a noncubic infinite well in which L_1, L_2, and L_3 are slightly different so that the excited levels are slightly split apart and the degeneracy is removed.

3-11 The Schrödinger Equation for Two Identical Particles

Our discussion of quantum mechanics has thus far been limited to situations in which a single particle moves in some force field characterized by a potential-energy function U. The most important physical problem of this type is the hydrogen atom, in which a single electron moves in the Coulomb potential of the proton nucleus. This problem, which we will consider in some detail in Chapter 4, is actually a two-body problem since the proton also moves in the field of the electron. However, the motion of the proton requires only a very small correction to the energy of the atom that is easily made in both classical and quantum mechanics. When we consider more complicated problems, such as the helium atom, we must apply quantum mechanics to two or more electrons moving in an external field. Such problems are complicated by the interaction of the electrons with each other and also by the fact that the electrons are identical.

The interaction of two electrons with each other is electromagnetic and is essentially the same as that expected classically for two charged particles. The Schrödinger equation for an atom with two or more electrons cannot be solved exactly, so approximation methods must be used. This is not very different from the situation in classical problems with three or more particles. However, the complications arising from the identity of electrons are purely quantum mechanical and have no classical counterpart. They are due to the fact that it is impossible to keep track of which electron is which. Classically, identical particles can be identified by their positions, which can be determined with unlimited accuracy. This is impossible quantum mechanically because of the uncertainty principle. Figure 3-23 offers a schematic illustration of the problem.

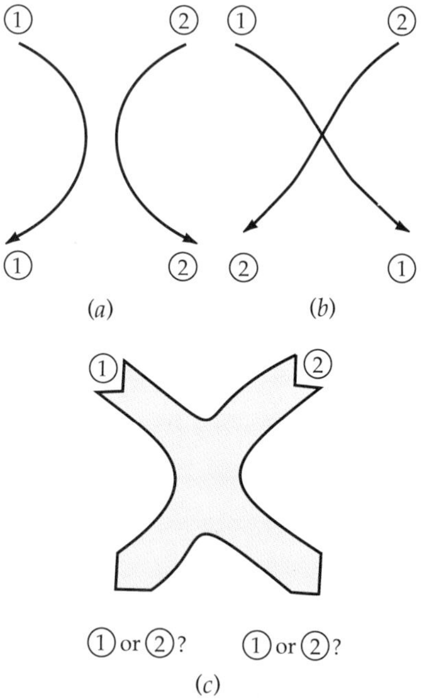

Figure 3-23 Two possible classical electron paths are shown in (a) and (b). The electrons can be distinguished classically. Because of the quantum-mechanical wave properties of the electrons, the paths are spread out, as indicated by the shaded region in (c). It is impossible to distinguish which electron is which after they separate.

Section 3-11 The Schrödinger Equation for Two Identical Particles

The indistinguishability of identical particles has important consequences. For instance, consider the very simple case of two identical, non-interacting particles in a one-dimensional infinite square well. The time-independent Schrödinger equation for two particles, each of mass m, is

$$-\frac{\hbar^2}{2m}\frac{\partial^2 \psi(x_1, x_2)}{\partial x_1^2} - \frac{\hbar^2}{2m}\frac{\partial^2 \psi(x_1, x_2)}{\partial x_2^2} + U\psi(x_1, x_2) = E\psi(x_1, x_2) \qquad 3\text{-}57$$

where x_1 and x_2 are the coordinates of the two particles. If the particles interact, the potential energy U contains terms with both x_1 and x_2 that cannot be separated into separate terms containing only x_1 or x_2. For example, the electrostatic repulsion of two electrons in one dimension is represented by the potential energy $ke^2/|x_2 - x_1|$. However, if the particles do not interact (as we are assuming here), we can write $U = U_1(x_1) + U_2(x_2)$. For the infinite square well, we need only solve the Schrödinger equation inside the well where $U = 0$, and require that the wave function be zero at the walls of the well. With $U = 0$, Equation 3-57 looks just like that for a particle in a two-dimensional well (Equation 3-52, with no z and with y replaced by x_2).

Solutions of this equation can be written in the form

$$\psi_{n,m} = \psi_n(x_1)\psi_m(x_2) \qquad 3\text{-}58$$

where ψ_n and ψ_m are the single-particle wave functions for a particle in an infinite well, and n and m are the quantum numbers of particles 1 and 2, respectively. For example, for $n = 1$ and $m = 2$, the wave function is

$$\psi_{1,2} = A \sin \frac{\pi x_1}{L} \sin \frac{2\pi x_2}{L} \qquad 3\text{-}59$$

The probability of finding particle 1 in dx_1 and particle 2 in dx_2 is $\psi_{n,m}^2(x_1, x_2)\, dx_1\, dx_2$, which is just the product of the separate probabilities $\psi_n^2(x_1)\, dx_1$ and $\psi_m^2(x_2)\, dx_2$. However, even though we have labeled the particles 1 and 2, we cannot distinguish which is in dx_1 and which is in dx_2 if they are identical. The mathematical descriptions of identical particles must be the same if we interchange the labels. The probability density $\psi^2(x_1, x_2)$ must therefore be the same as $\psi^2(x_2, x_1)$:

$$\psi^2(x_2, x_1) = \psi^2(x_1, x_2) \qquad 3\text{-}60$$

Equation 3-60 is satisfied if $\psi(x_2, x_1)$ is either **symmetric** or **antisymmetric** on the exchange of particles. That is, either

$$\psi(x_2, x_1) = \psi(x_1, x_2) \qquad \text{symmetric} \qquad 3\text{-}61$$

or

$$\psi(x_2, x_1) = -\psi(x_1, x_2) \qquad \text{antisymmetric} \qquad 3\text{-}62$$

Note that the wave functions given by Equations 3-58 and 3-59 are neither symmetric or antisymmetric. If we interchange x_1 and x_2 in these wave functions, we get a different wave function, which implies that the particles can be distinguished.

We can find symmetric and antisymmetric wave functions that are solutions of the Schrödinger equation by adding or subtracting $\psi_{n,m}$ and $\psi_{m,n}$. Adding them, we obtain

$$\psi_S = A'[\psi_n(x_1)\psi_m(x_2) + \psi_n(x_2)\psi_m(x_1)] \qquad \text{symmetric}$$

and subtracting them, we obtain

$$\psi_A = A'[\psi_n(x_1)\psi_m(x_2) - \psi_n(x_2)\psi_m(x_1)] \qquad \text{antisymmetric}$$

For example, the symmetric and antisymmetric wave functions for the first excited state of two identical particles in an infinite square well would be

$$\psi_S = A'\left[\sin\frac{\pi x_1}{L}\sin\frac{2\pi x_2}{L} + \sin\frac{\pi x_2}{L}\sin\frac{2\pi x_1}{L}\right]$$

and

$$\psi_A = A'\left[\sin\frac{\pi x_1}{L}\sin\frac{2\pi x_2}{L} - \sin\frac{\pi x_2}{L}\sin\frac{2\pi x_1}{L}\right]$$

There is an important difference between antisymmetric and symmetric wave functions. If $n = m$, the antisymmetric wave function is identically zero for all values of x_1 and x_2 whereas the symmetric wave function is not. Thus, if the wave function describing two identical particles is antisymmetric, the quantum numbers n and m of two particles cannot be the same. This is an example of the **Pauli exclusion principle,** which was first stated by Wolfgang Pauli for electrons in an atom:

Pauli exclusion principle

No two electrons in an atom can have the same quantum numbers.

(As we will see in Chapter 4, the state of an electron in an atom is described by four quantum numbers, one associated with each space coordinate and one associated with electron spin, which will be discussed in that chapter.) It is found that electrons, protons, neutrons, and some other particles have antisymmetric wave functions and obey the Pauli exclusion principle. These particles are called **fermions.** Other particles, such as α particles, deuterons, photons, and mesons, have symmetric wave functions and do not obey the Pauli exclusion principle. These particles are called **bosons.**

Summary

1. The state of a particle, such as an electron, is described by its wave function Ψ, which is the solution of the Schrödinger wave equation. The absolute square of the wave function $|\Psi|^2$ measures the probability of finding the particle in some region of space.

2. A harmonic wave of a single angular frequency ω and wave number k can represent an electron that is completely unlocalized and can be anywhere in space. In terms of the angular frequency and wave number, the de Broglie equations are

$$E = \hbar\omega$$

and

$$p = \hbar k$$

A localized electron can be represented by a wave packet, which is a group of waves of nearly equal frequencies and wavelengths. The wave packet moves with a group velocity

$$V_g = \frac{d\omega}{dk}$$

which equals the velocity of the electron.

3. Wave–particle duality leads to the uncertainty principle, which states that the product of the uncertainty in a measurement of the position of a particle and the uncertainty in a measurement of its momentum must be greater than $h/4\pi$, where h is Planck's constant:

$$\Delta x \, \Delta p \geq \frac{h}{4\pi} = \frac{1}{2}\hbar$$

Similarly, the uncertainty in the energy ΔE is related to the time interval Δt required to measure the energy by

$$\Delta E \, \Delta t \geq \tfrac{1}{2}\hbar$$

An important consequence of the uncertainty principle is that a particle confined in space has a minimum energy called the zero-point energy.

4. Light, electrons, neutrons, and all other carriers of momentum and energy exhibit both wave and particle properties. Everything propagates like a wave, exhibiting diffraction and interference, but exchanges energy in discrete lumps like a particle. Because the wavelengths of macroscopic objects are so small, diffraction and interference are not observed. Also, when a macroscopic amount of energy is exchanged, so many quanta are involved that the particle nature of the energy is not evident.

5. The wave function $\Psi(x, t)$ obeys the time-dependent Schrödinger equation

$$-\frac{\hbar^2}{2m}\frac{\partial^2 \Psi(x, t)}{\partial x^2} + U(x)\Psi(x, t) = i\hbar \frac{\partial \Psi(x, t)}{\partial t}$$

For any wave function that describes a particle in a state of definite energy, the time-dependent Schrödinger equation can be simplified by writing the wave function in the form

$$\Psi(x, t) = \psi(x)e^{-i\omega t}$$

where $\psi(x)$ is a function of x only. This leads to the time-independent Schrödinger equation

$$-\frac{\hbar^2}{2m}\frac{d^2\psi(x)}{dx^2} + U(x)\psi(x) = E\psi(x)$$

In addition to satisfying the Schrödinger equation, a wave function $\psi(x)$ must be continuous and must have a continuous first derivative $d\psi/dx$. Since the probability of finding an electron somewhere must be 1, the wave function must obey the normalization condition

$$\int_{-\infty}^{\infty} |\psi|^2 \, dx = 1$$

This condition implies the boundary condition that ψ must approach 0 as x approaches $\pm\infty$. Such boundary conditions lead to the quantization of energy.

6. A one-dimensional box of length L is described by the infinite square-well potential

$$U(x) = 0 \quad 0 < x < L$$
$$U(x) = \infty \quad x < 0 \text{ or } x > L$$

The wavelength for a particle confined in such a box obeys the standing-wave condition

$$n\frac{\lambda}{2} = L$$

This results in the energy of the particle being quantized to the values

$$E_n = n^2 E_1$$

where E_1 is the ground-state energy given by

$$E_1 = \frac{h^2}{8mL^2}$$

The number n is called a quantum number.

The wave functions for a particle in a box are given by

$$\psi_n = \sqrt{\frac{2}{L}} \sin \frac{n\pi x}{L}$$

7. An electron in a stationary state can be pictured as a cloud of charge with charge density proportional to $|\psi|^2$.

8. When the quantum numbers of a system are very large, quantum calculations and classical calculations agree—a result known as Bohr's correspondence principle.

9. In a finite well of height U_0, there is a finite number of allowed energies, which are slightly less than the corresponding energies in an infinite well.

10. The expectation value of x is defined by

$$\langle x \rangle = \int x |\Psi(x, t)|^2 \, dx$$

The expectation value of x is the same as the average value of x that we would expect to obtain from a measurement of the positions of a large number of particles with the same wave function $\Psi(x, t)$.

For a particle in a state of definite energy, the expectation value of x is

$$\langle x \rangle = \int x |\psi(x)|^2 \, dx$$

The expectation value of any function $f(x)$ is given by

$$\langle f(x) \rangle = \int f(x) |\psi(x)|^2 \, dx$$

11. When the potential changes abruptly over a small distance, a particle may be reflected even though $E > U(x)$. A particle may also penetrate a region in which $E < U(x)$. Reflection and penetration of de Broglie waves are similar to those for other kinds of waves.

12. When more than one wave function is associated with the same energy level, the energy level is said to be degenerate. Degeneracy arises because of spatial symmetry.

13. A wave function that describes two identical particles must be either symmetric or antisymmetric when the coordinates of the particles are exchanged. Fermions, which include electrons, protons, and neutrons, are described by antisymmetric wave functions and obey the Pauli exclusion principle, which states that no two particles can have the same quantum number. Bosons, which include α particles, deuterons, photons, and mesons, have symmetric wave functions and do not obey the Pauli exclusion principle.

Scanning Tunneling Microscopy

Ellen D. Williams
University of Maryland

All scientists believe that matter consists of atoms. However, the evidence for the existence of atoms is mostly indirect. There are only three known methods for actually imaging individual atoms and thus studying their behavior directly. These are transmission electron microscopy, field ion microscopy, and scanning tunneling microscopy. Of these three, the last is the most recently developed and the most versatile. Gert Binnig and Heinrich Rohrer were awarded the Nobel Prize in Physics in 1986 for the development of this technique, jointly with Ernst Ruska who was honored for his work on the development of electron microscopy.

To understand scanning tunneling microscopy (STM), we first need to understand a little bit about the behavior of electrons in metals (or any other electrically conducting material). A metal consists of a large number of atoms, which are held together by the electrostatic forces acting between the electrons and the nuclei of the atoms. Most of the electrons are bound tightly to the individual nuclei, just as in the case of isolated atoms. However, the electrons that are farthest from the nuclei feel a relatively weak electrostatic attraction, and thus are free to wander about in the space between the nuclei. These are called the conduction electrons because they are the ones that carry (or conduct) electric current. It is a pretty good approximation to treat these electrons as if they are moving in a nearly constant attractive potential, so that they behave very much like particles in a one-dimensional box. We draw an energy-level diagram similar to Figure 3-7 for the conduction electrons as shown in Figure 1.

Continued

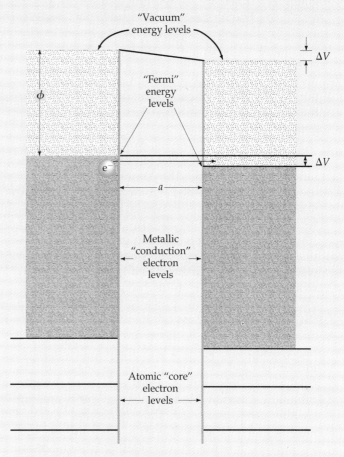

Figure 1 Energy-level diagram for the conduction electrons in two metals separated by a distance a. The electrons move as essentially free particles within the metal. The most weakly bound electrons have the highest energy (the Fermi energy E_F) and are held within the solid by a potential-energy barrier known as the work function ϕ. In order to induce a measurable electron tunneling current, a small voltage difference ΔV is applied between the two metals. This offsets their energy levels as shown, so that electrons can flow from the occupied states near E_F on the more negative metal (on the left in the figure) to the unoccupied states just above E_F on the more positive metal (on the right in the figure).

Ellen Williams was born in Oshkosh, Wisconsin (so had a connection with Paul Tipler right from the start, see page ix). After completing her BS in chemistry at Michigan State University, she earned her doctorate at the California Institute of Technology, studying the properties of thin layers of molecules on surfaces. Since then, she has been at the Department of Physics and Astronomy at the University of Maryland, where her work in surface physics has continued. This has focused on investigation of the structures and properties of very thin layers of metals on silicon. Enhanced understanding of this is crucial to the development of ultrafast microchip technology and other applications. The use of scanning tunneling microscopy has been of central importance in her research.

Because there are large numbers of levels ($\sim 10^{23}$), they overlap to provide a continuous distribution of states available to the conduction electrons. Only the lower energy levels are occupied by electrons. The energy of the most weakly bound electrons is called the Fermi energy (E_F). The electrons at the Fermi energy are held in the metal by an energy barrier of about 5 eV, the work function (ϕ) described in Section 2-2. Classically, these electrons can never leave the metal unless they are given the energy necessary to go over this potential barrier. Quantum mechanically, however, electrons near the Fermi energy can tunnel through the potential barrier, as discussed in Sections 3-4 to 3-9. By placing two pieces of metal close to one another, as shown in Figure 1, a finite square-well barrier like that shown in Figure 3-18 can be created. The probability for the electrons at the Fermi energy to tunnel through the barrier is proportional to $e^{-\alpha a}$, where a is the distance separating the two pieces of metal and α depends on the barrier height (in this case the work function), as discussed in Section 3-9. As explained below, this exponential dependence of the transmission probability on separation is what makes STM possible.

The mechanism of STM is illustrated in Figure 2. If a pointed metal probe is placed sufficiently close to a sample and a small voltage (say ΔV about 10 mV) is applied between the probe and sample, then electron tunneling can occur. The net flow of electrons can be measured as a tunneling current, which is proportional to the transmission probability. If we then scan the probe back and forth above the sample, any bumps on the sample surface will change the separation. Because of the exponential relationship between the separation and the transmission probability, changes in the separation as small as 0.01 nm result in measurable changes in the tunneling current. Measurement of the tunneling current while scanning thus generates a topographic map of the surface. Thus, in principle, it is possible to measure the topography of the surface using STM. In practice, formidable experimental problems arise in trying to image individual atoms on the surface. The challenges lie in three general areas: vibrations, tip sharpness, and position control.

Vibrations are important because the separation between the sample and probe must be very small. For a work function of around 5 eV, a separation of only a few nanometers (comparable to the size of atoms) is needed. For such a small separation, a minor perturbation such as vibrations set up by a sneeze can jam the probe right into the sample, ruining the experiment. The most common source

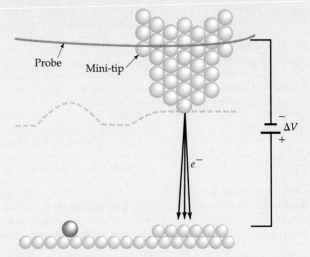

Figure 2 Schematic illustration of the path of a probe (dashed line) scanned across a surface while maintaining constant tunneling current. If the probe is very large (as illustrated by the solid line), tunneling occurs over a large area, and atomic features cannot be resolved. However, if the probe has a mini-tip of atomic dimensions, then tunneling occurs into a small area, allowing very small features (even individual atoms) to be imaged.

of vibrations is motion of the floor, which typically has an amplitude of about 1 μm—a thousand times larger than the tip–sample separation that must be maintained. Thus, very careful engineering is required to make the instrument rigid and to isolate it from these external disturbances.

The second problem—probe sharpness—determines how small a structure can be imaged on a surface. Electrochemical etching can be used to sharpen the end of a metal wire to a radius of about 1 μm (1000 nm). A probe with such a large surface area would allow tunneling to occur over a large region of the sample surface. In order to resolve small features such as atoms, it is necessary to have a probe comparable in size to the features. In trying to fabricate such a probe, we are rewarded for not being able to do a perfect job in making micron-radius probes. The metal wires that we polish electrochemically are rough on an atomic scale: Their surfaces bear many mini-tips like the one illustrated in Figure 2. The end of such a mini-tip will present a single atom (perhaps a few) close to the surface. The exponential dependence of transmission probability on separation then guarantees that tunneling will occur preferentially from the end of the mini-tip.

The third problem in STM is that of position control. How is it possible to move the probe

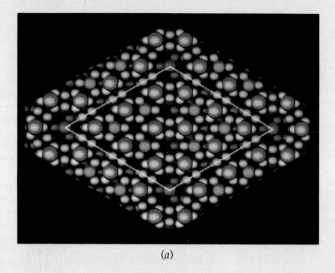

(a)

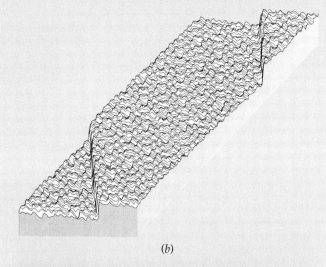

(b)

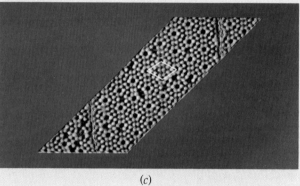

(c)

around with controllable displacements of less than 0.1 nm? The answer lies in a special type of material known as piezoelectric ceramic. This material expands and contracts when an external voltage is applied to electrodes on opposite faces. Typically, expansions are on the order of a few tenths of a nanometer per applied volt. As a result, a probe attached to a piece of piezoelectric ceramic can be moved with great precision by application of external voltages.

When Binnig and Rohrer first demonstrated that these challenges could be surmounted, it generated tremendous excitement because it opened the possibility of answering fundamental questions about the properties of surfaces, as well as a wealth of potential practical applications. The power of STM is illustrated in Figure 3, where a model of the atomic structure of atoms on the surface of silicon is compared with STM images of the real surface. The data from an STM scan of the surface consist of values of the surface height versus position. This can be immediately presented in the form of a line scan, as shown by the dashed line in Figure 2. The line scan image of Figure 3b is easier to visualize if the data are represented by a gray scale as in Figure 3c. Here, the height at each point is represented by the intensity of color—ranging from white for the highest points to black for the lowest—showing a striking correspondence to the atomic model in Figure 3a. The deep holes correspond to the positions of missing atoms in the model, and the bright spots are due to the atoms that protrude above the average surface plane. There are also two lines where the surface abruptly changes height in this image. These surface steps are important in practical processes such as crystal growth and microfabrication.

In addition to fundamental studies of the phys-

Figure 3 Scanning tunneling microscope used to image atoms and steps on a silicon surface. (a) A model of the atoms on a Si surface. The red circles represent atoms that protrude highest above the surface, the blue circles are atoms in a lower layer, and the gray circles are atoms in a still lower layer. The white diamond shows the repeating unit of the structure. The length of each side of this unit cell is 2.7 nm. (b) The traces of height versus position (line scans) obtained by an STM over an area of approximately 10 nm by 35 nm. (c) The data shown in (b) presented in a gray-scale representation, allowing the imaging of the highest layer of atoms [see red circles in the atomic model of (a)] to be immediately recognized. The unit cell of the structure is indicated for comparison with (a).

ics of atoms at surfaces, STM has a range of potentially practical applications, in part because STM is quite insensitive to its microscopic environment. For the description of tunneling presented above, the material in the gap between the sample and probe is not too important. Tunneling microscopes operate in vacuum, air, liquid helium, oil, water, and even in electrolytic solutions. This makes it possible to

Continued

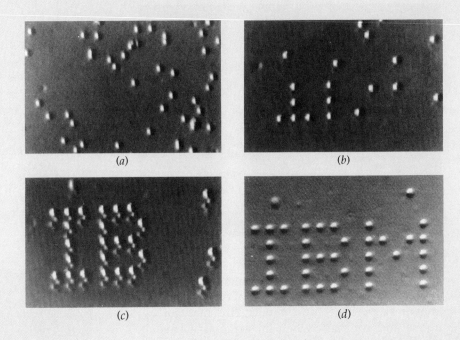

Figure 4 A sequence of STM images taken during the construction of a patterned array of xenon atoms on a nickel surface at a temperature of 4 K. Xenon atoms were allowed to stick randomly on the surface from the gas phase (upper left). The STM tip was then used to "pull" the atoms one by one across the surface to spell out the name of the company that sponsored the development of STM.

apply the STM to such important problems as imaging DNA in a biological environment and observing the surfaces of battery electrodes while they are operating. Variations of STM have also been developed capable of imaging samples that are not conductors (atomic force microscopy) and of imaging the magnetic properties at surfaces. Perhaps the most stunning possibility is that of using STM to write with atomic resolution. Features a few nanometers wide have been written by using the probe to scratch or dent the surface directly or by using the tunneling current to heat the surface. However, the ultimate limit of resolution has been demonstrated by using the probe to pull *individual atoms* of xenon around on a surface to spell out a message, as illustrated in Figure 4.

Scanning tunneling microscopy is a practical demonstration of quantum mechanics and an illustration that understanding of basic concepts of physics can yield tremendous gains in advanced technology. It is also an object lesson in the long-term and often unforeseeable benefits that accrue from developing fundamental ideas. The scientists who first explored the physical possibility of tunneling during the early part of this century would be amazed and delighted to see its application in STM.

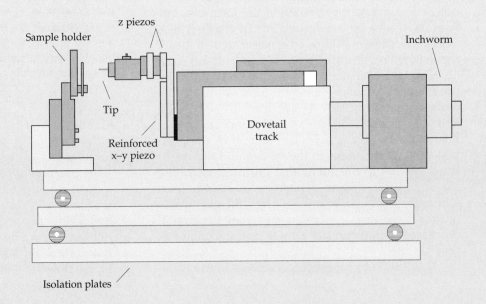

Figure 5 Schematic diagram of an STM. The sample holder is rigidly mounted to the top plate in a stack of isolation plates. The tip is fixed onto the x, y, z scanning piezos. In order to position the tip so tunneling can occur, these piezos are mounted on a heavy block that can slide on a dovetail track. The block is then pushed by an electronically controlled "walker" device called an inchworm, which can move forward and back in steps of 4 nm.

Suggestions for Further Reading

Everhart, Thomas E., and Thomas L. Hayes: "The Scanning Electron Microscope," *Scientific American*, January 1972, p. 54.

This article describes how the interaction between a beam of high-energy electrons and matter is used by the scanning electron microscope to create an image of three-dimensional appearance.

Hey, Tony, and Patrick Walters: *The Quantum Universe*, Cambridge University Press, Cambridge, 1987.

This "coffee-table quantum mechanics" book examines some of the many fields in which quantum theory has had an impact.

Rae, Alastair: *Quantum Physics: Illusion or Reality?*, Cambridge University Press, Cambridge, 1986.

This small book seeks to explain to the general reader the conceptual problems raised by quantum physics.

Shimony, Abner: "The Reality of the Quantum World," *Scientific American*, January 1988, p. 46.

This article explains why "common sense" (including Einstein's!) is unsuccessful in predicting the outcome of some recent experiments.

Review

A. Objectives: After studying this chapter, you should:

1. Know the properties of wave packets and, in particular, the relations between Δx and Δk, and between $\Delta \omega$ and Δt.

2. Be able to discuss the uncertainty principle and some of its consequences.

3. Be able to discuss particle–wave duality.

4. Be able to write the time-dependent and time-independent Schrödinger equations.

5. Be able to solve the time-independent Schrödinger equation for the infinite square-well problem and discuss how energy quantization arises.

6. Be able to sketch $\psi(x)$ and $\psi^2(x)$ for the infinite and finite square-well potentials.

7. Be able to draw an energy-level diagram for the infinite square-well potential.

8. Be able to state Bohr's correspondence principle.

9. Know how expectation values are calculated from the wave function for a particular state.

10. Be able to discuss qualitatively the reflection and transmission of waves at a barrier.

11. Know how quantum numbers arise in the solution of the Schrödinger equation in more than one dimension.

12. Be able to discuss the general features of the solution of the Schrödinger equation for two particles in an infinite square well.

B. Define, explain, or otherwise identify:

Probability distribution function
Group velocity
Uncertainty principle
Zero-point energy
Wave–particle duality
Classical particle
Classical wave
Time-dependent Schrödinger equation
Time-independent Schrödinger equation
Normalization condition
Infinite square-well potential
Boundary condition
Quantum number
Bohr's correspondence principle
Expectation value
Reflection coefficient
Transmission coefficient
Barrier penetration
Scanning tunneling electron microscope
Degeneracy
Symmetric wave function
Antisymmetric wave function
Pauli exclusion principle
Fermion
Boson

C. True or false: If the statement is true, explain why it is true. If it is false, give a counterexample.

1. The velocity of an electron is the same as the phase velocity of the wave describing it.

2. It is impossible in principle to know precisely the position of an electron.

3. A particle that is confined to some region of space cannot have zero energy.

4. All phenomena in nature are adequately described by classical wave theory.

5. The Schrödinger equation follows from Newton's laws of motion.

6. Boundary conditions on the wave function lead to energy quantization.

7. The expectation value of a quantity is the value that you expect to measure.

8. The penetration of a barrier by a wave has no physical significance.

9. Bosons do not obey the Pauli exclusion principle.

Problems

Level I

3-1 The Electron Wave Function

1. A 100-g rigid sphere of radius 1 cm has a kinetic energy of 2 J and is confined to move in a force-free region between two rigid walls separated by 50 cm. (a) What is the probability of finding the center of the sphere exactly midway between the two walls? (b) What is the probability of finding the center of the sphere between the 24.9- and 25.1-cm marks?

2. For the sphere described in Problem 1, make a graph of the classical probability density $P(x)$ as a function of position between the walls.

3-2 Electron Wave Packets

3. The wave function describing a state of an electron confined to move along the x axis is given at time zero by $\psi(x, 0) = Ae^{-x^2/4\sigma^2}$. Find the probability of finding the electron in a region dx centered at (a) $x = 0$, (b) $x = \sigma$, and (c) $x = 2\sigma$. (d) Where is the electron most likely to be found?

4. At time zero, three finite waves have the form $\cos kx$ with the wave numbers given by $k_1 - \delta$, k_1, and $k_1 + \delta$. The three waves are chopped so that there respectively are 5, 6, and 7 wavelengths in the region of space $-L < x < L$. The amplitude of the wave for $k = k_1$ is twice the amplitude of the other two. They are in phase at $x = 0$ but interfere destructively at the ends of the wave packet. Make a careful graph of each of the waves and of the resultant of the three chopped wave trains.

5. What is the approximate value of $\Delta x \, \Delta k$ for the wave packet formed in Problem 4?

6. An electron that is not localized in space is described by the wave function $\psi = A \sin(kx - \omega t)$. The kinetic energy of the electron is 1 keV. Find k and ω.

3-3 The Uncertainty Principle

7. Suppose a one-dimensional wave packet represents the state of a 1-ng particle moving along the x axis. The velocity of the wave packet is 10^{-2} m/s with a statistical spread of $\pm 10^{-4}$ m/s. What is the minimum corresponding statistical spread in the position predicted for the particle?

8. If the uncertainty in the position of a wave packet representing the state of a quantum system particle is equal to its de Broglie wavelength, how does the uncertainty in momentum compare with the value of the momentum of the particle?

9. Suppose that the wave function describing the state of an electron predicted a statistical spread in the velocity of 10^{-5} m/s. What is the corresponding statistical spread in the position of the electron?

10. If an excited state of an atom is known to have a lifetime of 10^{-7} s, what is the statistical spread in the energy of photons emitted by such atoms in the spontaneous decay to the ground state?

11. A wave pulse of frequency f_0 has a duration of Δt. It travels with speed v and occupies a region of space $\Delta x = v \, \Delta t$ as shown in Figure 3-24. Let N be the approximate number of waves in Δx. (a) How are N, f_0, and Δt related? (b) What is the approximate uncertainty in the wave number k?

Figure 3-24 Problem 11.

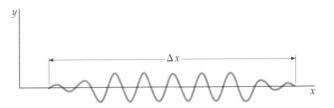

3-4 Wave–Particle Duality

12. What is the de Broglie wavelength of a neutron with speed 10^6 m/s?

13. Suppose you have a spherical object of mass 4 g moving at 100 m/s. What size aperture is necessary for the object to show diffraction? Show that no common objects would be small enough to squeeze through such an aperture.

14. A neutron has a kinetic energy of 10 MeV. What size object is necessary to observe neutron diffraction effects? Is there anything in nature of this size that could serve as a target to demonstrate the wave nature of 10-MeV neutrons?

15. What is the de Broglie wavelength of an electron accelerated from rest through a potential difference of 200 volts? What are some common targets that could demonstrate the wave nature of such an electron?

3-5 The Schrödinger Equation

16. Show that the wave function $\Psi(x, t) = Ae^{kx-\omega t}$ does not satisfy the time-dependent Schrödinger equation.

17. Show that $\Psi(x, t) = Ae^{i(kx-\omega t)}$ satisfies both the time-dependent Schrödinger equation and the classical wave equation (Equation 3-24).

18. In a region of space, a particle has a wave function given by $\psi(x) = Ae^{-x^2/2L^2}$, and energy $\hbar^2/2mL^2$, where L is some length. (a) Find the potential energy as a function of x, and sketch U versus x. (b) What is the classical potential that has this dependence?

19. (a) For Problem 18, find the kinetic energy as a function of x. (b) Show that $x = L$ is the classical turning point. (c) The potential energy of a simple harmonic oscillator in terms of its angular frequency ω is given by $U = \frac{1}{2}m\omega^2 x^2$. Compare this with your answer to part (a) of Problem 18, and show that the total energy for this wave function can be written $E = \frac{1}{2}\hbar\omega$.

3-6 A Particle in a Box

20. A particle of mass m is confined to a tube of length L.

(a) Use the uncertainty relationship to estimate the smallest possible energy. (b) Assume that the inside of the tube is a force-free region and that the particle makes elastic reflections at the tube ends. Use Schrödinger's equation to find the ground-state energy for the particle in the tube. Compare the answer to that of part (a).

21. (a) What is the wavelength associated with the particle of Problem 20 if the particle is in its ground state? (b) What is the wavelength if the particle is in its second excited state (quantum number $n = 3$)? (c) Use de Broglie's relationship to find the magnitude for the momentum of the particle in its ground state. (d) Show that $p^2/2m$ gives the correct energy for the ground state of this particle in the box.

22. Sketch the wave function $\psi(x)$ and the probability distribution $\psi^2(x)$ for the state $n = 4$ of the infinite square-well potential given by Equation 3-32.

23. A particle is in a box of size L. Calculate the ground-state energy if (a) the particle is a proton and $L = 0.1$ nm, a typical size for a molecule; (b) the particle is a proton and $L = 1$ fm, a typical size for a nucleus.

24. A particle is in the ground state of an infinite square-well potential given by Equation 3-32. Find the probability of finding the particle in the interval $\Delta x = 0.002\, L$ at (a) $x = L/2$, (b) $x = 2L/3$, and (c) $x = L$. (Since Δx is very small, you need not do any integration.)

25. Do Problem 24 for a particle in the first excited state ($n = 2$) of an infinite square-well potential.

26. Do Problem 24 for a particle in the second excited state ($n = 3$) of an infinite square-well potential.

27. A mass of 10^{-6} g is moving with a speed of about 10^{-1} cm/s in a box of length 1 cm. Treating this as a one-dimensional infinite square-well potential, calculate the approximate value of the quantum number n.

28. (a) For the classical particle of Problem 27, find Δx and Δp, assuming that $\Delta x/L = 0.01$ percent and $\Delta p/p = 0.01$ percent. (b) What is $(\Delta x\, \Delta p)/\hbar$?

3-7 A Particle in a Finite Square Well

29. Sketch (a) the wave function and (b) the probability distribution for the $n = 4$ state for the finite square-well potential.

30. Sketch (a) the wave function and (b) the probability distribution for the $n = 5$ state for the finite square-well potential.

3-8 Expectation Values

31. Find (a) $\langle x \rangle$ and (b) $\langle x^2 \rangle$ for the first excited state ($n = 2$) in an infinite square-well potential.

32. Find (a) $\langle x \rangle$ and (b) $\langle x^2 \rangle$ for the second excited state ($n = 3$) in an infinite square-well potential.

33. (a) Show that the classical probability distribution function for a particle in a one-dimensional infinite square-well potential of length L is given by $P(x) = 1/L$. (b) Use your result in (a) to find $\langle x \rangle$ and $\langle x^2 \rangle$ for a classical particle in such a well.

3-9 Reflection and Transmission of Electron Waves: Barrier Penetration

34. A free particle of mass m with wave number k_1 is traveling to the right. At $x = 0$, the potential jumps from zero to U_0 and remains at this value for positive x. (a) If the total energy is $E = \hbar^2 k_1^2/2m = 2U_0$, what is the wave number k_2 in the region $x > 0$? Express your answer in terms of k_1 and in terms of U_0. (b) Calculate the reflection coefficient R at the potential step. (c) What is the transmission coefficient T? (d) If one million particles with wave number k_1 are incident upon the potential step, how many particles are expected to continue along in the positive x direction? How does this compare with the classical prediction?

35. Suppose that the potential jumps from zero to $-U_0$ at $x = 0$ so that the free particle speeds up instead of slowing down. The wave number for the incident particle is again k_1, and the total energy is $2U_0$. (a) What is the wave number for the particle in the region of positive x? (b) Calculate the reflection coefficient R at $x = 0$. (c) What is the transmission coefficient T? (d) If one million particles with wave number k_1 are incident upon the potential step, how many particles are expected to continue along in the positive x direction? How does this compare with the classical prediction?

36. Work Problem 34 for the case in which the energy of the incident particle is $1.01 U_0$ instead of $2U_0$.

Section 3-10 The Schrödinger Equation in Three Dimensions

37. A particle is confined to a three-dimensional box that has sides L_1, $L_2 = 2L_1$, and $L_3 = 3L_1$. Give the quantum numbers n_1, n_2, n_3 that correspond to the lowest ten energy levels of this box.

38. Give the wave functions for the lowest ten energy levels of the particle in Problem 37.

39. (a) Repeat Problem 37 for the case $L_2 = 2L_1$ and $L_3 = 4L_1$. (b) What quantum numbers correspond to degenerate energy levels?

40. Give the wave functions for the lowest ten energy levels of the particle in Problem 39.

41. A particle moves in a potential well given by $U(x, y, z) = 0$ for $-L/2 < x < L/2$, $0 < y < L$, and $0 < z < L$, and $U = \infty$ outside these ranges. (a) Write an expression for the ground-state wave function for this particle. (b) How do the allowed energies compare with those for a box having $U = 0$ for $0 < x < L$, rather than for $-L/2 < x < L/2$?

Section 3-11 The Schrödinger Equation for Two Identical Particles

42. Show that Equation 3-59 satisfies Equation 3-57 with $U = 0$, and find the energy of this state.

43. What is the ground-state energy of ten noninteracting bosons in a one-dimensional box of length L?

44. What is the ground-state energy of ten noninteracting fermions, such as neutrons, in a one-dimensional box of length L? (Because of the quantum number associated with spin to be discussed in Chapter 4, each spatial state can hold two neutrons.)

Level II

45. A proton is in an infinite square-well potential given by Equation 3-32 with $L = 1$ fm. (a) Find the ground-state energy in MeV. (b) Make an energy-level diagram for this system. Calculate the wavelength of the photon emitted for the transitions (c) $n = 2$ to $n = 1$, (d) $n = 3$ to $n = 2$, and (e) $n = 3$ to $n = 1$.

46. Suppose that the bottom of the infinite potential well of Equation 3-32 is changed from zero potential to $+U_0$. Solve the time-independent Schrödinger equation, and find the wave functions and allowed energies.

47. A particle is in the ground state of an infinite square-well potential given by Equation 3-32. Calculate the probability that the particle will be found in the region (a) $0 < x < \frac{1}{2}L$, (b) $0 < x < \frac{1}{3}L$, and (c) $0 < x < \frac{3}{4}L$.

48. (a) Show that for large n, the fractional difference in energy between state n and state $n + 1$ for a particle in an infinite square well is given approximately by

$$\frac{E_{n+1} - E_n}{E_n} \approx \frac{2}{n}$$

(b) What is the approximate percentage energy difference between the states $n_1 = 1000$ and $n_2 = 1001$? (c) Comment on how this result is related to Bohr's correspondence principle.

49. A particle is in the first excited state ($n = 2$) of an infinite square-well potential given by Equation 3-32. (a) Sketch $\psi^2(x)$ versus x for this state. (b) What is the expectation value $\langle x \rangle$ for this state? (c) What is the probability of finding the particle in some small region dx centered at $x = \frac{1}{2}L$? (d) Are your answers for (b) and (c) contradictory? If not, explain.

50. A particle of mass m moves in a region in which the potential energy is constant $U(x) = U_0$. (a) Show that neither $\Psi(x, t) = A \sin(kx - \omega t)$ nor $\Psi(x, t) = A \cos(kx - \omega t)$ satisfies the time-dependent Schrödinger equation. (Hint: If $C_1 \sin \phi + C_2 \cos \phi = 0$ for all values of ϕ, then C_1 and C_2 must be zero.) (b) Show that $\Psi(x, t) = A[\cos(kx - \omega t) + i \sin(kx - \omega t)] = Ae^{i(kx-\omega t)}$ does satisfy the time-independent Schrödinger equation providing that k, U_0, and ω are related by Equation 3-28.

51. Repeat Problem 47 for a particle in the first excited state of the infinite square-well potential.

52. For the square-well wave functions

$$\psi_n(x) = \sqrt{2/L} \sin(n\pi x/L) \quad n = 1, 2, 3, \ldots$$

corresponding to an infinite square well of length L, show that

$$\langle x^2 \rangle = \frac{L^2}{3} - \frac{L^2}{2n^2\pi^2}$$

53. A 10-eV electron is incident on a potential barrier of height 25 eV and width of 1 nm. (a) Use Equation 3-51 to calculate the order of magnitude of the probability that the electron will tunnel through the barrier. (b) Repeat your calculation for a width of 0.1 nm.

54. The standard deviation of measurements of the position of a particle is defined as the square root of the average value of the square of $(x - \langle x \rangle)$:

$$\sigma_x = \sqrt{\langle (x - \langle x \rangle)^2 \rangle}$$

Show that this can be written

$$\sigma_x = \sqrt{\langle x^2 \rangle - \langle x \rangle^2}$$

55. Use the result of Problem 54 to calculate σ_x for a classical particle in a box for which the probability distribution is $P(x) = 1/L$ for $0 < x < L$, and $P(x) = 0$ for $x < 0$ and for $x > L$.

56. (a) Use the results of Problems 52 and 54 to calculate σ_x for a particle in an infinite square-well potential for a general state n. (b) Evaluate your result for $n = 1$. (c) Show that for large n, your result approaches that for a classical particle found in Problem 55.

57. Use Equation 3-51 to calculate the order of magnitude of the probability that a proton will tunnel out of a nucleus in one collision with the nuclear barrier if it has energy 6 MeV below the top of the potential barrier and the barrier thickness is 10^{-15} m.

58. The minimum kinetic energy of a particle confined to a region of space of length L satisfies the inequality given in Equation 3-23. The ground-state energy is obtained from Schrödinger's equation applied to a particle in a box, where the potential energy may be taken to be zero inside the box. Show that $p^2/2m$ is the ground-state energy, and show that Equation 3-23 is satisfied.

59. Quantum mechanics predicts that any particle localized in space has a nonzero velocity and consequently can never be at rest. Consider a Ping-Pong ball of diameter 2 cm and mass 2 g that can move back and forth in a box of length 2.001 cm. Hence, the space in which the ball moves is only 0.001 cm in length. (a) What is the minimum speed of the Ping-Pong ball according to Schrödinger's equation? (b) What is the period of one oscillation?

60. A particle of mass m is in an infinite square-well potential given by

$$U = \infty \quad x < -\tfrac{1}{2}L$$
$$U = 0 \quad -\tfrac{1}{2}L < x < +\tfrac{1}{2}L$$
$$U = \infty \quad +\tfrac{1}{2}L < x$$

Since this potential is symmetric about the origin, the probability density $\psi^2(x)$ must also be symmetric. (a) Show that this implies that either $\psi(-x) = \psi(x)$ or $\psi(-x) = -\psi(x)$. (b) Show that the proper solutions of the time-independent Schrödinger equation can be written

$$\psi(x) = \sqrt{\frac{2}{L}} \cos \frac{n\pi x}{L} \quad n = 1, 3, 5, 7, \ldots$$

and

$$\psi(x) = \sqrt{\frac{2}{L}} \sin \frac{n\pi x}{L} \quad n = 2, 4, 6, 8, \ldots$$

(c) Show that the allowed energies are the same as those for the infinite square well given by Equation 3-32. (d) Sketch the wave functions and probability distributions for the ground state and the first excited state.

61. Calculate $\langle x \rangle$ and $\langle x^2 \rangle$ for the ground state of the infinite square-well potential in Problem 60.

62. A particle moves in a potential given by $U(x) = A|x|$. Without attempting to solve the Schrödinger equation, sketch the wave function for (a) the ground-state energy of a particle inside this potential and (b) the first excited state for this potential.

63. Calculate $\langle x \rangle$ and $\langle x^2 \rangle$ for the first excited state of the infinite square-well potential in Problem 60.

64. A particle moves freely in the two-dimensional region defined by $0 \leq x \leq L$ and $0 \leq y \leq L$. (a) Find the wave function satisfying Schrödinger's equation. (b) Find the corresponding energies. (c) Find the lowest two states that are degenerate. Give the quantum numbers for this case. (d) Find the lowest three states that have the same energy. Give the quantum numbers for the three states having the same energy.

65. What is the next energy level above that found in Problem 64 for a particle in a two-dimensional square box for which the degeneracy is greater than 2?

66. In Problem 18 the wave function $\psi_0(x) = Ae^{-x^2/2L^2}$ represents the ground-state energy of a harmonic oscillator. (a) Show that $\psi_1 = L\, d\psi_0(x)/dx$ is also a solution of Schrödinger's equation. (b) What is the energy of this new state? (c) From a look at the nodes of this wave function, how would you classify this excited state?

Level III

67. For the wave functions in Equation 3-39, show that for any two functions $\psi_n(x)$ and $\psi_m(x)$

$$\int \psi_n(x)\psi_m(x)\, dx = 0$$

unless the integer $n = m$. This is the important orthogonality property of the wave functions of quantum mechanics.

68. A particle of mass m near the earth's surface at $z = 0$ can be described by the potential energy

$$U = mgz \quad z > 0$$
$$U = \infty \quad z < 0$$

For some positive value of total energy E, indicate the classically allowed region on a sketch of $U(z)$ versus z. Sketch also the kinetic energy versus z. The Schrödinger equation for this problem is quite difficult to solve. Using arguments similar to those in Section 3-7 about the curvature of the wave function as given by the Schrödinger equation, sketch your "guesses" for the shape of the wave function for the ground state and the first two excited states.

69. Use the Schrödinger equation to show that the expectation value of the kinetic energy of a particle is given by

$$\langle K \rangle = \int_{-\infty}^{+\infty} \psi(x) \left(-\frac{\hbar^2}{2m} \frac{d^2\psi(x)}{dx^2} \right) dx$$

70. Show that Equations 3-49 and 3-50 imply that the transmission coefficient for particles of energy E incident on a step barrier $U_0 < E$ is given by

$$T = \frac{4k_1 k_2}{(k_1 + k_2)^2} = \frac{4r}{(1+r)^2}$$

where $r = k_2/k_1$.

71. (a) Show that for the case of a particle of energy E incident on a step barrier $U_0 < E$, the wave numbers k_1 and k_2 are related by

$$\frac{k_2}{k_1} = r = \sqrt{1 - (U_0/E)}$$

Use this and the results of Problem 70 to calculate the transmission coefficient T and the reflection coefficient R for the case (b) $E = 1.2 U_0$, (c) $E = 2.0 U_0$, and (d) $E = 10.0 U_0$.

72. Consider the one-dimensional, classical wave equation, Equation B-1 in Appendix B. Substitute the trial function $\Psi(x, t) = \psi(x)f(t)$ into the classical wave equation to obtain

$$\frac{1}{\psi(x)} \frac{d^2\psi(x)}{dx^2} = \frac{1}{v^2 f(t)} \frac{d^2 f(t)}{dt^2} \qquad 3\text{-}63$$

In Equation 3-63, the variables x and t are separated. Since the left side is a function of x only, it cannot depend on t. Similarly, the right side cannot depend on x. Therefore, both sides must equal some constant. (b) Set each side of Equation 3-63 equal to the constant $-k^2$, and show that the solution for $f(t)$ can be expressed as $e^{\pm i\omega t}$, where $\omega = kv$. (c) Show that the time-independent equation for x is given by

$$\frac{d^2\psi(x)}{dx^2} + k^2\psi = 0$$

73. (a) Repeat the use of the method of separation of variables, used in Problem 72, on Schrödinger's time-dependent equation to obtain

$$-\frac{\hbar^2}{2m} \frac{d^2\psi(x)/dx^2}{\psi(x)} + U(x) = i\hbar \frac{df/dt}{f} \qquad 3\text{-}64$$

(b) Since the left side of Equation 3-64 does not vary with x, the right side also cannot vary with x. Similarly, neither side can vary with t; thus, they both must equal the same constant E. Show that this implies that $f(t)$ is given by

$$f(t) = e^{-i\omega t}$$

where $\omega = E/\hbar$. Use the de Broglie relation to argue that E must be the total energy. (c) Use the left-hand side of Equation 3-64 to obtain Equation 3-30.

Chapter 4

Atoms

A scanning-tunneling microscope image of iodine atoms (pink) absorbed on platinum. The yellow pocket reflects the gap where an iodine atom has been dislodged.

Slightly more than 100 different elements have been discovered, 92 of which are found in nature. Each is characterized by an atom that contains a number of protons Z, an equal number of electrons, and a number of neutrons N. The number of protons Z is called the **atomic number.** The lightest atom, hydrogen (H), has $Z = 1$; the next lightest, helium (He), has $Z = 2$; the next lightest, lithium (Li), has $Z = 3$; and so forth. Nearly all the mass of the atom is concentrated in a tiny nucleus, which contains the protons and neutrons. The nuclear radius is typically about 1 to 10 fm (1 fm = 10^{-15} m). The distance between the nucleus and the electrons is about 0.1 nm = 100,000 fm. This distance determines the "size" of the atom.

The chemical and physical properties of an element are determined by the number and arrangement of the electrons in the atom. Because each proton has a positive charge $+e$, the nucleus has a total positive charge $+Ze$. The electrons are negatively charged $(-e)$, so they are attracted to the nucleus and repelled by each other. The electrons are arranged in shells. The first shell has up to two electrons. The second shell, about four times farther out from the nucleus than the first, can contain up to eight electrons. The third shell, about nine times farther out than the first, can contain up to eighteen electrons. This shell structure accounts for the periodic nature of the properties of the elements shown in the periodic table (Appendix F). Elements with a single electron in an outer shell (hydrogen, lithium, sodium, etc.) or those with a single vacancy in an outer shell (fluorine, chlorine, bromine, etc.) are very active chemically and readily combine to form molecules. Those with completely filled outer shells (helium, neon, argon, etc.) are more or less chemically inert. The calculation of the electron configurations of the atoms and the determination of the resulting chemical properties were great triumphs of quantum mechanics in the 1920s.

Since electrons and protons have equal but opposite charges and there are an equal number of electrons and protons in an atom, atoms are electrically neutral. Atoms that lose or gain one or more electrons are then electrically charged and are called *ions*. Atoms with just one electron in an outer shell (such as sodium, which has eleven electrons) tend to lose it readily and become positive ions; those lacking just one electron for a complete shell tend to gain an electron to become negative ions (for example, chlorine). Atoms bond together to form molecules, such as H_2O, or solids. This bonding involves only the outer electrons called **valence electrons.** In a molecule or solid, the separation of the atomic nuclei is about one atomic diameter, which is of the order of 0.1 nm.

In this chapter, we will apply our knowledge of quantum mechanics from Chapter 3 to give a qualitative description of the simplest atom, the hydrogen atom. We will then discuss qualitatively the structure of other atoms and the periodic table of the elements.

4-1 Quantum Theory of the Hydrogen Atom

Despite its spectacular successes, the Bohr model of the hydrogen atom described in Section 2-5 had many shortcomings. There was no justification for the postulates of stationary states or the quantization of angular momentum other than the fact that these postulates led to energy levels that agreed with spectroscopic data. Furthermore, the Bohr model gave no information about the intensities of the spectral lines, and attempts to apply it to more complicated atoms had little success. The quantum-mechanical theory resolved these difficulties. The stationary states of the Bohr model correspond to the standing-wave solutions of the Schrödinger wave equation. Energy quantization is a direct consequence of the frequency quantization that results from standing waves and the de Broglie relation $E = hf$. The quantized energies resulting from the standing-wave solutions of the Schrödinger equation agree with those obtained from the Bohr model and with experiment. The quantization of angular momentum that had to be postulated in the Bohr model is predicted by the quantum theory.

In quantum theory, the electron is described by its wave function ψ. The absolute square of the electron wave function $|\psi|^2$ gives the probability of finding the electron in some region of space. Boundary conditions on the wave function lead to the quantization of the wavelengths and frequencies and thereby to the quantization of the electron energy.

We can treat the hydrogen atom as a stationary nucleus, the proton, that has a single particle, an electron, moving with kinetic energy $p^2/2m$ and potential energy $U(r)$ due to the electrostatic attraction of the proton:

$$U(r) = -k\frac{Ze^2}{r} \qquad 4\text{-}1$$

We include the factor Z, which is 1 for hydrogen, so that we can apply our results to other, one-electron atoms, such as ionized helium He^+, for which $Z = 2$. The time-independent Schrödinger equation for a particle of mass m moving in three dimensions is given by Equation 3-52:

$$-\frac{\hbar^2}{2m}\left(\frac{\partial^2 \psi}{\partial x^2} + \frac{\partial^2 \psi}{\partial y^2} + \frac{\partial^2 \psi}{\partial z^2}\right) + U\psi = E\psi \qquad 4\text{-}2$$

Since the potential energy $U(r)$ depends only on the radial distance $r = \sqrt{x^2 + y^2 + z^2}$, the problem is most conveniently treated using the spherical coordinates r, θ, and ϕ, which are related to the rectangular coordinates x, y, and z by

$$z = r \cos \theta$$
$$x = r \sin \theta \cos \phi \qquad 4\text{-}3$$
$$y = r \sin \theta \sin \phi$$

These relations are shown in Figure 4-1. The transformation of the three-dimensional Schrödinger equation into spherical coordinates is straightforward but involves much tedious calculation, which we will omit. The result is

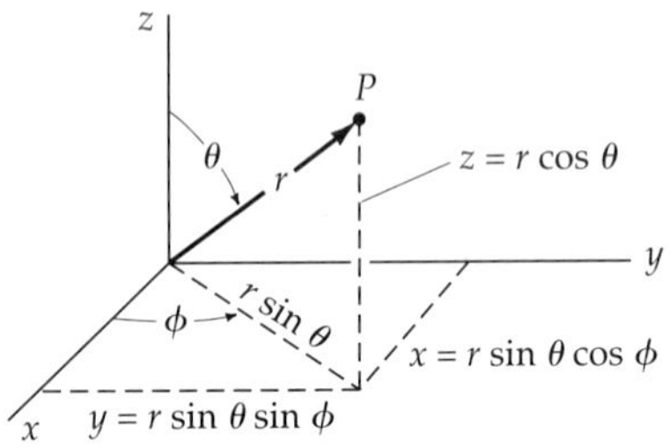

$$-\frac{\hbar^2}{2m} \frac{1}{r^2} \frac{\partial}{\partial r}\left(r^2 \frac{\partial \psi}{\partial r}\right)$$
$$-\frac{\hbar^2}{2mr^2}\left[\frac{1}{\sin \theta} \frac{\partial}{\partial \theta}\left(\sin \theta \frac{\partial \psi}{\partial \theta}\right) + \frac{1}{\sin^2 \theta} \frac{\partial^2 \psi}{\partial \phi^2}\right] + U\psi = E\psi \qquad 4\text{-}4$$

Despite the formidable appearance of this equation, it was not difficult for Schrödinger to solve because it is similar to other partial differential equations that arise in classical physics and such equations had been thoroughly studied. We will not solve this equation but merely discuss qualitatively some of the interesting features of the wave functions that satisfy it.

The first step in the solution of a partial differential equation such as Equation 4-4 is to separate the variables by writing the wave function $\psi(r, \theta, \phi)$ as a product of functions of each single variable:

$$\psi(r, \theta, \phi) = R(r) f(\theta) g(\phi) \qquad 4\text{-}5$$

Figure 4-1 Geometric relations between spherical and rectangular coordinates.

where R depends only on the radial coordinate r, f depends only on θ, and g depends only on ϕ. When this form of $\psi(r, \theta, \phi)$ is substituted into Equation 4-4, the partial differential equation can be transformed into three ordinary differential equations, one for $R(r)$, one for $f(\theta)$, and one for $g(\phi)$. The potential energy $U(r)$ appears only in the equation for $R(r)$, which is called the **radial equation.** The particular form of $U(r)$ given in Equation 4-1 therefore has no effect on the solutions of the equations for $f(\theta)$ and $g(\phi)$. These solutions are applicable to any central-field problem, that is, to any problem in which the potential energy depends only on r.

As we saw in Section 3-10, the requirement that the wave function be well behaved so that it is continuous and can be normalized introduces three quantum numbers, each associated with one of the three variables. The solution of the Schrödinger equation in spherical coordinates leads to three quantum numbers that are labeled n, ℓ, and m. The quantum numbers n_1, n_2, and n_3 that we found for a particle in a three-dimensional square well were independent of one another, but the quantum numbers associated with wave functions in spherical coordinates are interdependent. The possible values of these quantum numbers are

Quantum numbers for the hydrogen atom

$$n = 1, 2, 3, \ldots$$
$$\ell = 0, 1, 2, \ldots, n-1 \qquad 4\text{-}6$$
$$m = -\ell, -\ell+1, -\ell+2, \ldots, +\ell$$

That is, n can be any integer; ℓ can be 0 or any integer up to $n - 1$; and m can have $2\ell + 1$ possible values, ranging from $-\ell$ to $+\ell$ in integral steps.

The number n is called the **principal quantum number.** It is associated with the dependence of the wave function on the distance r and therefore with the probability of finding the electron at various distances from the nucleus. The quantum numbers ℓ and m are associated with the angular momentum of the electron and with the angular dependence of the electron wave function. The quantum number ℓ is called the **orbital quantum number.** The orbital angular momentum L of the electron is related to ℓ by

$$L = \sqrt{\ell(\ell+1)}\,\hbar \qquad 4\text{-}7$$

The quantum number m is called the **magnetic quantum number.** It is related to the component of the angular momentum in some direction. Ordinarily, all directions are equivalent, but one particular direction can be specified by placing the atom in a magnetic field. If the z direction is chosen for the magnetic field, the z component of the angular momentum of the electron is given by

$$L_z = m\hbar \qquad 4\text{-}8$$

If we measure the angular momentum of the electron in units of $\hbar$, we see that the angular momentum is quantized to the value $\sqrt{\ell(\ell+1)}$ units and that its component along any direction can have only the $2\ell+1$ values ranging from $-\ell$ to $+\ell$ units. Figure 4-2 shows a vector-model diagram illustrating the possible orientations of the angular-momentum vector for $\ell = 2$. Note that only specific values of θ are allowed; that is, the directions in space are quantized.

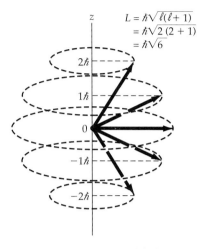

Figure 4-2 Vector-model diagram illustrating the possible values of the z component of the angular-momentum vector for the case $\ell = 2$.

Example 4-1

If an atom has an angular momentum characterized by the quantum number $\ell = 2$, what are the possible values of L_z, and what is the smallest possible angle between **L** and the z axis?

The possible values of L_z are $m\hbar$, where the $2\ell + 1 = 5$ values of m are -2, -1, 0, $+1$, and $+2$. The magnitude of **L** is $L = \sqrt{\ell(\ell+1)}\,\hbar = \sqrt{6}\,\hbar$. From Figure 4-1, the angle between **L** and the z axis is given by

$$\cos\theta = \frac{L_z}{L} = \frac{m\hbar}{\sqrt{\ell(\ell+1)}\,\hbar} = \frac{m}{\sqrt{\ell(\ell+1)}}$$

The smallest angle occurs when $m = +\ell$ or $-\ell$, which for $\ell = 2$ gives $\cos\theta = 2/\sqrt{6} = 0.816$ or $\theta = 35.3°$. We note the somewhat strange result that the angular-momentum vector cannot lie along the z axis. This is related to an uncertainty relation for angular momentum that implies that no two components of the angular momentum can be precisely known except when $\ell = 0$.

Exercise

An atom has an angular momentum characterized by the quantum number $\ell = 4$. What are the possible values of m? (Answer: -4, -3, -2, -1, 0, 1, 2, 3, 4)

The allowed energies of the hydrogen atom and other one-electron atoms that result from the solution of the Schrödinger equation with $U(r)$ given by Equation 4-1 are given by

$$E_n = -\frac{Z^2 E_0}{n^2} \qquad n = 1, 2, 3, \ldots \qquad 4\text{-}9 \quad \textit{Energy levels for hydrogen}$$

where

$$E_0 = \frac{k^2 e^4 m_e}{2\hbar^2} \approx 13.6 \text{ eV}$$

These energies are the same as in the Bohr model. Note that the energy is negative, indicating that the electron is bound to the nucleus (thus the term *bound state*), and that it depends only on the principal quantum number n. The fact that the energy does not depend on ℓ is a peculiarity of the inverse-square force and holds only for the hydrogen atom. For more complicated atoms having several electrons, the interaction of the electrons leads to a dependence of the energy on the orbital quantum number. In general, the lower the value of ℓ, the lower the energy for such atoms. Since, in general, there is no preferred direction in space, the energy for any atom does not depend on the magnetic quantum number m, which is related to the z component of the angular momentum. The energy does depend on m if the atom is in a magnetic field.

Figure 4-3 shows an energy-level diagram for hydrogen. This diagram is similar to Figure 2-14, except that the states with the same value of n but different values of ℓ are shown separately. These states (called *terms*) are referred to by giving the value of n along with a code letter: S for $\ell = 0$, P for $\ell = 1$, D for $\ell = 2$, and F for $\ell = 3$. (These code letters are remnants of spectroscopists' descriptions of various spectral lines as *sharp, principal, diffuse,* and *fundamental*. For values greater than 3, the letters follow alphabetically; thus, G is used for $\ell = 4$ and so forth.) When an atom makes a transition from one allowed energy state to another, electromagnetic radiation in the form of a photon is emitted or absorbed. Such transitions result in spectral lines that are characteristic of the atom. The transitions obey the **selection rules**

$$\Delta m = 0 \text{ or } \pm 1$$
$$\Delta \ell = \pm 1$$

4-10

These selection rules are related to the conservation of angular momentum and to the fact that the photon itself has an intrinsic angular momentum that

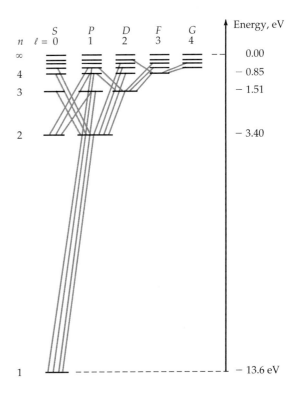

Figure 4-3 Energy-level diagram for hydrogen. The diagonal lines show transitions that involve emission or absorption of radiation and obey the selection rule $\Delta \ell = \pm 1$. States with the same value of n but different values of ℓ have the same energy $-E_0/n^2$, where $E_0 = 13.6$ eV as in the Bohr model.

has a maximum component along any axis of $1\hbar$. The wavelengths of the spectral lines emitted by hydrogen (and by other atoms) are related to the energy levels by the Bohr formula

$$hf = \frac{hc}{\lambda} = E_i - E_f \qquad 4\text{-}11$$

where E_i and E_f are the energies of the initial and final states.

4-2 The Hydrogen-Atom Wave Functions

The wave functions that are solutions of the Schrödinger equation are characterized by the quantum numbers and are written $\psi_{n\ell m}$. For any given value of n, there are n possible values of ℓ ($\ell = 0, 1, \ldots, n - 1$), and for each value of ℓ, there are $2\ell + 1$ possible values of m. Since the energy depends only on n, there are generally many different wave functions that correspond to the same energy, except at the lowest energy level, for which $n = 1$ and therefore ℓ and m must be 0. As discussed previously, the origins of this degeneracy are the $1/r$ dependence of the potential energy and the fact that, in the absence of any external fields, there is no preferred direction in space.

In the lowest energy state, the ground state, the principal quantum number n has the value 1, ℓ is 0, and m is 0. The energy is -13.6 eV, the same as in the Bohr model, but the angular momentum is zero rather than $1\hbar$ as in the Bohr model. The wave function for the ground state is

$$\psi_{100} = C_{100} e^{-Zr/a_0}$$

where

$$a_0 = \frac{\hbar^2}{m_e k e^2} = 0.0529 \text{ nm}$$

is the first Bohr radius and C_{100} is a constant that is determined by normalization. In three dimensions, the normalization condition is

$$\int \psi^2 \, dV = 1$$

where dV is a volume element and the integration is performed over all space. In spherical coordinates, the volume element (Figure 4-4) is

$$dV = (r \sin \theta \, d\phi)(r \, d\theta) \, dr = r^2 \sin \theta \, dr \, d\theta \, d\phi$$

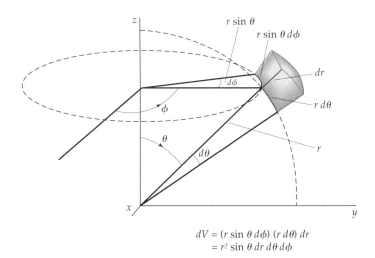

$dV = (r \sin \theta \, d\phi) \, (r \, d\theta) \, dr$
$= r^2 \sin \theta \, dr \, d\theta \, d\phi$

Figure 4-4 Volume element in spherical coordinates.

We integrate over all space by integrating over ϕ from $\phi = 0$ to $\phi = 2\pi$, over θ from $\theta = 0$ to $\theta = \pi$, and over r from $r = 0$ to $r = \infty$. The normalization condition is thus

$$\int \psi^2 \, dV = \int_0^\infty \int_0^\pi \int_0^{2\pi} \psi^2 r^2 \sin\theta \, d\phi \, d\theta \, dr$$

$$= \int_0^{2\pi} d\phi \int_0^\pi \sin\theta \, d\theta \int_0^\infty \psi^2 r^2 \, dr$$

$$= \int_0^{2\pi} d\phi \int_0^\pi \sin\theta \, d\theta \int_0^\infty C_{100}^2 r^2 e^{-2Zr/a_0} \, dr = 1$$

Since there is no θ or ϕ dependence in ψ_{100}, the integration over the angles gives 4π. From a table of integrals, we obtain

$$\int_0^\infty r^2 e^{-2Zr/a_0} \, dr = \frac{a_0^3}{4Z^3}$$

Then

$$4\pi C_{100}^2 \left(\frac{a_0^3}{4Z^3}\right) = 1$$

and

$$C_{100} = \frac{1}{\sqrt{\pi}} \left(\frac{Z}{a_0}\right)^{3/2}$$

The normalized ground-state wave function is thus

$$\psi_{100} = \frac{1}{\sqrt{\pi}} \left(\frac{Z}{a_0}\right)^{3/2} e^{-Zr/a_0} \qquad 4\text{-}12$$

The probability of finding the electron in the volume dV is $\psi^2 \, dV$. The probability density ψ^2 is illustrated in Figure 4-5. Note that this probability density is spherically symmetric; that is, it depends only on r and not on θ or ϕ. The probability density is maximum at the origin. We are more often interested in the probability of finding the electron at some radial distance r between r and $r + dr$. This radial probability $P(r) \, dr$ is the probability density ψ^2 times the volume of the spherical shell of thickness dr, which is $dV =$

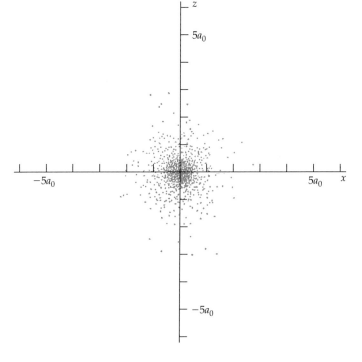

Figure 4-5 Computer-generated dot picture of the probability density ψ^2 for the ground state of hydrogen. The quantity $e\psi^2$ can be thought of as the electron charge density in the atom. The density is spherically symmetric, is greatest at the origin, and decreases exponentially with r.

(a)

(b) ⊢4μ⊣

(a) The tip of a tungsten probe used in a scanning-tunneling microscope (STM). Some probes have tips only one atom wide. During operation, a small current flows from the surface of the material that is being imaged onto the probe tip. The current results from electrons, on the surface of the material, tunneling through the potential difference between the surface and the probe. Because the tunneling current decreases exponentially with distance, it is very sensitive to changes in that distance. (See essay, *Scanning Tunneling Microscopy*, at the end of Chapter 3.) (b) The microcantilever of an atomic-force microscope (AFM). Repulsion between electrons on the surface that is being imaged and electrons on the cantilever causes the cantilever to deflect. The degree of deflection can be closely calibrated to the distance the cantilever is from the surface. Both this image and that in part *a* were taken using a scanning electron microscope (SEM).

$4\pi r^2\, dr$, at a distance r. The probability of finding the electron in the range from r to $r + dr$ is thus $P(r)\, dr = \psi^2 4\pi r^2\, dr$, and the **radial probability density** is

$$P(r) = 4\pi r^2 \psi^2 \qquad \text{4-13} \quad \textit{Radial probability density}$$

For the hydrogen atom in the ground state, the radial probability density is

$$P(r) = 4\pi r^2 \psi^2 = 4\pi r^2 C_{100}^2 e^{-2Zr/a_0} = 4\left(\frac{Z}{a_0}\right)^3 r^2 e^{-2Zr/a_0} \qquad \text{4-14}$$

Figure 4-6 shows the radial probability density $P(r)$ as a function of r. The maximum value of $P(r)$ occurs at $r = a_0/Z$, which for $Z = 1$ is the first Bohr radius. In contrast to the Bohr model in which the electron stays in a well-defined orbit at $r = a_0$, we see that it is possible for the electron to be found at any distance from the nucleus. However, the most probable distance is a_0 (assuming $Z = 1$), and the chance of finding the electron at a much different distance is small. It is sometimes useful to think of the electron in an atom as a charged cloud of charge density $\rho = e\psi^2$, but we must remember that an electron is always observed to interact with matter or radiation as a single charge.

In the first excited state, $n = 2$ and ℓ can be either 0 or 1. For $\ell = 0$, $m = 0$, and we again have a spherically symmetric wave function, this time given by

$$\psi_{200} = C_{200}\left(2 - \frac{Zr}{a_0}\right)e^{-Zr/2a_0} \qquad \text{4-15}$$

For $\ell = 1$, m can be +1, 0, or −1. The corresponding wave functions are

$$\psi_{210} = C_{210}\frac{Zr}{a_0}e^{-Zr/2a_0}\cos\theta \qquad \text{4-16}$$

$$\psi_{21\pm 1} = C_{211}\frac{Zr}{a_0}e^{-Zr/2a_0}\sin\theta\, e^{\pm i\phi} \qquad \text{4-17}$$

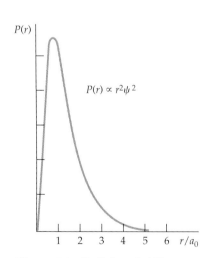

Figure 4-6 Radial probability density $P(r)$ versus r/a_0 for the ground state of the hydrogen atom. $P(r)$ is proportional to $r^2\psi^2$. The value of r for which $P(r)$ is maximum is the most probable distance $r = a_0$.

In general, the wave functions for $m \neq 0$ are proportional to $e^{im\phi}$ and are therefore not real. As we discussed in Chapter 3, the probability density for complex functions is given by $|\psi|^2$. We may write $|\psi|^2$ as

$$|\psi|^2 = \psi^*\psi \qquad 4\text{-}18$$

where ψ^* is the complex conjugate of ψ, which is obtained from ψ by replacing i by $-i$ wherever it appears.[†] Since $(e^{im\phi})^* = e^{-im\phi}$ and $(e^{im\phi})^*(e^{im\phi}) = e^{-im\phi}e^{im\phi} = 1$, the probability densities do not depend on m or ϕ even though the wave functions do.

Figure 4-7 shows the probability density $\psi^*\psi$ for $n = 2$, $\ell = 0$, $m = 0$ (Figure 4-7a); for $n = 2$, $\ell = 1$, $m = 0$ (Figure 4-7b); and for $n = 2$, $\ell = 1$, $m = \pm 1$ (Figure 4-7c). An important feature of these plots is that the electron cloud is spherically symmetric for $\ell = 0$ and is not spherically symmetric for $\ell \neq 0$. These angular distributions of the electron charge density depend only on the values of ℓ and m and not on the radial part of the wave function. Similar charge distributions for the valence electrons of more complicated atoms play an important role in the chemistry of molecular bonding, which we will discuss in Chapter 5.

Figure 4-7 Computer-generated dot picture of the probability densities $|\psi|^2$ for the electron in the $n = 2$ states of hydrogen. (a) For $\ell = 0$, $|\psi|^2$ is spherically symmetric. (b) For $\ell = 1$ and $m = 0$, $|\psi|^2$ is proportional to $\cos^2 \theta$. (c) For $\ell = 1$ and $m = +1$ or -1, $|\psi|^2$ is proportional to $\sin^2 \theta$.

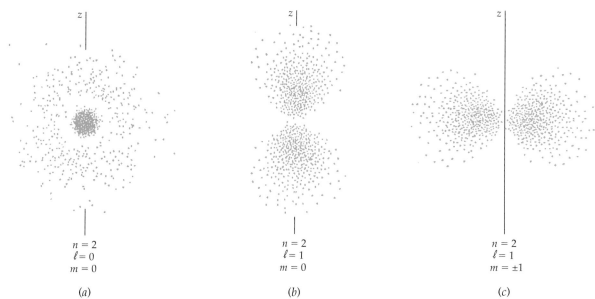

Figure 4-8 shows the probability of finding the electron at a distance r as a function of r for $n = 2$ when $\ell = 1$ and when $\ell = 0$. We can see from the figure that the probability distribution depends on ℓ as well as on n. In particular, we can see that for very small values of r ($r < a_0$), the S state ($\ell = 0$) has a larger probability density than does the P state ($\ell = 1$). This will be important when we consider atoms with more than one electron.

For $n = 1$, we found that the most likely distance between the electron and the nucleus is a_0, the first Bohr radius, whereas for $n = 2$, $\ell = 1$, it is $4a_0$. These are the orbital radii for the first and second Bohr orbits (Equation 2-22). For $n = 3$ (and $\ell = 2$),[‡] the most likely distance between the electron and nucleus is $9a_0$, the radius of the third Bohr orbit.

[†] Every complex number can be written in the form $z = a + bi$, where a and b are real numbers and $i = \sqrt{-1}$. The magnitude or absolute value of z is defined as $\sqrt{a^2 + b^2}$. The complex conjugate of z is $z^* = a - bi$, so $z^*z = (a - bi)(a + bi) = a^2 + b^2 = |z|^2$.

[‡] The correspondence with the Bohr model is closest for the maximum value of ℓ, which is $n - 1$.

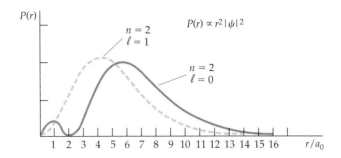

Figure 4-8 Radial probability density $P(r)$ versus r/a_0 for the $n = 2$ states of hydrogen. For $\ell = 1$, $P(r)$ is maximum at the Bohr value $r = 2^2 a_0$. For $\ell = 0$, there is a maximum near this value and a smaller submaximum near the origin.

Example 4-2

For the ground state of the hydrogen atom, find the probability of finding the electron in the range $\Delta r = 0.02 a_0$ at (a) $r = a_0$ and (b) $r = 2a_0$.

Because the range Δr is so small, the variation in the radial probability density $P(r)$ can be neglected. The probability of finding the electron in Δr at $r = a_0$ is then

$$P(r)\, dr \approx P(r)\, \Delta r$$

(a) Using Equation 4-14 for $P(r)$, with $Z = 1$, we obtain for $r = a_0$

$$P(r)\, \Delta r = 4\left(\frac{1}{a_0}\right)^3 r^2 e^{-2r/a_0}\, \Delta r = 4\left(\frac{1}{a_0}\right)^3 a_0^2 e^{-2}(0.02 a_0) = 0.0108$$

Thus, there is about a 1 percent chance of finding the electron in this range at $r = a_0$.

(b) For $r = 2a_0$, we obtain

$$P(r)\, \Delta r = 4\left(\frac{1}{a_0}\right)^3 r^2 e^{-2r/a_0}\, \Delta r = 4\left(\frac{1}{a_0}\right)^3 4a_0^2 e^{-4}(0.02 a_0) = 0.00586$$

4-3 Magnetic Moments and Electron Spin

When a spectral line of hydrogen is viewed under high resolution, it is found to consist of two closely spaced lines. (The spectral lines for other elements may consist of more than two closely spaced lines.) This splitting of a line into two or more closely spaced lines is called **fine structure**. To explain this fine structure and to clear up a major difficulty with the quantum-mechanical explanation of the periodic table (which will be discussed in the next section), W. Pauli suggested in 1925 that the electron has an additional quantum number that can take on just two values. In the same year, S. Goudsmit and G. Uhlenbeck, graduate students at Leiden, suggested that this fourth quantum number was the z component m_s of an intrinsic angular momentum of the electron called **electron spin**. In the particle model, the electron is pictured as a spinning ball that orbits the nucleus (Figure 4-9), much as the earth rotates about its axis as it revolves around the sun. If this intrinsic spin angular momentum is to be described by a quantum number s, like the orbital quantum number ℓ, we expect the z component to have $2s + 1$ possible values just as there are $2\ell + 1$ possible z components of the orbital angular momentum. If m_s is to have just two possible values, s must be $\frac{1}{2}$. Then m_s can be either $m_s = -s = -\frac{1}{2}$ or $m_s = +s = +\frac{1}{2}$ corresponding to the z components of intrinsic electron spin of $+\frac{1}{2}\hbar$ and $-\frac{1}{2}\hbar$.

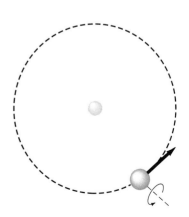

Figure 4-9 The electron can be pictured as a spinning ball that orbits the nucleus somewhat like the spinning earth orbits the sun.

One consequence of electron spin is that the electron possesses an intrinsic magnetic moment—a result to be expected since a spinning charge is equivalent to a set of current loops. The magnetic moment[†] $\boldsymbol{\mu}$ of a rotating charged system is related to its angular momentum by:

$$\boldsymbol{\mu} = \frac{q}{2m_q}\mathbf{L} \qquad 4\text{-}19$$

where m_q is the mass of the particle that has a charge q. Substituting $q = -e$ and $m_q = m_e$ for the electron, we have

$$\boldsymbol{\mu} = -\frac{e}{2m_e}\mathbf{L} \qquad 4\text{-}20$$

Applying this equation to the orbital angular momentum of the electron in the hydrogen atom, we have for the magnitude and z component of the magnetic moment

$$\mu = \frac{e}{2m_e}L = \frac{e}{2m_e}\sqrt{\ell(\ell+1)}\,\hbar = \sqrt{\ell(\ell+1)}\,\mu_B \qquad 4\text{-}21$$

and

$$\mu_z = -\frac{e}{2m_e}m\hbar = -\frac{e\hbar}{2m_e}m = -m\mu_B \qquad 4\text{-}22$$

where μ_B is the Bohr magneton, which has the value

$$\mu_B = \frac{e\hbar}{2m_e} = 9.27 \times 10^{-24}\text{ J/T} = 5.79 \times 10^{-5}\text{ eV/T}$$

We thus see that the quantization of angular momentum implies the quantization of magnetic moment.

Since the electron has an intrinsic spin angular momentum characterized by the quantum numbers s and m_s, we would expect it to have an intrinsic magnetic moment given by Equation 4-21 and a z component of its magnetic moment given by Equation 4-22, with ℓ replaced by s and m replaced by m_s. In particular, we would expect the z component of the intrinsic magnetic moment of the electron to be $\pm\frac{1}{2}\mu_B$. However, the measured value is twice this expected value; that is, the electron has an intrinsic magnetic moment of one Bohr magneton, not one-half Bohr magneton. It is customary to write the relation between the z component of any type of angular momentum J_z and the z component of the magnetic moment μ_z as

$$\mu_z = -g\mu_B\frac{J_z}{\hbar} \qquad 4\text{-}23$$

where g, called the **gyromagnetic ratio,** has the values $g_\ell = 1$ for orbital angular momentum and $g_s = 2$ for spin. Precise measurements indicate that $g_s = 2.00232$. This result and the phenomenon of electron spin itself was predicted by P. Dirac in 1927. He combined special relativity and quantum mechanics into a relativistic wave equation called the **Dirac equation.** The fact that the intrinsic magnetic moment of the electron is approximately twice what we would expect makes it clear that the simple model of the electron as a spinning ball should not be taken literally. Like the Bohr model

---

[†] Because the magnetic quantum number is designated by the symbol m, we will now use μ for the magnetic moment.

of the atom, the simple classical picture is useful for describing quantum-mechanical calculations, and it often provides useful guidelines as to what to expect from an experiment.

The interaction of the magnetic moments associated with the orbital angular momentum and the spin angular momentum of the electron gives rise to a small splitting of the energy levels of hydrogen and other atoms, which is responsible for the fine-structure splitting of the spectral lines. Since the spin angular momentum can have two possible orientations relative to some direction in space, the electron's intrinsic spin magnetic moment can have two possible orientations relative to the magnetic moment associated with the orbital angular momentum. These two orientations are usually described by saying that the two magnetic moments can be parallel or antiparallel to each other. Because of the interaction of the magnetic moments, the energies of these two orientations are slightly different, resulting in the splitting of each energy level in the hydrogen atom (except those with $\ell = 0$) into two nearly equal energy levels.

The addition of the electron-spin quantum number completes the quantum-mechanical description of the hydrogen atom. The wave functions for the electron in the hydrogen atom are characterized by four quantum numbers: n, ℓ, m, and m_s.

Question

1. The Bohr theory and the Schrödinger theory of the hydrogen atom give the same results for energy levels. Discuss the advantages and disadvantages of each model.

4-4 The Stern–Gerlach Experiment

In addition to explaining the fine structure and the periodic table, the proposal of electron spin explained an interesting experiment first performed by O. Stern and W. Gerlach in 1922. In this experiment, atoms from an oven are collimated and sent through a magnet whose poles are shaped so that the magnetic field B_z is not constant in space but increases slightly with z (Figure 4-10). The behavior of a magnetic moment in a magnetic field can be visualized by considering a small bar magnet (Figure 4-11). When the magnet is

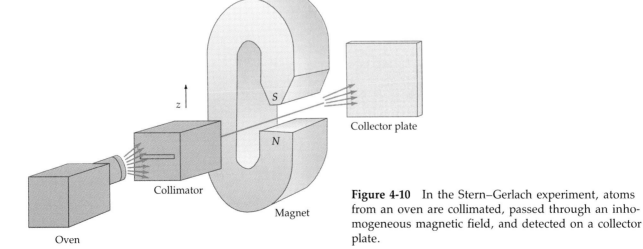

Figure 4-10 In the Stern–Gerlach experiment, atoms from an oven are collimated, passed through an inhomogeneous magnetic field, and detected on a collector plate.

Figure 4-11 Bar-magnet model of a magnetic moment. (a) In an external magnetic field, the moment experiences a torque that tends to align it with the field. (b) If the magnet is spinning, the torque causes the spin axis to precess about the direction of the external field.

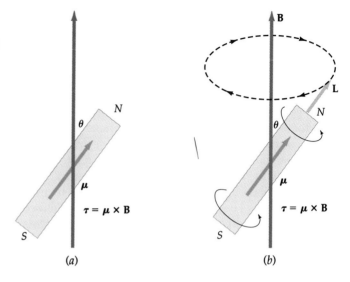

placed in a uniform external magnetic field **B**, there is a torque $\boldsymbol{\tau} = \boldsymbol{\mu} \times \mathbf{B}$ that tends to align the magnet with the direction of the field **B**. If the magnet is spinning about its axis, the effect of the torque is to make the spin axis precess about the direction of the external field. If the magnetic field is not uniform, the force on one pole of the magnet will be greater or less than that on the other pole, depending on the orientation of the magnet. Figure 4-12 illustrates the effects of such a field on three bar magnets of different orientations. In addition to the torque, which merely causes the magnetic moment to precess about the field direction, there is a net force in the positive or negative z direction given by

$$F_z = \mu_z \frac{dB_z}{dz} \qquad 4\text{-}24$$

This force deflects the magnet up or down by an amount that depends on the inhomogeneity of the magnetic field and on the z component of the magnetic moment.

Figure 4-12 In an inhomogeneous magnetic field, a bar magnet experiences a net force that depends on the magnet's orientation. Here the force is stronger near the south pole of the large magnet, so B_z increases in the z direction.

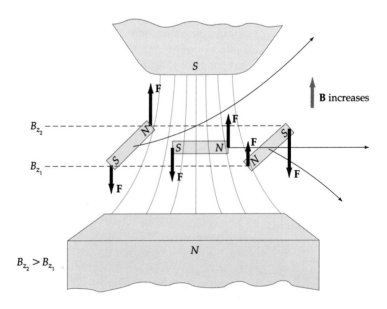

Classically, one would expect a continuum of deflections corresponding to the continuum of possible orientations of the magnetic moments. However, since the magnetic moment of an atom is quantized, quantum mechanics predicts that the z component of the magnetic moment can have

$2j + 1$ values, where j is a quantum number associated with the total angular momentum of the atom that results from the combination of the orbital and spin angular momenta of the electrons in the atom. This experiment thus measures the angular momentum of the atom. If $j = 0$, corresponding to zero angular momentum for the atom, the magnetic moment will be zero and there will be no deflection of the atoms. If $j = 1$, there will be three possible orientations of the z component of the magnetic moment, and the original beam of atoms will be split into three beams. If $j = \frac{1}{2}$, the original beam will be split into two beams corresponding to the two possible orientations of the magnetic moment.

This experiment was performed by Stern and Gerlach in 1922 using silver atoms and by Phipps and Taylor in 1927 using hydrogen atoms. In each case, the beam was split into two beams as shown in Figure 4-13. Since the ground state of hydrogen has $\ell = 0$, we would expect no splitting were it not for electron spin. The net angular momentum of the hydrogen atom in its ground state is just the spin angular momentum of the electron.[†] The splitting of the beam of hydrogen atoms into two beams confirms the result that the spin angular momentum of the electron can have just two orientations corresponding to the quantum number $s = \frac{1}{2}$. The quantization of the orientation of the magnetic moment of the electron to either of two directions in space is called **space quantization.**

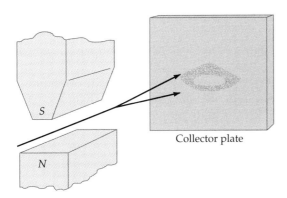

Collector plate

Figure 4-13 Results of the Stern–Gerlach experiment. The atomic beam is split into two beams, indicating that the magnetic moments of the atoms are quantized to two orientations in space. The shape of the upper line is due to the greater inhomogeneity of the magnetic field near the upper pole face.

Exercise
A beam of atoms is split into five lines when passing though an inhomogeneous magnetic field. What is the quantum number associated with the angular momentum of the atoms?   (Answer: 2)

4-5 Addition of Angular Momenta and the Spin–Orbit Effect

In general, an electron in an atom has both orbital angular momentum **L** characterized by the quantum number ℓ and spin angular momentum **S** characterized by the quantum number s. Analogous classical systems that have two kinds of angular momentum are the earth, which is spinning about its axis of rotation in addition to revolving about the sun, or a precess-

---
†The nucleus of an atom also has angular momentum and, therefore, a magnetic moment. However, the mass of the nucleus is about 2000 times that of the electron for hydrogen and is greater still for other atoms. From Equation 4-19, we expect the magnetic moment of the hydrogen nucleus to be on the order of $1/2000$ of a Bohr magneton since m_q is now the mass of a proton rather than an electron.

ing gyroscope, which has angular momentum of precession in addition to its spin. Classically, the total angular momentum **J** is given by

$$\mathbf{J} = \mathbf{L} + \mathbf{S} \qquad 4\text{-}25$$

This is an important quantity because the net torque on a system equals the rate of change of the total angular momentum, and in the case of central forces, the total angular momentum is conserved. For a classical system, the magnitude of the total angular momentum **J** can have any value between $L + S$ and $L - S$. We have seen that angular momentum in quantum mechanics is more complicated. Both **L** and **S** are quantized, and their directions are restricted. Quantum mechanics also limits the possible values of the total angular momentum **J**. For an electron with orbital angular momentum characterized by the quantum number ℓ and spin $s = \frac{1}{2}$, the total angular momentum **J** has the magnitude $\sqrt{j(j+1)}\,\hbar$, where the quantum number j can be either

$$j = \ell + \tfrac{1}{2}$$

or $\qquad\qquad\qquad\qquad\qquad\qquad\qquad\qquad\qquad\qquad 4\text{-}26$

$$j = \ell - \tfrac{1}{2} \qquad \ell \neq 0$$

(For $\ell = 0$, the total angular momentum is simply the spin, so $j = \frac{1}{2}$.) In Figure 4-14, vector diagrams illustrate the two possible combinations $j = \frac{3}{2}$ and $j = \frac{1}{2}$ for the case of $\ell = 1$. The lengths of the vectors are proportional to $\sqrt{\ell(\ell+1)}\,\hbar$, $\sqrt{s(s+1)}\,\hbar$, and $\sqrt{j(j+1)}\,\hbar$. The spin and orbital angular momenta are said to be parallel when $j = \ell + s$ and antiparallel when $j = \ell - s$.

Equation 4-26 is a special case of a more general rule for combining two angular momenta that is useful when dealing with more than one particle. For example, there are two electrons in helium, each with spin, orbital, and total angular momenta. The general rule is

If $\mathbf{J}_1$ is one angular momentum (orbital, spin, or a combination) and $\mathbf{J}_2$ is another, the resulting total angular momentum $\mathbf{J} = \mathbf{J}_1 + \mathbf{J}_2$ has a magnitude $\sqrt{j(j+1)}\,\hbar$, where j can have any of the values

$$j = j_1 + j_2,\ j_1 + j_2 - 1,\ \ldots,\ |j_1 - j_2| \qquad 4\text{-}27$$

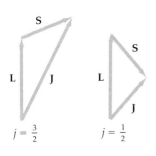

Figure 4-14 Vector diagrams illustrating the addition of orbital and spin angular momentum for the case $\ell = 1$ and $s = \frac{1}{2}$. There are two possible values of the quantum number for the total angular momentum, $j = \ell + s = \frac{3}{2}$ and $j = \ell - s = \frac{1}{2}$.

The z component of the total angular momentum can then have any of the $2j + 1$ values

$$m_j = -j,\ -j + 1,\ \ldots,\ +j$$

Example 4-3

Two electrons each have zero orbital angular momentum. What are the possible quantum numbers for the total angular momentum of the two-electron system?

In this case, $j_1 = j_2 = \frac{1}{2}$. Equation 4-27 then gives two possible results, $j = 1$ and $j = 0$. These results are commonly called parallel spins and antiparallel spins. For $j = 1$, the quantum number m_j can be -1, 0, or $+1$; for $j = 0$, m_j must be 0. Thus the $j = 1$ state is commonly referred to as the **triplet state** and the $j = 0$ state is referred to as the **singlet state**.

Example 4-4

An electron in an atom has orbital angular momentum L_1 with quantum number $\ell_1 = 2$ and a second electron has orbital angular momentum L_2 with quantum number $\ell_2 = 3$. What are the possible quantum numbers ℓ for the total orbital angular momentum $L = L_1 + L_2$?

Since $\ell_1 + \ell_2 = 2 + 3 = 5$ and $|\ell_1 - \ell_2| = |2 - 3| = 1$, the possible values of ℓ are 5, 4, 3, 2, and 1.

In spectroscopic notation, the total angular-momentum quantum number of an atomic state is written as a subscript after the code letter describing the orbital angular momentum. For example, the ground state of hydrogen is written $1S_{1/2}$, where the number preceding the code letter indicates the value of n. The $n = 2$ states can have either $\ell = 0$ or $\ell = 1$, and the $\ell = 1$ state can have either $j = \frac{3}{2}$ or $j = \frac{1}{2}$. These states are thus denoted by $2S_{1/2}$, $2P_{3/2}$, and $2P_{1/2}$. Thus the code denotes $n(\ell)_j$.

Atomic states with the same values of n and ℓ but different values of j have slightly different energies because of the interaction between the spin of the electron and its orbital motion. This effect is called the **spin–orbit effect**. The resulting splitting of the spectral line, such as the splitting that results from the transitions $2P_{3/2} \to 2S_{1/2}$ and $2P_{1/2} \to 2S_{1/2}$ in hydrogen, is called **fine-structure splitting**. We can understand the spin–orbit effect qualitatively from the simple Bohr-model picture shown in Figure 4-15. In this figure, the electron moves in a circular orbit with speed v around a fixed proton. The orbital angular momentum L is up. In the frame of reference of the electron, the proton is moving in a circle around it, thus constituting a circular loop of current that produces a magnetic field B at the position of the electron. The direction of B is up, parallel to L. The potential energy of a magnetic moment in a magnetic field depends on its orientation and is given by:

$$U = -\boldsymbol{\mu}\cdot\mathbf{B} = -\mu_z B \qquad 4\text{-}28$$

The potential energy is lowest when the magnetic moment is parallel to B and highest when it is antiparallel. Since the magnetic moment of the electron is directed opposite to its spin (because the electron has a negative charge), the spin–orbit energy is highest when the spin is parallel to B and thus to L. The energy of the $2P_{3/2}$ state in hydrogen, in which L and S are parallel, is therefore slightly higher than the $2P_{1/2}$ state, in which L and S are antiparallel (Figure 4-16).

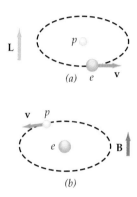

Figure 4-15 (*a*) An electron moving about a proton with angular momentum L up. (*b*) The magnetic field B seen by the electron due to the apparent (relative) motion of the proton is also up. When the electron spin is parallel to L, the magnetic moment is antiparallel to L and B, so the spin–orbit energy is at its greatest.

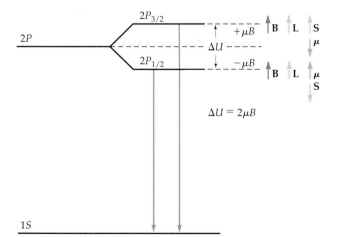

Figure 4-16 Fine-structure energy-level diagram. On the left, the levels in the absence of a magnetic field are shown. The effect of an applied field is shown on the right. Because of the spin–orbit interaction, the magnetic field splits the $2P$ level into two energy levels, with the $j = \frac{3}{2}$ level having slightly greater energy than the $j = \frac{1}{2}$ level. The spectral line due to the transition $2P \to 1S$ is therefore split into two lines of slightly different wavelengths.

Example 4-5

The fine-structure splitting of the $2P_{3/2}$ and $2P_{1/2}$ levels in hydrogen is 4.5×10^{-5} eV. If the $2p$ electron sees an internal magnetic field **B**, the spin–orbit energy splitting will be of the order of $\Delta E = 2\mu_B B$, where μ_B is the Bohr magneton. From this, estimate the magnetic field that the $2p$ electron in hydrogen experiences.

We have

$$\Delta E = 2\mu_B B = 4.5 \times 10^{-5} \text{ eV}$$

$$B = \frac{4.5 \times 10^{-5} \text{ eV}}{2\mu_B} = \frac{4.5 \times 10^{-5} \text{ eV}}{2(5.79 \times 10^{-5} \text{ eV/T})} = 0.39 \text{ T}$$

When an atom is placed in an external magnetic field **B**, the energy of the atomic state characterized by the angular-momentum quantum number j is split into $2j + 1$ energy levels corresponding to the $2j + 1$ possible values of the z component of the angular momentum and therefore to the $2j + 1$ possible values of the z component of the magnetic moment. This splitting of the energy levels in the atom gives rise to a splitting of the spectral lines emitted by the atom. The splitting of the spectral lines of an atom placed in an external magnetic field was discovered by P. Zeeman and is known as the **Zeeman effect**.

4-6 The Periodic Table

In the quantum-mechanical theory, the state of each electron in atoms with more than one electron is described by the quantum numbers n, ℓ, m, and m_s. The energy of the electron is determined mainly by the principal quantum number n (which is related to the radial dependence of the wave function) and by the orbital angular-momentum quantum number ℓ. Generally, the lower the values of n and ℓ, the lower the energy. The dependence of the energy on ℓ is due to the interaction of the electrons in the atom with each other. In hydrogen, of course, there is only one electron, and the energy is independent of ℓ. The specification of n and ℓ for each electron in an atom is called the **electron configuration**. Customarily, ℓ is specified according to the same code used to label the states of the hydrogen atom rather than by its numerical value. The code is

Code	s	p	d	f	g	h
ℓ value	0	1	2	3	4	5

Capital letters are used to specify atomic states, and lowercase letters are used for individual electron states. The n values are sometimes referred to as shells, which are identified by another letter code: $n = 1$ denotes the K shell; $n = 2$, the L shell; and so on.

An important principle that governs the electron configuration of atoms is the **Pauli exclusion principle**:

> No two electrons in an atom can be in the same quantum state; that is, no two electrons can have the same set of values for the quantum numbers n, ℓ, m, and m_s.

As we saw in Section 3-11, the exclusion principle is related to the fact that the wave function describing the atom must be antisymmetric in exchange of any two electrons. Using the exclusion principle and the restrictions on the quantum numbers discussed in the previous section (n is an integer, ℓ is an integer that ranges from 0 to $n - 1$, m can have $2\ell + 1$ values from $-\ell$ to ℓ in integral steps, and m_s can be either $+\frac{1}{2}$ or $-\frac{1}{2}$), we can understand much of the structure of the periodic table. We have already discussed the lightest element, hydrogen, which has just one electron. In the ground (lowest energy) state, the electron has $n = 1$ and $\ell = 0$, with $m = 0$ and $m_s = +\frac{1}{2}$ or $-\frac{1}{2}$. We call this a 1s electron. The 1 signifies that $n = 1$, and the s signifies that $\ell = 0$.

As electrons are added to make the heavier atoms, the electrons go into those states that will give the lowest total energy consistent with the Pauli exclusion principle.

Helium ($Z = 2$)

The next element after hydrogen is helium ($Z = 2$), which has two electrons. In the ground state, both electrons are in the K shell with $n = 1$, $\ell = 0$, and $m = 0$; one electron has $m_s = +\frac{1}{2}$ and the other has $m_s = -\frac{1}{2}$. This configuration is lower in energy than any other two-electron configuration. The resultant spin of the two electrons is zero. Since the orbital angular momentum is also zero, the total angular momentum is zero. The electron configuration for helium is written $1s^2$. The 1 signifies that $n = 1$, the s signifies that $\ell = 0$, and the 2 signifies that there are two electrons in this state. Since ℓ can be only 0 for $n = 1$, these two electrons fill the K ($n = 1$) shell. The energy required to remove an electron from an atom is called the **ionization energy.** The ionization energy is the binding energy of the last electron placed in the atom. For helium, the ionization energy is 24.6 eV, which is relatively large. Helium is therefore basically inert.

Example 4-6

Calculate the energy of interaction of the two electrons in the ground state of the helium atom and use it to find the average separation of the two electrons.

If the electrons did not interact, the energy of each electron in helium would be given by Equation 4-9 with $Z = 2$. For the ground state, $n = 1$ and

$$E_1 = -\frac{Z^2 E_0}{n^2} = -\frac{(2)^2(13.6 \text{ eV})}{(1)^2} = -54.4 \text{ eV}$$

The total energy of the two electrons in the ground state would be twice this, $2(-54.4 \text{ eV}) = -108.8 \text{ eV}$. If one electron were removed, the energy of the other electron would be -54.4 eV. Thus, the energy needed to remove one electron would be 54.4 eV, which would be the ionization energy. Since the measured ionization energy is 24.6 eV, the energy of the ground state of helium must be $-(54.4 \text{ eV} + 24.6 \text{ eV}) = -79.0$ eV. This is 29.8 eV higher than -108.8 eV, so the energy of interaction of the two electrons in the ground state of helium is 29.8 eV.

The energy of interaction of two electrons a distance r apart is the potential energy $U = ke^2/r$. Setting this equal to 29.8 eV, we obtain

$$U = \frac{ke^2}{r} = 29.8 \text{ eV}$$

Since we know that $ke^2/a_0 = 13.6$ eV, it is convenient to multiply this equation by a_0/a_0 before solving it for r. We then obtain

$$U = \frac{ke^2}{r}\frac{a_0}{a_0} = \frac{ke^2}{a_0}\frac{a_0}{r} = (13.6 \text{ eV})\frac{a_0}{r} = 29.8 \text{ eV}$$

$$r = \left(\frac{13.6 \text{ eV}}{29.8 \text{ eV}}\right)a_0 = 0.456 a_0 = 0.024 \text{ nm}$$

where we have used $a_0 = 0.0529$ nm. This separation is approximately equal to the radius of the first Bohr orbit for an electron in helium, which is $r_1 = a_0/Z = \tfrac{1}{2}a_0$.

Lithium ($Z = 3$)

The next element, lithium, has three electrons. Since the K shell is completely filled with two electrons, the third electron must go into a higher energy shell. The next lowest energy shell after $n = 1$ is the $n = 2$ or L shell. The outer electron is much farther from the nucleus than are the two inner, $n = 1$ electrons. It is most likely to be found at the radius of the second Bohr orbit, which is four times the radius of the first Bohr orbit. (The radii of the Bohr orbits, as given by Equation 2-22, are proportional to n^2.)

The nuclear charge is partially screened from the outer electron by the two inner electrons. Recall that the electric field outside a spherically symmetric charge density is the same as if all the charge were at the center of the sphere. If the outer electron were completely outside of the charge cloud of the two inner electrons, the electric field it would see would be that of a single charge $+e$ at the center due to the nuclear charge of $+3e$ and the charge $-2e$ of the inner electron cloud. However, the outer electron does not have a well-defined orbit; instead, it is itself a charge cloud that penetrates the charge cloud of the inner electrons to some extent. Because of this penetration, the effective nuclear charge $Z'e$ is somewhat greater than $+1e$. The energy of the outer electron at a distance r from a point charge $+Z'e$ is

$$E = -\frac{1}{2}\frac{kZ'e^2}{r} \qquad \text{4-29}$$

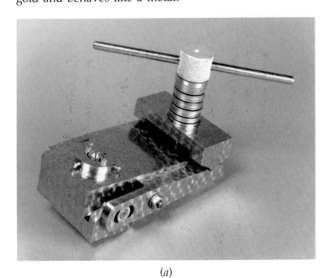

(a) A diamond anvil cell, in which the facets of two diamonds (about 1 mm² each) are used to compress a sample substance, subjecting it to very high pressure.
(b) Samarium monosulfide (SmS) is normally a black, dull-looking semiconductor. When it is subjected to pressure above 7000 atm, an electron from the $4f$ state is dislocated into the $5d$ state. The resulting compound glitters like gold and behaves like a metal.

(a)

(b)

(This is Equation 2-19 with the nuclear charge $+Ze$ replaced by $+Z'e$.) The greater the penetration of the inner electron cloud, the greater the effective nuclear charge $Z'e$ and the lower the energy. Because the penetration is greater for lower ℓ values (see Figure 4-8), the energy of the outer electron in lithium is lower for the s state ($\ell = 0$) than for the p state ($\ell = 1$). The electron configuration of lithium in the ground state is therefore $1s^2 2s$. The ionization energy of lithium is only 5.39 eV. Because its outer electron is so loosely bound to the atom, lithium is very active chemically. It behaves like a "one-electron atom," and its spectrum is similar to that of hydrogen.

Beryllium ($Z = 4$)

The fourth electron has the least energy in the $2s$ state. There can be two electrons with $n = 2$, $\ell = 0$, and $m = 0$ because of the two possible values for the spin quantum number m_s. The configuration of beryllium is thus $1s^2 2s^2$.

Boron to Neon ($Z = 5$ to $Z = 10$)

Since the $2s$ subshell is filled, the fifth electron must go into the next available (lowest energy) subshell, which is the $2p$ subshell, with $n = 2$ and $\ell = 1$. Since there are three possible values of m ($+1$, 0, and -1) and two values of m_s for each value of m, there can be six electrons in this subshell. The electron configuration for boron is $1s^2 2s^2 2p$. The electron configurations for the elements carbon ($Z = 6$) to neon ($Z = 10$) differ from that for boron only in the number of electrons in the $2p$ subshell. The ionization energy increases with Z for these elements, reaching the value of 21.6 eV for the last element in the group, neon. Neon has the maximum number of electrons allowed in the $n = 2$ shell. Its electron configuration is $1s^2 2s^2 2p^6$. Because of its very high ionization energy, neon, like helium, is basically chemically inert. The element just before neon, fluorine, has a "hole" in the $2p$ subshell; that is, it has room for one more electron. It readily combines with elements such as lithium that have one outer electron. Lithium, for example, will donate its single outer electron to the fluorine atom to make an F$^-$ ion and a Li$^+$ ion. These ions then bond together to form a molecule of lithium fluoride.

Sodium to Argon ($Z = 11$ to $Z = 18$)

The eleventh electron must go into the $n = 3$ shell. Since this electron is very far from the nucleus and from the inner electrons, it is weakly bound in the sodium ($Z = 11$) atom. The ionization energy of sodium is only 5.14 eV. Sodium therefore combines readily with atoms such as fluorine. With $n = 3$, the value of ℓ can be 0, 1, or 2. Because of the lowering of the energy due to penetration of the electron shield formed by the other ten electrons (similar to that discussed for lithium) the $3s$ state is lower than the $3p$ or $3d$ states. This energy difference between subshells of the same n value becomes greater as the number of electrons increases. The electron configuration of sodium is $1s^2 2s^2 2p^6 3s^1$. As we move to elements with higher values of Z, the $3s$ subshell and then the $3p$ subshell fill. These two subshells can accommodate $2 + 6 = 8$ electrons. The configuration of argon ($Z = 18$) is $1s^2 2s^2 2p^6 3s^2 3p^6$. One might expect the nineteenth electron to go into the third subshell (the d subshell with $\ell = 2$), but the penetration effect is now so strong that the energy of the next electron is lower in the $4s$ subshell than in the $3d$ subshell. There is thus another large energy difference between the eighteenth and nineteenth electrons, and so argon, with its full $3p$ subshell, is basically stable and inert.

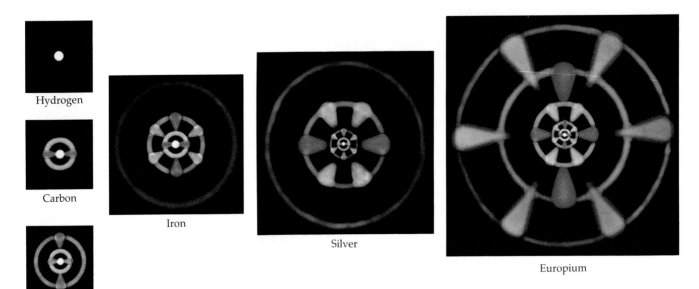

Hydrogen

Carbon

Silicon

Iron

Silver

Europium

A schematic depiction of the electron configurations in atoms. The spherically symmetric s states can contain 2 electrons and are colored white and blue. The dumbbell-shaped p states can contain up to 6 electrons and are colored orange. The d states can contain up to 10 electrons and are colored yellow-green. The f states can contain up to 14 electrons and are colored purple.

Elements with $Z > 18$

The nineteenth electron in potassium ($Z = 19$) and the twentieth electron in calcium ($Z = 20$) go into the $4s$ rather than the $3d$ subshell. The electron configurations of the next ten elements, scandium ($Z = 21$) through zinc ($Z = 30$), differ only in the number of electrons in the $3d$ shell, except for chromium ($Z = 24$) and copper ($Z = 29$), each of which has only one $4s$ electron. These ten elements are called **transition elements.** Since their chemical properties are mainly due to their $4s$ electrons, they are quite similar chemically.

Figure 4-17 shows a plot of the first ionization energy (the energy needed to remove one electron from an atom) versus Z for $Z = 1$ to $Z = 60$. The peaks in ionization energy at $Z = 2, 10, 18, 36,$ and 54 mark the closing of a shell or subshell. Table 4-1 (see pp. 142–143) gives the electron configurations of all the elements.

Figure 4-17 First ionization energy versus Z for $Z = 1$ to $Z = 60$. This energy is the binding energy of the last electron in the atom. The binding energy increases with Z until a shell is closed at $Z = 2, 10, 18, 36,$ and 54. Elements with a closed shell plus one outer electron, such as sodium ($Z = 11$), have very low binding energies because the outer electron is very far from the nucleus and is shielded by the inner core electrons.

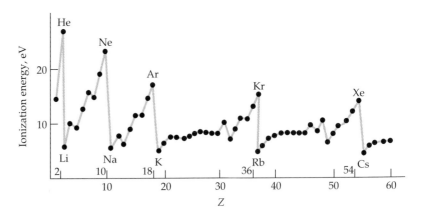

Questions

2. Why is the energy of the $3s$ state considerably lower than that of the $3p$ state in sodium, whereas in hydrogen these states have essentially the same energy?

3. Discuss the evidence from the periodic table of the need for a fourth quantum number. How would the properties of helium differ if there were only three quantum numbers, n, ℓ, and m?

4-7 Optical and X-Ray Spectra

Optical Spectra

When an atom is in an excited state (that is, when one or more of its electrons is in an energy state above the ground state), the electrons make transitions to lower energy states and, in doing so, emit electromagnetic radiation. The frequency of this radiation is related to the initial and final energy states of the electron by the Bohr formula (Equation 2-17), $f = (E_i - E_f)/h$, where E_i and E_f are the initial and final energies and h is Planck's constant. The wavelength of the radiation is, of course, related to the frequency by $\lambda = c/f$. An atom can be excited to a higher energy state by bombarding it with a beam of electrons in a glass tube with a high voltage across it. Since the excited energy states of an atom are discrete rather than continuous, only certain wavelengths are emitted. The spectral lines at these wavelengths constitute the emission spectrum of the atom.

To understand atomic spectra, we need to understand the excited states of an atom. The situation for an atom with many electrons is, in general, much more complicated than that of hydrogen with its one electron. An excited state of the atom may involve a change in the state of any one of its electrons or even two or more of its electrons. Fortunately, in most cases, an excited state of an atom involves the excitation of just one of the electrons in the atom. The energies of excitation of the outer, valence electrons of an atom are of the order of a few electron volts. Transitions involving these electrons result in photons in or near the visible or **optical spectrum.** (Recall that the energies of visible photons range from about 1.8 eV to 3 eV.) Excitation energies can often be calculated using a simple model in which the atom is pictured as a single electron plus a stable core consisting of the nucleus and the other, inner electrons. This model works particularly well for the alkali metals—lithium, sodium, potassium, rubidium, and cesium—the elements in the first column of the periodic table.

Figure 4-18 shows an energy-level diagram for the optical transitions in sodium, whose electrons consist of a neon core plus one outer electron. Since the spin angular momentum of the core adds up to zero, the spin of each state of sodium is $\frac{1}{2}$. Because of the spin–orbit effect, the states with $j = \ell - \frac{1}{2}$ have a slightly lower energy than those with $j = \ell + \frac{1}{2}$. Each state (except for the S states) is therefore a doublet. The doublet splitting is very small and is not evident with the energy scale used in the figure. The states are labeled using spectroscopic notation in which the superscript 2 before the letter code indicates that the state is a doublet. Thus, $^2P_{3/2}$ (read as "doublet P three halves") denotes a state in which $\ell = 1$ and $j = \frac{3}{2}$. (The S states are customarily labeled as if they were doublet even though they are not.)

In the first excited state, the outer electron is excited from the $3s$ level to the $3p$ level, which is about 2.1 eV above the ground state. The energy difference between the $P_{3/2}$ and $P_{1/2}$ states due to the spin–orbit effect is about 0.002 eV. Transitions from these states to the ground state give the familiar yellow doublet of sodium:

$$3p(^2P_{1/2}) \rightarrow 3s(^2S_{1/2}) \quad \lambda = 589.59 \text{ nm}$$
$$3p(^2P_{3/2}) \rightarrow 3s(^2S_{1/2}) \quad \lambda = 588.99 \text{ nm}$$

It is important to distinguish between the doublet energy states and doublet spectral lines. All transitions beginning or ending on an S state give doublet lines because they involve one doublet state and one singlet state. [The selection rule $\Delta \ell = \pm 1$ (Equation 4-10) rules out transitions between two S states.] There are four possible energy differences between two doublet

A neon sign outside a Chinatown restaurant in Paris. Neon atoms in the tube are excited by an electron current passing through the tube. The excited neon atoms emit light in the visible range as they decay toward their ground states. The colors of neon signs result from the characteristic red-orange spectrum of neon plus the color of the glass tube itself.

Table 4-1 Electron configurations of the atoms in their ground states. For some of the rare-earth elements ($Z = 57$ to 71) and the heavy elements ($Z > 89$) the configurations are not firmly established.

Z		Element	Shell: K n: 1 ℓ: s	L 2 s p	M 3 s p d	N 4 s p d f	O 5 s p d f	P 6 s p d	Q 7 s
1	H	hydrogen	1						
2	He	helium	2						
3	Li	lithium	2	1					
4	Be	beryllium	2	2					
5	B	boron	2	2 1					
6	C	carbon	2	2 2					
7	N	nitrogen	2	2 3					
8	O	oxygen	2	2 4					
9	F	fluorine	2	2 5					
10	Ne	neon	2	2 6					
11	Na	sodium	2	2 6	1				
12	Mg	magnesium	2	2 6	2				
13	Al	aluminum	2	2 6	2 1				
14	Si	silicon	2	2 6	2 2				
15	P	phosphorus	2	2 6	2 3				
16	S	sulfur	2	2 6	2 4				
17	Cl	chlorine	2	2 6	2 5				
18	Ar	argon	2	2 6	2 6				
19	K	potassium	2	2 6	2 6 .	1			
20	Ca	calcium	2	2 6	2 6 .	2			
21	Sc	scandium	2	2 6	2 6 1	2			
22	Ti	titanium	2	2 6	2 6 2	2			
23	V	vanadium	2	2 6	2 6 3	2			
24	Cr	chromium	2	2 6	2 6 5	1			
25	Mn	manganese	2	2 6	2 6 5	2			
26	Fe	iron	2	2 6	2 6 6	2			
27	Co	cobalt	2	2 6	2 6 7	2			
28	Ni	nickel	2	2 6	2 6 8	2			
29	Cu	copper	2	2 6	2 6 10	1			
30	Zn	zinc	2	2 6	2 6 10	2			
31	Ga	gallium	2	2 6	2 6 10	2 1			
32	Ge	germanium	2	2 6	2 6 10	2 2			
33	As	arsenic	2	2 6	2 6 10	2 3			
34	Se	selenium	2	2 6	2 6 10	2 4			
35	Br	bromine	2	2 6	2 6 10	2 5			
36	Kr	krypton	2	2 6	2 6 10	2 6			
37	Rb	rubidium	2	2 6	2 6 10	2 6 . .	1		
38	Sr	strontium	2	2 6	2 6 10	2 6 . .	2		
39	Y	yttrium	2	2 6	2 6 10	2 6 1 .	2		
40	Zr	zirconium	2	2 6	2 6 10	2 6 2 .	2		
41	Nb	niobium	2	2 6	2 6 10	2 6 4 .	1		
42	Mo	molybdenum	2	2 6	2 6 10	2 6 5 .	1		
43	Tc	technetium	2	2 6	2 6 10	2 6 6 .	1		
44	Ru	ruthenium	2	2 6	2 6 10	2 6 7 .	1		
45	Rh	rhodium	2	2 6	2 6 10	2 6 8 .	1		
46	Pd	palladium	2	2 6	2 6 10	2 6 10 .	.		
47	Ag	silver	2	2 6	2 6 10	2 6 10 .	1		
48	Cd	cadmium	2	2 6	2 6 10	2 6 10 .	2		
49	In	indium	2	2 6	2 6 10	2 6 10 .	2 1		
50	Sn	tin	2	2 6	2 6 10	2 6 10 .	2 2		
51	Sb	antimony	2	2 6	2 6 10	2 6 10 .	2 3		
52	Te	tellurium	2	2 6	2 6 10	2 6 10 .	2 4		

Table 4-1 (Continued)

Z		Element	Shell: K n: 1 ℓ: s	L 2 s p	M 3 s p d	N 4 s p d f	O 5 s p d f	P 6 s p d	Q 7 s
53	I	iodine	2	2 6	2 6 10	2 6 10 .	2 5		
54	Xe	xenon	2	2 6	2 6 10	2 6 10 .	2 6		
55	Cs	cesium	2	2 6	2 6 10	2 6 10 .	2 6 . .	1	
56	Ba	barium	2	2 6	2 6 10	2 6 10 .	2 6 . .	2	
57	La	lanthanum	2	2 6	2 6 10	2 6 10 .	2 6 1 .	2	
58	Ce	cerium	2	2 6	2 6 10	2 6 10 1	2 6 1 .	2	
59	Pr	praseodymium	2	2 6	2 6 10	2 6 10 3	2 6 . .	2	
60	Nd	neodymium	2	2 6	2 6 10	2 6 10 4	2 6 . .	2	
61	Pm	promethium	2	2 6	2 6 10	2 6 10 5	2 6 . .	2	
62	Sm	samarium	2	2 6	2 6 10	2 6 10 6	2 6 . .	2	
63	Eu	europium	2	2 6	2 6 10	2 6 10 7	2 6 . .	2	
64	Gd	gadolinium	2	2 6	2 6 10	2 6 10 7	2 6 1 .	2	
65	Tb	terbium	2	2 6	2 6 10	2 6 10 9	2 6 . .	2	
66	Dy	dysprosium	2	2 6	2 6 10	2 6 10 10	2 6 . .	2	
67	Ho	holmium	2	2 6	2 6 10	2 6 10 11	2 6 . .	2	
68	Er	erbium	2	2 6	2 6 10	2 6 10 12	2 6 . .	2	
69	Tm	thulium	2	2 6	2 6 10	2 6 10 13	2 6 . .	2	
70	Yb	ytterbium	2	2 6	2 6 10	2 6 10 14	2 6 . .	2	
71	Lu	lutetium	2	2 6	2 6 10	2 6 10 14	2 6 1 .	2	
72	Hf	hafnium	2	2 6	2 6 10	2 6 10 14	2 6 2 .	2	
73	Ta	tantalum	2	2 6	2 6 10	2 6 10 14	2 6 3 .	2	
74	W	tungsten (wolfram)	2	2 6	2 6 10	2 6 10 14	2 6 4 .	2	
75	Re	rhenium	2	2 6	2 6 10	2 6 10 14	2 6 5 .	2	
76	Os	osmium	2	2 6	2 6 10	2 6 10 14	2 6 6 .	2	
77	Ir	iridium	2	2 6	2 6 10	2 6 10 14	2 6 7 .	2	
78	Pt	platinum	2	2 6	2 6 10	2 6 10 14	2 6 9 .	1	
79	Au	gold	2	2 6	2 6 10	2 6 10 14	2 6 10 .	1	
80	Hg	mercury	2	2 6	2 6 10	2 6 10 14	2 6 10 .	2	
81	Tl	thallium	2	2 6	2 6 10	2 6 10 14	2 6 10 .	2 1	
82	Pb	lead	2	2 6	2 6 10	2 6 10 14	2 6 10 .	2 2	
83	Bi	bismuth	2	2 6	2 6 10	2 6 10 14	2 6 10 .	2 3	
84	Po	polonium	2	2 6	2 6 10	2 6 10 14	2 6 10 .	2 4	
85	At	astatine	2	2 6	2 6 10	2 6 10 14	2 6 10 .	2 5	
86	Rn	radon	2	2 6	2 6 10	2 6 10 14	2 6 10 .	2 6	
87	Fr	francium	2	2 6	2 6 10	2 6 10 14	2 6 10 .	2 6 .	1
88	Ra	radium	2	2 6	2 6 10	2 6 10 14	2 6 10 .	2 6 .	2
89	Ac	actinium	2	2 6	2 6 10	2 6 10 14	2 6 10 .	2 6 1	2
90	Th	thorium	2	2 6	2 6 10	2 6 10 14	2 6 10 .	2 6 2	2
91	Pa	protactinium	2	2 6	2 6 10	2 6 10 14	2 6 10 1	2 6 2	2
92	U	uranium	2	2 6	2 6 10	2 6 10 14	2 6 10 3	2 6 1	2
93	Np	neptunium	2	2 6	2 6 10	2 6 10 14	2 6 10 4	2 6 1	2
94	Pu	plutonium	2	2 6	2 6 10	2 6 10 14	2 6 10 6	2 6 .	2
95	Am	americium	2	2 6	2 6 10	2 6 10 14	2 6 10 7	2 6 .	2
96	Cm	curium	2	2 6	2 6 10	2 6 10 14	2 6 10 7	2 6 1	2
97	Bk	berkelium	2	2 6	2 6 10	2 6 10 14	2 6 10 8	2 6 1	2
98	Cf	californium	2	2 6	2 6 10	2 6 10 14	2 6 10 10	2 6 .	2
99	Es	einsteinium	2	2 6	2 6 10	2 6 10 14	2 6 10 11	2 6 .	2
100	Fm	fermium	2	2 6	2 6 10	2 6 10 14	2 6 10 12	2 6 .	2
101	Md	mendelevium	2	2 6	2 6 10	2 6 10 14	2 6 10 13	2 6 .	2
102	No	nobelium	2	2 6	2 6 10	2 6 10 14	2 6 10 14	2 6 .	2
103	Lw	lawrencium	2	2 6	2 6 10	2 6 10 14	2 6 10 14	2 6 1	2

Figure 4-18 Energy-level diagram for sodium. The diagonal lines show observed optical transitions, with wavelengths given in nanometers. The energy of the ground state has been chosen as the zero point for the scale on the left.

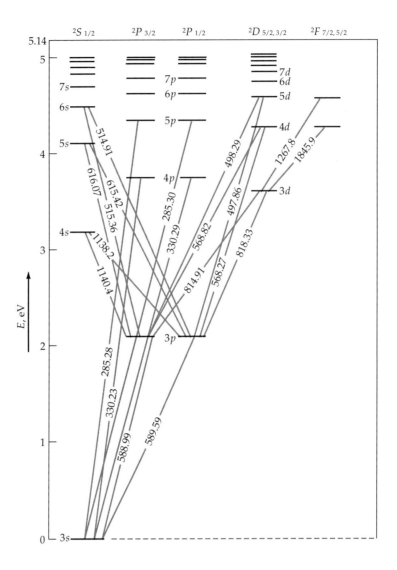

states. One of these is ruled out by a selection rule on j, which is

$$\Delta j = \pm 1 \text{ or } 0 \qquad \text{allowed}$$

but

$$j = 0 \rightarrow j = 0 \qquad \text{forbidden}$$

Transitions between doublet states therefore result in triplet spectral lines. The energy levels and optical spectra of the other alkali metals are similar to those for sodium.

The optical spectra for elements that have two outer electrons, such as helium, beryllium, and magnesium, are considerably more complex because of the interaction between the two outer electrons.

X-Ray Spectra

The energy needed to excite an inner, core electron, for example, an electron in the $n = 1$ state (K shell), is much greater than that needed to excite an outer, valence electron. An inner electron cannot be excited to any of the filled states, such as the $n = 2$ states in sodium, because of the Pauli exclusion principle. Therefore, the energy required to excite an inner electron to an unoccupied state is typically of the order of several thousand electron volts (keV). An inner electron can be excited by bombarding the atom with a high-energy electron beam in, for example, an x-ray tube. If an electron is knocked out of the $n = 1$ (K) shell, there is a vacancy left in this shell. This vacancy can be filled by an electron from the *L* shell or a higher shell that makes a transition to the *K* shell. The photons emitted by electrons making such transitions have energies of the order of 1 keV. They constitute the **characteristic x-ray spectrum** of an element, appearing as sharp peaks in the continuous x-ray spectrum of the element, as shown for molybdenum in Figure 4-19. Spectral lines arising from transitions that end at the $n = 1$ (K) shell make up the *K* series of the characteristic x-ray spectrum of an element. For instance, the K_α line in the figure arises from transitions from the $n = 2$ (*L*) shell to the $n = 1$ (*K*) shell, and the K_β line arises from transitions from the $n = 3$ shell to the $n = 1$ shell. A second series, the *L* series, is produced by transitions from higher energy states to a vacated place in the $n = 2$ (*L*) shell.

We can use the Bohr theory to calculate the approximate frequencies of characteristic x-ray spectra. According to the Bohr model, the energy of a single electron in a state *n* is given by Equation 4-9:

$$E_n = -Z^2 \frac{13.6 \text{ eV}}{n^2}$$

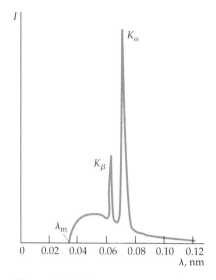

Figure 4-19 X-ray spectrum of molybdenum. The sharp peaks labeled K_α and K_β are characteristic of the element. The cutoff wavelength λ_m is independent of the target element and is related to the voltage *V* of the x-ray tube by $\lambda_m = hc/eV$.

Since for any atom other than hydrogen there are two electrons in the innermost shell, the *K* shell, the effective charge seen by one of the electrons is less than *Ze* because of the shielding due to the other electron. If the effective charge is $(Z - 1)e$, the energy of an electron in the *K* shell is given by this equation when $n = 1$ and *Z* is replaced by $Z - 1$:

$$E_1 = -(Z - 1)^2 (13.6 \text{ eV})$$

The energy of an electron in state *n* (provided that the effective charge is the same) is given by

$$E_n = -(Z - 1)^2 \frac{13.6 \text{ eV}}{n^2} \qquad 4\text{-}30$$

When an electron from state *n* drops into the vacated state in the $n = 1$ shell, a photon of energy $E_n - E_1$ is emitted. The wavelength of this photon is

$$\lambda = \frac{hc}{E_n - E_1} = \frac{hc}{(Z - 1)^2 (13.6 \text{ eV})(1 - 1/n^2)} \qquad 4\text{-}31$$

In 1913, the English physicist H. Moseley measured the wavelengths of the characteristic x-ray spectra for about 40 elements. From his data he was able to determine the atomic number *Z* for each element.

Example 4-7

Calculate the wavelength of the K_α x-ray line for molybdenum ($Z = 42$) and compare it with the value $\lambda = 0.0721$ nm measured by Moseley.

The K_α line corresponds to a transition from $n = 2$ to $n = 1$. The wavelength is given by Equation 4-31 with $Z = 42$ and $n = 2$:

$$\lambda = \frac{hc}{(41)^2 (13.6 \text{ eV})(1 - \frac{1}{4})} = \frac{1240 \text{ eV} \cdot \text{nm}}{(41)^2 (13.6 \text{ eV})(3/4)} = 0.0723 \text{ nm}$$

This result is in good agreement with the measured value.

Questions

4. Would you expect the optical spectrum of potassium to be like that of hydrogen or that of helium?

5. Would you expect the optical spectrum of beryllium to be like that of hydrogen or that of helium?

4-8 Absorption, Scattering, and Stimulated Emission

Information about the energy levels of an atom is usually obtained from the radiation emitted when the atom makes a transition from an excited state to a state of lower energy. We can also obtain information about such energy levels from the absorption spectrum. When atoms are irradiated with a continuous spectrum of radiation, the transmitted radiation shows dark lines corresponding to absorption of light at discrete wavelengths. The absorption spectra of atoms were the first line spectra observed. Since at normal temperatures atoms and molecules are in either their ground states or low-lying excited states, absorption spectra are usually simpler than emission spectra. For example, only those lines corresponding to the Lyman emission series are seen in the absorption spectrum of atomic hydrogen because nearly all the atoms are initially in their ground states.

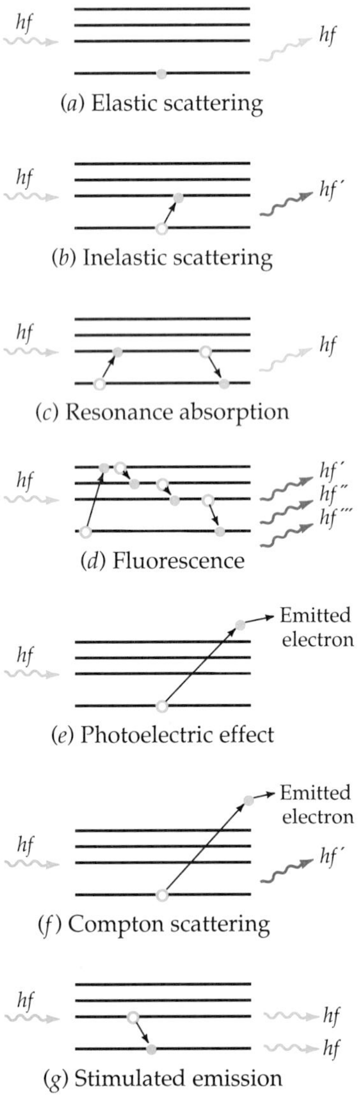

Figure 4-20 illustrates several interesting phenomena in addition to absorption that can occur when a photon is incident on an atom. In Figure 4-20a, the energy of the incoming photon is too small to excite the atom to one of its excited states, so the atom remains in its ground state and the photon is said to be scattered. Since the incoming and outgoing or scattered photons have the same energy the scattering is said to be elastic. If the wavelength of the incident light is large compared with the size of the atom, the scattering can be described in terms of classical electromagnetic theory and is called **Rayleigh scattering** after Lord Rayleigh, who worked out the theory in 1871. The probability of Rayleigh scattering varies as $1/\lambda^4$. This means that blue light is scattered much more readily than red light, which accounts for the bluish color of the sky. The removal of blue light by Raleigh scattering also accounts for the reddish color of the transmitted light seen in sunsets.

Figure 4-20b shows **inelastic scattering,** which occurs when the incident photon has enough energy to cause the atom to make a transition to an excited state. The energy of the scattered photon hf' is related to the energy of the incident photon hf by

$$hf' = hf - \Delta E$$

where ΔE is the excitation energy, which is the difference between the energy of the ground state and the energy of the excited state. Inelastic scattering of light from molecules was first observed by the Indian physicist C. V. Raman and is therefore often referred to as **Raman scattering.**

In Figure 4-20c, the energy of the incident photon is just equal to the difference in energy between the ground state and the first excited state of the atom. The atom makes a transition to its first excited state and then after a short delay makes a transition back to its ground state with the emission of a photon whose energy is equal to that of the incident photon. This multistep process is called **resonance absorption.** The phase of the emitted photon is not correlated with the phase of the incident photon. The emission of a photon as an atom spontaneously makes a transition to a lower state is called **spontaneous emission.**

In Figure 4-20d, the energy of the incident photon is great enough to excite the atom to one of its higher excited states. The atom then loses its energy by spontaneous emission as it makes one or more transitions to lower energy states. A common example occurs when the atom is excited by ultraviolet light and emits visible light as it returns to its ground state. This process is called **fluorescence.** Since the lifetime of a typical excited atomic energy state is of the order of 10^{-8} s, this process appears to occur instantaneously. However, some excited states have much longer lifetimes—of the

Figure 4-20 Phenomena that can occur when a photon is incident on an atom.

(a) Elastic scattering
(b) Inelastic scattering
(c) Resonance absorption
(d) Fluorescence
(e) Photoelectric effect
(f) Compton scattering
(g) Stimulated emission

(a)

(b)

order of milliseconds or occasionally seconds or even minutes. Such a state is called a **metastable state. Phosphorescent materials** have metastable states, and so emit light long after the original excitation.

Figure 4-20e illustrates the photoelectric effect, in which the absorption of the photon ionizes the atom by causing the emission of an electron. Figure 4-20f illustrates Compton scattering, which occurs if the energy of the incident photon is much greater than the ionization energy. Note that in Compton scattering, a photon is emitted, whereas in the photoelectric effect, the photon is absorbed with none emitted.

Figure 4-20g illustrates **stimulated emission.** This process occurs if the atom or molecule is initially in an excited state of energy E_2, and the energy of the incident photon is equal to $E_2 - E_1$, where E_1 is the energy of a lower state or the ground state. In this case, the oscillating electromagnetic field associated with the incident photon stimulates the excited atom or molecule which then emits a photon in the same direction as the incident photon and in phase with it. In spontaneous emission, the phase of the light from one atom is unrelated to that from another atom, so the resulting light is incoherent. However, in stimulated emission, the phase of the light emitted from one atom is related to that emitted by every other atom, so the resulting light is coherent. As a result, interference of the light from different atoms can be observed.

In addition to being excited by incident radiation, atoms may be excited to higher energy states by collisions with electrons or with other atoms.

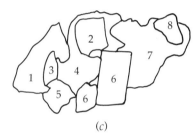
(c)

A collection of minerals in (a) daylight and (b) ultraviolet light (sometimes called "black light"). Identified by number in the schematic (c), they are 1 powellite, 2 willemite, 3 scheelite, 4 calcite, 5 calcite and willemite composite, 6 optical calcite, 7 willemite, 8 opal. The change in color is due to the minerals fluorescing under the ultraviolet light. In the case of optical calcite, both fluorescence and phosphorescence are occurring.

4-9 The Laser

The **laser** (*l*ight *a*mplification by *s*timulated *e*mission of *r*adiation) is a device that produces a strong beam of coherent photons by stimulated emission. Consider a system of atoms that have a ground-state energy E_1 and an excited-state energy E_2. If these atoms are irradiated by photons of energy $E_2 - E_1$, those atoms in the ground state can absorb a photon and make the transition to state E_2, whereas those atoms already in the excited state may be stimulated to decay back to the ground state. The relative probabilities of absorption and stimulated emission were first worked out by Einstein, who showed them to be equal. Ordinarily, at normal temperatures, nearly all the atoms will initially be in the ground state, so absorption will be the main effect. To produce more stimulated-emission transitions than absorption transitions, we must arrange to have more atoms in the excited state than in the ground state. This condition is called **population inversion.** It can be achieved if the excited state is a metastable state. Population inversion is often obtained by a method called **optical pumping** in which atoms are "pumped" up to energy levels of energy greater than E_2 by the absorption of an intense auxiliary radiation. The atoms then decay down to state E_2 by either spontaneous emission or by nonradiative transitions such as those due to collisions.

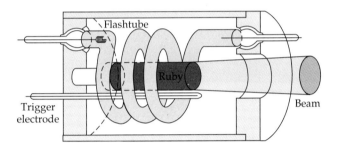

Figure 4-21 Schematic diagram of the first ruby laser.

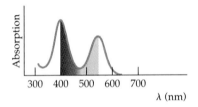

Figure 4-22 Absorption versus wavelength for Cr^{3+} in ruby. Ruby appears red because of the strong absorption of green and blue light by the chromium ions.

Figure 4-21 shows a schematic diagram of the first laser, a ruby laser built by Theodore Maiman in 1960. It consists of a small rod of ruby (a few centimeters long) surrounded by a helical gaseous flashtube. The ends of the ruby rod are flat and perpendicular to the axis of the rod. Ruby is a transparent crystal of Al_2O_3 with a small amount (about 0.05 percent) of chromium. It appears red because the chromium ions (Cr^{3+}) have strong absorption bands in the blue and green regions of the visible spectrum as shown in Figure 4-22. The energy levels of chromium that are important for the operation of a ruby laser are shown in Figure 4-23.

When the flashtube is fired, there is an intense burst of light lasting a few milliseconds. Absorption excites many of the chromium ions to the bands of energy levels indicated by the shading in Figure 4-23. The chromium ions then relax, giving up their energy to the crystal in nonradiative transitions and drop down to a pair of metastable states labeled E_2 in the figure. These metastable states are about 1.79 eV above the ground state. If

Figure 4-23 Energy levels in a ruby laser. To make the population of the metastable states greater than that of the ground state, the ruby crystal is subjected to intense radiation (in the green and blue wavelengths). This excites atoms from the ground state to the bands of energy levels indicated by the shading, from which they decay to the metastable states by nonradiative transitions.

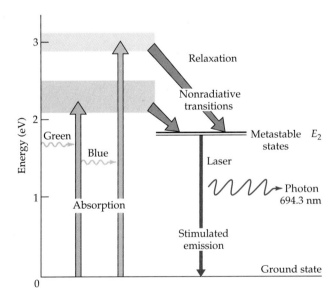

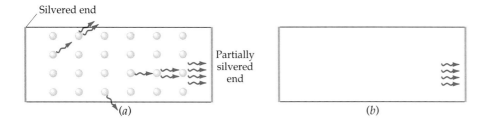
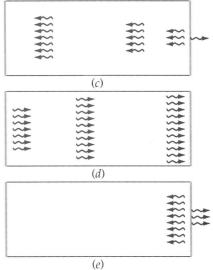

the flash is intense enough, more atoms will make the transition to the states E_2 than remain in the ground state. As a result, the populations of the ground state and the metastable states become inverted. When some of the atoms in the states E_2 decay to the ground state by spontaneous emission, they emit photons of energy 1.79 eV and wavelength 694.3 nm. Some of these photons stimulate other excited atoms to emit photons of the same energy and wavelength.

In the ruby laser, both ends of the crystal are silvered such that one end is almost totally reflecting (about 99.9 percent) and the other end is only partially reflecting (about 99 percent) so that some of the beam is transmitted. If the ends are parallel, standing waves are set up, and an intense beam of coherent light emerges through the partially silvered end. Figure 4-24 illustrates the buildup of the beam inside the laser. When photons traveling parallel to the axis of the crystal strike the silvered ends, all are reflected from the back face and most are reflected from the front face, with a few escaping through the partially silvered front face. During each pass through the crystal, the photons stimulate more and more atoms so that the photon beam builds up and an intense beam is emitted.

Modern ruby lasers generate intense light beams with energies ranging from 50 J to 100 J in pulses lasting a few milliseconds. The beam can have a diameter as small as 1 mm and an angular divergence as small as 0.25 milliradian to about 7 milliradians.

In 1961, the first successful operation of a continuous helium–neon gas laser was announced by Ali Javan, W. R. Bennet, Jr., and D. R. Herriott. Figure 4-25 shows a schematic diagram of the type of helium–neon laser commonly used for physics demonstrations. It consists of a gas tube containing 15 percent helium gas and 85 percent neon gas. A totally reflecting flat mirror is mounted at one end of the gas tube and a partially reflecting concave mirror is placed at the other end. The concave mirror focuses parallel light at the flat mirror and also acts as a lens that transmits part of the light so that it emerges as a parallel beam.

Figure 4-24 Buildup of photon beam in a laser. In (a), some of the atoms spontaneously emit photons, some of which travel to the right and stimulate other atoms to emit photons parallel to the axis of the crystal. In (b), four photons strike the partially silvered right face of the laser. In (c), one photon has been transmitted and the others have been reflected. As these photons traverse the laser crystal, they stimulate other atoms to emit photons and the beam builds up. By the time the beam reaches the right face again in (d), it comprises many photons. In (e), some of these photons are transmitted and the rest are reflected.

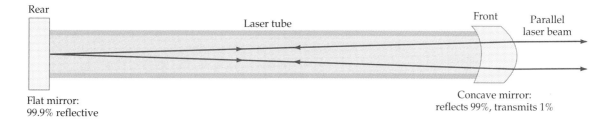

Figure 4-25 Schematic drawing of a helium–neon laser. The use of a concave mirror rather than a second plane mirror makes the alignment of the mirrors less critical than it is for the ruby laser. The concave mirror on the right also serves as a lens that focuses the emitted light into a parallel beam.

(a) Beams from a krypton and an argon laser, split into their component wavelengths. In these gas lasers, krypton and argon atoms have been stripped of multiple electrons, forming positive ions. The light-emitting energy transitions occur when excited electrons in the ions decay from one upper energy level to another. Here, several energy transitions are occuring at once, each corresponding to emitted light of a different wavelength. (b) Fluorescent organic compounds called dyes, dissolved in a solvent, are used in tuneable dye lasers. An external light source excites the dye molecules, which decay to their ground state in a series of steps, emitting light during some of the transitions. The dye molecules (which are large molecules, with multiple ring structures) have a range of vibrational and rotational, as well as electronic, energy states. The energy levels are closely enough spaced to allow a near continuum of possible transitions. In the laser, the dyes are contained in adjustable resonant cavities. Light emitted by the dye oscillates in the cavity, which is tuned to amplify only certain wavelengths.

(a)

(b)

Population inversion is achieved somewhat differently in the helium–neon laser than in the ruby laser. Figure 4-26 shows the energy levels of helium and neon that are important for operation of the laser. (The complete energy-level diagrams for helium and neon are considerably more complicated.) Helium has an excited energy state $E_{2,\text{He}}$ that is 20.61 eV above its ground state. Helium atoms are excited to state $E_{2,\text{He}}$ by an electric discharge. Neon has an excited state $E_{3,\text{Ne}}$ that is 20.66 eV above its ground state. This is just 0.05 eV above the first excited state of helium. The neon atoms are excited to state $E_{3,\text{Ne}}$ by collisions with excited helium atoms. The kinetic energy of the helium atoms provides the extra 0.05-eV energy needed to excite the neon atoms. There is another excited state of neon $E_{2,\text{Ne}}$ that is 18.70 eV above its ground state and 1.96 eV below state $E_{3,\text{Ne}}$. Since state $E_{2,\text{Ne}}$ is normally unoccupied, population inversion between states $E_{3,\text{Ne}}$ and $E_{2,\text{Ne}}$ is obtained immediately. The stimulated emission that occurs between these states results in photons of energy 1.96 eV and wavelength 632.8 nm, which produces a bright red light. After stimulated emission, the atoms in state $E_{2,\text{Ne}}$ decay to the ground state by spontaneous emission.

Figure 4-26 Energy levels of helium and neon that are important for the helium–neon laser. The helium atoms are excited by electrical discharge to an energy state 20.61 eV above the ground state. They collide with neon atoms, exciting some neon atoms to an energy state 20.66 eV above the ground state. Population inversion is thus achieved between this level and one 1.96 eV below it. The spontaneous emission of photons of energy 1.96 eV stimulates other atoms in the upper state to emit photons of energy 1.96 eV.

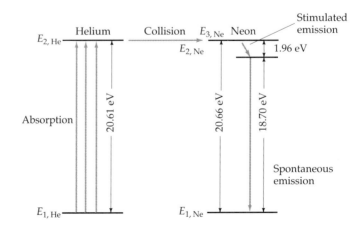

(c)

(d)

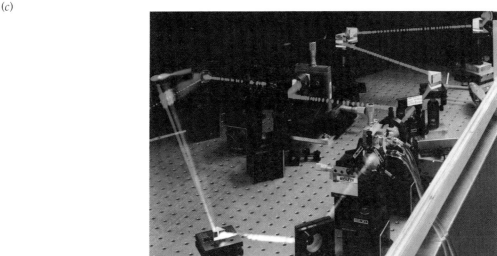

(e)

(c) The titanium-sapphire laser is presently state-of-the-art in tuneable lasers. The blue beam is laser light acting as an optical pump for the titanium-sapphire laser, whose cover is removed. Vibrational energy levels in the titanium atoms embedded in the sapphire crystal are superimposed on the normal energy levels of the aluminum and oxygen atoms that compose the sapphire. Sapphire atoms are excited to higher energy states by absorbing the light from the optical pump. They emit red light and return to a lower energy state corresponding to the lowest vibrational energy mode of the crystal. Because the electronic energy transitions can terminate anywhere in the band of the finely spaced vibrational states, the light is emitted in a broad range of wavelengths. This light is diverted to a tuning cavity where a particular wavelength can be extracted (for instance, by adjusting the angle of a diffraction grating). (d) A semiconductor laser, shown for scale in the eye of a needle. *pn* Junction semiconductors (Chapter 6) are constructed by doping a crystal with small amounts of an impurity element. When an external voltage is applied across the semiconductor, free electrons from the impurity element combine with holes in the crystal lattice and light energy is released. In the cleaved-coupled cavity design shown here, opposite ends of the semiconductor crystal are cleaved to form reflective facets. The reflected light itself is partially absorbed by electron-hole pairs, causing additional electron-hole combinings and thereby amplifying the net emitted light. (e) A femtosecond pulsed laser. By a technique known as "modelocking," different excited modes within a laser's cavity can be made to interfere with one another and create a series of ultrashort pulses, picoseconds long, that correspond to the time it takes light to bounce back and forth once within the cavity. It has been possible to compress such pulses even further, once they have left the laser. Ultrashort pulses have been used as probes to study the behavior of molecules during chemical reactions.

Note that there are four energy levels involved in the helium–neon laser, whereas the ruby laser involved only three levels. In a three-level laser, population inversion is difficult to achieve because more than half the atoms in the ground state must be excited. In a four-level laser, population inversion is easily achieved because the state after stimulated emission is not the ground state but an excited state that is normally unpopulated.

A laser beam is coherent, very narrow, and intense. Its coherence makes the laser beam useful in the production of holograms, which we discuss next. The precise direction and small angular spread of the beam make it useful as a surgical tool for destroying cancer cells or reattaching a detached retina. Lasers are also used by surveyors for precise alignment over large distances. Distances can be accurately measured by reflecting a laser pulse from a mirror and measuring the time the pulse takes to travel to the mirror and back. The distance to the moon has been measured to within a few centimeters using a mirror placed on the moon for that purpose. Laser beams are also used in fusion research. An intense laser pulse is focused on tiny pellets of deuterium–tritium in a combustion chamber. The beam heats the pellets to temperatures of the order of 10^8 K in a very short time, causing the deuterium and tritium to fuse and release energy.

Laser technology is advancing so fast that it is possible to mention only a few of the recent developments. In addition to the ruby laser, there are many other solid-state lasers with output wavelengths ranging from about 170 nm to about 3900 nm. Lasers that generate more than 1 kW of continuous power have been constructed. Pulsed lasers can now deliver nanosecond pulses of power exceeding 10^9 W. Various gas lasers can now produce beams of wavelengths ranging from the far infrared to the ultraviolet. Semiconductor lasers (also known as diode lasers or junction lasers) the size of a pinhead can develop 200 mW of power. Liquid lasers using chemical dyes can be tuned over a range of wavelengths (about 70 nm for continuous lasers and more than 170 nm for pulsed lasers). A relatively new laser, the free-electron laser, extracts light energy from a beam of free electrons moving through a spatially varying magnetic field. The free-electron laser has the potential for very high power and high efficiency and can be tuned over a large range of wavelengths. There appears to be no limit to the variety and uses of modern lasers.

Question

6. Why is helium needed in a helium–neon laser? Why not just use neon?

Holograms

An interesting application of lasers is the production of a three-dimensional photograph called a **hologram.** In an ordinary photograph, the intensity of reflected light from an object is recorded on a film. When the film is viewed by transmitted light, a two-dimensional image is produced. In a hologram, a beam from a laser is split into two beams, a reference beam and an object beam. The object beam reflects from the object to be photographed and the interference pattern between it and the reference beam is recorded on a photographic film. This can be done because the laser beam is coherent so that the relative phase difference between the reference beam and object beam can be kept constant during the exposure. The interference fringes on the film act as a diffraction grating. When the film is illuminated with a laser, a three-dimensional replica of the object is produced.

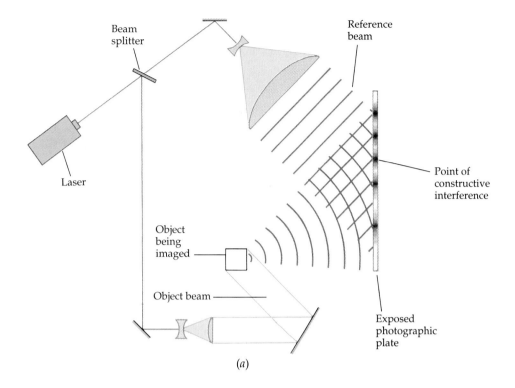

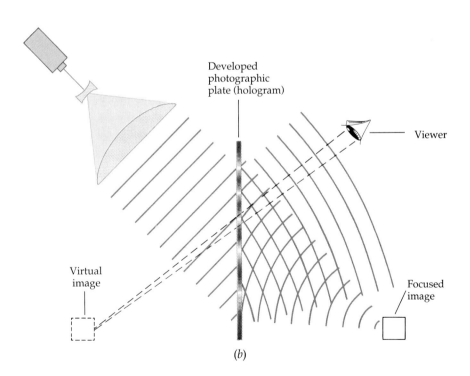

(a) The production of a hologram. The interference pattern produced by the reference beam and object beam is recorded on a photographic film. (b) When the film is developed and illuminated by coherent laser light, a three-dimensional image is seen. Holograms that you see on credit cards or postage stamps, called rainbow holograms, are more complicated. A horizontal strip of the original hologram is used to make a second hologram. The three-dimensional image can be seen as the viewer moves from side to side, but if viewed with laser light, the image disappears when the viewer's eyes move above or below the slit image. When viewed with white light, the image is seen in different colors as the viewer moves in the vertical direction.

(a) A technician produces a hologram of a statuette at the University of Strasbourg. When the glass plate is later illuminated by laser light, the statuette appears as a three-dimensional image. (b) and (c) Two views of the hologram "Digital." Note that different parts of the circuit board appear behind the front magnifying lens. (d) A holograph emulsion magnified 100 times. (e) A "head-up display" uses a holographic projection of important information from the airplane's control panel, so the pilot can view the runway and the control panel at the same time.

(a)

(b)

(c)

(d)

(e)

Summary

1. In quantum theory, the hydrogen atom is described by a wave function, the square of which gives the probability of finding the electron in a given region of space. The wave function is characterized by four quantum numbers:

$$n = 1, 2, 3, \ldots$$
$$\ell = 0, 1, \ldots, n - 1$$
$$m = -\ell, -\ell + 1, \ldots, +\ell$$
$$m_s = +\tfrac{1}{2} \text{ or } -\tfrac{1}{2}$$

   The energy of the hydrogen atom depends only on the principal quantum number n and is the same as in the Bohr model. In the ground state, $n = 1$, $\ell = 0$, and $m = 0$, and the probability distribution is spherically symmetric, with the electron most likely to be found near the first Bohr radius. It is convenient to think of the electron as a charged cloud with a charge density proportional to the probability distribution.

2. In multielectron atoms, the energy of the electron is determined mainly by the principal quantum number n (which is related to the radial dependence of the wave function) and by the orbital angular-momentum quantum number ℓ. Generally, the lower the values of n and ℓ, the lower the energy. The specification of n and ℓ for all the electrons in an atom is called the electron configuration. Customarily, ℓ is specified by a letter code rather than by its numerical value. The code is

Code	s	p	d	f	g	h
ℓ value	0	1	2	3	4	5

3. An important principle that governs the electron configurations of atoms is the Pauli exclusion principle. It states that no two electrons in an atom can be in the same quantum state; that is, no two electrons can have the same set of values for the quantum numbers n, ℓ, m, and m_s. Using the exclusion principle, we can understand much of the structure of the periodic table of the elements.

4. Atomic spectra consist of optical spectra and x-ray spectra. Optical spectra can be understood in terms of transitions between energy levels of a single outer electron moving in the field of the nucleus and core electrons of the atom. Characteristic x-ray spectra result from the excitation of an inner core electron and the subsequent filling of the resulting vacancy by another electron in the atom.

5. Stimulated emission occurs if an atom is initially in an excited state and a photon of energy equal to the excitation energy is incident on the atom. The oscillating electromagnetic field of the incident photon stimulates the excited atom to emit another photon in the same direction and in phase with the incident photon. The operation of a laser depends on population inversion, in which there are more atoms in an excited state than in the ground state or a lower state. A laser produces an intense, coherent, and narrow beam of photons.

Trapped Atoms and Laser Cooling

D. J. Wineland
National Institute of Standards and Technology,
Boulder, Colorado

Physicists are often interested in investigating the energy-level structure of atoms with high accuracy. Perhaps the most important reason for this is the desire to test with precision the theories that predict this energy structure. If a deviation, no matter how small, occurs between the theoretically predicted structure and the experimental measurements, the theory must be modified. What is most interesting is to find the difference to be explained by some new physical effect which must then be incorporated into the theory.

Spectroscopy is an experimental procedure for measuring the energy-level structure of atoms. It involves measuring the energy differences between atomic states by observing the frequency f (or equivalently, the wavelength) of the radiation that is emitted or absorbed in transitions between atomic states. The energy difference between two particular states is $E_2 - E_1 = hf$, where h is Planck's constant (Equation 2-17). The energy levels of the atom can then be constructed from series of such energy differences.

David Wineland has been a staff physicist at the National Institute of Standards and Technology in Boulder, Colorado since 1975. After receiving his bachelor's degree from Berkeley he attended Harvard University where he received his PhD in 1970. From 1970 to 1975, he worked at the University of Washington. There, he learned about trapping techniques from Hans Dehmelt who shared the 1989 Physics Nobel Prize for the development of ion traps. When not in the lab, he enjoys bicycling and skiing.

Measurements of energy differences ($E_2 - E_1$) are always accompanied by some measurement uncertainty ΔE. In the case of absorption or emission spectra, the uncertainty in energy corresponds to the width of spectral lines: the smaller the uncertainty ΔE, the smaller the range of frequencies spanned, that is, the narrower the lines. Furthermore, the highest resolution (smallest ΔE) can be obtained when the time for transition between states Δt is as long as possible. This is a consequence of the time-energy uncertainty relation, which states that the time over which an energy is measured Δt and the uncertainty of that energy ΔE are related by $\Delta E \Delta t \sim \hbar$ (Section 3-3). In practice, Δt cannot be made arbitrarily long because processes other than emission or absorption—such as collisions with other atoms—can intervene and significantly alter the state of the atom. Such an intervention cuts short the time the atom has to emit or absorb in its undisturbed state, and therefore cuts short the time available for measuring the energy of such an emission or absorption.

The atoms in a sample being analyzed (for example, helium atoms emerging from a small hole in a helium reservoir) are normally moving relative to the radiation source (which is fixed in the laboratory), so the apparent frequencies of light emitted by the atoms are shifted by the doppler effect (Section 1-6). ^{4}He atoms at room temperature move with speeds of about 1.5×10^5 cm/s; therefore the apparent emission frequency of an atom moving toward the radiation source would be shifted to a higher value by the fraction $v/c = 5 \times 10^{-6}$. (The frequency of absorption would be shifted to a lower value by the same amount.) This is a significant amount in very high resolution studies.

The energy resolution can be increased (ΔE made small) and the doppler shift reduced if the atoms are confined to a region of space. By holding the atoms, we can make the time for transition Δt large and ΔE small. As the atoms are held in a localized region, their velocity and associated doppler shift average to zero. The trick in this procedure is to hold the atoms in such a way that when they strike the walls of the container, the atomic states are not distorted to such a degree that the measurement is rendered inaccurate. Various methods are employed to do this; for atomic or molecular ions we can use special configurations of electric and magnetic fields that act on the charge of the ion to pro-

This essay is a publication of the National Institute of Standards and Technology; not subject to U.S. copyright.

Figure 1 Schematic representation of the electrode configuration for the Penning or Paul trap. Electrode surfaces are figures of revolution about the z axis and are made as close as possible to be equipotentials of the quadratic potential ϕ discussed in the text.

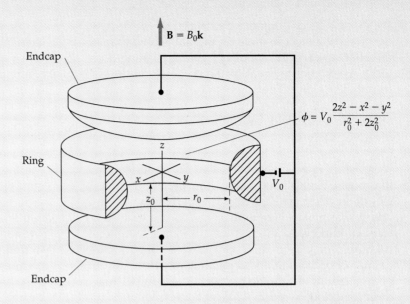

Ion Trapping

Several kinds of electromagnetic "traps" are used to confine ions. Cyclotrons and synchrotrons are used for accelerating and storing high-energy elementary particles. Here we'll briefly describe one kind of trap for atomic spectroscopy. This trap, shown in Figure 1, uses static electric and magnetic fields and is usually called a Penning trap, after F. M. Penning who reported the basic device in 1936.

Assume we are interested in confining positively charged ions, for example Be^+ or Hg^+, where one electron has been removed from the neutral atom. If we put a positive voltage V_0 on the two endcap electrodes and leave the central ring at ground potential, the ions experience a force in a direction toward the xy plane; in other words, they are confined as to how far they can travel up or down in the z direction. Unfortunately, this configuration of electrodes causes the ions to feel a radial electric force away from the z axis of the trap, so they accelerate rapidly into the ring electrode. This difficulty can be overcome if we superpose a static magnetic field $\mathbf{B}_0$ whose field lines are parallel to the z axis. This axial magnetic field acting with the radial electric force causes the ions to orbit about the z axis of the trap and confines the ions in all three dimensions.

vide confinement. It turns out that these electric and magnetic fields perturb the ion's internal structure very little, so that they do not limit the accuracy of the observation. The "walls" of the container provided by the fields are, in effect, very soft.

Let us analyze quantitatively the motion of an ion in a common type of Penning trap. One that is often used has electrode surfaces that conform to equipotentials of the potential (in cartesian coordinates)

$$\phi = V_0(2z^2 - x^2 - y^2)/(r_0^2 + 2z_0^2)$$

This is provided, to a good approximation, by the electrode shapes shown in Figure 2. The particular shape is that generated by rotating a hyperbola around the z axis. From this potential, we can find

Continued

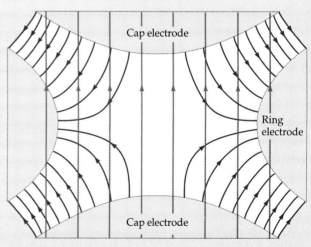

Figure 2 Electric (red) and magnetic (blue) field lines of trap shown in Figure 1.

the electric field along the z axis as $E_z = -d\phi/dz \propto -z$. Thus the electric force on the ion is always in a direction toward the xy plane, and the strength of the force is proportional to the distance the ion is from the xy plane. The magnetic force on the ion can act only in a direction perpendicular to the direction of the magnetic field, meaning that in this case the magnetic force will act only in the xy plane. So the only force on the ion in the z direction is the electric force. Since $F_z \propto -z$, we know that the motion of the ion must be simple harmonic. It is in fact this property of an electrode shaped as a hyperboloid of revolution that makes it a popular one to use in an ion trap. In particular, the fact that the ion's motion is harmonic (in the z direction) means that its frequency of oscillation is independent of its amplitude of oscillation. From the properties of simple harmonic motion, we can work out the frequency of oscillation in the z direction, f_z, and find

$$f_z^2 = qV_0/\pi^2 m(r_0^2 + 2z_0^2)$$

where q and m are the ion's charge and mass respectively. To get an idea of the magnitude of this frequency, assume some typical values of parameters encountered in the laboratory. For $V_0 = 1$ V, $m \approx 9$ atomic mass units (e.g., ^{9}Be$^+$ ions), $r_0 = z_0\sqrt{2} = 1$ cm, we find $f_z \approx 74$ kHz.

A charged particle in a magnetic field executes circular motion around the field lines at the cyclotron frequency $f_c = qB/2\pi m$. However the addition of the potential ϕ gives rise to a radial electric field whose strength is proportional to the distance of the ion from the z axis. As this radial electric field is perpendicular to the magnetic field, the center of the ions' cyclotron orbits drifts in a direction perpendicular to the magnetic and electric field lines. For the cylindrical geometry of the trap, this drift motion is a circular orbit about the z axis. Therefore the overall motion in the direction normal to the trap axis is a composite of circular motion at approximately the cyclotron frequency (which is actually shifted slightly by the presence of the radial electric field) and a circular motion around the z axis due to the crossed electric and magnetic fields. The latter is usually called the magnetron motion. The energy extracted from this motion in a "magnetron" tube is what provides heat in a microwave oven. The complete motion of the ion in the Penning trap is summarized in Figure 3a, b.

If the ions collide with background gas atoms, the radius of the magnetron motion gradually increases until the ions strike the ring electrode and stick to it. To prevent this, the trap apparatus is nor-

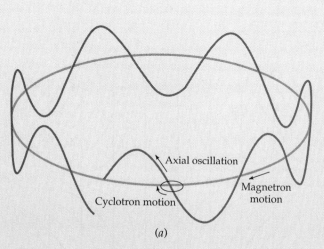
(a)

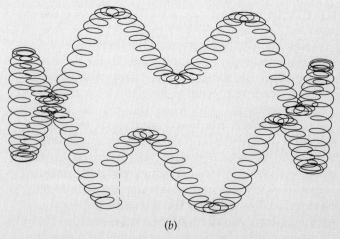
(b)

Figure 3 (a) Three modes of motion of the ion in the trap combine to yield a complex trajectory (b). The ion revolves rapidly in the cyclotron orbit; at the same time the center of the cyclotron orbit follows the much larger circular path of the magnetron motion; all the while the ion is oscillating along the axis of the trap.

Figure 4 Electrodes for a small Paul trap mounted on a penny to give the size. In the experiment, these electrodes are mounted inside a quartz vacuum enclosure where the vacuum is about 10^{-8} Pa or about 10^{-13} atm. The electrodes on either end are held at ground potential and a potential of about 500 V which oscillates sinusoidally at a frequency around 20 MHz is applied to the center ring electrode. The quartz vacuum enclosure allows the ultraviolet light scattered from the ion to pass and be photographed.

mally installed in an evacuated envelope (for example, a sealed glass tube) where the pressure inside the envelope is on the order of 10^{-8} Pa (about 10^{-13} atm). Under these conditions, the ions can be stored in the trap for many days.

Another kind of ion trap uses the same electrode shapes as the Penning trap but confines ions by the action of an oscillating electric potential $V_0 \cos \omega t$ applied between the endcaps and ring electrode. This trap is called the Paul trap after Wolfgang Paul who proposed it in the early 1950s (Figure 4). Figure 5 shows an ultraviolet photograph of a single Hg^+ ion confined in a miniature Paul trap. The high degree of localization obtained with the trap allows this kind of photograph to be made.

With both kinds of traps, ions, electrons, and even more exotic particles such as positrons or antiprotons can be confined for such long times that the resolution of energy measurements is no longer limited by the residence time of the ions in the trap.

Various types of traps can be used for neutral atoms. These provide trapping by electric and magnetic fields but (because of the overall charge neutrality of the atom) the fields must act on the electric or magnetic dipole moment of the atom.

Continued

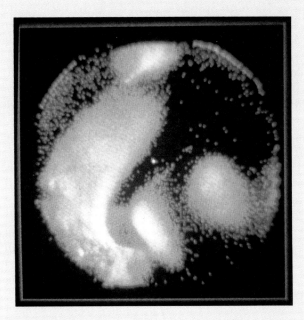

Figure 5 False-color image of a *single* Hg^+ ion (small isolated dot near the center) stored in the Paul trap shown in Figure 4. The other shapes are due to light reflected from the trap electrodes. The inner diameter of the ring electrode in this trap was about 0.9 mm. The ion is actually confined to a region of space much smaller than indicated by the size of the dot but the lens used in this experiment limited the resolution obtained.

High Accuracy Atomic Spectroscopy

Accuracies of energy measurements in certain atomic spectroscopy experiments are now better than 1 part in 10^{13}. This means that when we measure the frequency of an atomic transition, we know all of the environmental perturbations on the atom to such an extent that we can tell you the frequency at which the atom would absorb the radiation if it were isolated and at rest in space. If the accuracy is 1 part in 10^{13}, the inaccuracy of our prediction of the frequency is only 10^{-13} of the value of the frequency. As a comparison, if we could measure the distance between two points on the east and west coasts of the United States to an accuracy of 1 part in 10^{13}, the error in our measurement would be about 5×10^{-5} cm (the distance of one wavelength of visible light).

With such high accuracies in atomic spectroscopy, we must consider a number of effects that are typically small enough that we do not ordinarily worry about them. One such effect is called the second-order doppler shift. It can be derived by expressing Equation 1-24a or 1-24b in a power series in V/c. The term proportional to V/c is the same as the classical expression for sound. The second-order doppler shift is the term proportional to $(V/c)^2$. This is the effect caused by relativistic time dilation; because the ions (or atoms) are moving with respect to our radiation source, which is stationary in the lab, time moves more slowly for them. Hence, when we measure the frequencies of their transitions, we measure a value which is slightly lower than the frequency we would measure if they were at rest. This is not a particularly big effect—for ^{9}Be$^+$ ions stored in a trap where the kinetic energies are near room temperature (about 300 K), the magnitude of this shift is fractionally $V^2/2c^2 \approx 5 \times 10^{-12}$. However, it is a difficult shift to measure accurately as it is hard to measure the ions' velocity distribution precisely. One approach to this problem is to reduce the temperature of the ions; an effective method is laser cooling.

Laser Cooling

We are familiar with the use of lasers to provide heat (e.g., laser surgery, welding, and inertial confinement fusion). However, as explained below, laser light has now also been used to cool small samples of trapped ions and atoms to very low temperatures—in some cases to about 1 μK. This cooling results from the mechanical momentum imparted to the atoms when they scatter light; by suitable arrangement of the laser beam's frequency and position, the atoms can be made to scatter light only when this scattering causes their momentum to be reduced.

That electromagnetic radiation can impart momentum to matter was known to James Clerk Maxwell in the late nineteenth century. Albert Einstein used the discrete momentum changes imparted to atoms by electromagnetic radiation in his theoretical studies of thermal equilibrium between radiation and matter. In 1933, Otto Frisch demonstrated experimentally the transfer of momentum from photons to atoms by deflecting a beam of sodium atoms with resonance radiation from a lamp. With the development of tunable lasers in the 1970s, such effects could be much more pronounced. Recently, the narrow spectral width of lasers has also allowed cooling of atoms by these mechanical forces. The simplest form of laser cooling, and the one most commonly used, is called doppler cooling. It relies on the high spectral purity of lasers, the fact that atoms tend to absorb light only at particular frequencies, and the frequency shift of the light (as viewed by the atom) due to the doppler effect.

First, suppose we use a tunable laser to measure the absorption spectrum of an atom near one of its optical transitions. If we could hold the atom stationary, the absorption would be strongest at a particular frequency f_0 and have a narrow range Δf over which it absorbs most strongly. For the transition of interest in ^{9}Be$^+$ ions, $f_0 \approx 10^{15}$ Hz and $\Delta f \approx$ 20 MHz.

Suppose we now release the atom and subject it to a laser beam coming from the left. Assume this laser has frequency f_L where $f_L < f_0$. If the atom moves to the left with velocity V, then in the atom's reference frame, the laser light appears to have a frequency approximately equal to $f_L(1 + V/c)$ due to the doppler frequency shift (Section 1-7). For a particular value of V, $f_L(1 + V/c) \approx f_0$, and the atom absorbs and re-emits photons at a high rate. On absorption, the photon's momentum is transferred to the atom and thereby reduces the atom's momentum by approximately h/λ where λ is the wavelength of the laser radiation (see Equation 2-29). However, the photon re-emission is spatially symmetric, so on the average, there is no net momentum imparted to the atom by re-emission. Hence, on the average, the atom's momentum is reduced by h/λ for each scattering event. This is really no different than in collisions of macroscopic bodies, since all we need to take care of is conservation of energy and momentum.

If, instead, the atom moves to the right, each scattering event increases the atom's momentum by h/λ. However, the scattering rate is much less for atoms moving to the right because the frequency of the radiation (in the atom's frame) is now $f_L(1 - V/c) < f_L < f_0$ and the laser appears to be tuned away from the atom's resonance. This asymmetry in the scattering rate, and in the accompanying transferred momentum, for atoms moving left or right, gives rise to a net cooling effect. If an atom is subjected to three mutually orthogonal, intersecting pairs of counterpropagating laser beams tuned to $f_L < f_0$, the atom feels a damping or cooling force independent of which direction it moves. Such a configuration has been called "optical molasses."

The randomness in the times of absorption and the randomness in the direction of photon re-emission act like random impulses on the atom, which counteract the cooling effect. These random impulses, which cause heating, reach a balance with the cooling when the effective temperature of the atoms reaches a minimum value equal to $h \Delta f/2k$ (k = Boltzmann's constant). For many atoms, this temperature is around 1 mK or less. At 1 mK, the second-order doppler shift of $^9Be^+$ ions is about 1.5×10^{-17}, so by the use of laser cooling we significantly reduce the perturbation to the measured frequency caused by the second-order doppler shift.

Other Applications of Trapping and Cooling

In the above, we have discussed how the techniques of atom trapping and cooling can be used for precision atomic spectroscopy. It appears that these techniques can also be used to advantage for other purposes. Some examples are briefly discussed here.

Atomic Clocks The regular oscillations or vibrations of atoms and molecules can be likened to the oscillations of the pendulum in a grandfather clock. To make a clock based on atoms, we can tune the frequency of a radiation source until we drive a particular transition in an atom with maximum probability. If we then count the oscillations of the radiation source and wait until a certain number of cycles have elapsed, we define a unit of time. The nice thing about atoms is that, as far as we know, all atoms of a particular kind (such as $^9Be^+$ ions) are the same. No matter where two people are in the universe, if they agree to synchronize their radiation sources to a particular transition in a particular atom, when they count a given number of cycles, the unit of time they measure will be the same independent of a direct comparison. This is to be contrasted to pendulum clocks where no matter how much care is taken in their construction, they will oscillate at slightly different frequencies as it is hard to make the length of the pendula the same. Eventually we hope that, with the aid of trapping and cooling techniques, we will be able to make a clock that will be accurate to 1 second over the age of the universe. Accurate clocks are very useful in satellite and deep-space navigation systems.

Collision Studies Atomic collision studies at extremely low energies are now possible. At very low temperatures, the atom's de Broglie wavelength is long and quantum-mechanical effects are very important in describing the collisions. If the atom's de Broglie wavelength is large compared to the attractive region near a material surface, the atom may experience only the repulsive part of the surface and elastically bounce rather than stick. This may help provide nearly ideal atom "boxes."

Atom Manipulation Optical forces, such as those used in laser cooling, have been used to slow neutral atoms, steer atomic beams, and make atom "traps." The traps provided by optical forces are typically shallow (trap depths corresponding to a few kelvins). Therefore atoms from an atomic beam can be first slowed by overlapping them with a counterpropagating laser beam whose cooling force can stop the atoms at the position of the trap. In one trap called "optical tweezers," which uses the forces of a focused laser beam, the trapped atoms can be moved to different spatial locations by simply moving the laser beam. Applications of atom traps may include storage and manipulation of atomic antimatter, which must avoid contact with ordinary matter to prevent annihilation.

Condensed Matter Collections of atomic ions contained in ion traps (for instance, Figure 6) can be viewed as plasmas. At the very low temperatures provided by laser cooling, the Coulomb potential energy between adjacent ions exceeds their kinetic energy and the ions show regular spatial structure. In Figure 7, this regular spatial structure takes the form of shells of

Continued

Figure 6 Photograph of the Penning trap used to confine the $^9Be^+$ ions shown in Figure 7. In this trap, which has cylindrical geometry (inner diameter of the cylinders ≈ 2.5 cm), the sections at either end have the function of the endcaps of the trap shown in Figure 1. They are at a positive potential with respect to the central electrodes. A uniform magnetic field from a superconducting magnet (not shown) is parallel to the axis of the cylinders. This trap is also mounted in a quartz vacuum enclosure. Photographs of the ions are made by viewing along the axis of the cylinders.

ions. If a sample of weakly interacting atoms (e.g., atomic hydrogen) is trapped and sufficiently cooled, it may be possible to observe a transition to a state where the wave functions of the atoms are all the same and occupy the same region of space. This phenomenon, which has yet to be observed, is called Bose–Einstein condensation.

These topics have been developed in more detail by P. Ekstrom and D. Wineland, "The Isolated Electron," *Scientific American*, August 1980, pp. 104–121; D. J. Wineland and W. M. Itano, "Laser Cooling," *Physics Today* vol. 40, no. 6, June 1987, pp. 34–40; W. D. Phillips, and H. J. Metcalf, "Cooling and Trapping Atoms," *Scientific American*, March 1987, pp. 50–56. J. J. Bollinger and D. J. Wineland, "Microplasmas," *Scientific American,* January 1990, pp. 124–130. C. Cohen-Tannoudji and W. D. Phillips, *Physics Today*, vol. 43, no. 10, October 1990, pp. 33–40.

Figure 7 Ultraviolet photograph of a small Be^+ ion plasma which has been stored in the Penning trap shown in Figure 6 and laser cooled to about 10 mK. This picture was taken by viewing the ion plasma along the z axis through one of the endcaps. At low temperatures, plasmas become "strongly coupled" and show spatial structure. Here, this structure takes the form of cylindrical shells, which have been partially illuminated by a laser beam. The diameter of the outer shell in this picture is about 150 μm.

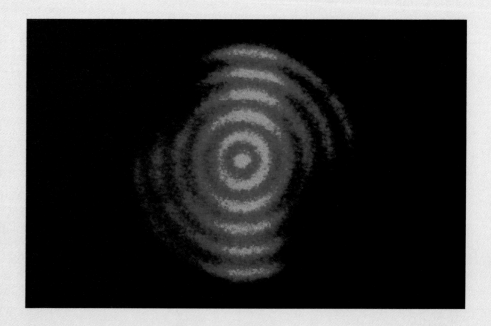

Suggestions for Further Reading

Leith, Emmett N., and Juris Upatnicks: "Photography by Laser," *Scientific American*, June 1965, p. 24.

The interference of coherent light produced by a laser is employed in wavefront reconstruction photography, more commonly known as holography.

Schawlow, Arthur L.: "Laser Light," *Scientific American*, September 1968, p. 120.

How lasers work and how laser light differs from ordinary light are discussed in this article.

Schewe, Phillip F.: "Lasers," *The Physics Teacher*, vol. 19, no. 8, 1981, p. 534.

This is an excellent and comprehensive exposition of the principles of laser operation, the types of lasers in use today, and application of laser light.

Vali, Victor: "Measuring Earth Strains by Laser," *Scientific American*, December 1969, p. 88.

One important modern use for laser interferometers is to measure small changes in the compression of the earth's crust, both steady and sudden, as in earthquakes.

Walker, Jearl: "The Amateur Scientist: The Colors Seen in the Sky Offer Lessons in Optical Scattering," *Scientific American*, January 1988, p. 102.

The scattering of light by molecules and dust particles explains blue skies, red sunsets, and other lesser-known atmospheric color phenomena.

Walker, Jearl: "The Amateur Scientist: The Spectra of Streetlights Illuminate Basic Principles of Quantum Mechanics," *Scientific American*, January 1984, p. 138.

This straightforward account of the development and application of quantum mechanics to the explanation of atomic spectra is illustrated with novel photographs.

Zare, Richard N.: "Laser Separation of Isotopes," *Scientific American*, February 1977, p. 86.

This article explains how laser irradiation of atoms or molecules in a beam can be useful in separating the isotopes of an element.

Review

A. Objectives: After studying this chapter, you should:

1. Know the origins of the quantum numbers n, ℓ, and m and the possible values of these numbers.

2. Know the relation between the quantum numbers n, ℓ, and m and the quantization of energy and angular momentum in the hydrogen atom.

3. Be able to compare the Schrödinger and Bohr models of the hydrogen atom.

4. Be able to sketch the wave function and probability distribution functions for the ground state of hydrogen.

5. Know the connection between magnetic moment and angular momentum, and be able to describe the Stern–Gerlach experiment.

6. Know the rules for the combination of angular momenta, and be able to discuss qualitatively the spin–orbit effect.

7. Be able to discuss the shell structure of atoms and the periodic table.

8. Be able to compare the optical and x-ray spectra of an atom.

9. Be able to describe the operation of a ruby laser and a helium–neon laser.

B. Define, explain, or otherwise identify:

Atomic number
Valence electron
Radial equation for hydrogen
Principal quantum number
Orbital quantum number
Magnetic quantum number
Selection rules
Radial probability density
Fine structure
Electron spin
Gyromagnetic ratio
Dirac equation
Stern–Gerlach experiment
Space quantization
Triplet state
Singlet state
Spin–orbit effect
Fine-structure splitting
Zeeman effect
Electron configuration
Pauli exclusion principle
Ionization energy
Transition elements
Optical spectrum
Characteristic x-ray spectrum
Rayleigh scattering
Inelastic scattering
Raman scattering
Resonance absorption
Spontaneous emission
Fluorescence
Metastable state
Phosphorescent materials
Stimulated emission
Laser
Population inversion
Optical pumping
Hologram

C. True or false: If the statement is true, explain why it is true. If it is false, give a counterexample.

1. No two electrons can be in the same quantum state.

2. Elements with one electron outside a closed shell have small ionization energies and are chemically active.

3. Visible light results from transitions involving only the outermost electrons in an atom.

4. Characteristic x rays result from transitions made by inner electrons.

Problems

Level I

4-1 Quantum Theory of the Hydrogen Atom

1. For $\ell = 1$, find (a) the magnitude of the angular momentum L and (b) the possible values of m. (c) Draw to scale a vector diagram showing the possible orientations of **L** with the z axis.

2. Work Problem 1 for $\ell = 3$.

3. If $n = 3$, (a) what are the possible values of ℓ? (b) For each value of ℓ in (a), list the possible values of m. (c) Using the fact that there are two quantum states for each value of ℓ and m because of electron spin, find the total number of electron states with $n = 3$.

4. Find the total number of electron states with (a) $n = 2$ and (b) $n = 4$. (See Problem 3.)

5. The moment of inertia of a phonograph record is about 10^{-3} kg·m². (a) Find the angular momentum $L = I\omega$ when it rotates at $\omega/2\pi = 33.3$ rev/min and (b) find the approximate value of the quantum number ℓ.

6. Find the minimum value of the angle θ between **L** and the z axis for (a) $\ell = 1$, (b) $\ell = 4$, and (c) $\ell = 50$.

7. What are the possible values of n and m if (a) $\ell = 3$, (b) $\ell = 4$, and (c) $\ell = 0$?

8. What are the possible values of n and ℓ if (a) $m = 0$, (b) $m = -1$, and (c) $m = 2$?

4-2 The Hydrogen-Atom Wave Functions

9. For the ground state of the hydrogen atom, find the values of (a) ψ, (b) ψ^2, and (c) the radial probability density $P(r)$ at $r = a_0$. Give your answers in terms of a_0.

10. (a) If spin is not included, how many different wave functions are there corresponding to the first excited energy level $n = 2$ for hydrogen? (b) List these functions by giving the quantum numbers for each state.

11. For the ground state of the hydrogen atom, find the probability of finding the electron in the range $\Delta r = 0.03 a_0$ at (a) $r = a_0$ and (b) $r = 2a_0$.

12. Show that the radial probability density for the $n = 2$, $\ell = 1$, $m = 0$ state of a one-electron atom can be written as

$$P(r) = A \cos^2\theta \, r^4 e^{-Zr/a_0}$$

where A is a constant.

13. The value of the constant C_{200} in Equation 4-15 is

$$C_{200} = \frac{1}{4\sqrt{2\pi}} \left(\frac{Z}{a_0}\right)^{3/2}$$

Find the values of (a) ψ, (b) ψ^2, and (c) the radial probability density $P(r)$ at $r = a_0$ for the state $n = 2$, $\ell = 0$, $m = 0$ in hydrogen. Give your answers in terms of a_0.

14. Find the probability of finding the electron in the range $\Delta r = 0.02 a_0$ at (a) $r = a_0$ and (b) $r = 2a_0$ for the state $n = 2$, $\ell = 0$, $m = 0$ in hydrogen. (See Problem 13 for the value of C_{200}.)

4-3 Magnetic Moments and Electron Spin;
4-4 The Stern–Gerlach Experiment

15. The potential energy of a magnetic moment in an external magnetic field is given by $U = -\boldsymbol{\mu}\cdot\mathbf{B}$. (a) Calculate the difference in energy between the two possible orientations of an electron in a magnetic field $\mathbf{B} = 0.600$ T **k**. (b) If these electrons are bombarded with photons of energy equal to this energy difference, "spin flip" transitions can be induced. Find the wavelength of the photons needed for such transitions. This phenomenon is called **electron-spin resonance**.

16. Calculate the force on an electron in an inhomogeneous magnetic field for which $dB_z/dz = 850$ T/m.

17. How many beams would be expected in a Stern–Gerlach experiment done with atoms for which the spin is zero but the orbital quantum number $\ell = 1$?

18. Consider the magnet orientations shown in Figure 4-12. Could the magnet shown in the middle of the figure be an electron? Explain.

19. A convenient unit for the magnetic moment of nuclei is the nuclear magneton $\mu_N = e\hbar/2m_p$, where m_p is the mass of the proton. Calculate the magnitude of the nuclear magneton in (a) joules per tesla and (b) electron volts per gauss.

4-5 Addition of Angular Momenta and the Spin–Orbit Effect

20. The total angular momentum of a hydrogen atom in a certain excited state has the quantum number $j = \frac{1}{2}$. What can you say about the orbital angular-momentum quantum number ℓ?

21. The total angular momentum of a hydrogen atom in a certain excited state has the quantum number $j = 1\frac{1}{2}$. What can you say about the orbital angular-momentum quantum number ℓ?

22. A hydrogen atom is in the state $n = 3$, $\ell = 2$. (a) What are the possible values of j? (b) What are the possible values of the magnitude of the total angular momentum including spin? (c) What are the possible z components of the total angular momentum?

23. A deuteron is a nucleus with one proton and one neutron, each having spin $\frac{1}{2}$. (a) What are the possible values of the total spin quantum number of the deuteron ($\ell = 0$)? (b) In the ground state, the deuteron has $\ell = 0$ and $s = 1$. What is the magnitude of the angular momentum of the deuteron? (c) Draw a vector diagram illustrating the spins of the proton, neutron, and deuteron, and find the angle between the spins of the neutron and proton.

24. List all the spectroscopic state designations in atomic hydrogen for $n = 2$ and $n = 4$, including the label for total angular momentum.

4-6 The Periodic Table

25. Write the electron configuration of (a) carbon and (b) oxygen.

26. Write the electron configuration of (a) aluminum and (b) chromium.

27. What element has the electron configuration (a) $1s^2 2s^2 2p^6 3s^2 3p^2$ and (b) $1s^2 2s^2 2p^6 3s^2 3p^6 4s^2$?

28. The properties of iron ($Z = 26$) and cobalt ($Z = 27$), which have adjacent atomic numbers, are similar, whereas the properties of neon ($Z = 10$) and sodium ($Z = 11$), which also have adjacent atomic numbers, are very different. Explain why.

29. In Figure 4-17, there are small dips in the ionization-energy curve at $Z = 31$ (gallium) and $Z = 49$ (indium) that are not labeled. Explain these dips using the electron configurations of these atoms given in Table 4-1.

30. Which of the following elements would you expect to have a ground state split by the spin–orbit interaction: Li, B, Na, Al, K, Cu, Ga, Ag? *Hint:* Use Table 4-1 to see which elements have $\ell = 0$ in their ground state and which do not.

31. If the outer electron in lithium moves in the $n = 2$ Bohr orbit, the effective nuclear charge would be $Z'e = 1e$, and the energy of the electron would be $-13.6 \text{ eV}/2^2 = -3.4 \text{ eV}$. However, the ionization energy of lithium is 5.39 eV, not 3.4 eV. Use this fact and Equation 4-29 to calculate the effective nuclear charge Z' seen by the outer electron in lithium. Assume that $r = 4a_0$ for the outer electron.

32. Give the possible values of the z component of the orbital angular momentum of (a) a d electron and (b) an f electron.

33. Separate the following six elements—potassium, calcium, titanium, chromium, manganese, and copper—into two groups of three each such that those in a group have similar properties.

4-7 Optical and X-Ray Spectra

34. The optical spectra of atoms with two electrons in the same outer shell are similar, but they are quite different from the spectra of atoms with just one outer electron because of the interaction of the two electrons. Separate the following elements into two groups such that those in each group have similar spectra: lithium, beryllium, sodium, magnesium, potassium, calcium, chromium, nickel, cesium, and barium.

35. Give the possible electron configurations for the first excited state of (a) hydrogen, (b) sodium, and (c) helium.

36. Which of the following elements should have optical spectra similar to that of hydrogen and which should have optical spectra similar to that of helium: Li, Ca, Ti, Rb, Ag, Cd, Ba, Hg, Fr, and Ra?

37. (a) Calculate the next two longest wavelengths after the K_α line in the K series of molybdenum. (b) What is the shortest wavelength in this series?

38. The wavelength of the K_α line for a certain element is 0.3368 nm. What is the element?

39. The wavelength of the K_α line for a certain element is 0.0794 nm. What is the element?

40. Calculate the wavelength of the K_α line of rhodium.

41. Calculate the wavelength of the K_α line of (a) magnesium ($Z = 12$) and (b) copper ($Z = 29$).

42. (a) Calculate the energy of the electron in the K shell for tungsten using $Z - 1$ for the effective nuclear charge. (b) The experimental result for this energy is 69.5 keV. Assume that the effective nuclear charge is $(Z - \sigma)$, where σ is the screening constant, and calculate σ from the experimental result for the energy.

4-8 Absorption, Scattering, and Stimulated Emission

There are no problems for this section.

4-9 The Laser

43. A pulse from a ruby laser has an average power of 10 MW and lasts 1.5 ns. (a) What is the total energy of the pulse? (b) How many photons are emitted in this pulse?

44. A helium–neon laser emits light of wavelength 632.8 nm and has a power output of 4 mW. How many photons are emitted per second by this laser?

45. A laser beam is aimed at the moon a distance 3.84×10^8 m away. The angular spread of the beam is given by the diffraction formula, $\sin \theta = 1.22\lambda/D$, where D is the diameter of the laser tube. Calculate the size of the beam on the moon for $D = 10$ cm and $\lambda = 600$ nm.

Level II

46. The angular momentum of the yttrium atom in the ground state is characterized by the quantum number $j = 1\frac{1}{2}$. How many lines would you expect to see if you could perform the Stern–Gerlach experiment with yttrium atoms?

47. Find the final state (or the final kinetic energy) of the hydrogen-atom electron if a photon of (a) 12.09 eV and (b) 20 eV is absorbed by the hydrogen atom when it is in the ground state.

48. The total energy of the electron of momentum p and mass m a distance r from the proton in the hydrogen atom is given by

$$E = \frac{p^2}{2m} - \frac{ke^2}{r}$$

where k is the Coulomb constant. If we assume that the minimum value of p^2 is $p^2 \approx (\Delta p)^2 = \hbar^2/r^2$, where Δp is the uncertainty in p and we have taken $\Delta r \sim r$ for the order of magnitude of the uncertainty in position, the energy is

$$E = \frac{\hbar^2}{2mr^2} - \frac{ke^2}{r}$$

Find the radius r_m for which this energy is a minimum, and calculate the minimum value of E in electron volts.

49. The L shell in an atom has $n = 2$ and $\ell = 1$. For atoms with eight or more electrons, there are two electrons in the K shell and six in the L shell. An electron in the L shell is thus shielded from the nuclear charge by the two electrons in the K shell and is partially shielded by the five other electrons in the L shell. There may also be some shielding due to the penetration of the wave functions of the electrons in outer shells. The frequencies for x rays in the L series involve transitions from some state $n_1 > 2$ to the state $n_2 = 2$. Equation 4-30 gives a good approximation for the initial and final electron energies of L-series x

rays if $(Z-1)$ in that equation is replaced by $(Z-7.4)$. Use this result to calculate the shortest wavelength in the L series for (a) molybdenum ($Z = 42$) and (b) zinc ($Z = 30$).

50. For $\ell = 2$, (a) what is the minimum value of $L_x^2 + L_y^2$? (b) What is the maximum value of $L_x^2 + L_y^2$? (c) What is the value of $L_x^2 + L_y^2$ for $\ell = 2$ and $m = 1$?

51. The L_α x-ray line arising from the transition from $n_1 = 3$ to $n_2 = 2$ for a certain element has a wavelength of 0.3617 nm. What is the element? (See Problem 49.)

52. A particle of mass m moves in a circle of radius r with speed v in the xy plane around the z axis. Consider the time when the particle is on the y axis. Multiply the uncertainty product $\Delta x\, \Delta p$ by r/r to obtain an uncertainty relation between the angle ϕ and the z component of the angular momentum.

53. (a) Calculate the energies in electron volts of the photons corresponding to the two yellow lines in the optical spectrum of sodium whose wavelengths are 589.0 nm and 589.6 nm. (b) The difference in energies of these photons equals the difference in energy ΔE between the $3P_{3/2}$ and $3P_{1/2}$ states in sodium. Calculate ΔE. (c) Estimate the magnetic field that the $3p$ electron in sodium experiences.

54. Find the minimum angle between the angular momentum $\mathbf{L}$ and the z axis for a general value of ℓ, and show that for large values of ℓ, $\theta_{\min} \approx 1/\sqrt{\ell}$.

55. The wavelengths of the photons emitted by potassium corresponding to transitions from the $4P_{3/2}$ and $4P_{1/2}$ states to the ground state are 766.41 nm and 769.90 nm. (a) Calculate the energies of these photons in electron volts. (b) The difference in the energies of these photons equals the difference in energy ΔE between the $4P_{3/2}$ and $4P_{1/2}$ states in potassium. Calculate ΔE. (c) Estimate the magnetic field that the $4p$ electron in potassium experiences.

56. If a classical system does not have a constant charge-to-mass ratio throughout the system, the magnetic moment can be written

$$\mu = g \frac{Q}{2M} L$$

where Q is the total charge, M is the total mass, and $g \neq 1$. (a) Show that $g = 2$ for a solid cylinder that spins about its axis and has a uniform charge on its cylindrical surface. (b) Show that $g = 2\frac{1}{2}$ for a solid sphere that has a ring of charge on its surface at the equator.

57. In a Stern–Gerlach experiment hydrogen atoms in their ground state move with speed $v_x = 14.5$ km/s. The magnetic field is in the z direction and its maximum gradient is given by $dB_z/dz = 600$ T/m. (a) Find the maximum acceleration of the hydrogen atoms. (b) If the region of the magnetic field extends over a distance $\Delta x = 75$ cm, and there is an additional 1.25 m from the edge of the field to the detector, find the maximum distance between the two lines on the detector.

58. The radial probability distribution function for a one-electron atom in its ground state can be written $P(r) = Cr^2 e^{-2Zr/a_0}$, where C is a constant. Show that $P(r)$ has its maximum value at $r = a_0/Z$.

59. Consider a system of two electrons, each with $\ell = 1$ and $s = \frac{1}{2}$. (a) Neglecting spin, find the possible values of the quantum number for the total orbital angular momentum $\mathbf{L} = \mathbf{L}_1 + \mathbf{L}_2$. (b) What are the possible values of the quantum number S for the total spin $\mathbf{S} = \mathbf{S}_1 + \mathbf{S}_2$? (c) Using the results of parts (a) and (b), find the possible quantum numbers j for the combination $\mathbf{J} = \mathbf{L} + \mathbf{S}$. (d) What are the possible quantum numbers j_1 and j_2 for the total angular momentum of each particle? (e) Use the results of part (d) to calculate the possible values of j from the combination of j_1 and j_2. Are these the same as in part (c)?

60. (a) Show that the radial probability distribution for the $n = 2$, $\ell = 1$ energy levels of a one-electron atom can be written $P(r) = Ar^4 e^{-Zr/a_0}$, where A depends on θ but not on r. (b) Show that $P(r)$ is maximum at $r = 4a_0/Z$.

61. The radius of a proton is about $R_0 = 10^{-15}$ m. The probability that the hydrogen-atom electron is inside the proton is

$$P = \int_0^{R_0} P(r)\, dr$$

where $P(r)$ is the radial probability density. Calculate this probability for the ground-state of hydrogen. *Hint:* Show that $e^{-2r/a_0} \approx 1$ for $r \leq R_0$ is valid for this calculation.

62. Show that the expectation value of r for the electron in the ground state of a one-electron atom is $\langle r \rangle = \frac{3}{2} a_0/Z$.

63. Show that the expectation value of the potential energy $U(r) = -kZe^2/r$ for the ground state of a one-electron atom is given by $\langle U(r) \rangle = -2Z^2 E_0$.

64. Show by direct substitution that the ground-state wave function for hydrogen given by Equation 4-12 satisfies the Schrödinger equation in spherical coordinates (Equation 4-4).

Level III

65. Show by direct substitution that the wave function ψ_{210} given by Equation 4-16 satisfies the Schrödinger equation in spherical coordinates (Equation 4-4).

66. Show that the number of states in the hydrogen atom for a given n is $2n^2$.

67. Find the two values of r for which $P(r)$ is a maximum for the $2s$ wave function of hydrogen.

68. Derive the expression for C_{200} given in Problem 13.

69. Calculate the probability that the electron in the ground state of a hydrogen atom is in the region $0 < r < a_0$.

Chapter 5

Molecules

Single atoms rarely occur in nature. Instead, atoms usually bond together to form molecules or solids. Molecules may exist as separate entities, as in gaseous oxygen (O_2) or nitrogen (N_2), or they may bond together to form liquids or solids. A molecule is the smallest constituent of a substance that retains the chemical properties of the substance. The study of the properties of molecules thus forms the basis for theoretical chemistry.

The application of quantum mechanics to molecular physics has been spectacularly successful in explaining the structure of molecules and the complexity of their spectra. It has also provided answers to such puzzling questions as why two hydrogen atoms join together to form a molecule but three hydrogen atoms do not. As in atomic physics, the quantum-mechanical calculations in molecular physics are very difficult, so much of our discussion will be qualitative.

There are two extreme views that we can take of a molecule. Consider, for example, gaseous hydrogen (H_2). We can think of it either as two hydrogen atoms somehow joined together or as a quantum-mechanical system consisting of two protons and two electrons. The latter view is more fruitful in this case because neither of the electrons in the H_2 molecule can be identified as belonging to either proton. Instead, the wave function for each elec-

Hemoglobin is an iron-containing protein found in red-blood cells that transports oxygen throughout the body. Shown here is a model of part of a molecule synthesized in the laboratory and designed to function, potentially, as artificial human hemoglobin. The notch in the center is the binding site for oxygen. The two hemispheres are compounds known as cyclodextrins made up of carbon, hydrogen, and oxygen. They encage a molecule formed of iron and a porphyrin, a nitrogen-containing organic compound found universally in protoplasm. The artificial hemoglobin has the interesting property that its subunits will spontaneously self-assemble themselves. Each individual subunit is held together by covalent chemical bonds, whereas the subunits themselves bind to one another by electrostatic and "water-avoiding" interactions. Though complicated molecules routinely assemble themselves at high speed in living organisms, chemists are only now beginning to duplicate this process in the test tube.

tron is spread out in space throughout the whole molecule. For more complicated molecules, however, an intermediate view is useful. For example, the nitrogen molecule N_2 has 14 protons and 14 electrons, but only two of the electrons take part in the bonding. The electron configuration of a nitrogen atom in the ground state is $1s^2 2s^2 2p^3$. Of the three electrons in the $2p$ state, two have their spins paired; that is, their spins are antiparallel, so their resultant spin is zero. The single electron with the unpaired spin is the most free to take part in the bonding of one nitrogen atom to another. We can therefore consider the N_2 molecule as consisting of two N^+ ions and two electrons that belong to the molecule as a whole. The molecular wave functions for these bonding electrons are called **molecular orbitals.** In many cases, molecular wave functions can be constructed by combining the atomic wave functions with which we are familiar.

5-1 Molecular Bonding

The two principal types of bonds responsible for the formation of molecules are the ionic bond and the covalent bond. Other types of bonds that are important in the bonding of liquids and solids are van der Waals bonds, metallic bonds, and hydrogen bonds. In many cases, bonding is a mixture of these mechanisms.

The Ionic Bond

The simplest type of bond is the **ionic bond,** which is found in most salts. Consider sodium chloride (NaCl) as an example. The sodium atom has one $3s$ electron outside a stable core. It takes just 5.14 eV to remove this electron from sodium (see Figure 4-17). The ionization energies for the other alkali metals are also low. The removal of one electron from sodium leaves a positive ion with a spherically symmetric, closed-shell electron core. Chlorine, on the other hand, is only one electron short of having a closed shell. The energy released by an atom's acquisition of one electron is called its **electron affinity,** which in the case of chlorine is 3.61 eV. The acquisition of one electron by chlorine results in a negative ion with a spherically symmetric, closed-shell electron core. Thus, the formation of a Na^+ ion and a Cl^- ion by the donation of one electron of sodium to chlorine requires only 5.14 eV − 3.61 eV = 1.53 eV at infinite separation. The electrostatic potential energy of the two ions when they are a distance r apart is $-ke^2/r$. When the separation of the ions is less than about 0.94 nm, the negative potential energy of attraction is of greater magnitude than the 1.53 eV of energy needed to create the ions. Thus, at separation distances less than 0.94 nm it is energetically favorable (that is, the total energy of the system is reduced) for the sodium atom to donate an electron to the chlorine atom to form NaCl.

Since the electrostatic attraction increases as the ions get closer together, it might seem that equilibrium could not exist. However, when the separation of the ions is very small, there is a strong repulsion that is quantum mechanical in nature and is related to the exclusion principle. This **exclusion-principle repulsion** is responsible for the repulsion of the atoms in all molecules (except H_2) no matter what the bonding mechanism is. (In H_2, the repulsion is simply that of the two positively charged protons.)

We can understand the exclusion-principle repulsion qualitatively as follows. When the ions are very far apart, the wave function for a core electron in one of the ions does not overlap that of any electron in the other ion. We can distinguish the electrons by the ion to which they belong. This means that electrons in the two ions can have the same quantum numbers because they occupy different regions of space. (Recall from our discussion in Section 3-11 that the exclusion principle is related to the fact that the wave function for two identical electrons is antisymmetric on the exchange of the electrons and that an antisymmetric wave function for two electrons with the same quantum numbers is zero if the space coordinates of the electrons are the same.) However, as the distance between the ions decreases, the wave functions of the core electrons begin to overlap; that is, the electrons in the two ions begin to occupy the same region of space. Because of the exclusion principle, some of these electrons must go into higher-energy quantum states. But energy is required to shift the electrons into higher-energy quantum states. This increase in energy when the ions are pushed closer together is equivalent to a repulsion of the ions. This is not a sudden process. The energy states of the electrons change gradually as the ions are brought together. A sketch of the potential energy of the Na^+ and Cl^- ions versus separation is shown in Figure 5-1. The energy is lowest at an equilibrium separation of about 0.236 nm. At smaller separations, the energy rises steeply as a result of the exclusion principle. The energy required to separate the ions and form neutral sodium and chlorine atoms is called the **dissociation energy**, which is about 4.26 eV for NaCl.

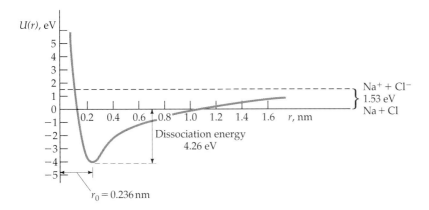

Figure 5-1 Potential energy for Na^+ and Cl^- ions as a function of separation distance r. The energy at infinite separation was chosen to be 1.53 eV, corresponding to the energy needed to form the ions from neutral atoms. The minimum energy for this curve is at the equilibrium separation $r_0 = 0.236$ nm for the ions in the molecule.

The equilibrium separation distance of 0.236 nm is for gaseous, diatomic NaCl, which can be obtained by evaporating solid NaCl. Normally, NaCl exists in a cubic crystal structure, with the Na^+ and Cl^- ions at the alternate corners of a cube. The separation of the ions in a crystal is somewhat larger, about 0.28 nm. Because of the presence of neighboring ions of opposite charge, the Coulomb energy per ion pair is lower when the ions are in a crystal.

Example 5-1

The electron affinity of fluorine is 3.45 eV, and the equilibrium separation of sodium fluoride (NaF) is 0.193 nm. (*a*) How much energy is needed to form Na^+ and F^- ions from neutral sodium and fluorine atoms? (*b*) What is the electrostatic potential energy of the Na^+ and F^- ions at their equilibrium separation? (*c*) The dissociation energy of NaF is 4.99 eV. What is the energy due to repulsion of the ions at the equilibrium separation?

(*a*) Since the energy needed to ionize sodium is 5.14 eV, the energy needed to form Na^+ and F^- ions from the neutral sodium and fluorine atoms is $5.14 \text{ eV} - 3.45 \text{ eV} = 1.69 \text{ eV}$.

(*b*) The electrostatic potential energy of Na^+ and F^- ions at their equilibrium separation (with $U = 0$ at infinite separation) is

$$U = -\frac{ke^2}{r} = -\frac{(8.99 \times 10^9 \text{ N·m}^2/\text{C}^2)(1.60 \times 10^{-19} \text{ C})^2}{1.93 \times 10^{-10} \text{ m}}$$

$$= -1.19 \times 10^{-18} \text{ J}$$

Converting this to electron volts, we obtain

$$U = -1.19 \times 10^{-18} \text{ J} \frac{1 \text{ eV}}{1.60 \times 10^{-19} \text{ J}} = -7.45 \text{ eV}$$

(c) If we choose the potential energy at infinity to be 1.69 eV (the energy needed to form the Na$^+$ and F$^-$ ions from neutral sodium and fluorine atoms), the electrostatic potential energy is

$$U = -\frac{ke^2}{r} + 1.69 \text{ eV}$$

At the equilibrium separation, this energy is $U = -7.45$ eV $+ 1.69$ eV $= -5.76$ eV. Since the dissociation energy is 4.99 eV, the energy due to repulsion of the Na$^+$ and F$^-$ ions at the equilibrium separation must be 5.76 eV $-$ 4.99 eV $= 0.77$ eV.

The Covalent Bond

A completely different mechanism, the **covalent bond,** is responsible for the bonding of identical or similar atoms to form such molecules as gaseous hydrogen (H$_2$), nitrogen (N$_2$), and carbon monoxide (CO). If we calculate the energy needed to form H$^+$ and H$^-$ ions by the transfer of an electron from one atom to the other and then add this energy to the electrostatic potential energy, we find that there is no separation distance for which the total energy is negative. The bond thus cannot be ionic. Instead, the attraction of two hydrogen atoms is an entirely quantum-mechanical effect. The decrease in energy when two hydrogen atoms approach each other is due to the sharing of the two electrons by both atoms. It is intimately connected with the symmetry properties of the wave functions of the electrons.

We can gain some insight into covalent bonding by considering a simple, one-dimensional, quantum-mechanics problem of two finite square wells. We first consider a single electron that is equally likely to be in either well. Since the wells are identical, symmetry requires that ψ^2 be symmetric about the midpoint between the wells. Then ψ must be either symmetric or antisymmetric. The two possibilities for the ground state are shown in Figure 5-2a for the case in which the wells are far apart and in Figure 5-2b for the case in which the wells are close together. An important feature of Figure 5-2b is that in the region between the wells the symmetric wave function is large and the antisymmetric wave function is small.

Now consider adding a second electron to the two wells. The total wave function for the two electrons must be antisymmetric on exchange of the electrons. (Note that exchanging the electrons in the wells is the same as exchanging the wells.) The total wave function for two electrons can be written as a product of a space part and a spin part.

To understand the symmetry of the total wave function, we must first digress and consider the symmetry of the spin part of the wave function. When the two electrons have parallel spins ($S = 1$), the spin state $\phi_{1,+1}$, corresponding to $m_S = +1$, can be written as

$$\phi_{1,+1} = \uparrow_1 \uparrow_2 \quad S = 1, m_S = +1 \quad \quad 5\text{-}1$$

where $\uparrow_1$ means that electron 1 is "up" ($m_1 = +\frac{1}{2}$) and $\uparrow_2$ means the same for electron 2. Similarly, the spin state for $m_S = -1$ can be written as

$$\phi_{1,-1} = \downarrow_1 \downarrow_2 \quad S = 1, m_S = -1 \quad \quad 5\text{-}2$$

where $\downarrow_1$ means that electron 1 is "down" ($m_1 = -\frac{1}{2}$). Note that both of these states are symmetric upon exchange of the electrons. The spin state

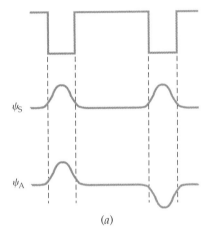

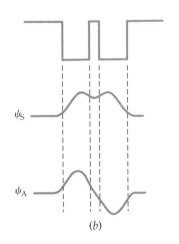

Figure 5-2 (a) Two square wells that are far apart. The electron wave function can be either symmetric (ψ_S) or antisymmetric (ψ_A) in space. The probability distributions and energies are the same for the two wave functions.
(b) Two square wells that are close together. The symmetric wave function is larger between the wells than the antisymmetric wave function.

corresponding to $S = 1$ and $m_S = 0$ is not quite so obvious. It turns out to be proportional to

$$\phi_{1,0} = \uparrow_1 \downarrow_2 + \uparrow_2 \downarrow_1 \qquad S = 1, m_S = 0 \qquad 5\text{-}3$$

This spin state is also symmetric upon exchange of the electrons. The spin state for two electrons with antiparallel spins ($S = 0$) is

$$\phi_{0,0} = \uparrow_1 \downarrow_2 - \uparrow_2 \downarrow_1 \qquad S = 0, m_S = 0 \qquad 5\text{-}4$$

This spin state is antisymmetric upon exchange of electrons.

We thus have the important result that the spin part of the wave function is symmetric for parallel spins and antisymmetric for antiparallel spins. To obtain a total wave function that is antisymmetric, we form the product of a symmetric space state and an antisymmetric spin state, or of an antisymmetric space state and a symmetric spin state. This gives us the following result:

> For the total wave function of two electrons to be antisymmetric, the space part of the wave function must be antisymmetric for parallel spins ($S = 1$) and symmetric for antiparallel spins ($S = 0$).

We can now consider the problem of two hydrogen atoms. Figure 5-3a shows a symmetric wave function ψ_S and an antisymmetric wave function ψ_A for two hydrogen atoms that are far apart, and Figure 5-3b shows the same two wave functions for two hydrogen atoms that are close together. The squares of these two wave functions are shown in Figure 5-3c. Note that the probability distribution $|\psi|^2$ in the region between the protons is large for the symmetric wave function and small for the antisymmetric wave function. Thus, when the spins of the two electrons are antiparallel, which means that the space part of the wave function is symmetric, the electrons are often found in the region between the protons, and the protons are bound together by the negatively charged electrons between them. The negatively charged electron cloud representing these electrons is concentrated in the space between the protons, as shown in the upper part of Figure 5-3c. Conversely, when the electron spins are parallel, which means that the space part of the wave function is antisymmetric, the electrons spend little time between the protons, and the atoms do not bind together to form a molecule. In this case, the electron cloud is not concentrated in the space between the protons, as shown in the lower part of Figure 5-3c.

The total electrostatic potential energy for the H$_2$ molecule consists of the positive energy of repulsion of the two electrons and the negative poten-

Figure 5-3 One-dimensional symmetric and antisymmetric wave functions for two hydrogen atoms (a) far apart and (b) close together. (c) Electron probability distributions $|\psi^2|$ for the wave functions in (b). For the symmetric wave function on the left, the electron charge density is large between the protons. This negative charge density holds the protons together in the hydrogen molecule H$_2$. For the antisymmetric wave functions on the right, the electron charge density is not large between the protons, and the atoms do not bond together to form a molecule.

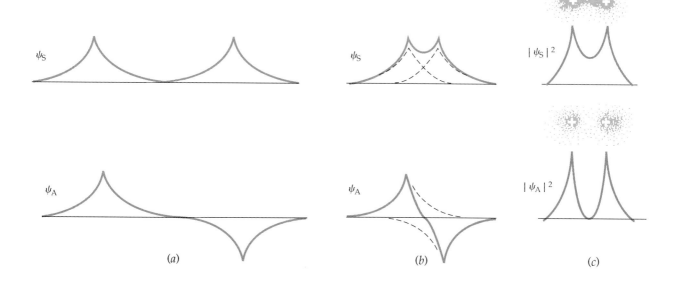

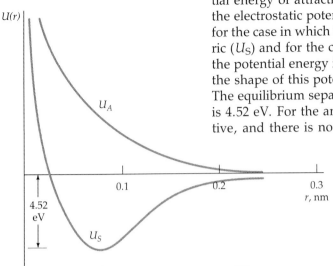

tial energy of attraction of each electron for each proton. Figure 5-4 shows the electrostatic potential energy for two hydrogen atoms versus separation for the case in which the space part of the electron wave function is symmetric (U_S) and for the case in which it is antisymmetric (U_A). We can see that the potential energy for the symmetric state is the lower of the two and that the shape of this potential-energy curve is similar to that for ionic bonding. The equilibrium separation for H_2 is $r_0 = 0.074$ nm, and the binding energy is 4.52 eV. For the antisymmetric state, the potential energy is never negative, and there is no bonding.

Figure 5-4 Potential energy versus separation for two hydrogen atoms. The curve labeled U_S is for a wave function with a symmetric space part, and the curve labeled U_A is for a wave function with an antisymmetric space part.

We can now see why three hydrogen atoms do not bond to form H_3. If a third hydrogen atom is brought near an H_2 molecule, the third electron cannot be in a 1s state and have its spin antiparallel to the spin of both of the other electrons. If this electron is in an antisymmetric space state with respect to exchange with one of the electrons, the repulsion of this atom is greater than the attraction of the other. As the three atoms are pushed together, the third electron is, in effect, forced into a higher quantum-energy state by the exclusion principle. The bond between two hydrogen atoms is called a **saturated bond** because there is no room for another electron. The two shared electrons essentially fill the 1s states of both atoms.

We can also see why two helium atoms do not normally bond together to form the He_2 molecule. There are no valence electrons that can be shared. The electrons in the closed shells are forced into higher energy states when the two atoms are brought together. At low temperatures or high pressures, helium atoms do bond together due to van der Waals forces, which we will discuss next. This bonding is so weak that at atmospheric pressure helium boils at 4 K, and it does not form a solid at any temperature unless the pressure is greater than about 20 atm.

When two identical atoms bond, as in O_2 or N_2, the bonding is purely covalent. However, the bonding of two dissimilar atoms is often a mixture of covalent and ionic bonding. Even in NaCl, the electron donated by sodium to chlorine has some probability of being at the sodium atom because its wave function does not suddenly fall to zero. Thus, this electron is partially shared in a covalent bond, although this bonding is only a small part of the total bond, which is mainly ionic.

A measure of the degree to which a bond is ionic or covalent can be obtained from the electric dipole moment of the molecule. For example, if the bonding in NaCl were purely ionic, the center of positive charge would be at the Na^+ ion and the center of negative charge would be at the Cl^- ion. The electric dipole moment would have the magnitude

$$p_{\text{ionic}} = er_0 \qquad 5\text{-}5$$

where r_0 is the equilibrium separation of the ions. Thus, the dipole moment of NaCl would be

$$p_{\text{ionic}} = er_0$$
$$= (1.60 \times 10^{-19} \text{ C})(2.36 \times 10^{-10} \text{ m}) = 3.78 \times 10^{-29} \text{ C·m}$$

The actual measured electric dipole moment of NaCl is

$$p_{\text{measured}} = 3.00 \times 10^{-29} \text{ C·m}$$

We can define the ratio of p_measured to p_ionic as the fractional amount of ionic bonding. For NaCl, this ratio is $3.00/3.78 = 0.79$. Thus, the bonding in NaCl is about 79 percent ionic.

Exercise

The equilibrium separation of HCl is 0.128 nm and its measured electric dipole moment is 3.60×10^{-30} C·m. What is the percentage of ionic bonding in HCl? (Answer: 18 percent)

The van der Waals Bond

Any two separated molecules will be attracted to one another by electrostatic forces called van der Waals forces. So will any two atoms that do not form ionic or covalent bonds. The **van der Waals bonds** formed because of these forces are much weaker than those already discussed. At high enough temperatures, these forces are not strong enough to overcome the ordinary thermal agitation of the atoms or molecules, but at sufficiently low temperatures, thermal agitation becomes negligible, and the van der Waals forces will cause virtually all substances to condense into a liquid and then a solid form. (Helium is the only element that does not solidify at any temperature at atmospheric pressure.)

The van der Waals forces arise from the interaction of the instantaneous electric dipole moments of the molecules. Figure 5-5 shows how two polar molecules—molecules with permanent electric dipole moments, such as H_2O—can bond. The electric field due to the dipole moment of one molecule orients the other molecule such that the two dipole moments attract.

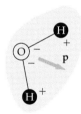
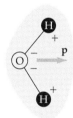

Figure 5-5 The bonding of H_2O molecules due to the attraction of the electric dipoles. The dipole moment of each molecule is indicated by **p**.

Nonpolar molecules also attract other nonpolar molecules via the van der Waals forces. Although nonpolar molecules have zero electric dipole moments on the average, they have instantaneous dipole moments that are generally not zero because of fluctuations in the positions of the charges. When two nonpolar molecules are near each other, the fluctuations in the instantaneous dipole moments tend to become correlated so as to produce attraction. This is illustrated in Figure 5-6.

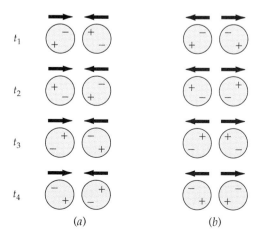

Figure 5-6 The van der Waals attraction of molecules with zero average dipole moments. (a) Possible orientations of the instantaneous dipole moments at different times leading to attraction. (b) Possible orientations leading to repulsion. The electric field of the instantaneous dipole moment of one molecule tends to polarize the other molecule. Thus, the orientations in (a) leading to attraction are much more likely than those in (b) leading to repulsion.

Figure 5-7 The DNA molecule.

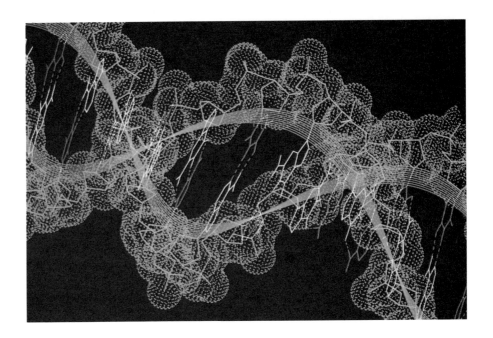

The Hydrogen Bond

Another bonding mechanism of great importance is the hydrogen bond. It often holds groups of molecules together and is responsible for the cross-linking that allows giant biological molecules and polymers to hold their fixed shapes. The well-known helical structure of DNA is due to hydrogen-bond linkages across turns of the helix (Figure 5-7). The hydrogen bond is formed by the sharing of a proton (the nucleus of the hydrogen atom) between two atoms, frequently two oxygen atoms. This sharing of a proton is similar to the sharing of electrons responsible for the covalent bond already discussed. It is facilitated by the small mass of the proton and by the absence of inner-core electrons in hydrogen.

The Metallic Bond

The bonding of atoms in a metal is different from the bonding of atoms in a molecule. In a metal, two atoms do not bond together by exchanging or sharing an electron to form a molecule. Instead, each valence electron is shared by many atoms. The bonding is thus distributed throughout the entire metal. A metal can be thought of as a lattice of positive ions held together by a "gas" of essentially free electrons that roam throughout the solid. In the quantum-mechanical picture, these free electrons form a cloud of negative charge density between the positively charged lattice ions that holds the ions together. In this respect, the metallic bond is somewhat similar to the covalent bond. However, with the metallic bond, there are far more than just two atoms involved, and the negative charge is distributed uniformly throughout the volume of the metal. The number of free electrons varies from metal to metal but is of the order of one per atom.

Questions

1. Why would you expect the separation distance between the two protons to be smaller in the H_2^+ ion than in the H_2 molecule?
2. Would you expect the NaCl molecule to be polar or nonpolar?
3. Would you expect the N_2 molecule to be polar or nonpolar?
4. Does neon occur naturally as Ne or Ne_2? Why?

5-2 Polyatomic Molecules

Molecules with more than two atoms range from such relatively simple molecules as water, which has a molecular mass of 18, to such giants as proteins, which can have molecular masses of hundreds of thousands or even a million. As with diatomic molecules, the structure of polyatomic molecules can be understood by applying basic quantum mechanics to the bonding of individual atoms. The bonding mechanisms for most polyatomic molecules are the covalent bond and the hydrogen bond. We will discuss only some of the simplest polyatomic molecules—H_2O, NH_3, and CH_4—to illustrate both the simplicity and complexity of the application of quantum mechanics to molecular bonding.

The basic requirement for the sharing of electrons in a covalent bond is that the wave functions of the valence electrons in the individual atoms must overlap as much as possible. As our first example, we will consider the water molecule. The ground-state configuration of the oxygen atom is $1s^2 2s^2 2p^4$. The $1s$ and $2s$ electrons are in closed-shell states and do not contribute to the bonding. The $2p$ shell has room for six electrons, two in each of the three space states corresponding to $\ell = 1$. In an isolated atom, we describe these space states by the hydrogen-like wave functions corresponding to $\ell = 1$ and $m = +1, 0,$ and -1. Since the energy is the same for these three space states, we could equally well use any linear combination of these wave functions. When an atom participates in molecular bonding, certain combinations of these atomic wave functions are important. These combinations are called the p_x, p_y, and p_z **atomic orbitals.** The angular dependence of these orbitals is

$$p_x \propto \sin\theta \cos\phi \qquad 5\text{-}6$$

$$p_y \propto \sin\theta \sin\phi \qquad 5\text{-}7$$

$$p_z \propto \cos\theta \qquad 5\text{-}8$$

The electron charge distribution is maximum along the x, y, or z axis for these orbitals as shown in Figure 5-8. When oxygen is in an H_2O molecule, maximum overlap of the electron wave functions occurs when two of the four $2p$ electrons are paired with their spins antiparallel in one of the orbitals, for example, p_z, one of the other electrons is in the p_x orbital, and the other electron is in the p_y orbital. Each of the unpaired electrons in the p_x and p_y orbitals forms a bond with the electron of a hydrogen atom as shown in Figure 5-9. Because of the repulsion of the two hydrogen atoms, the angle between the O–H bonds is actually greater than 90°. The effect of this repulsion can be calculated, and the result is in agreement with the measured angle of 104.5°.

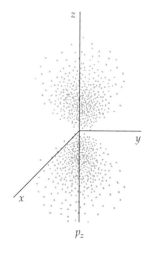

Figure 5-8 Computer-generated dot plot illustrating the spatial dependence of the electron charge distribution in the p_x, p_y, and p_z atomic orbitals.

Figure 5-9 Electron charge distribution in the H₂O molecule.

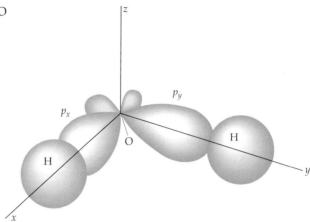

Similar reasoning leads to an understanding of the bonding in NH$_3$. In the ground state, nitrogen has three electrons in the $2p$ state. When these three electrons are in the p_x, p_y, and p_z atomic orbitals, they bond to the electrons of hydrogen atoms. Again, because of the repulsion of the hydrogen atoms, the angles between the bonds are somewhat larger than 90°.

The bonding of carbon atoms is somewhat more complicated. Carbon forms a wide variety of different types of molecular bonds, leading to a great diversity in the kinds of organic molecules that contain carbon. The ground-state configuration of carbon is $1s^2 2s^2 2p^2$. From our previous discussion, we might expect carbon to be divalent, with the two $2p$ electrons forming bonds at approximately 90°. However, one of the most important features of the chemistry of carbon is that tetravalent carbon compounds, such as CH$_4$, are overwhelmingly favored.

The observed valence of 4 for carbon comes about in an interesting way. One of the first excited states of carbon occurs when a $2s$ electron is excited to a $2p$ state, giving a configuration of $1s^2 2s^1 2p^3$. In this excited state, we can have four unpaired electrons, one each in the $2s$, $2p_x$, $2p_y$, and $2p_z$ atomic orbitals. We might expect there to be three similar bonds corresponding to the three p orbitals and one different bond corresponding to the s orbital. However, when carbon forms tetravalent bonds, these four atomic orbitals become mixed and form four new *equivalent* molecular orbitals called **hybrid orbitals**. This mixing of atomic orbitals, called **hybridization**, is probably the most important new feature involved in the physics of complex molecular bonds. Figure 5-10 shows the tetrahedral structure of the methane molecule (CH$_4$), and Figure 5-11 shows the structure of the ethane molecule (CH$_3$–CH$_3$), which is similar to two joined methane molecules in which one of the C–H bonds is replaced with a C–C bond.

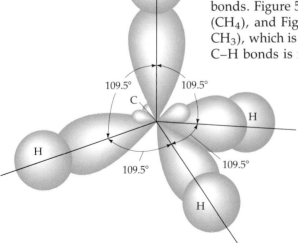

Figure 5-10 Electron charge distribution in the CH$_4$ molecule (methane).

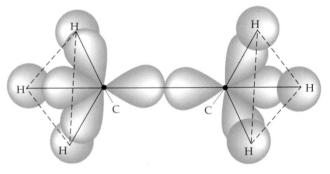

Figure 5-11 Electron charge distribution in the CH$_3$–CH$_3$ molecule (ethane).

Carbon orbitals can also hybridize with the s, p_x, and p_y orbitals combining to form three hybrid orbitals in the xy plane with 120° bonds and the p_z orbital remaining unmixed. An example of this configuration is graphite, in which the bonds in the xy plane provide the strongly layered structure characteristic of the material. Structures of some common molecules are shown in Table 5-1 on pages 178–179.

5-3 Energy Levels and Spectra of Diatomic Molecules

As is the case with an atom, a molecule often emits electromagnetic radiation when it makes a transition from an excited energy state to a state of lower energy. Conversely, a molecule can absorb radiation and make a transition from a lower energy state to a higher energy state. The study of molecular emission and absorption spectra thus provides us with information about the energy states of molecules. For simplicity, we will consider only diatomic molecules here.

The energy of a molecule can be conveniently separated into three parts: electronic due to the excitation of the electrons of the molecule, vibrational due to the oscillations of the atoms of the molecule, and rotational due to the rotation of the molecule about its center of mass. The magnitudes of these energies are sufficiently different that they can be treated separately. The energies due to the electronic excitations of a molecule are of the order of magnitude of 1 eV, the same as for the excitation of an atom. The energies of vibration and rotation are much smaller than this.

Rotational Energy Levels

Figure 5-12 shows a simple schematic model of a diatomic molecule consisting of a mass m_1 and a mass m_2 separated by a distance r and rotating about its center of mass. Classically, the kinetic energy of rotation is

$$E = \tfrac{1}{2} I \omega^2 \qquad 5\text{-}9$$

where I is the moment of inertia and ω is the angular frequency of rotation. If we write this in terms of the angular momentum $L = I\omega$, we have

$$E = \frac{(I\omega)^2}{2I} = \frac{L^2}{2I} \qquad 5\text{-}10$$

The solution of the Schrödinger equation for rotation leads to quantization of the angular momentum with values given by

$$L^2 = \ell(\ell + 1)\hbar^2 \qquad \ell = 0, 1, 2, \ldots \qquad 5\text{-}11$$

where ℓ is the **rotational quantum number.** This is the same quantum condition on angular momentum that holds for the orbital angular momentum of an electron in an atom. Note, however, that L in Equation 5-10 refers to the angular momentum of the entire molecule rotating about its center of mass. The energy levels of a rotating molecule are therefore given by

$$E = \frac{\ell(\ell + 1)\hbar^2}{2I} = \ell(\ell + 1) E_{0r} \qquad \ell = 0, 1, 2, \ldots \qquad 5\text{-}12$$

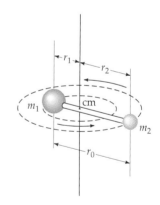

Figure 5-12 Diatomic molecule rotating about an axis through its center of mass.

Rotational energy levels

Table 5-1 Some Common Molecules

ESTRADIOL $C_{18}H_{24}O_2$

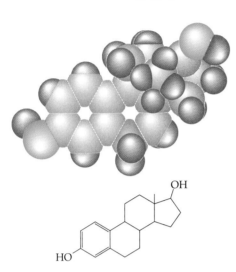

Estradiol is one of the principal female sex hormones. It is released at puberty and decreases in abundance at menopause.

CHOLESTEROL $C_{27}H_{46}O$

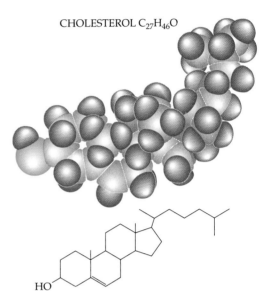

Cholesterol is produced in the liver and plays an essential role in metabolism in the body.

2-FURYLMETHANETHIOL C_5H_6OS

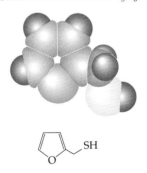

2-Furylmethanethiol is one of the molecules responsible for the aroma of coffee.

CAFFEINE $C_8H_{10}O_2N_4$

Caffeine is a stimulant found in coffee, tea, and soft drinks.

CITRIC ACID $C_6H_8O_7$

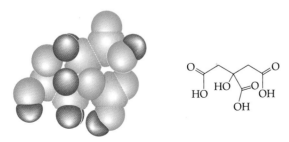

Citric acid is found in fruits such as lemons, grapefruit, and oranges.

FRUCTOSE $C_6H_{12}O_6$

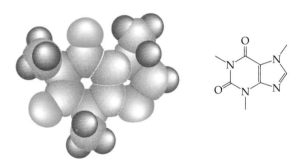

Fructose is a sugar found in honey and many fruits. It is also the sugar that powers the motion of sperm.

Section 5-3 Energy Levels and Spectra of Diatomic Molecules 179

MORPHINE $C_{17}H_{19}O_3N$

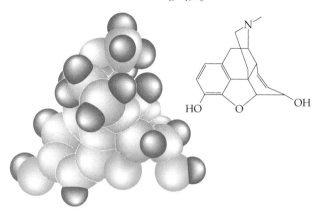

Morphine is the principal component of opium, which acts on the central nervous system and can induce addiction.

VANILLIN $C_8H_8O_3$

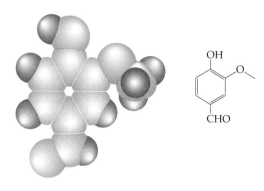

Vanillin is the essential component of oil of vanilla used to flavor many foods.

PELARGONIDIN $C_{15}H_{11}O_5$

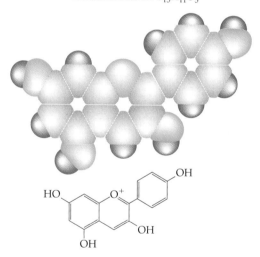

Pelargonidin is one of the molecules responsible for the red color of raspberries, strawberries, and apples.

para-HYDROXYPHENOL-2-BUTANONE $C_{10}H_{12}O_2$

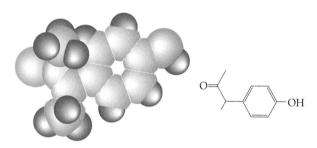

Para-hydroxyphenol-2-butanone is the molecule chiefly responsible for the aroma of raspberries.

METHYL CYANOACRYLATE $C_5H_5O_2N$

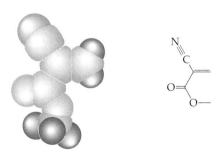

Methyl cyanoacrylate is the substance used in *Super Glue*.

TRINITROTOLUENE $C_7H_5O_6N_3$

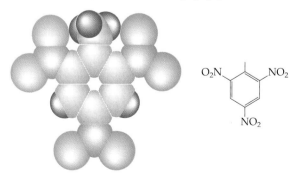

When the oxygen atoms of trinitrotoluene are nudged slightly, they combine with the carbon and hydrogen atoms to produce CO_2, H_2O, and gaseous N_2—thus converting the compact molecule into a suddenly expanding gas and giving TNT its explosiveness.

Characteristic rotational energy

where E_{0r} is the characteristic rotational energy of a particular molecule, which is inversely proportional to its moment of inertia:

$$E_{0r} = \frac{\hbar^2}{2I} \qquad 5\text{-}13$$

A measurement of the rotational energy of a molecule from its rotational spectrum can be used to determine the moment of inertia of the molecule, which can then be used to find the separation of the atoms in the molecule. The moment of inertia about an axis through the center of mass of a diatomic molecule (see Figure 5-12) is

$$I = m_1 r_1^2 + m_2 r_2^2$$

Using $m_1 r_1 = m_2 r_2$, which relates the distances r_1 and r_2 from the atoms to the center of mass, and $r_0 = r_1 + r_2$ for the separation of the atoms, we can write the moment of inertia as (see Problem 22)

$$I = \mu r_0^2 \qquad 5\text{-}14$$

where μ, called the **reduced mass**, is

Reduced mass

$$\mu = \frac{m_1 m_2}{m_1 + m_2} \qquad 5\text{-}15$$

If the masses are equal ($m_1 = m_2 = m$), as in H_2 and O_2, the reduced mass $\mu = \tfrac{1}{2}m$ and

$$I = \tfrac{1}{2} m r_0^2 \qquad 5\text{-}16$$

A unit of mass convenient for discussing atomic and molecular masses is the **unified mass unit** u, which is defined as one-twelfth the mass of the neutral carbon-12 (^{12}C) atom. The mass of one ^{12}C atom is thus 12 u. The mass of an atom in unified mass units is therefore numerically equal to the molar mass of the atom in grams. The unified mass unit is related to the gram and kilogram by

$$1 \text{ u} = \frac{1 \text{ g}}{N_A} = \frac{10^{-3} \text{ kg}}{6.0221 \times 10^{23}} = 1.6606 \times 10^{-27} \text{ kg} \qquad 5\text{-}17$$

where N_A is Avogadro's number.

Example 5-2

Find the reduced mass of the HCl molecule.

From the periodic table in Appendix F, the mass of the hydrogen atom is 1.01 u and that of the chlorine atom is 35.5 u. The reduced mass of HCl is therefore

$$\mu = \frac{m_1 m_2}{m_1 + m_2} = \frac{(1.01 \text{ u})(35.5 \text{ u})}{1.01 \text{ u} + 35.5 \text{ u}} = 0.982 \text{ u}$$

Note that the reduced mass is less than the mass of either atom in the molecule and that it is approximately equal to the mass of the hydrogen atom. When one atom of a diatomic molecule is much more massive than the other, the center of mass of the molecule is approximately at the center of the more massive atom, and the reduced mass is approximately equal to the mass of the lighter atom.

Example 5-3

Estimate the characteristic rotational energy of an O_2 molecule, assuming that the separation of the atoms is 0.1 nm.

If r_0 is the separation of the atoms in the oxygen molecule, the moment of inertia is

$$I = \tfrac{1}{2} m r_0^2$$

where m is the mass of an oxygen atom. Substituting this into Equation 5-13, we obtain

$$E_{0r} = \frac{\hbar^2}{m r_0^2}$$

Using $m = 16 \text{ u} = 16(1.66 \times 10^{-27} \text{ kg}) = 2.66 \times 10^{-26} \text{ kg}$ and $r_0 = 10^{-10}$ m, we obtain

$$E_{0r} = \frac{\hbar^2}{m r_0^2} = \frac{(1.05 \times 10^{-34} \text{ J·s})^2}{(2.66 \times 10^{-26} \text{ kg})(10^{-10} \text{ m})^2}$$

$$= 4.14 \times 10^{-23} \text{ J} = 2.59 \times 10^{-4} \text{ eV}$$

We can see from Example 5-3 that the rotational energy levels are several orders of magnitude smaller than those due to electron excitation, which have energies of the order of 1 eV or higher. Transitions within a given set of rotational energy levels yield photons in the far infrared region of the electromagnetic spectrum. Note that the rotational energies are small compared with the typical thermal energy kT at normal temperatures. For $T = 300$ K, for example, kT is about 2.6×10^{-2} eV. Thus, at ordinary temperatures, a molecule can be easily excited to the lower rotational energy levels by collisions with other molecules. But such collisions cannot excite the molecule to its electronic energy levels above the ground state.

Vibrational Energy Levels

The quantization of energy in a simple harmonic oscillator was one of the first problems solved by Schrödinger in his paper proposing his wave equation. Solving the Schrödinger equation for a simple harmonic oscillator gives

$$E_\nu = (\nu + \tfrac{1}{2}) h f \qquad \nu = 0, 1, 2, 3, \ldots \qquad \text{5-18}$$

Vibrational energy levels

where f is the frequency of the oscillator and ν is the **vibrational quantum number**.* An interesting feature of this result is that the energy levels are equally spaced with intervals equal to hf. The frequency of vibration of a diatomic molecule can be related to the force exerted by one atom on the other. Consider two objects of mass m_1 and m_2 connected by a spring of force constant K. The frequency of oscillation of this system can be shown to be (see Problem 34)

$$f = \frac{1}{2\pi} \sqrt{\frac{K}{\mu}} \qquad \text{5-19}$$

where μ is the reduced mass given by Equation 5-15. The effective force constant of a diatomic molecule can thus be determined from a measurement of the frequency of oscillation of the molecule.

---

*We use ν (the Greek letter nu) here rather than n so as not to confuse the vibrational quantum number with the principal quantum number n for electronic energy levels.

A selection rule on transitions between vibrational states (of the same electronic state) requires that ν can change only by ± 1, so the energy of a photon emitted by such a transition is hf and the frequency is f, the same as the frequency of vibration. There is a similar selection rule that ℓ must change by ± 1 for transitions between rotational states. A typical measured frequency of a transition between vibrational states is 5×10^{13} Hz, which gives for the order of magnitude of vibrational energies

$$E \sim hf = (4.14 \times 10^{-15} \text{ eV·s})(5 \times 10^{13} \text{ s}^{-1}) = 0.2 \text{ eV}$$

Note that a typical vibrational energy is about 1000 times greater than the typical rotational energy E_{0r} of the O_2 molecule we found in Example 5-3 and about 8 times greater than the typical thermal energy $kT = 0.026$ eV at $T = 300$ K.

Example 5-4

The frequency of vibration of the CO molecule is 6.42×10^{13} Hz. What is the effective force constant for this molecule?

Using 12 u for the mass of the carbon atom and 16 u for the mass of the oxygen atom, we obtain for the reduced mass

$$\mu = \frac{m_1 m_2}{m_1 + m_2} = \frac{(12 \text{ u})(16 \text{ u})}{12 \text{ u} + 16 \text{ u}} = 6.86 \text{ u}$$

Multiplying by 1.66×10^{-27} kg/u to convert to SI units, we obtain

$$\mu = (6.86 \text{ u})(1.66 \times 10^{-27} \text{ kg/u}) = 1.14 \times 10^{-26} \text{ kg}$$

Solving Equation 5-19 for the effective force constant K, we obtain

$$K = (2\pi f)^2 \mu = 4\pi^2 (6.42 \times 10^{13} \text{ Hz})^2 (1.14 \times 10^{-26} \text{ kg})$$
$$= 1.86 \times 10^3 \text{ N/m}$$

Emission Spectra

Figure 5-13 shows schematically some electronic, vibrational, and rotational energy levels of a diatomic molecule. The vibrational levels are labeled with the quantum number ν and the rotational levels are labeled with ℓ. The lower vibrational levels are evenly spaced, with $\Delta E = hf$. For higher vibra-

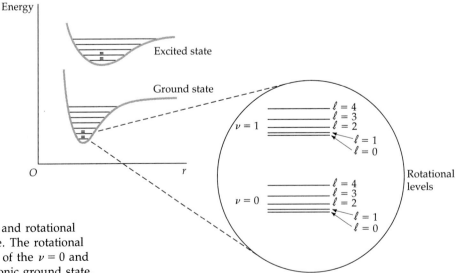

Figure 5-13 Electronic, vibrational, and rotational energy levels of a diatomic molecule. The rotational levels are shown in an enlargement of the $\nu = 0$ and $\nu = 1$ vibrational levels of the electronic ground state.

tional levels, the approximation that the vibration is simple harmonic is not valid and the levels are not quite evenly spaced. Note that the potential-energy curves representing the force between the two atoms in the molecule do not have exactly the same shape for the electronic ground and excited states. This implies that the fundamental frequency of vibration f is different for different electronic states. For transitions between vibrational states of different electronic states, the selection rule $\Delta \nu = \pm 1$ does not hold. Such transitions result in the emission of photons of wavelengths in or near the visible spectrum, so the emission spectrum of a molecule for electronic transitions is also sometimes called the optical spectrum.

The spacing of the rotational levels increases with increasing values of ℓ. Since the energies of rotation are so much smaller than those of vibrational or electronic excitation of a molecule, molecular rotation shows up in optical spectra as a fine splitting of the spectral lines. When the fine structure is not resolved, the spectrum appears as bands as shown in Figure 5-14a. Close inspection of these bands reveals that they have a fine structure due to the rotational energy levels, as shown in the enlargement in Figure 5-14b.

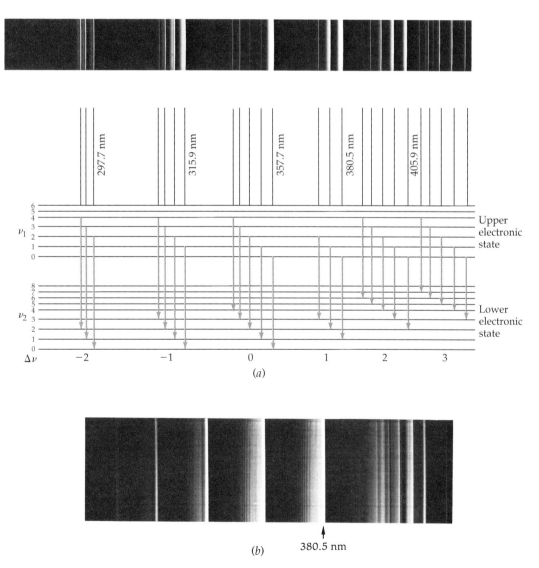

Figure 5-14 Part of the emission spectrum of N_2. (a) These components of the band are due to transitions between the vibrational levels of two electronic states, as indicated in the diagram. (b) An enlargement of part of a shows that the apparent lines in a are in fact band heads with structure caused by rotational levels.

Absorption Spectra

Much molecular spectroscopy is done using infrared absorption techniques in which only the vibrational and rotational energy levels of the ground-state electronic level are excited. For ordinary temperatures, the vibrational energies are sufficiently large in comparison with the thermal energy kT that most of the molecules are in the lowest vibrational state $\nu = 0$, for which the energy is $E_0 = \frac{1}{2}hf$. The transition from $\nu = 0$ to $\nu = 1$ is the predominant transition in absorption. The rotational energies, however, are sufficiently less than kT that the molecules are distributed among several rotational energy states. If the molecule is originally in a vibrational state characterized by $\nu = 0$ and a rotational state characterized by the quantum number ℓ, its initial energy is

$$E_\ell = \tfrac{1}{2}hf + \ell(\ell + 1)E_{0r} \qquad 5\text{-}20$$

where E_{0r} is given by Equation 5-13. From this state, two transitions are permitted by the selection rules. For a transition to the next highest vibrational state $\nu = 1$ and a rotational state characterized by $\ell + 1$, the final energy is

$$E_{\ell+1} = \tfrac{3}{2}hf + (\ell + 1)(\ell + 2)E_{0r} \qquad 5\text{-}21$$

For a transition to the next highest vibrational state and to a rotational state characterized by $\ell - 1$, the final energy is

$$E_{\ell-1} = \tfrac{3}{2}hf + (\ell - 1)\ell E_{0r} \qquad 5\text{-}22$$

The energy differences are

$$\Delta E_{\ell \to \ell+1} = E_{\ell+1} - E_\ell = hf + 2(\ell + 1)E_{0r} \qquad 5\text{-}23$$

where $\ell = 0, 1, 2, \ldots$, and

$$\Delta E_{\ell \to \ell-1} = E_{\ell-1} - E_\ell = hf - 2\ell E_{0r} \qquad 5\text{-}24$$

where $\ell = 1, 2, 3, \ldots$. (In Equation 5-24, ℓ begins at $\ell = 1$ because from $\ell = 0$ only the transition $\ell \to \ell + 1$ is possible.) Figure 5-15 illustrates these transitions. The frequencies of these transitions are given by

$$f_{\ell \to \ell+1} = \frac{\Delta E_{\ell \to \ell+1}}{h} = f + \frac{2(\ell + 1)E_{0r}}{h} \qquad \ell = 0, 1, 2, \ldots \qquad 5\text{-}25$$

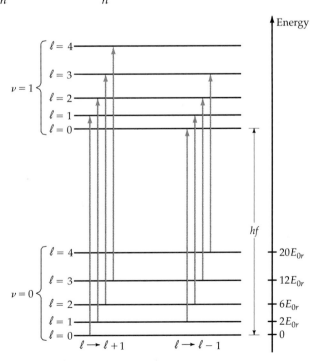

Figure 5-15 Absorptive transitions between the lowest vibrational states $\nu = 0$ and $\nu = 1$ in a diatomic molecule. These transitions obey the selection rule $\Delta \ell = \pm 1$ and fall into two bands. The energies of the $\ell \to \ell + 1$ band are $hf + 2E_{0r}$, $hf + 4E_{0r}$, $hf + 6E_{0r}$, and so forth, whereas the energies of the $\ell \to \ell - 1$ band are $hf - 2E_{0r}$, $hf - 4E_{0r}$, $hf - 6E_{0r}$, and so forth.

and

$$f_{\ell \to \ell-1} = \frac{\Delta E_{\ell \to \ell-1}}{h} = f - \frac{2\ell E_{0r}}{h} \qquad \ell = 1, 2, 3, \ldots \qquad 5\text{-}26$$

The frequencies for the transitions $\ell \to \ell + 1$ are thus $f + 2(E_{0r}/h)$, $f + 4(E_{0r}/h)$, $f + 6(E_{0r}/h)$, and so forth; those corresponding to the transition $\ell \to \ell - 1$ are $f - 2(E_{0r}/h)$, $f - 4(E_{0r}/h)$, $f - 6(E_{0r}/h)$, and so forth. We thus expect the absorption spectrum to contain frequencies equally spaced by $2E_{0r}/h$ except for a gap of $4E_{0r}/h$ at the vibrational frequency f as shown in Figure 5-16. A measurement of the position of the gap gives f, and a measurement of the spacing of the absorption peaks gives E_{0r}, which is inversely proportional to the moment of inertia of the molecule.

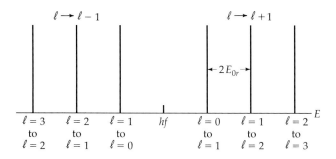

Figure 5-16 Expected absorption spectrum of a diatomic molecule. The right branch corresponds to the transitions $\ell \to \ell + 1$ and the left branch to the transitions $\ell \to \ell - 1$. The lines are equally spaced by $2E_{0r}$. The energy midway between the branches is hf, where f is the frequency of vibration of the molecule.

Figure 5-17 shows the absorption spectrum of HCl. The double-peak structure results from the fact that chlorine occurs naturally in two isotopes, ^{35}Cl and ^{37}Cl, which have different moments of inertia. If all the rotational levels were equally populated initially, we would expect the intensities of each absorption line to be equal. However, the population of a rotational level ℓ is proportional to the degeneracy of the level, that is, to the number of states with the same value of ℓ, which is $2\ell + 1$, and to the Boltzmann factor $e^{-E/kT}$, where E is the energy of the state. For low values of ℓ, the population increases slightly because of the degeneracy factor, whereas for higher values of ℓ, the population decreases because of the Boltzmann factor. The intensities of the absorption lines therefore increase with ℓ for low values of ℓ and then decrease with ℓ for high values of ℓ, as can be seen from the figure.

Figure 5-17 Absorption spectrum of the diatomic molecule HCl. The double-peak structure results from the two isotopes of chlorine, ^{35}Cl (abundance 75.5 percent) and ^{37}Cl (abundance 24.5 percent). The intensities of the peaks vary because the population of the initial state depends on ℓ.

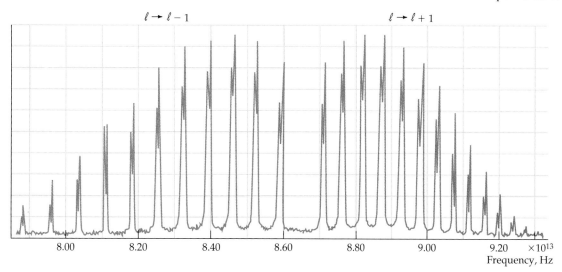

Questions

5. How does the effective force constant calculated for the CO molecule in Example 5-4 compare with the force constant of an ordinary spring?

6. Why can an atom absorb radiation only from the ground state whereas a diatomic molecule can absorb radiation from many different rotational states?

Summary

1. Bonding mechanisms for atoms and molecules include ionic, covalent, van der Waals, hydrogen, and metallic bonds. Ionic bonds result when an electron is transferred from one atom to another, resulting in a positive ion and a negative ion that bond together. The covalent bond is a quantum-mechanical effect that arises from the sharing of one or more electrons by identical or similar atoms. The van der Waals bonds are weak bonds that result from the interaction of the instantaneous electric dipole moments of molecules. The hydrogen bond results from the sharing of a hydrogen atom by other atoms. In the metallic bond, the positive lattice ions of the metal are held together by a cloud of negative charge comprised of free electrons.

2. A diatomic molecule formed from two identical atoms, such as O_2, must bond by covalent bonding. The bonding of two nonidentical atoms is often a mixture of covalent and ionic bonding. The percentage of ionic bonding can be found from the ratio of the measured electric dipole moment to the ionic electric dipole moment defined by

$$p_{\text{ionic}} = er_0$$

where r_0 is the equilibrium separation of the ions.

3. The shapes of such polyatomic molecules as H_2O and NH_3 can be understood from the spatial distribution of the atomic-orbital or molecular-orbital wave functions. The tetravalent nature of the carbon atom is a result of the hybridization of the $2s$ and $2p$ atomic orbitals.

4. The rotational energies of a diatomic molecule are quantized to the values

$$E_\ell = \frac{\ell(\ell+1)\hbar^2}{2I} = \ell(\ell+1)E_{0r} \qquad \ell = 0, 1, 2, \ldots$$

where

$$E_{0r} = \frac{\hbar^2}{2I}$$

and I is the moment of inertia of the molecule. The moment of inertia of a diatomic molecule is related to the equilibrium separation r_0 by

$$I = \mu r_0^2$$

where

$$\mu = \frac{m_1 m_2}{m_1 + m_2}$$

is the reduced mass.

5. The vibrational energies of a diatomic molecule are quantized to the values

$$E_\nu = (\nu + \tfrac{1}{2})hf \qquad \nu = 0, 1, 2, 3, \ldots$$

where f is the frequency of vibration of the molecule, which is related to the effective force constant K by

$$f = \frac{1}{2\pi}\sqrt{\frac{K}{\mu}}$$

6. The optical spectra of molecules have a band structure due to transitions between rotational levels. Information about the structure and bonding of a molecule can be found from its rotational and vibrational absorption spectrum involving transitions from one vibrational–rotational level to another. These transitions obey the selection rules

$$\Delta \nu = \pm 1$$
$$\Delta \ell = \pm 1$$

Suggestions for Further Reading

Atkins, P. W.: *Molecules*, Scientific American Library, A Division of HPHLP, New York, 1987.

This book illustrates the structure and describes some properties of a selection of molecules.

Nassau, Kurt: *The Physics and Chemistry of Color: The Fifteen Causes of Color*, John Wiley & Sons, New York, 1983.
Nassau, Kurt: "The Causes of Color," *Scientific American*, October 1980, p. 124.

This fascinating book and article discuss mechanisms of color production in such objects as hot bodies, neon signs, organic dyes, gemstones, metals, and semiconductors.

Pauling, Linus, and Roger Hayward: *The Architecture of Molecules*, W. H. Freeman and Company, San Francisco and London, 1964.

This is a picture book illustrating the relative positions of the various atoms in some common molecules and solids, with commentary by one of the men most responsible for our modern understanding of the chemical bond.

Ronn, Avigdor M.: "Laser Chemistry," *Scientific American*, May 1979, p. 114.

This article explains how laser light of the correct frequency can efficiently induce the electronic or molecular transitions required for a desired chemical reaction to occur.

Review

A. Objectives: After studying this chapter, you should:

1. Be able to list the ways atoms can bond together and describe each briefly.

2. Be able to compare ionic and covalent bonding.

3. Know the general shapes of the H_2O and NH_3 molecules and understand the origin of these shapes.

4. Be able to describe the general features of the energy-level diagram for a diatomic molecule and discuss the vibrational–rotational spectrum.

5. Be able to discuss why only certain lines in the emission spectrum of a molecule are seen in the absorption spectrum.

B. Define, explain, or otherwise identify:

Molecular orbitals
Ionic bond
Electron affinity
Exclusion-principle repulsion
Dissociation energy
Covalent bond
Saturated bond
The van der Waals bond
Hydrogen bond
Metallic bond
Atomic orbital

Hybrid orbital
Hybridization
Rotational quantum number
Reduced mass
Unified mass unit
Vibrational quantum number

188 Chapter 5 Molecules

C. True or false: If the statement is true, explain why it is true. If it is false, give a counterexample.

1. Ionic bonds are formed by atoms whose outer electron shells are filled.

2. The repulsive force between the two oxygen atoms in the O_2 molecule is due to the Coulomb repulsion of the nuclei of the atoms.

3. The dissociation energy of NaCl is the energy needed to move the Na^+ and Cl^- ions to an infinite separation.

4. Covalent bonding is a rare occurrence in nature.

5. The energies of vibration and rotation of a molecule are usually much greater than the energy of electronic excitation.

Problems

Level I

5-1 Molecular Bonding

1. Calculate the separation of Na^+ and Cl^- ions for which the potential energy is -1.53 eV.

2. The dissociation energy of Cl_2 is 2.48 eV. Consider the formation of an NaCl molecule by the reaction

$$Na + \tfrac{1}{2}Cl_2 \rightarrow NaCl$$

Is this reaction endothermic (requiring energy) or exothermic (giving off energy)? How much energy per molecule is required or given off?

3. What kind of bonding mechanism would you expect for (a) the KCl molecule, (b) the O_2 molecule, and (c) copper atoms in a solid?

4. What type of bonding mechanism would you expect for (a) NaF, (b) KBr, and (c) N_2?

5. The dissociation energy is sometimes expressed in kilocalories per mole. (a) Find the relation between electron volts per molecule and kilocalories per mole. (b) Find the dissociation energy of molecular NaCl in kilocalories per mole.

6. What kind of bonding mechanism would you expect for (a) KF, (b) NO, and (c) silver atoms in a solid?

7. The equilibrium separation of the atoms in the HF molecule is 0.0917 nm, and its measured electric dipole moment is 6.40×10^{-30} C·m. What percentage of the bonding is ionic?

8. Repeat Problem 7 for CsCl, for which the equilibrium separation is 0.291 nm and the measured electric dipole moment is 3.48×10^{-29} C·m.

9. The equilibrium separation of CsF is 0.2345 nm. If its bonding is 70 percent ionic, what should its measured electric dipole moment be?

5-2 Polyatomic Molecules

10. Using Table 4-1, find other elements with the same subshell electron configuration in the two outermost orbitals as carbon. Would you expect the same type of hybridization for these elements as occurs for carbon?

5-3 Energy Levels and Spectra of Diatomic Molecules

11. The characteristic rotational energy E_{0r} for the N_2 molecule is 2.48×10^{-4} eV. From this, find the separation distance of the nitrogen atoms in N_2.

12. Explain why the moment of inertia of a diatomic molecule increases slightly with increasing angular momentum.

13. Show that when one atom in a diatomic molecule is much more massive than the other, the reduced mass is approximately equal to the mass of the lighter atom.

14. Calculate the reduced mass in unified mass units for (a) H_2, (b) N_2, (c) CO, and (d) HCl.

15. The separation of the oxygen atoms in O_2 is actually slightly greater than the 0.1 nm assumed in Example 5-3, and the characteristic energy of rotation E_{0r} is 1.78×10^{-4} eV rather than the result obtained in that example. Use this value to calculate the separation distance of the oxygen atoms in O_2.

Level II

16. We are often interested in finding the quantity ke^2/r in electron volts when r is given in nanometers. Show that $ke^2 = 1.44$ eV·nm.

17. The equilibrium separation of the K^+ and Cl^- ions in KCl is about 0.267 nm. (a) Calculate the potential energy of attraction of the ions, assuming them to be point charges at this separation. (b) The ionization energy of potassium is 4.34 eV and the electron affinity of chlorine is 3.61 eV. Find the dissociation energy for KCl, neglecting any energy of repulsion (see Figure 5-1). (c) The measured dissociation energy is 4.43 eV. What is the energy due to repulsion of the ions at the equilibrium separation?

18. Indicate the mean value of r for two vibration levels on the potential-energy curve for a diatomic molecule, and show that because of the asymmetry in the curve, r_{av} increases with increasing vibrational energy, which is why solids expand when heated.

19. (a) Calculate the potential energy of attraction between the Na^+ and Cl^- ions of NaCl at the equilibrium separation $r_0 = 0.236$ nm, and compare this result with the dissociation energy given in Figure 5-1. (b) What is the energy due to repulsion of the ions at the equilibrium separation?

20. Show that the reduced mass is smaller than either mass in a diatomic molecule.

21. The equilibrium separation of the K^+ and F^- ions in KF is about 0.217 nm. (a) Calculate the potential energy of attraction of the ions, assuming them to be point charges at this separation. (b) The ionization energy of potassium is 4.34 eV and the electron affinity of fluorine is 3.45 eV. Find the dissociation energy of KF, neglecting any energy of repulsion. (c) The measured dissociation energy is 5.07 eV. Calculate the energy due to repulsion of the ions at the equilibrium separation.

22. Derive Equations 5-14 and 5-15 for the moment of inertia in terms of the reduced mass of a diatomic molecule.

23. The effective force constant for the HF molecule is 970 N/m. Find the frequency of vibration for this molecule.

24. The frequency of vibration of the NO molecule is 5.63×10^{13} Hz. Find the effective force constant for NO.

25. Use the equilibrium separation of the K^+ and Cl^- ions given in Problem 17 and the reduced mass of KCl to calculate the characteristic rotational energy E_{0r} of KCl.

26. The central frequency for the absorption band of HCl shown in Figure 5-17 is at $f = 8.66 \times 10^{13}$ Hz, and the absorption peaks are separated by about $\Delta f = 6 \times 10^{11}$ Hz. Use this information to find (a) the lowest vibrational energy for HCl, (b) the moment of inertia of HCl, (c) the equilibrium separation of the atoms.

27. Calculate the effective force constant for HCl from its reduced mass and the fundamental vibrational frequency obtained from Figure 5-17.

28. The potential energy between two atoms in a molecule can often be described rather well by the Lenard–Jones potential, which can be written

$$U = U_0 \left[\left(\frac{a}{r}\right)^{12} - 2\left(\frac{a}{r}\right)^6 \right]$$

where U_0 and a are constants. (a) Find the interatomic separation r_0 in terms of a for which the potential energy is minimum. (b) Find the corresponding value of U_{min}. (c) Use Figure 5-4 to obtain numerical values for r_0 and U_0 for the H_2 molecule. Express your answers in nanometers and electron volts.

29. Make a plot of the potential energy $U(r)$ versus the internuclear separation r for the H_2 molecule. Use the Lenard–Jones potential determined in Problem 28 and plot each term separately, together with the total $U(r)$.

30. In this problem, you are to find how the van der Waals force between a polar and a nonpolar molecule depends on the distance between the molecules. Let the dipole moment of the polar molecule be in the x direction and the nonpolar molecule be a distance x away. (a) How does the electric field due to an electric dipole depend on the distance x? (b) Use the facts that the potential energy of an electric dipole of moment p in an electric field $\mathbf{E}$ is $U = -\mathbf{p} \cdot \mathbf{E}$ and that the induced dipole moment of the nonpolar molecule is proportional to $\mathbf{E}$ to find how the potential energy of interaction of the two molecules depends on separation distance. (c) Using $F_x = -dU/dx$, find the x dependence of the force between the two molecules.

31. Find the dependence of the force between two polar molecules on separation distance. (See Problem 30.)

32. Use the infrared absorption spectrum of HCl given in Figure 5-17 to obtain (a) the characteristic rotational energy E_{0r} in electron volts and (b) the vibrational frequency f and the vibrational energy hf in electron volts.

33. For a molecule such as CO, which has a permanent electric dipole moment, radiative transitions obeying the selection rule $\Delta \ell = \pm 1$ between two rotational energy levels of the same vibrational energy state are allowed. (That is, the selection rule $\Delta \nu = \pm 1$ does not hold.) (a) Find the moment of inertia of CO for which $r_0 = 0.113$ nm, and calculate the characteristic rotational energy E_{0r} in electron volts. (b) Make an energy-level diagram for the rotational levels for $\ell = 0$ to $\ell = 5$ for some vibrational level. Label the energies in electron volts, starting with $E = 0$ for $\ell = 0$. (c) Indicate on your diagram transitions that obey $\Delta \ell = -1$ and calculate the energy of the photons emitted. (d) Find the wavelength of the photon emitted for each transition in (c). In what region of the electromagnetic spectrum are these photons?

Level III

34. Two objects of mass m_1 and m_2 are attached to a spring of force constant K and equilibrium length r_0. (a) Show that when m_1 is moved a distance Δr_1 from its equilibrium position, the force exerted by the spring is

$$F = -K \frac{m_1 + m_2}{m_2} \Delta r_1$$

(b) Show that the frequency of vibration is $f = (1/2\pi)\sqrt{K/\mu}$, where μ is the reduced mass.

35. (a) Calculate the reduced mass for the $H^{35}Cl$ and $H^{37}Cl$ molecules and the fractional difference $\Delta \mu / \mu$. (b) Show that the mixture of isotopes in HCl leads to a fractional difference in the frequency of a transition from one rotational state to another given by $\Delta f/f = -\Delta \mu/\mu$. (c) Compute $\Delta f/f$ and compare your result with Figure 5-17.

36. In calculating the rotational energy levels of a diatomic molecule, we did not consider rotation of the molecule about the line joining the atoms. (a) Estimate the moment of inertia of the H_2 molecule about this line. (b) Use your results for (a) to estimate the typical rotational energy E_{0r} for rotation about the line joining the atoms. (c) Compare your answer in (b) with the typical thermal energy kT at $T = 300$ K.

Chapter 6

Solids

Molten tin solidifies in a pattern of tree-shaped crystals called dendrites, as it cools under controlled circumstances.

The many and varied properties of solids have intrigued us for centuries. Technological developments involving metals and alloys have shaped the courses of civilizations, and the symmetry and beauty of naturally occurring, large single crystals have consistently captured our imaginations. However, the origins of the physical properties of solids were not understood at all until the development of quantum mechanics. The application of quantum mechanics to solids has provided the basis for much of the technological progress of modern times. We will briefly study some aspects of the structure of solids and then concentrate on their electrical properties.

6-1 The Structure of Solids

In our everyday world, we see matter in three phases: gases, liquids, and solids. In a gas, the average distance between two molecules is large compared with the size of a molecule. The molecules have little influence on one another except during their frequent but brief collisions. In a liquid or solid, the molecules are close together and exert forces on one another that are comparable to the forces that bind atoms into molecules. In a liquid, the molecules form temporary short-range bonds that are continually broken and reformed due to the thermal kinetic energy of the molecules. The strength of these bonds depends on the type of molecule. For example, the bonds between helium atoms are the very weak van der Waals bonds, and helium does not liquefy at atmospheric pressure unless the temperature is 4.2 K or lower.

If a liquid is cooled slowly so that the kinetic energy of its molecules is reduced slowly, the molecules may arrange themselves in a regular crystalline array, producing the maximum number of bonds and leading to a minimum potential energy. However, if the liquid is cooled rapidly so that its internal energy is removed before the molecules have a chance to arrange themselves, the solid formed is often not crystalline but instead resembles a snapshot of the liquid. Such a solid is called an **amorphous solid.** It displays short-range order but not the long-range order (over many molecular diameters) that is characteristic of a crystal. Glass is a typical amorphous solid. A characteristic of the long-range ordering of a crystal is that it has a well-defined melting point, whereas an amorphous solid merely softens as its temperature is increased. Many materials may solidify into either an amorphous or a crystalline state depending on how they are prepared. Others exist only in one form or the other. Most common solids are polycrystalline; that is, they consist of many single crystals that meet at grain boundaries. The size of a single crystal is typically a fraction of a millimeter. However, large single crystals do occur naturally and can be produced artificially. We will discuss only simple crystalline solids in this chapter.

The most important property of a single crystal is the symmetry and regularity of its structure. It can be thought of as having a single unit structure that is repeated throughout the crystal. This smallest unit of a crystal is called the **unit cell.** The structure of the unit cell depends on the type of bonding between the atoms, ions, or molecules in the crystal. If more than one kind of atom is present, the structure will also depend on the relative sizes of the atoms. The bonding mechanisms are those discussed in Chapter 5—ionic, covalent, metallic, hydrogen, and van der Waals. Figure 6-1 shows the structure of the ionic crystal sodium chloride (NaCl). The Na^+ and Cl^- ions are spherically symmetric, and the Cl^- ion is approximately twice as large as the Na^+ ion. The minimum potential energy for this crystal occurs when an ion of either kind has six nearest neighbors of the other kind. This structure is called *face-centered-cubic (fcc)*. Note that the Na^+ and Cl^- ions in solid NaCl are *not* paired into NaCl molecules.

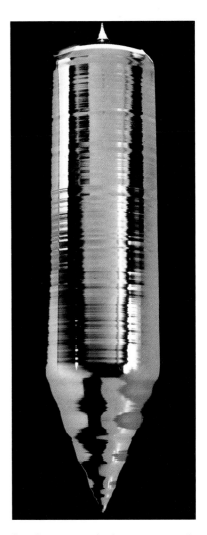

Synthetic crystal silicon is created beginning with a raw material containing silicon (for instance, common beach sand), purifying out the silicon, and melting it. From a seed crystal, the molten silicon grows into a cylindrical crystal, such as the one shown here. The crystals (typically about 1.3 m long) are formed under highly controlled conditions to ensure that they are flawless, and sliced into thousands of thin wafers, onto which the layers of an integrated circuit are etched.

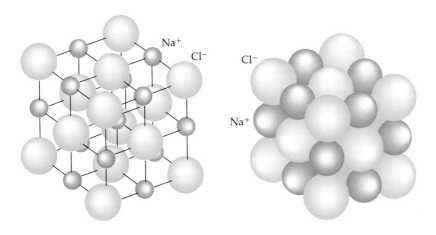

Figure 6-1 Face-centered-cubic structure of the NaCl crystal.

(a)   (b)

(a) The hexagonal symmetry of a snowflake arises from a hexagonal symmetry in its lattice of hydrogen and oxygen atoms. (b) NaCl (salt) crystals, magnified about 30 times. The crystals are built up from a cubic lattice of sodium and chloride ions. In the absence of impurities, an exact cubic crystal is formed. This (false-color) scanning electron micrograph shows that in practice the basic cube is often disrupted by dislocations, giving rise to crystals with a wide variety of shapes. The underlying cubic symmetry, though, remains evident. (c) A crystal of quartz (SiO_2, silicon dioxide), the most abundant and widespread mineral on earth. If molten quartz is allowed to solidify without crystallizing, it will form glass. (d) A soldering iron tip, ground down to reveal the copper core within its iron sheath. Visible in the iron is its underlying microcrystalline structure.

The net attractive part of the potential energy of an ion in a crystal can be written

$$U_{att} = -\alpha \frac{ke^2}{r} \qquad 6\text{-}1$$

where r is the separation distance between neighboring ions (which is 0.281 nm for the Na^+ and Cl^- ions in crystalline NaCl); and α, called the **Madelung constant,** depends on the geometry of the crystal. If only the 6 nearest neighbors of each ion were important, α would be 6. However, in addition to the 6 neighbors of the opposite charge at a distance r, there are 12 ions of the same charge at a distance $\sqrt{2}r$, 8 ions of opposite charge at a distance $\sqrt{3}r$, and so on. The Madelung constant is thus an infinite sum:

$$\alpha = 6 - \frac{12}{\sqrt{2}} + \frac{8}{\sqrt{3}} - \cdots \qquad 6\text{-}2$$

The result* for face-centered-cubic structures is $\alpha = 1.7476$.

When Na^+ and Cl^- ions are very close together, they repel each other because of the overlap of their electrons and the exclusion-principle repulsion discussed in Section 5-1. A simple empirical expression for the potential energy associated with this repulsion that works fairly well is

$$U_{rep} = \frac{A}{r^n}$$

---

*A large number of terms is needed to calculate the Madelung constant accurately because the sum converges very slowly.

(c)

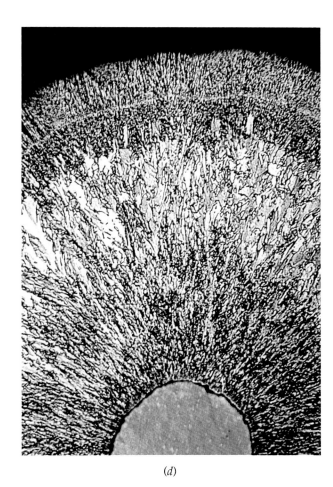
(d)

where A and n are constants. The total potential energy of an ion is then

$$U = -\alpha \frac{ke^2}{r} + \frac{A}{r^n} \qquad 6\text{-}3$$

The equilibrium separation $r = r_0$ is that at which the force $F = -dU/dr$ is zero. Differentiating and setting $dU/dr = 0$ at $r = r_0$, we obtain

$$A = \frac{\alpha k e^2 r_0^{n-1}}{n} \qquad 6\text{-}4$$

Exercise

Derive Equation 6-4.

The total potential energy can thus be written

$$U = -\alpha \frac{ke^2}{r_0}\left[\frac{r_0}{r} - \frac{1}{n}\left(\frac{r_0}{r}\right)^n\right] \qquad 6\text{-}5$$

At $r = r_0$, we have

$$U(r_0) = -\alpha \frac{ke^2}{r_0}\left(1 - \frac{1}{n}\right) \qquad 6\text{-}6$$

If we know the equilibrium separation r_0, the value of n can be found approximately from the **dissociation energy** of the crystal, which is the energy needed to break up the crystal into atoms.

Example 6-1

Calculate the equilibrium spacing r_0 for NaCl from the measured density of NaCl, which is $\rho = 2.16$ g/cm^3.

We consider each ion to occupy a cubic volume of side r_0. The mass of 1 mol of NaCl is 58.4 g, which is the sum of the atomic masses of sodium and chlorine. The ions occupy a volume of $2N_A r_0^3$, where $N_A = 6.02 \times 10^{23}$ is Avogadro's number. The density is thus related to r_0 by

$$\rho = \frac{m}{V} = \frac{m}{2N_A r_0^3}$$

Then

$$r_0^3 = \frac{m}{2N_A \rho} = \frac{58.4 \text{ g}}{2(6.02 \times 10^{23})(2.16 \text{ g/cm}^3)} = 2.24 \times 10^{-23} \text{ cm}^3$$

$$r_0 = 2.82 \times 10^{-8} \text{ cm} = 0.282 \text{ nm}$$

The measured dissociation energy of NaCl is 770 kJ/mol. Using 1 eV = 1.602×10^{-19} J, and the fact that 1 mol of NaCl contains N_A pairs of ions, we can express the dissociation energy in electron volts per ion pair. The conversion between electron volts per ion pair and kilojoules per mole is

$$1 \frac{\text{eV}}{\text{ion pair}} \times \frac{6.022 \times 10^{23} \text{ ion pairs}}{\text{mol}} \times \frac{1.602 \times 10^{-19} \text{ J}}{1 \text{ eV}}$$

The result is

$$1 \frac{\text{eV}}{\text{ion pair}} = 96.47 \frac{\text{kJ}}{\text{mol}} \qquad 6\text{-}7$$

Thus 770 kJ/mol = 7.98 eV per ion pair. Substituting -7.98 eV for $U(r_0)$, 0.282 nm for r_0, and 1.75 for α in Equation 6-6, we can solve for n. The result is $n = 9.35 \approx 9$.

Most ionic crystals, such as LiF, KF, KCl, KI, and AgCl, have a face-centered-cubic structure. Some elemental solids that have this structure are silver, aluminum, gold, calcium, copper, nickel, and lead.

Figure 6-2 shows the structure of CsCl, which is called the *body-centered-cubic (bcc) structure*. In this structure, each ion has eight nearest neighbor ions of the opposite charge. The Madelung constant for these crystals is 1.7627. Elemental solids with this structure include barium, cesium, iron, potassium, lithium, molybdenum, and sodium.

Figure 6-3 shows another important crystal structure: the *hexagonal close-packed (hcp) structure*. It is obtained by stacking identical spheres, such as bowling balls. In the first layer, each ball touches six others; thus, the name *hexagonal*. In the next layer, each ball fits into a triangular depression of the first layer. In the third layer, each ball fits into a triangular depression

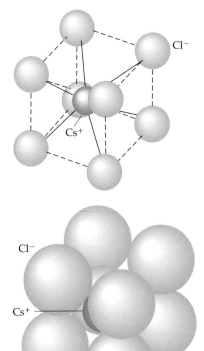

Figure 6-2 Body-centered-cubic structure of the CsCl crystal.

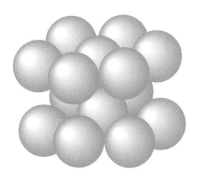

Figure 6-3 Hexagonal close-packed crystal structure.

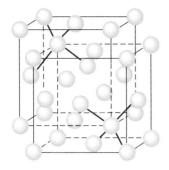

Figure 6-4 Diamond crystal structure. This structure can be considered to be a combination of two interpenetrating face-centered-cubic structures.

of the second layer, so it lies directly over a ball in the first layer. Elemental solids with *hcp* structure include beryllium, cadmium, cerium, magnesium, osmium, and zinc.

In some solids with covalent bonding, the crystal structure is determined by the directional nature of the bonds. Figure 6-4 illustrates the diamond structure of carbon, in which each atom is bonded to four others as a result of hybridization, which we discussed in Section 5-2. This is also the structure of germanium and silicon.

Questions

1. Why is r_0 different for solid NaCl than for the diatomic molecule?
2. Why would you not expect NaCl to have a *hcp* structure?

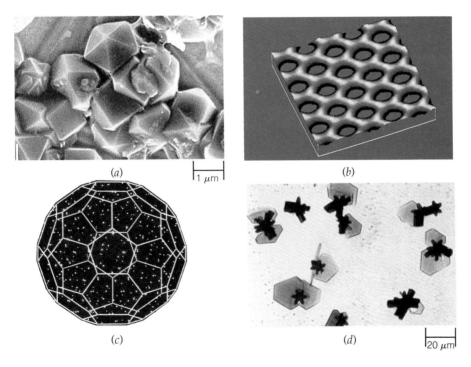

(a) $\overline{1\ \mu m}$ (b)

(c) (d) $\overline{20\ \mu m}$

Carbon exists in three well-defined crystalline forms: diamond, graphite, and fullerenes (short for "buckminsterfullerenes") —the third of which was predicted and discovered only a few years ago. The forms differ in how the carbon atoms are packed together in a lattice. A fourth form of carbon, in which no well-defined crystalline form exists, is common charcoal. (*a*) Synthetic diamonds, magnified about 50,000 times. In diamond, each carbon atom is centered in a tetrahedron of four other carbon atoms. The strength of these bonds accounts for the hardness of a diamond. (*b*) An atomic-force micrograph of graphite. In graphite, carbon atoms are arranged in sheets, each sheet made up of atoms in hexagonal rings. The sheets slide easily across one another, a property that allows graphite to function as a lubricant. (*c*) A single sheet of carbon rings can be closed on itself if certain rings are allowed to be pentagonal, instead of hexagonal. A computer-generated image of the smallest such structure, C_{60}, is shown here. Each of the 60 vertices corresponds to a carbon atom; 20 of the faces are hexagons and 12 are pentagons. The same geometric pattern is encountered in a soccer ball. (*d*) Fullerene crystals, in which C_{60} molecules are close-packed. The smaller crystals tend to form thin brownish platelets; larger crystals are usually rod-like in shape. Fullerenes exist in which more than 60 carbon atoms appear. In the crystals shown here, about one-sixth of the molecules are C_{70}.

6-2 The Classical Free-Electron Theory of Metals

A microscopic model of electric conduction was first proposed by P. Drude in 1900 and was developed by Hendrik A. Lorentz about 1909. This model, now called **the classical model of electric conduction**, successfully predicts Ohm's law and relates the conductivity and resistivity of conductors to the motion of free electrons within them.

In this model, a metal is pictured as a regular three-dimensional lattice of ions with a large number of electrons that are free to move throughout the whole metal. In copper, for example, there is about 1 free electron per copper atom. In the absence of an electric field, the free electrons move about the metal much like gas molecules move about in a container. The free electrons are in thermal equilibrium with the lattice ions with which they make collisions. When there is no electric field in the wire, these electrons move in random directions with relatively high speeds due to their thermal energy. Since the velocity vectors of the electrons are randomly oriented, the average velocity due to this thermal energy is zero. When an electric field is applied, for example, by connecting the wire to a battery that applies a potential difference along the wire, the free electrons experience a momentary acceleration due to the force $-e\mathbf{E}$. The electrons acquire a small velocity in the direction opposite the field but the kinetic energy acquired is quickly dissipated by collisions with the fixed ions in the wire. The electrons are then again accelerated by the field. The net result of this repeated acceleration and dissipation of energy is that the electrons have a small **drift velocity** opposite to the electric field superimposed on their large random thermal velocity. The motion of the free electrons in a metal is somewhat similar to that of the molecules of a gas such as air. In still air, the molecules move with large instantaneous velocities between collisions, but the average velocity is zero. When there is a breeze, the air molecules have a small drift velocity in the direction of the breeze superimposed on the much larger instantaneous velocity.

The current in a wire is the amount of charge passing through a cross-sectional area A per unit time. We can relate the current I to the number of electrons per unit volume n, the drift velocity v_d, the charge on the electron e and the cross-sectional area A. We will assume that each electron moves with a drift velocity v_d. In a time Δt all the particles in the volume $Av_d \Delta t$, shaded in Figure 6-5, pass through the area element. Since the number of electrons per unit volume is n, the number of particles in this volume is $nAv_d \Delta t$, and the total charge is

$$\Delta Q = enAv_d \Delta t$$

The current is thus

$$I = \frac{\Delta Q}{\Delta t} = neAv_d \qquad 6\text{-}8$$

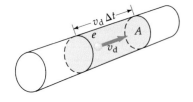

Figure 6-5 In time Δt, all the charges in the shaded volume pass through A. If there are n charge carriers per unit volume, each with charge e, the total charge in this volume is $\Delta Q = nev_dA \Delta t$, where v_d is the drift velocity of the charge carriers. The total current is then $I = \Delta Q/\Delta t = nev_dA$.

We can get an idea of the order of magnitude of the drift velocity for electrons in a conducting wire by putting typical magnitudes into Equation 6-8.

Example 6-2

What is the drift velocity of electrons in a typical copper wire of radius 0.815 mm carrying a current of 1 A?

If we assume one free electron per copper atom, the density of free electrons is the same as the density of atoms n_a, which is related to the mass density ρ, Avogadro's number N_A, and the molar mass M by

$$n_a = \frac{\rho N_A}{M}$$

For copper $\rho = 8.92$ g/cm^3, and $M = 63.5$ g/mol. Then

$$n_a = \frac{(8.93 \text{ g/cm}^3)(6.02 \times 10^{23} \text{ atoms/mol})}{63.5 \text{ g/mol}} = 8.47 \times 10^{22} \text{ atoms/cm}^3$$

The density of electrons is then

$$n = 8.47 \times 10^{22} \text{ electrons/cm}^3 = 8.47 \times 10^{28} \text{ electrons/m}^3$$

The drift velocity is therefore

$$v_d = \frac{I}{Ane} = \frac{1 \text{ C/s}}{\pi(0.00815 \text{ m})^2(8.47 \times 10^{28} \text{ m}^{-3})(1.6 \times 10^{-19} \text{ C})}$$

$$\approx 2.89 \times 10^{-5} \text{ m/s}$$

We see that typical drift velocities are of the order of 0.01 mm/s which is quite small.

According to Ohm's law, the current in a conducting wire segment is proportional to the voltage drop across the segment

$$I = \frac{V}{R}$$

where the resistance R does not depend on the current I or the voltage drop V. It is proportional to the length of the wire segment L and inversely proportional to the cross-sectional area A

$$R = \rho\frac{L}{A}$$

where ρ is the resistivity. For a uniform electric field E, the voltage drop across a segment of length L is $V = EL$. Substituting $\rho L/A$ for R, and EL for V, we can write Ohm's law

$$I = \frac{V}{R} = \frac{EL}{\rho L/A} = \frac{1}{\rho} EA \qquad 6\text{-}9$$

For materials that obey Ohm's law, the resistivity ρ must be independent of E.

The objective of the classical theory of conduction is to find an expression for ρ in terms of the properties of metals. In the presence of an electric field, a free electron experiences a force of magnitude eE. If this were the only force acting on the electron the electron would have an acceleration eE/m_e and its velocity would steadily increase. However, Ohm's law implies that there is a steady-state situation in which the drift velocity of the electron is proportional to the field E because the current I is proportional to E (Equation 6-9) and also to v_d (Equation 6-8). In the classical model, it is assumed that a free electron is accelerated for a short time and then makes a collision with a lattice ion. After the collision, the velocity of the electron is assumed to be completely unrelated to that before the collision. The justification for this assumption is that the drift velocity is very small compared with the random thermal velocity.

The average speed of electrons due to their thermal motion can be calculated from the Maxwell–Boltzmann distribution function (see Appendix C). The result is the same as that for ideal-gas molecules with the electron mass m_e replacing the mass of a molecule. It is related to the absolute temperature T by (Equation C-11)

$$v_{av} = \sqrt{\frac{8kT}{\pi m_e}} \qquad 6\text{-}10$$

For example, at $T = 300$ K, the average speed is

$$v_{av} = \sqrt{\frac{8(1.38 \times 10^{-23} \text{ J/K})(300 \text{ K})}{\pi(9.11 \times 10^{-31} \text{ kg})}} = 1.08 \times 10^5 \text{ m/s}$$

which is much greater than the drift velocity for electrons in a metal that we calculated in Example 6-2.

Let τ be the average time before an electron, picked at random, makes its next collision. Because the collisions are random, this time does not depend on the time elapsed since the electron's last collision. If we look at an electron immediately after it makes a collision, the average time before its next collision will be τ. Thus τ, called the **collision time**, is the average time between collisions. It is also the average time since the *last* collision of an electron picked at random.* The drift velocity is the average velocity of an electron picked at random. Since the acceleration is eE/m_e, the drift velocity is

$$v_d = \frac{eE}{m_e}\tau \qquad 6\text{-}11$$

The average distance the electron travels between collisions is called the **mean free path** λ. It is the product of the average speed v_{av} and the average time between collisions τ.

$$\lambda = v_{av}\,\tau$$

In terms of the mean free path, the drift velocity is

$$v_d = \frac{eE\lambda}{m_e v_{av}}$$

Using this result in Equation 6-8, we obtain

$$I = neAv_d = \frac{ne^2\lambda}{m_e v_{av}}EA \qquad 6\text{-}12$$

Comparing this with Ohm's law, Equation 6-9, we have for the resistivity

Resistivity in terms of the average speed and mean free path

$$\rho = \frac{m_e v_{av}}{ne^2\lambda} \qquad 6\text{-}13$$

We can relate the mean free path to the size of the ions in the metal lattice. Consider one electron moving with speed v through a region of stationary ions (Figure 6-6). Since this speed is related to its thermal energy, it is essentially unaffected by collisions. Assuming that the size of the electron is negligible, it will collide with an ion if it comes within a distance r from the center of the ion where r is the radius of the ion. In some time t, the electron moves a distance vt and collides with every ion in the the cylindrical tube of volume $\pi r^2 vt$ surrounding the path of the electron. (After each collision, the direction of the electron changes so the path is really a zigzag one.) The number of ions in this volume and hence the number of collisions in the time t is $n_a \pi r^2 vt$ where n_a is the number of ions per unit volume. The total path length divided by the number of collisions is the mean free path

$$\lambda = \frac{vt}{n_a \pi r^2 vt} = \frac{1}{n_a \pi r^2} \qquad 6\text{-}14$$

Example 6-3

Estimate the mean free path for electrons in copper.

From Equation 6-14, the mean free path λ is related to the number of copper ions per unit volume n_a and the radius of a copper ion r by

$$\lambda = \frac{1}{n_a \pi r^2}$$

---

*It is tempting but incorrect to think that if τ is the average time between collisions, the average time since its last collision is $\frac{1}{2}\tau$ rather than τ. If you find this confusing, you may take comfort in the fact that Drude used the incorrect result $\frac{1}{2}\tau$ in his original work.

The number of copper ions per unit volume was calculated in Example 6-2 to be $n_a = 8.47 \times 10^{22}$ ions/cm^3. Using $r \approx 10^{-10}$ m $= 10^{-8}$ cm for the radius of a copper ion, we obtain for an estimate of the mean free path of the electrons in copper

$$\lambda = \frac{1}{(8.47 \times 10^{22} \text{ cm}^{-3})\pi(10^{-8} \text{ cm})^2} \approx 4 \times 10^{-8} \text{ cm}$$

$$= 4 \times 10^{-10} \text{ m} = 0.4 \text{ nm}$$

Example 6-4

Using the result of Example 6-3 and the average velocity calculated from Equation 6-10, calculate the resistivity for copper at $T = 300$ K.

From Equation 6-13, the resistivity is given by

$$\rho = \frac{m_e v_{av}}{ne^2 \lambda}$$

Using $m_e = 9.11 \times 10^{-31}$ kg, $v_{av} = 1.08 \times 10^5$ m/s, $n = 8.47 \times 10^{28}$ electrons/m^3, $e = 1.6 \times 10^{-19}$ C, and $\lambda = 4 \times 10^{-10}$ m, we obtain

$$\rho = \frac{(9.11 \times 10^{-31} \text{ kg})(1.08 \times 10^5 \text{ m/s})}{(8.47 \times 10^{28} \text{ electrons/m}^3)(1.6 \times 10^{-19} \text{ C})^2(4 \times 10^{-10} \text{ m})}$$

$$= 1.13 \times 10^{-7} \text{ }\Omega\cdot\text{m}$$

This is about 6.6 times greater than the measured value of 1.7×10^{-8} $\Omega\cdot$m.

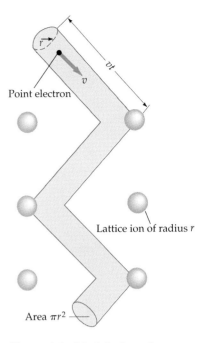

Figure 6-6 Model of an electron moving through the lattice ions in copper. The electron, which is considered to be a point, collides with a lattice ion if it comes within a distance r of the center of the ion, where r is the radius of the ion. If the electron has speed v, it collides with all the ions in the cylindrical volume $\pi r^2 v t$ in time t.

With the average speed given by Equation 6-10 and the mean free path by Equation 6-14, the resistivity (Equation 6-13) has been expressed in terms of the properties of metals, which was the objective of the classical theory of conduction. According to Ohm's law, the resistivity is independent of the electric field E. The quantities in Equation 6-13 that might depend on the electric field are the average speed v_{av} and the mean free path λ. As we have seen, the drift velocity is very much smaller than the average thermal speed of the electrons in equilibrium with the lattice ions. Thus the electric field has essentially no effect on the average speed of the electrons. The mean free path of the electrons depends on the size of the lattice ion and on the density of the ions, neither of which depends on the electric field E. Thus the classical model predicts Ohm's law with the resistivity as given by Equation 6-13.

Although successful in predicting Ohm's law, the classical theory of conduction has several defects. We saw from Example 6-4 that the magnitude of the resistivity calculated from Equation 6-12 is about 6 times the measured value at $T = 300$ K. The temperature dependence of ρ is also not correct. Experimentally, the resistivity varies linearly with temperature over a wide range of temperatures. The temperature dependence of resistivity in Equation 6-13 is given completely by the mean speed v_{av}, which according to Equation 6-10 is proportional to $\sqrt{T}$. Thus this calculation does not give a linear dependence on temperature. Finally, the classical model says nothing about why some materials are conductors, others insulators, and still others semiconductors.

In the quantum mechanical theory of electrical conduction, which is discussed in Section 6-4, the resistivity is again given by Equation 6-13, but the average speed and the mean free path are interpreted in terms of quantum theory. We will see that the average speed is not proportional to $\sqrt{T}$, but is approximately independent of T because the electron speeds do not obey the Maxwell–Boltzmann distribution law, but instead obey a quantum mechanical distribution law called the Fermi–Dirac distribution. Also, in the quantum mechanical calculation of the mean free path the wave nature of the electron (Chapter 2) is important. The collision of an electron with a lattice ion is not similar to the collision of a baseball and a tree, but instead involves the scattering of an electron wave by a regularly spaced lattice.

Heat Conduction and Heat Capacity

Good conductors of electricity are also good conductors of heat. The classical theory assumes that this is because the electron gas is mainly responsible for heat conduction in metals. This theory, which we will not discuss in detail, adequately describes the qualitative features of heat conduction in metals, but its quantitative predictions do not agree with experiment. In particular, the contribution of the electron gas to the heat capacity of a conductor, expected from classical theory, is not observed.

According to the equipartition theorem (Appendix C), the average energy per molecule of a system is $\frac{1}{2}kT$ per degree of freedom, where k is Boltzmann's constant and T is the absolute temperature. (Each coordinate, velocity component, or angular velocity component that appears as a squared term in the expression for the energy is considered to be a degree of freedom.) In a simple model, a solid is considered to consist of a regular array of atoms in which each of the atoms has a fixed equilibrium position and is connected by springs to its neighbors. Each atom can vibrate in the x, y, and z directions. Its total energy is thus

$$E = \tfrac{1}{2}mv_x^2 + \tfrac{1}{2}mv_y^2 + \tfrac{1}{2}mv_z^2 + \tfrac{1}{2}Kx^2 + \tfrac{1}{2}Ky^2 + \tfrac{1}{2}Kz^2 \qquad 6\text{-}15$$

where K is the effective force constant of the springs. Thus for oscillation in three dimensions, there are 6 degrees of freedom and the equipartition theorem predicts that the average energy is $6 \times \tfrac{1}{2}kT$ per atom, and $6 \times \tfrac{1}{2}RT$ per mole, where R is the gas constant. The average energy per atom is thus

$$E_{av} = 6 \times \tfrac{1}{2}kT = 3kT$$

and the internal energy of a mole of a solid is

$$U = 6 \times \tfrac{1}{2}RT = 3RT$$

The molar heat capacity is then

$$C_{mv} = \frac{dU}{dT} = 3R$$

The internal energy $3RT$ is the energy associated with lattice vibrations as described by Equation 6-15. If the solid is an insulator, these vibrations are the exclusive source of its internal energy. On the other hand, if the solid is a metal, there will be additional energy associated with the electron gas in it. If such a gas has a Maxwellian distribution, the average kinetic energy of the electrons would be $\tfrac{3}{2}kT$, corresponding to an additional energy of $\tfrac{3}{2}RT$ per mole. Thus the molar heat capacity of a metal would be

$$C_{mv} = (3R)_{\text{lattice vibrations}} + (\tfrac{3}{2}R)_{\text{electron gas}} = \tfrac{9}{2}R \qquad 6\text{-}16$$

which is $\tfrac{3}{2}R$ greater than that of an insulator. This is not observed. The molar heat capacity of metals is very nearly $3R$. At high temperature it is slightly greater, but the increase is nowhere near the value of $\tfrac{3}{2}R$ predicted by the classical theory. The increase is, in fact, proportional to the temperature, and at $T = 300$ K it is only about $0.02R$.

6-3 The Fermi Electron Gas

One of the difficulties of the classical free-electron theory of metals is related to the assumption that the average energy of the electrons is $\frac{3}{2}kT$. This result follows from the equipartition theorem, which applies to any system of particles that obey the Maxwell–Boltzmann energy distribution. However, because of the exclusion principle, the energy distribution of the free electrons in a metal is not even approximately given by the Maxwell–Boltzmann distribution, and the average energy is not $\frac{3}{2}kT$. We will consider first the energy distribution of electrons at temperature $T = 0$, which can be calculated rather easily and is a very good approximation to the distribution at other temperatures, even those as high as several thousand kelvins.

The Energy Distribution at $T = 0$

Classically, all the electrons in a conductor should have zero kinetic energy at $T = 0$. As the conductor is heated, the lattice ions acquire an average kinetic energy of $\frac{3}{2}kT$, which is imparted to the electron gas by the collisions between the electrons and the ions. In equilibrium, the electrons would be expected to have a mean kinetic energy of $\frac{3}{2}kT$.

Since the electrons are confined to the space occupied by the metal, it is clear from the uncertainty principle that, even at $T = 0$, an electron cannot have zero kinetic energy. Furthermore, the exclusion principle prevents more than two electrons (with opposite spins) from being in the lowest energy level. At $T = 0$, we expect the electrons to have the lowest energies consistent with the exclusion principle. It is instructive to consider first a one-dimensional model.

Consider N electrons in a one-dimensional infinite square-well potential of length L. The allowed energies were calculated in Chapter 3 and are given by Equation 3-39:

$$E_n = n^2 E_1 \qquad n = 1, 2, 3, \ldots \qquad \text{6-17}$$

where the energy of the lowest level E_1 (Equation 3-40) is

$$E_1 = \frac{h^2}{8m_e L^2} \qquad \text{6-18}$$

We can put two electrons with opposite spins in level $n = 1$, two in level $n = 2$, and so on. The N electrons will thus fill up $N/2$ levels from $n = 1$ to $n = N/2$. The energy of the last filled (or half-filled) level is called the **Fermi energy at $T = 0$**. We can calculate this Fermi energy E_F for N electrons by setting $n = N/2$ in Equation 6-17:

$$E_F = E_{N/2} = \left(\frac{N}{2}\right)^2 E_1 = \frac{h^2}{32m_e}\left(\frac{N}{L}\right)^2 = \frac{(hc)^2}{32m_e c^2}\left(\frac{N}{L}\right)^2 \qquad \text{6-19}$$

where we have multiplied the numerator and denominator by c^2 to simplify numerical calculations. We can see that the Fermi energy is a function of the number of electrons per unit length, which is the number density in one dimension. The number density of free electrons in copper, if we assume one free electron per atom, is $8.47 \times 10^{22}/\text{cm}^3$. In one dimension, this corresponds to

$$\frac{N}{L} = (8.47 \times 10^{22}/\text{cm}^3)^{1/3} = 4.39 \times 10^7/\text{cm} = 4.39/\text{nm}$$

The Fermi energy is then

$$E_F = \frac{(hc)^2}{32m_e c^2}\left(\frac{N}{L}\right)^2 = \frac{(1240 \text{ eV·nm})^2 (4.39/\text{nm})^2}{32(5.11 \times 10^5 \text{ eV})} = 1.82 \text{ eV}$$

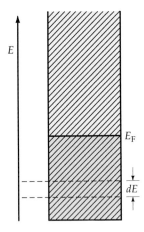

Figure 6-7 Energy levels in a one-dimensional square well. The Fermi energy E_F at $T = 0$ is the energy of the highest occupied level. The levels are so closely spaced they can be assumed to be continuous. The number of states between E and $E + dE$ is $g(E)\, dE$, where $g(E)$ is the density of states.

The average energy is the total energy divided by the number of particles:

$$E_{av} = \frac{1}{N} \sum_{n=1}^{N/2} 2n^2 E_1$$

Since $N/2 \gg 1$, we can approximate the sum using an integral:

$$\sum_{n=1}^{N/2} n^2 \approx \int_0^{N/2} n^2\, dn = \frac{1}{3}\left(\frac{N}{2}\right)^3$$

Thus

$$E_{av} = \frac{2E_1}{N} \frac{1}{3}\left(\frac{N}{2}\right)^3$$

$$= \frac{1}{3}\left(\frac{N}{2}\right)^2 E_1 = \frac{1}{3} E_F \qquad 6\text{-}20$$

Our one-dimensional calculation thus gives an average energy of about $\frac{1}{3}(1.82 \text{ eV}) \approx 0.6$ eV. The temperature at which the average energy would be 0.6 eV for a one-dimensional Maxwell–Boltzmann distribution, obtained from $\frac{1}{2}kT = 0.6$ eV, is about 14,000 K.

Since the energy states are so close together, we can assume that they are continuous (Figure 6-7). Let $n(E)\, dE$ be the number of particles with energies between E and $E + dE$. We can write this distribution function as

Energy distribution function

$$n(E)\, dE = g(E)\, dE\, F \qquad 6\text{-}21$$

where $g(E)\, dE$ is the number of states in dE, and F is the **Fermi factor**, which is the probability that a state will be occupied. At $T = 0$, the Fermi factor is 1 for states of energy less than E_F and 0 for states of energy greater than E_F:

Fermi factor at $T = 0$

$$F = 1 \quad E < E_F$$
$$F = 0 \quad E > E_F \qquad 6\text{-}22$$

The quantity $g(E)$ in Equation 6-21 is called the **density of states**. In one dimension, it is given by

$$g(E) = 2\frac{dn}{dE}$$

where $E = n^2 E_1$ and the 2 is for the two spin states per space state. Then

$$dE = 2nE_1\, dn = 2\left(\frac{E}{E_1}\right)^{1/2} E_1\, dn = 2(EE_1)^{1/2}\, dn$$

and

$$g(E) = E_1^{-1/2} E^{-1/2} \qquad 6\text{-}23$$

The energy distribution function is then

$$n(E) = E_1^{-1/2} E^{-1/2} F$$

In three dimensions, it is more difficult to count the number of states, so we will just give the results. The Fermi energy in three dimensions at $T = 0$ is given by

$$E_F = \frac{h^2}{8m_e}\left(\frac{3N}{\pi V}\right)^{2/3} = \frac{(hc)^2}{8m_ec^2}\left(\frac{3N}{\pi V}\right)^{2/3} \qquad 6\text{-}24a$$

Fermi energy in three dimensions at T = 0

where V is the volume of the metal. As in one dimension, the Fermi energy depends on the number density N/V. Table 6-1 lists the free-electron number densities and the calculated Fermi energies at $T = 0$ for several metals.

Table 6-1 Free-Electron Number Densities and Fermi Energies at $T = 0$ for Selected Elements

Element	N/V, electrons per cm^3	E_F, eV
Al aluminum	18.1×10^{22}	11.7
Ag silver	5.86×10^{22}	5.50
Au gold	5.90×10^{22}	5.53
Cu copper	8.47×10^{22}	7.04
Fe iron	17.0×10^{22}	11.2
K potassium	1.4×10^{22}	2.11
Li lithium	4.70×10^{22}	4.75
Mg magnesium	8.60×10^{22}	7.11
Mn manganese	16.5×10^{22}	11.0
Na sodium	2.65×10^{22}	3.24
Sn tin	14.8×10^{22}	10.2
Zn zinc	13.2×10^{22}	9.46

Example 6-5

Calculate the Fermi energy at $T = 0$ for copper.

Using Equation 6-24a, we obtain for the Fermi energy

$$E_F = \frac{(1240 \text{ eV·nm})^2}{8(5.11 \times 10^5 \text{ eV})}\left(\frac{3N}{\pi V}\right)^{2/3} = (0.365 \text{ eV·nm}^2)\left(\frac{N}{V}\right)^{2/3} \qquad 6\text{-}24b$$

From Table 6-1, the number density of electrons in copper is $8.47 \times 10^{22}/\text{cm}^3 = 84.7/\text{nm}^3$. Then

$$E_F = (0.365 \text{ eV·nm}^2)(84.7/\text{nm}^3)^{2/3} = 7.04 \text{ eV}$$

Note that the Fermi energy is much greater than kT at ordinary temperatures.

Exercise

Use Equation 6-24b to calculate the Fermi energy at $T = 0$ for gold. (Answer: 5.53 eV)

For the three-dimensional case, we again write the energy distribution as $g(E)F$, where F is the Fermi factor given by Equation 6-22, and $g(E)$ is the density of states in three dimensions, which is given by

$$g(E) = \frac{3N}{2}E_F^{-3/2}E^{1/2} \qquad 6\text{-}25$$

Density of states in three dimensions

The average energy at $T = 0$ is calculated from

$$E_{av} = \frac{\int_0^{E_F} E g(E)\, dE}{\int_0^{E_F} g(E)\, dE} = \frac{1}{N} \int_0^{E_F} E g(E)\, dE$$

The result is

Average energy at $T = 0$

$$E_{av} = \tfrac{3}{5} E_F \qquad 6\text{-}26$$

Exercise

Using Equation 6-25, show that $\int_0^{E_F} g(E)\, dE = N$.

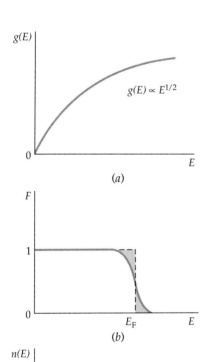

Figure 6-8 The Fermi energy distribution function $n(E)$ shown in (c) is the product of the density of states $g(E)$ shown in (a) and the Fermi factor F shown in (b). The dashed curves in (b) and (c) show the Fermi factor and energy distribution at $T = 0$. At higher temperatures, some electrons with energies near the Fermi energy are excited, as indicated by the shaded regions in (b) and (c).

The Energy Distribution at Temperature T

At temperatures greater than $T = 0$, some electrons will gain energy and occupy higher-energy states. The Fermi factor will then be slightly different from that given by Equation 6-22. However, an electron cannot move to a higher or lower state unless it is unoccupied. Since the kinetic energy of the lattice ions is of the order of kT, electrons cannot gain much more energy than kT in collisions with the lattice. Therefore, only those electrons with energies within about kT of the Fermi energy can gain energy as the temperature is increased. At 300 K, kT is only 0.026 eV, so the exclusion principle prevents all but a very few electrons near the top of the energy distribution from gaining energy through random collisions with the lattice ions. Figure 6-8 shows the density of states, the Fermi factor, and the product of these, which gives the energy distribution.

The **Fermi temperature** T_F is defined by

$$kT_F = E_F \qquad 6\text{-}27$$

For temperatures much lower than the Fermi temperature, the average energy of the lattice ions will be much less than the Fermi energy, and the electron energy distribution will not differ greatly from that at $T = 0$.

Example 6-6

Find the Fermi temperature for copper.

Using $E_F = 7.04$ eV and $k = 1.38 \times 10^{-23}$ J/K $= 8.62 \times 10^{-5}$ eV/K in Equation 6-27, we obtain

$$T_F = \frac{E_F}{k} = \frac{7.04 \text{ eV}}{8.62 \times 10^{-5} \text{ eV/K}} = 81{,}700 \text{ K}$$

We can see from this example that the Fermi temperature of copper is much greater than any temperature T for which copper remains a solid.

The complete quantum-mechanical distribution function for electrons at any temperature is called the **Fermi–Dirac distribution**. The energy distribution function $n(E)$ is given by Equation 6-21. The Fermi factor for temperatures much less than the Fermi temperature is sketched in Figure 6-8b. At temperatures above $T = 0$, there is no energy below which all states are full and above which all states are empty, so we must alter our definition of the Fermi energy:

The Fermi energy E_F at any temperature T is defined as that energy at which the probability of a state being occupied is $\frac{1}{2}$; that is, it is the energy at which the Fermi factor F has the value $\frac{1}{2}$.

Fermi energy at temperature T

For all but extremely high temperatures, the difference between the Fermi energy at temperature T and that at $T = 0$ is very small because only those electrons within about kT of the Fermi energy can be excited to higher energy states. The Fermi factor F at temperature T is given by

$$F = \frac{1}{e^{(E-E_F)/kT} + 1} \qquad 6\text{-}28$$

Fermi factor at temperature T

We can see from Equation 6-28 that as $T \to 0$, $e^{(E-E_F)/kT} \to 0$ for $E < E_F$ and $e^{(E-E_F)/kT} \to \infty$ for $E > E_F$, which is consistent with Equation 6-22. We can also see that for those few electrons with energies much greater than the Fermi energy, the Fermi factor approaches $e^{(E_F-E)/kT}$, which is proportional to $e^{-E/kT}$. Thus, the high-energy tail of the Fermi–Dirac energy distribution decreases as $e^{-E/kT}$, just like the classical Maxwell–Boltzmann energy distribution. In this energy region, there are many unoccupied energy states and few electrons, so the Pauli exclusion principle is not important and the distribution approaches the classical distribution. This result is important because it applies to the conduction electrons in semiconductors, as will be discussed in Sections 6-5 and 6-6.

Contact Potential

When two different metals are placed in contact, a potential difference develops between them. Figure 6-9a shows the energy levels for two metals with different Fermi energies E_{F1} and E_{F2} and different work functions ϕ_1 and ϕ_2. When the metals are in contact, the total energy of the system is lowered if electrons near the boundary move from the metal with the higher Fermi energy into the metal with the lower Fermi energy until the Fermi energies of the two metals are the same, as shown in Figure 6-9b. When equilibrium is established, the metal with the lower initial Fermi energy is negatively charged and that with the higher initial Fermi energy is positively charged, so there is a potential difference between them. This potential difference, called the **contact potential**, equals the difference in the work functions of the two metals divided by the electronic charge e:

$$V_{\text{contact}} = \frac{\phi_1 - \phi_2}{e} \qquad 6\text{-}29$$

Table 6-2 lists the work functions for several metals.

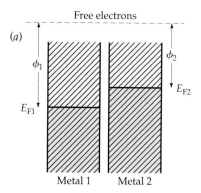
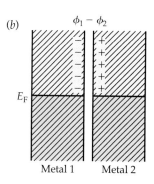

Figure 6-9 (a) Energy levels for two different metals with different Fermi energies and work functions. (b) When the metals are in contact, electrons flow from the metal that initially has the higher Fermi energy to the metal that initially has the lower Fermi energy until the Fermi energies are equal.

Table 6-2 Work Functions for Some Metals	
Metal	ϕ, eV
Ag silver	4.7
Au gold	4.8
Ca calcium	3.2
Cu copper	4.1
K potassium	2.1
Mn manganese	3.8
Na sodium	2.3
Ni nickel	5.2

Example 6-7

The threshold wavelength for the photoelectric effect is 271 nm for tungsten and 262 nm for silver. What is the contact potential developed when silver and tungsten are placed in contact?

From Equation 2-5, the work function ϕ is related to the threshold wavelength λ_t by

$$\phi = \frac{hc}{\lambda_t}$$

The work function for tungsten ϕ_W is thus

$$\phi_W = \frac{hc}{\lambda_t} = \frac{1240 \text{ eV} \cdot \text{nm}}{271 \text{ nm}} = 4.58 \text{ eV}$$

and that for silver is

$$\phi_{Ag} = \frac{1240 \text{ eV} \cdot \text{nm}}{262 \text{ nm}} = 4.73 \text{ eV}$$

The contact potential is thus

$$V_{\text{contact}} = \frac{\phi_{Ag} - \phi_W}{e} = 4.73 \text{ V} - 4.58 \text{ V} = 0.15 \text{ V}$$

6-4 Quantum Theory of Electrical Conduction

With two relatively simple but important quantum-mechanical modifications of the classical free-electron theory, we can understand the electrical conduction of metals. First, we must replace the classical Maxwell–Boltzmann energy distribution with the Fermi–Dirac distribution of energies in the electron gas as discussed in the previous section. Second, we must consider the effect of the wave properties of electrons on the scattering of electrons by the lattice ions. We will discuss this modification qualitatively.

We might expect that most of the electrons would not participate in the conduction of electricity because of the exclusion principle, but this is not the case because the electric field accelerates all the electrons together. Figure 6-10 shows the Fermi factor in one dimension versus velocity for an ordinary temperature (such as $T = 300$ K) that is small compared with T_F. The factor is approximately 1 for speeds v_x in the range $-u_F < v_x < u_F$, where the Fermi speed u_F is the speed corresponding to the Fermi energy:

$$u_F = \sqrt{\frac{2E_F}{m_e}} \qquad \qquad 6\text{-}30$$

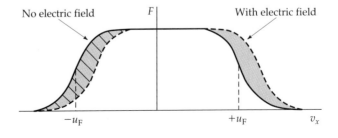

Figure 6-10 Fermi factor versus velocity in one dimension with no electric field (solid) and with an electric field in the $+x$ direction (dashed). The difference is greatly exaggerated.

Example 6-8

Calculate the Fermi speed for copper.

Using $E_F = 7.04$ eV, we obtain

$$u_F = \sqrt{\frac{2(7.04 \text{ eV})}{9.11 \times 10^{-31} \text{ kg}} \left(\frac{1.6 \times 10^{-19} \text{ J}}{1 \text{ eV}}\right)} = 1.57 \times 10^6 \text{ m/s}$$

The dashed curve in Figure 6-10 shows the Fermi factor after the electric field has been acting for some time t. Although all of the electrons have been shifted to higher velocities, the net effect is equivalent to shifting only the electrons near the Fermi level. We can use the classical equation for the resistivity (Equation 6-13) if we use the Fermi speed u_F in place of v_{av}:

$$\rho = \frac{m_e u_F}{ne^2 \lambda} \qquad 6\text{-}31$$

We now have two problems. First, since the Fermi speed u_F is approximately independent of temperature, the resistivity given by Equation 6-31 is independent of temperature unless the mean free path depends on it. The second problem concerns magnitudes. As mentioned earlier, the classical expression for resistivity using v_{av} calculated from the Maxwell–Boltzmann distribution gives values about 6 times too large at $T = 300$ K. Since the Fermi speed u_F is about 16 times the Maxwell–Boltzmann value of v_{av}, the magnitude of ρ predicted by Equation 6-31 will be more than 6 times greater than the experimentally determined value.

The resolution of both of these problems lies in the way the value of the mean free path is calculated. If we use u_F from Equation 6-30 and the experimental value for the resistivity, $\rho \approx 1.7 \times 10^{-8}$ Ω·m, we can solve Equation 6-31 for the mean free path. The result is $\lambda \approx 40$ nm, which is about 100 times the value of 0.4 nm calculated from Equation 6-14 using $r \approx 0.1$ nm as the radius of the copper ion. The reason for this large discrepancy between the classical calculation of the mean free path and the "experimental" result calculated from Equation 6-31 is that the wave nature of the electron must be taken into account. The collision of an electron with a lattice ion is not similar to the collision of a baseball and a tree. Instead, it involves the scattering of the electron wave by the regularly spaced lattice. Detailed calculations show that there is no scattering of the electron waves by a *perfectly* ordered crystal, and that the mean free path is infinite. The scattering of electron waves arises because of imperfections in the crystal lattice, most commonly due to impurities, or thermal vibrations of the lattice ions.

In Equation 6-14 for the classical mean free path, the quantity πr^2 can be thought of as the area A of the lattice ion as seen by an electron. The mean free path is then given by

$$\lambda = \frac{1}{nA} \qquad 6\text{-}32$$

Figure 6-11a depicts the classical picture in which the lattice ions have an area $A = \pi r^2$. According to the quantum-mechanical theory of electron scattering, however, the area A has nothing to do with the size of the lattice ions. Instead, it depends on deviations of the lattice ions from a perfectly ordered array. Figure 6-11b depicts the quantum picture in which the lattice ions are points that have no size but present an area $A = \pi r_0^2$, where r_0 is the amplitude of thermal vibrations, to the electrons. This area is proportional to the square of the amplitude of the vibrations. From our knowledge of simple

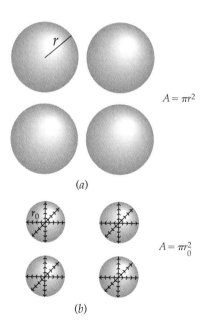

Figure 6-11 (a) Classical picture of the lattice ions as spherical balls of radius r that present an area πr^2 to the electrons. (b) Quantum-mechanical picture of the lattice ions as points that are vibrating in three dimensions. The area presented to the electrons is πr_0^2, where r_0 is the amplitude of oscillation of the ions.

harmonic motion, we know that the energy of vibration is also proportional to the square of the amplitude. Thus, the effective area A is proportional to the energy of vibration of the lattice ions. From the equipartition theorem, we know that the energy of vibration is proportional to kT, where k is Boltzmann's constant. Thus, the mean free path λ is proportional to $1/T$, so the resistivity is proportional to T in agreement with experiment.

The effective area A due to thermal vibrations can be calculated, and it turns out to be about 100 times smaller at $T = 300$ K than the actual area πr^2 of a lattice ion, thus giving values for the resistivity that are in agreement with experiment. We see, therefore, that the free-electron model of metals gives a good account of electrical conduction if the classical mean speed v_{av} is replaced by the Fermi speed u_F and if the collisions between electrons and the lattice ions are interpreted in terms of the scattering of electron waves, for which only deviations from a perfectly ordered lattice are important.

The presence of impurities in a metal also causes deviations from perfect regularity in the crystal lattice. The effects of impurities on resistivity are approximately independent of temperature. The resistivity of a metal containing impurities can be written $\rho = \rho_t + \rho_i$, where ρ_t is the resistivity due to the thermal motion of the lattice ions and ρ_i is the resistivity due to impurities. Figure 6-12 shows typical resistance-versus-temperature curves for metals with impurities. As the temperature approaches zero, ρ_t approaches zero and the resistivity approaches the constant ρ_i due to impurities.

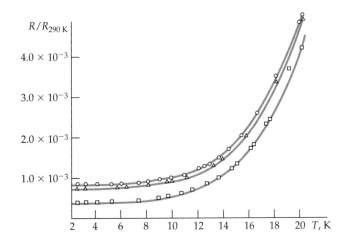

Figure 6-12 Relative resistance versus temperature for three samples of sodium. The three curves have the same temperature dependence but different magnitudes because of differing amounts of impurities in the samples.

Heat Conduction and Heat Capacity

Quantum-mechanical modifications of the theory of heat conduction in metals also bring that theory into good agreement with experimental measurements. By replacing the Maxwell–Boltzmann energy distribution of the electrons with the Fermi–Dirac distribution, we can understand why the contribution of the electron gas to the heat capacity of a metal is much less than the value $\frac{3}{2}R$ predicted by classical theory. At $T = 0$, the average energy of the electrons is $\frac{3}{5}E_F$, so the total energy is $E = \frac{3}{5}NE_F$. At some temperature T, only those electrons with energies near the Fermi energy can be excited by random collisions with the lattice ions, which have an average energy of the order of kT. The fraction of the electrons that are excited is of the order of kT/E_F, and their energy is increased from that at $T = 0$ by an amount that is of the order of kT. We can thus express the energy of N electrons at temperature T as

$$E = \frac{3}{5}NE_F + \alpha N \frac{kT}{E_F} kT \qquad 6\text{-}33$$

where α is some constant that we expect to be of the order of 1 if our reasoning is correct. The calculation of α involves the complete Fermi–Dirac distribution at an arbitrary temperature T and is quite difficult. The result is $\alpha = \pi^2/4$. Using this result and $E_F/k = T_F$, the Fermi temperature, we obtain for the contribution of the electron gas to the heat capacity at constant volume

$$C_v = \frac{dU}{dT} = 2\alpha Nk \frac{kT}{E_F} = \frac{\pi^2}{2} nR \frac{T}{T_F}$$

For one mole, $Nk = R$, the gas constant. The molar heat capacity at constant volume is then

$$C_{mv} = \frac{\pi^2}{2} R \frac{T}{T_F} \qquad 6\text{-}34$$

We can see that because of the large value of T_F, the contribution of the electron gas is a small fraction of R at ordinary temperatures. Because $T_F = 81{,}700$ K for copper, the molar heat capacity of the electron gas at $T = 300$ K is

$$C_{mv} = \frac{\pi^2}{2} \left(\frac{300 \text{ K}}{81{,}700} \right) R \approx 0.02R$$

which is in good agreement with experiment.

6-5 Band Theory of Solids

We have seen that, if the electron gas is treated as a Fermi gas and the electron–lattice collisions are treated as the scattering of electron waves, the free-electron model gives a good account of the thermal and electrical properties of conductors. However, this simple model gives no indication of why one material is a conductor and another is an insulator. Resistivities vary enormously between insulators and conductors. For example, the resistivity of a typical insulator, such as quartz, is of the order of 10^{16} $\Omega\cdot$m, whereas that of a typical conductor is of the order of 10^{-8} $\Omega\cdot$m. To understand why some materials conduct and others do not, we must refine the free-electron model and consider the effect of the lattice on the electron energy levels.

We begin by considering the energy levels of the individual atoms as they are brought together. As we have seen, the allowed energy levels in an isolated atom are often far apart. For example, in hydrogen, the energy for $n = 1$ is -13.6 eV and that for $n = 2$ is $(-13.6$ eV$)/4 = -3.4$ eV. Let us consider two identical hydrogen atoms, and focus our attention on one particular energy level, such as $n = 2$. When the atoms are far apart, the energy of this level is the same for each atom. As the atoms are brought close together, the energy of this level for each atom changes because of the influence of the other atom. As a result, the $n = 2$ level splits into two levels of slightly different energies for the two-atom system.

If we have N identical atoms, a particular energy level in the isolated atom splits into N different, nearly equal energy levels when the atoms are close together. Figure 6-13 shows the energy splitting of the 1s and 2s energy levels for six atoms as a function of the separation of the atoms. In a macroscopic solid, N is very large—of the order of Avogadro's number—so each energy level splits into a very large number of levels called a **band**. Because the number of levels in the band is so large, the levels are spaced almost continuously within the band. There is a separate band of levels for each particular energy level of the isolated atom. The bands may be widely separated in energy, they may be close together, or they may even overlap, depending on the kind of atom and the type of bonding in the solid.

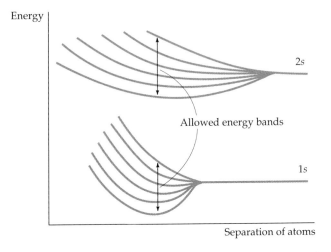

Figure 6-13 Energy splitting of the 1s and 2s energy levels for six atoms as a function of the separation of the atoms.

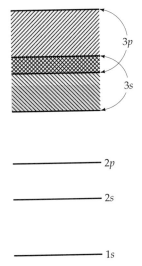

Figure 6-14 Energy band structure of sodium. The empty 3p band overlaps the half-filled 3s band. Just above the filled states are many empty states into which electrons can be excited by an electric field, so sodium is a conductor.

We can now understand why some solids are conductors and others are insulators. Consider sodium. There is room for two electrons in the 3s state of each atom, but each separate sodium atom has only one 3s electron. Therefore, when N sodium atoms are bound in a solid, the 3s energy band is only half filled. In addition, the empty 3p band overlaps the 3s band. The allowed energy bands of sodium are shown schematically in Figure 6-14. The occupied levels are shaded. We can see that many allowed energy states are available just above the filled ones, so the valence electrons can easily be raised to a higher-energy state by an electric field. Accordingly, sodium is a good conductor. Magnesium, on the other hand, has two 3s electrons, so the 3s band is filled. However, like sodium, the empty 3p band overlaps the 3s band, so magnesium is also a conductor.

The band structure of an ionic crystal, such as NaCl, is quite different. The energy bands arise from the energy levels of the Na^+ and Cl^- ions. Both of these ions have a closed-shell configuration, so the highest occupied band in NaCl is completely full. The next allowed band, which is empty, arises from the excited states of Na^+ and Cl^-. There is a large energy gap between the filled band and this empty band. A typical electric field applied to NaCl will be too weak to excite an electron from the upper energy levels of the filled band across the large gap into the lower energy levels of the empty band, so NaCl is an insulator. When an applied electric field is sufficiently strong to cause an electron to be excited to the empty band, the phenomenon is called dielectric breakdown.

Figure 6-15 shows four possible kinds of band structures for a solid. The band occupied by the outermost, valence electrons is called the **valence band**. The lowest band in which there are unoccupied states is called the **conduction band**. In sodium, the valence band is only half filled, so the valence band is also the conduction band. In magnesium, the filled 3s band and the empty 3p bands overlap, forming a combined valence–conduction band that is only partially filled, so magnesium is also a conductor.

The band structure for a conductor such as copper is shown in Figure 6-15a. The lower bands are filled with the inner electrons of the atoms. According to the Pauli exclusion principle, no more electrons can occupy levels in these bands. The uppermost band that contains electrons is only about half full. Because these electrons are the electrons that form the metallic bond, this band is the valence band. In the normal state, at low temperatures, the lower half of the valence band is filled and the upper half is empty, so this band is also the conduction band. At higher temperatures, a few of the electrons are in the higher energy states of this band because of thermal excitation, but there are still many unfilled energy states above the filled ones. When an electric field is established in the conductor, the electrons in the conduction band are accelerated, which means that their energy is increased. This is consistent with the Pauli exclusion principle because there are many empty energy states just above those occupied by electrons in this band. These electrons are thus the conduction electrons.

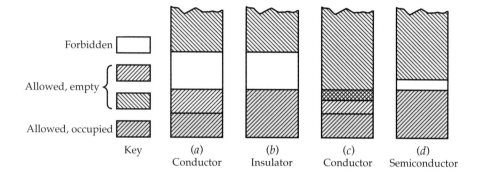

Figure 6-15 Four possible band structures for a solid. (a) A typical conductor. The valence band is only partially full, so electrons can be easily excited to nearby energy states. (b) A typical insulator. There is a forbidden band with a large energy gap between the filled valence band and the conduction band. (c) A conductor in which the allowed energy bands overlap. (d) A semiconductor. The energy gap between the filled valence band and the conduction band is very small, so some electrons are excited to the conduction band at normal temperatures, leaving holes in the valence band.

Figure 6-15b shows the band structure for a typical insulator. At $T = 0$ K, the highest energy band that contains electrons is completely full. The next energy band containing empty energy states, the conduction band, is separated from the last filled band by an energy gap. At $T = 0$, the conduction band is empty. At ordinary temperatures, a few electrons can be excited to states in this band, but most cannot be because the energy gap is large compared with the energy an electron might obtain by thermal excitation, which on the average is of the order of $kT \approx 0.026$ eV at $T = 300$ K. This energy gap is sometimes referred to as the **forbidden energy band.** Very few electrons can be thermally excited to the nearly empty conduction band, even at fairly high temperatures. When an electric field is established in the solid, electrons cannot be accelerated because there are no empty energy states at nearby energies. We describe this by saying that there are no free electrons. The small conductivity that is observed is due to the very few electrons that are thermally excited into the nearly empty conduction band.

In Figure 6-15c the valence and conduction bands overlap. A material such as magnesium with this type of band structure is a conductor.

In some materials the energy gap between the top filled band and the empty conduction band is very small, as shown in Figure 6-15d. At $T = 0$, there are no electrons in the conduction band and the material is an insulator. At ordinary temperatures, there are an appreciable number of electrons in the conduction band due to thermal excitation. Such a material is called an **intrinsic semiconductor.** In the presence of an electric field, the electrons in the conduction band can be accelerated because there are empty states nearby. Also, for each electron in the conduction band there is a vacancy, or hole, in the nearly filled valence band. In the presence of an electric field, electrons in this band can be excited to a vacant energy level. This contributes to the electric current and is most easily described as the motion of a hole in the direction of the field and opposite the motion of the electrons. The hole thus acts like a positive charge. An analogy of a two-lane, one-way road with one lane full of parked cars and the other empty may help to visualize the conduction of holes. If a car moves out of the filled lane into the empty lane, it can move ahead freely. As the other cars move up to occupy the space left, the empty space propagates backwards in the direction opposite the motion of the cars. Both the forward motion of the car in the nearly empty lane and the backward propagation of the empty space contribute to a net forward propagation of the cars.

An interesting characteristic of semiconductors is that the conductivity increases (and the resistivity decreases) as the temperature increases, which is contrary to the case for normal conductors. The reason is that as the temperature is increased, the number of free electrons is increased because there are more electrons in the conduction band. The number of holes in the valence band is also increased of course. In semiconductors, the effect of the increase in the number of charge carriers, both electrons and holes, exceeds the effect of the increase in resistivity due to the increased scattering of the electrons by the lattice ions due to thermal vibrations. Semiconductors therefore have a negative temperature coefficient of resistivity.

A typical intrinsic semiconductor is silicon, which has four valence electrons in the $n = 3$ shell. In a solid crystal of silicon, each atom forms a covalent bond with four neighboring atoms and shares one of its valence electrons with each neighbor, as illustrated schematically in Figure 6-16. Figure 6-17 shows the energy level structures of carbon, silicon, and germanium. The $2s$ and $2p$ states of carbon, the $3s$ and $3p$ states of silicon, and $4s$ and $4p$ states of germanium have room for eight electrons. These states split into two hybrid states, each with room for four electrons. The lower hybrid state is filled with the valence electrons and the upper one is empty. At the separation of about 0.15 nm for carbon, the energy gap between these states is about 7 eV, so carbon is an insulator. For silicon and germanium, the separation is about 0.24 nm and the energy gap is about 1 eV, so these elements are intrinsic semiconductors.

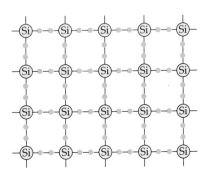

Figure 6-16 A two-dimensional schematic illustration of solid silicon. Each atom forms a covalent bond with four neighbors, sharing one of its four valence electrons with each neighbor.

Figure 6-17 Splitting of the $2s$ and $2p$ states of carbon, the $3s$ and $3p$ states of silicon, or the $4s$ and $4p$ states of germanium versus the separation of the atoms. The energy gap between the filled states and empty states is about 7 eV for carbon but only about 1 eV for silicon and germanium.

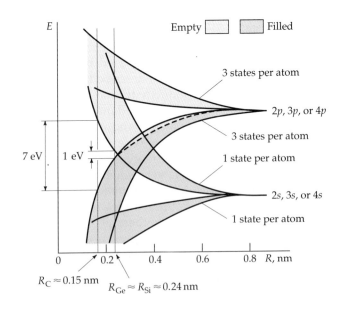

6-6 Impurity Semiconductors

Most semiconductor devices, such as the semiconductor diode and the transistor, make use of **impurity semiconductors,** which are created through the controlled addition of certain impurities to intrinsic semiconductors. This process is called **doping.** Figure 6-18a is a schematic illustration of silicon doped with a small amount of arsenic such that arsenic atoms replace a few of the silicon atoms in the crystal lattice. Arsenic has five electrons in its valence shell rather than the four of silicon. Four of these electrons take part in covalent bonds with the four neighboring silicon atoms, and the fifth electron is very loosely bound to the atom. This extra electron occupies an energy level that is just slightly below the conduction band in the solid, and it is easily excited into the conduction band, where it can contribute to electrical conduction.

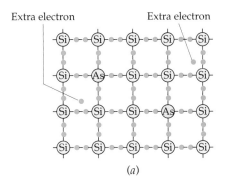
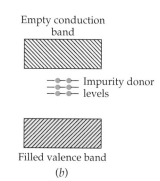

Figure 6-18 (a) A two-dimensional schematic illustration of silicon doped with arsenic. Because arsenic has five valence electrons, there is an extra, weakly bound electron that is easily excited to the conduction band, where it can contribute to electrical conduction. (b) Band structure of an *n*-type semiconductor such as silicon doped with arsenic. The impurity atoms provide filled energy levels that are just below the conduction band. These levels donate electrons to the conduction band.

The effect on the band structure of a silicon crystal achieved by doping it with arsenic is shown in Figure 6-18b. The levels shown just below the conduction band are due to the extra electrons of the arsenic atoms. These levels are called **donor levels** because they donate electrons to the conduction band without leaving holes in the valence band. Such a semiconductor is called an *n*-type semiconductor because the major charge carriers are *negative* electrons. The conductivity of a doped semiconductor can be controlled by controlling the amount of impurity added. The addition of just one part per million can increase the conductivity by several orders of magnitude.

Another type of impurity semiconductor can be made by replacing a silicon atom with a gallium atom that has 3 electrons in its valence level (Figure 6-19a). The gallium atom accepts electrons from the valence band to complete its four covalent bonds, thus creating a hole in the valence band. The effect on the band structure of silicon achieved by doping it with gallium is shown in Figure 6-19b. The empty levels shown just above the valence band are due to the holes from the ionized gallium atoms. These levels are called **acceptor levels** because they accept electrons from the filled valence band when these electrons are thermally excited to a higher energy state.

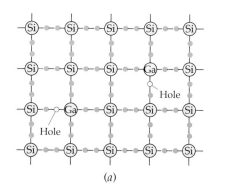
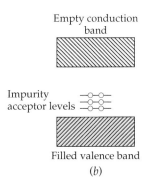

Figure 6-19 (a) A two-dimensional schematic illustration of silicon doped with gallium. Because gallium has only three valence electrons, there is a hole in one of its bonds. As electrons move into the hole, the hole moves about, contributing to the conduction of electrical current. (b) Band structure of a *p*-type semiconductor such as silicon doped with gallium. The impurity atoms provide empty energy levels just above the filled valence band that accept electrons from the valence band.

This creates holes in the valence band that are free to propagate in the direction of an electric field. Such a semiconductor is called a *p*-type **semiconductor** because the charge carriers are *positive* holes. The fact that conduction is due to the motion of holes can be verified by the Hall effect.

6-7 Semiconductor Junctions and Devices

Semiconductor devices such as diodes and transistors make use of *n*-type and *p*-type semiconductors joined together as shown in Figure 6-20. In practice, the two types of semiconductors are often a single silicon crystal doped with donor impurities on one side and acceptor impurities on the other. The region in which the semiconductor changes from a *p*-type to an *n*-type is called a **junction.**

When an *n*-type and a *p*-type semiconductor are placed in contact, the initially unequal concentrations of electrons and holes result in the diffusion of electrons across the junction from the *n* side to the *p* side and holes from the *p* side to the *n* side until equilibrium is established. The result of this diffusion is a net transport of positive charge from the *p* side to the *n* side. Unlike the case when two different metals are in contact, the electrons cannot travel very far from the junction region because the semiconductor is not a particularly good conductor. The diffusion of electrons and holes therefore creates a double layer of charge at the junction similar to that on a parallel-plate capacitor. There is thus a potential difference V across the junction, which tends to inhibit further diffusion. In equilibrium, the *n* side with its net positive charge will be at a higher potential than the *p* side with its net negative charge. In the junction region, there will be very few charge carriers of either type, so the junction region has a high resistance. Figure 6-21 shows the energy level diagram for a *pn* junction. The junction region is also called the **depletion region** because it has been depleted of charge carriers.

A semiconductor with a *pn* junction can be used as a simple diode rectifier. In Figure 6-22, an external potential difference has been applied across the junction by connecting a battery and resistor to the semiconductor. When the positive terminal of the battery is connected to the *p* side of the junction, as shown in Figure 6-22a, the diode is said to be **forward biased.** Forward biasing lowers the potential across the junction. The diffusion of electrons and holes is thereby increased as they attempt to reestablish equilibrium, resulting in a current in the circuit. If the positive terminal of the battery is connected to the *n* side of the junction as shown in Figure 6-22b, the diode is said to be **reverse biased.** Reverse biasing tends to increase the potential difference across the junction, thereby further inhibiting diffusion. Figure 6-23 shows a plot of current versus voltage for a typical semiconductor junction. Essentially, the junction conducts only in one direction, the same as a vacuum-tube diode. Junction diodes have replaced vacuum diodes in nearly all applications except when a very high current is required.

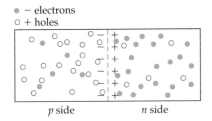

Figure 6-20 A *pn* junction. Because of the difference in their concentrations, holes diffuse from the *p* side to the *n* side and electrons diffuse from the *n* side to the *p* side. As a result, there is a double layer of charge at the junction with the *p* side being negative and the *n* side being positive.

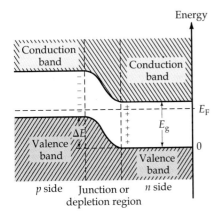

Figure 6-21 Electron energy levels for an unbiased *pn* junction.

Figure 6-22 A *pn*-junction diode. (a) Forward-biased *pn* junction. The applied potential difference enhances the diffusion of holes from the *p* side to the *n* side and electrons from the *n* side to the *p* side, resulting in a current I. (b) Reverse-biased *pn* junction. The applied potential difference inhibits the further diffusion of holes and electrons, so there is no current.

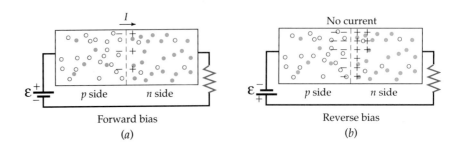

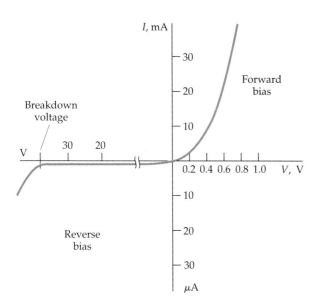

Figure 6-23 Current versus applied voltage across a *pn* junction. Note the different scales for the forward and reverse bias conditions.

Note that the current in Figure 6-23 suddenly increases in magnitude at extreme values of reverse bias. In such large electric fields, electrons are stripped from their atomic bonds and accelerated across the junction. These electrons, in turn, cause others to break loose. This effect is called **avalanche breakdown**. Although such a breakdown can be disastrous in a circuit where it is not intended, the fact that it occurs at a sharp voltage value makes it of use in a special voltage reference standard known as a **Zener diode**.

An interesting effect that we can discuss only qualitatively occurs if both the *n* side and *p* side of a *pn*-junction diode are so heavily doped that the donors on the *n* side provide so many electrons that the lower part of the conduction band is practically filled and the acceptors on the *p* side accept so many electrons that the upper part of the valence band is nearly empty. Figure 6-24a shows the energy-level diagram for this situation. Because the depletion region is now so narrow, electrons can easily penetrate the potential barrier across the junction. This flow of electrons is called a **tunneling current**, and such a heavily doped diode is called a **tunnel diode**.

At equilibrium with no bias, there is an equal tunneling current in each direction. When a small bias voltage is applied across the junction, the energy-level diagram is as shown in Figure 6-24b, and the tunneling of electrons from the *n* side to the *p* side is increased whereas that in the opposite direction is decreased. This tunneling current in addition to the usual current due to diffusion results in a considerable net current. When the bias voltage is increased slightly, the energy-level diagram is as shown in Figure 6-24c, and the tunneling current is decreased. Although the diffusion current is increased, the net current is decreased. At large bias voltages, the

Figure 6-24 Electron energy levels for a heavily doped *pn*-junction tunnel diode. (*a*) With no bias voltage, some electrons tunnel in each direction. (*b*) With a small bias voltage, the tunneling current is enhanced in one direction, making a sizable contribution to the net current. (*c*) With further increases in the bias voltage, the tunneling current decreases dramatically.

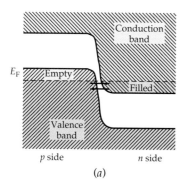
(a)

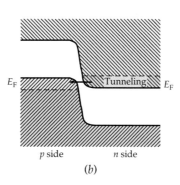
(b)

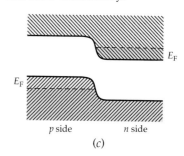
(c)

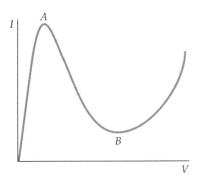

Figure 6-25 Current versus applied voltage for a tunnel diode. Up to point A, an increase in the bias voltage enhances tunneling. Between points A and B, an increase in the bias voltage inhibits tunneling. After point B, the tunneling is negligible, and the diode behaves like an ordinary pn-junction diode.

Figure 6-26 A pn-junction semiconductor as a solar cell. When light strikes the p-type region, electron–hole pairs are created, resulting in a current through the load resistance R_L.

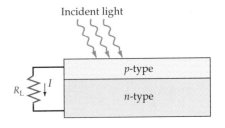

tunneling current is completely negligible, and the total current increases with increasing bias voltage due to diffusion as in an ordinary pn-junction diode. Figure 6-25 shows the current-versus-voltage curve for a tunnel diode. Such diodes are used in electric circuits because of their very fast response time. When operated near the peak in the current-versus-voltage curve, a small change in bias voltage results in a large change in the current.

Another use for the pn-junction semiconductor is the **solar cell**, which is illustrated schematically in Figure 6-26. When a photon of energy greater than the gap energy (1.1 eV in silicon) strikes the p-type region, it can excite an electron from the valence band into the conduction band, leaving a hole in the valence band. This region is already rich in holes. Some of the electrons created by the photons will recombine with holes, but some will migrate to the junction. From there they are accelerated into the n-type region by the electric field between the double layer of charge. This creates an excess negative charge in the n-type region and excess positive charge in the p-type region. The result is a potential difference between the two regions, which in practice is about 0.6 V. If a load resistance is connected across the two regions, a charge flows through the resistance. Some of the incident light energy is thus converted into electrical energy. The current in the resistor is proportional to the number of incident photons, which is in turn proportional to the intensity of the incident light.

There are many other applications of semiconductors with pn junctions Particle detectors called **surface-barrier detectors** consist of a pn-junction semiconductor with a large reverse bias so that there is ordinarily no current. When a high-energy particle, such as an electron, passes through the semiconductor, it creates many electron–hole pairs as it loses energy. The resulting current pulse signals the passage of the particle. **Light-emitting diodes (LEDs)** are pn-junction semiconductors with a large forward bias that produces a large excess concentration of electrons on the p side and holes on the n side of the junction. Under these conditions, the diode emits light as the electrons and holes recombine. This is essentially the reverse of the process that occurs in a solar cell, in which electron–hole pairs are created by the absorption of light. LEDs are commonly used as displays for digital watches and calculators.

A light-emitting diode (LED).

Transistors

The transistor, invented in 1948 by William Shockley, John Bardeen, and Walter H. Brattain, has revolutionized the electronics industry and our everyday world. A simple junction transistor consists of three distinct semiconductor regions called the **emitter,** the **base,** and the **collector.** The base is a very thin region of one type of semiconductor sandwiched between two regions of the opposite type. The emitter semiconductor is much more heavily doped than either the base or the collector. In an npn transistor, the emitter and collector are n-type semiconductors and the base is a p-type semiconductor; in a pnp transistor, the base is an n-type semiconductor and the emitter and collector are p-type semiconductors. The emitter, base, and

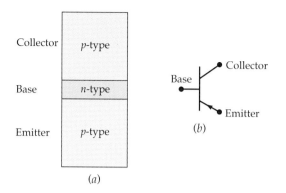

Figure 6-27 A *pnp* transistor. The heavily doped emitter emits holes that pass through the thin base to the collector. (*b*) Symbol for a *pnp* transistor in a circuit. The arrow points in the direction of the conventional current, which is the same as that of the emitted holes.

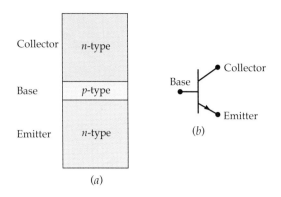

Figure 6-28 An *npn* transistor. The heavily doped emitter emits electrons that pass through the thin base to the collector. (*b*) Symbol for an *npn* transistor. The arrow points in the direction of the conventional current, which is opposite the direction of the emitted electrons.

collector behave somewhat similarly to the cathode, grid, and plate in a vacuum-tube triode, except that in a *pnp* transistor it is holes that are emitted rather than electrons.

Figures 6-27 and 6-28 show, respectively, a *pnp* transistor and an *npn* transistor with the symbols used to represent each transistor in circuit diagrams. We see that a transistor consists of two *pn* junctions. We will discuss the operation of a *pnp* transistor. The operation of an *npn* transistor is similar.

In normal operation, the emitter–base junction is forward biased, and the base–collector junction is reverse biased, as shown in Figure 6-29. The heavily doped *p*-type emitter emits holes that flow across the emitter–base junction into the base. Because the base is very thin, most of these holes flow across the base into the collector. This flow constitutes a current I_c from the emitter to the collector. However, some of the holes recombine in the base producing a positive charge that inhibits the further flow of current. To prevent this, some of the holes that do not reach the collector are drawn off the base as a base current I_b in a circuit connected to the base. In Figure 6-29, therefore, I_c is almost but not quite equal to I_e, and I_b is much smaller than either I_c or I_e. It is customary to express I_c as

$$I_c = \beta I_b \qquad 6\text{-}35$$

where β is called the **current gain** of the transistor. Transistors can be designed to have values of β as low as 10 or as high as several hundred.

Figure 6-30 shows a simple *pnp* transistor used as an amplifier. A small time-varying input voltage v_s is connected in series with a bias voltage V_{eb}.

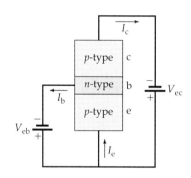

Figure 6-29 A *pnp* transistor biased for normal operation. Holes from the emitter can easily diffuse across the base, which is only tens of nanometers thick. Most of the holes flow to the collector, producing the current I_c.

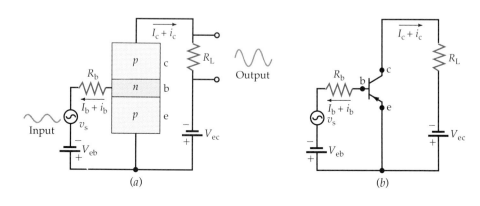

Figure 6-30 (*a*) A *pnp* transistor used as an amplifier. A small change i_b in the base current results in a large change i_c in the collector current. Thus a small signal in the base circuit results in a large signal across the load resistor R_L in the collector circuit. (*b*) The same circuit as (*a*) with the conventional symbol for the transistor.

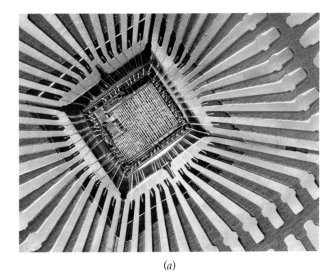
(a)

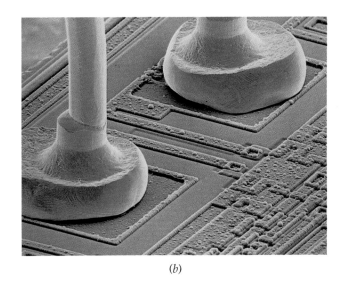
(b)

Integrated circuits (often called ICs or chips) combine "active" electronic devices (transistors and diodes) with "passive" ones (capacitors and resistors) on a single semiconductor crystal. (a) This particular chip, shown connected to 44 conductor leads, has an actual size of 6.4 mm square. It is used to format digitized voice and data signals, so they can share a single transmission line. (b) A scanning electron micrograph showing two conductor leads precision bonded to the edge of a chip (magnification: ×163).

The base current is then the sum of a steady current I_b produced by the bias voltage V_{eb} and a varying current i_b due to the signal voltage v_s. Because v_s may at any instant be either positive or negative, the bias voltage V_{eb} must be large enough to ensure that there is always a forward bias on the emitter–base junction. The collector current will consist of two parts: a direct current $I_c = \beta I_b$ and an alternating current $i_c = \beta i_b$. We thus have a current amplifier in which the time-varying output current i_c is β times the input current i_b. In such an amplifier, the steady currents I_c and I_b, although essential to the operation of the transistor, are usually not of interest. The input signal voltage v_s is related to the base current by Ohm's law:

$$i_b = \frac{v_s}{R_b + r_b} \qquad 6\text{-}36$$

where r_b is the internal resistance of the transistor between the base and emitter. Similarly, the collector current i_c produces a voltage v_L across the output or load resistance R_L given by

$$v_L = i_c R_L \qquad 6\text{-}37$$

Using Equation 6-35, we have

$$i_c = \beta i_b = \beta \frac{v_s}{R_b + r_b}$$

The output voltage is thus related to the input voltage by

$$v_L = \beta \frac{R_L}{R_b + r_b} v_s \qquad 6\text{-}38$$

The ratio of the output voltage to the input voltage is the **voltage gain** of the amplifier:

$$\text{Voltage gain} = \frac{v_L}{v_s} = \beta \frac{R_L}{R_b + r_b} \qquad 6\text{-}39$$

A typical amplifier, such as that in a tape player, has several transistors similar to the one in Figure 6-30 connected in series so that the output of one transistor serves as the input for the next. Thus, the very small voltage produced by the passage of the magnetized tape past the pickup heads controls the large amounts of power required to drive the loudspeakers. The power delivered to the speakers is supplied by the sources of direct voltage connected to each transistor.

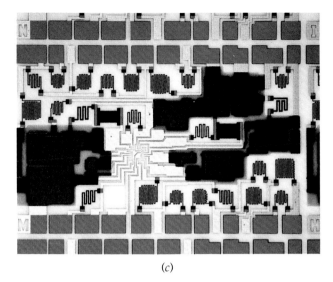

(c)

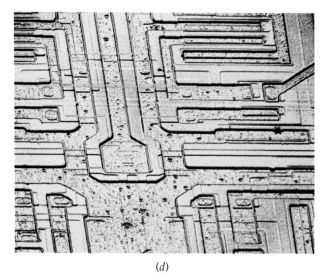

(d)

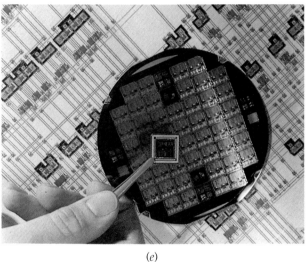

(e)

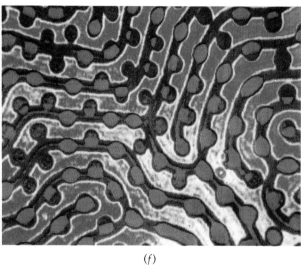

(f)

(c) Capacitors (orange blocks), resistors (brown blocks and meandering black lines), and conductors (gold lines) on a ceramic base, formed here by metal films only a few tenths of a micrometer thick. No means have been found to directly fabricate inductors (the other passive circuit component) on ICs; they are simulated with other circuitry or appended to a chip as discrete components. (d) A scanning electron micrograph of metal oxide silicon (MOS) transistors in patterned layers (magnification ×106). MOS transistors are manufactured by heating an original silicon wafer to about 1000°C, causing a layer of silicon dioxide (SiO_2) to form on its surface. This is coated with a photoresist and exposed to light through a mask. Unexposed (masked) windows of photoresist are etched away with a developer, exposing the silicon dioxide, which is etched away with acid. The exposed (unmasked) areas are resistant to the developer and are not affected. The wafer is again heated and this time doped, via a diffusion process, with a p-type impurity, forming pn junctions in the n-type silicon. The chip is covered with a contact metal (typically aluminum), which bonds to the SiO_2 that has reformed in windows while the chip was heated and doped. The contact metal itself is patterned in a final photo-etching process. Entire microchips are fabricated by an elaboration, using many masks, of this process. (e) The chip in the tweezers holds 150,000 transistors. Beneath it is a 4-inch wide silicon wafer, awaiting dicing, on which a group of chips have been fabricated simultaneously. In the background is a detail of the "stare plot," the layout of the chip's circuits.
(f) Magnetized domains ("bubbles"), blue in this video micrograph, flow along channels in a thin-film garnet memory crystal. Magnetic bubble memory chips are the integrated-circuit analog to magnetic recording tape and disks. The bubbles (actually, cylinders seen in cross section) are created when the garnet is placed between two permanent magnets. They represent regions whose magnetic polarity points in a direction opposite to that of the surrounding crystal. An additional external magnetic field manipulates the position of the bubbles. (Garnet is easy to magnetize, up or down, along a particular axis—but hard to magnetize perpendicular to that axis. This property is necessary for the formation and movement of bubbles.) Storage sites for bubbles are established using a layer of ferromagnetic material deposited on the surface of the crystal; the presence or absence of a bubble at a site can be used to represent a bit of data.

A fixed pattern of light and dark is established in a photorefractive crystal, for instance by letting two laser beams of the same frequency interfere within the crystal.

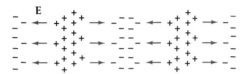

Responding to the electric fields of the light, free electrons or holes move away from bright regions of the crystal. The resulting build-up of charge creates a static electric field within the crystal.

The static electric field acts on the atoms of the crystal lattice. The positions of the atoms shift and the lattice becomes distorted. Distortions in the lattice cause light to propagate faster in some regions of the crystal and slower in others: that is, the refractive index of the crystal is altered.

(a)

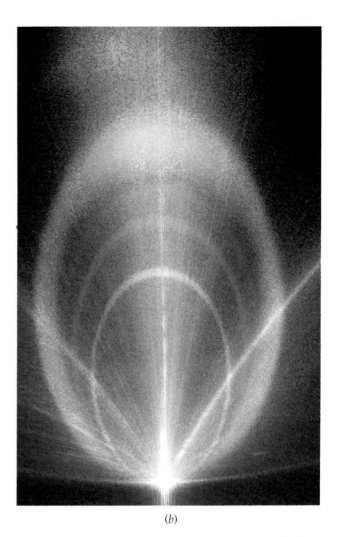

(b)

Photorefractivity. Light incident on a certain class of materials called "photorefractive crystals" causes a rearrangement of their crystal lattice, which in turn alters the optical properties of that lattice. This behavior is due to impurities or defects in the photorefractive crystal. These defects are a source of free electrons or holes that migrate away from illuminated areas in the crystal. The resulting rearrangement of free charge gives rise to electric fields that act on the atoms in the crystal lattice, distorting its shape. This process is summarized in schematic (a). Here, the incident light is from two interfering laser beams. The interference pattern of these beams is translated into a periodic pattern in the crystal lattice itself. Such a pattern is sometimes referred to as a "refractive-index grating." Typically, the grating pattern will be phase shifted with respect to the interference pattern of the light beams. One of the beams will tend to interfere constructively with light scattered by the grating, the other destructively. As a result, the first beam will be intensified passing through the crystal, and the second beam will be reduced. Figure (b) shows the result of the "beam-fanning effect" occurring in the photorefractive crystal barium titanate (BaTiO$_3$). Multicolored incident laser beams are partially scattered by defects in the crystal. An interference pattern arises between the incident and scattered beams, and the resulting deformation of the crystal lattice changes the lattice's index of refraction. The incident beams are refracted to a different angle and a new interference pattern arises, causing a further change in the index of refraction.

Beam fanning lies at the heart of new devices called "phase-conjugate mirrors." Such mirrors produce beams that retrace the path of any incident beam exactly, with the leading edge of the incident wavefront appearing last in the phase-conjugate beam. (In reflection from an ordinary mirror the leading edge of the incident wavefront emerges first.) Incident light undergoing beam fanning in a photorefractive material may be preferentially swept into a corner of the crystal where it is twice internally reflected. The path of such light forms a loop, in which the exiting beam is phase conjugate to the incoming beam. Such a configuration is shown in (c) inside a crystal of barium titanate. The volume refractive-index grating generated in the crystal by these loops produces the phase-conjugate beam. A phase-conjugate beam can be used to restore its incident beam if the incident beam has been overlaid with noise, or in other ways distorted.

Mutually reinforcing beams can give rise to spectacular effects. For instance, if a laser beam is sent

Continued

(c)

(d)

separately into each of two crystals positioned, say, a meter apart, a third beam of light will, after a few seconds, spontaneously arise between the two crystals—connecting them. Light scattered by each crystal is phase conjugated by the other crystal. Such a phase-conjugate beam is aimed directly at the other crystal and so contributes to a growing beam connecting the two.

Refractive index gratings, if formed from a reference beam interfering with a beam scattered from an object, are effectively holograms stored in the photorefractive material. Such holograms are often referred to as being "real-time," since they can be generated and erased continuously in a crystal. The use of applied light to control the transmission of light in crystals with impurities is generally reminiscent of the use of applied voltages to control electrical conduction and electrical properties of semiconductors. The hope thus arises that an "integrated optics" technology can be developed that might supersede the microelectronics of integrated circuits. Early prototypes along these lines have, in fact, been developed—one of which is shown in (d).

6-8 Superconductivity

There are some materials for which the resistivity is zero below a certain temperature, called the **critical temperature** T_c. This phenomenon, called **superconductivity,** was discovered in 1911 by the Dutch physicist H. Kamerlingh Onnes. Figure 6-31 shows his plot of the resistance of mercury versus temperature. The critical temperature for mercury is 4.2 K. The critical temperature varies from material to material, but below this temperature the resistance of the material is zero. Critical temperatures for other superconducting elements range from less than 0.1 K for hafnium and iridium to 9.2 K for niobium. The critical temperatures of various superconducting materials are given in Table 6-3 on page 227. In the presence of a magnetic field B, the critical temperature is lower than it is when there is no field. As the magnetic field increases, the critical temperature decreases. If the magnetic field is greater than some critical field B_c, superconductivity does not exist at any temperature.

Many metallic compounds are also superconductors. For example the superconducting alloy Nb_3Ge, discovered in 1973, has a critical temperature of 23.2 K, which was the highest known until 1986. Despite the cost and inconvenience of refrigeration with expensive liquid helium, which boils at 4.2 K, many superconducting magnets were built using such materials, because such magnets produce no heat.

The conductivity of a superconductor cannot be defined since its resistance is zero. There can be a current in a superconductor even when the electric field in the superconductor is zero. Indeed, steady currents have

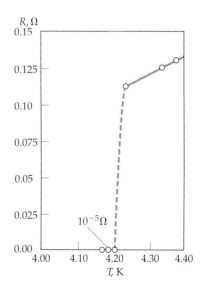

Figure 6-31 Plot by Kamerlingh Onnes of the resistance of mercury versus temperature, showing sudden decrease at the critical temperature $T = 4.2$ K.

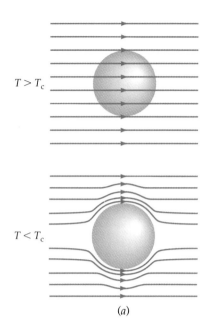
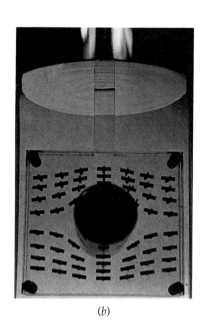

Figure 6-32 (a) The Meissner effect in a superconducting sphere cooled in a constant applied magnetic field. As the temperature drops below the critical temperature T_c, the magnetic-field lines are expelled from the sphere. (b) Demonstration of the Meissner effect. A superconducting tin cylinder is situated with its axis perpendicular to a horizontal magnetic field. The directions of the field lines near the cylinder are indicated by weakly magnetized compass needles mounted in a Lucite sandwich so that they are free to turn.

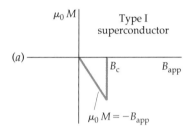
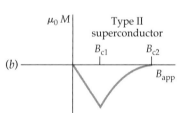

Figure 6-33 Plots of μ_0 times the magnetization M versus applied magnetic field for type I and type II superconductors. (a) In a type I superconductor, the resultant magnetic field is zero below a critical applied field B_c because the field due to induced currents on the surface of the superconductor exactly cancels the applied field. Above the critical field, the material is a normal conductor and the magnetization is too small to be seen on this scale. (b) In a type II superconductor, the magnetic field starts to penetrate the superconductor at a field B_{c1}, but the material remains superconducting up to the field B_{c2} after which it becomes a normal conductor.

been observed to persist for years without apparent loss in superconducting rings in which there was no electric field.

Consider a superconducting material that is originally at a temperature greater than the critical temperature and is in the presence of a small external magnetic field $B < B_c$. We now cool the material below the critical temperature so that it becomes superconducting. Since the resistance is now zero, there can be no emf in the superconductor. Thus, from Faraday's law, the magnetic field in the superconductor cannot change. We therefore expect from classical physics that the magnetic field in the superconductor will remain constant. However, it is observed experimentally that when a superconductor is cooled below the critical temperature in an external magnetic field, the magnetic-field lines are expelled from the superconductor and thus the magnetic field inside the superconductor is zero (Figure 6-32). This effect was discovered by Meissner and Ochsenfeld in 1933 and is now known as the **Meissner effect.** The mechanism by which the magnetic-field lines are expelled is an induced superconducting current on the surface of the superconductor. The magnetic levitation shown on page 223 results from the repulsion between the permanent magnet producing the external field and the magnetic field produced by the currents induced in the superconductor. Only certain superconductors called **type I superconductors** exhibit the complete Meissner effect. Figure 6-33a shows a plot of the magnetization M times μ_0 versus the applied magnetic field B_{app} for a type I superconductor. For a magnetic field less than the critical field B_c, the magnetic field $\mu_0 M$ induced in the superconductor is equal and opposite to the external magnetic field; that is, the superconductor is a perfect diamagnet. The values of B_c for type I superconductors are always too small for such materials to be useful in the coils of a superconducting magnet.

Other materials, known as **type II superconductors,** have a magnetization curve similar to that in Figure 6-33b. Such materials are usually alloys or metals that have large resistivities in the normal state. Type II superconductors exhibit the electrical properties of superconductors except for the Meissner effect up to the critical field B_{c2}, which may be several hundred times the typical values of critical fields for type I superconductors. For example, the alloy Nb_3Ge has a critical field $B_{c2} = 34$ T. Such materials can be used for high-field superconducting magnets. Below the critical field B_{c1}, the behavior of a type II superconductor is the same as that of a type I superconductor. In the region between fields B_{c1} and B_{c2}, the superconductor is said to be in a vortex state.

The BCS Theory

It had been recognized for some time that superconductivity is due to a collective action of the conducting electrons. In 1957, John Bardeen, Leon Cooper, and Bob Schrieffer published a successful theory of superconductivity now known as the **BCS theory**. According to this theory, the electrons in a superconductor are coupled in pairs at low temperatures. The coupling comes about because of the interaction between electrons and the crystal lattice. One electron interacts with the lattice and perturbs it. The perturbed lattice interacts with another electron in such a way that there is an attraction between the two electrons that at low temperatures can exceed the Coulomb repulsion between them. The electrons form a bound state called a **Cooper pair**. The electrons in a Cooper pair have opposite spins and equal and opposite linear momenta. They thus form a system with zero spin and zero momentum. Each Cooper pair may be considered as a single particle with zero spin. Such a particle does not obey the Pauli exclusion principle, so any number of Cooper pairs may be in the same quantum state with the same energy. In the ground state of a superconductor (at $T = 0$), all the electrons are in Cooper pairs and all the Cooper pairs are in the same energy state. In the superconducting state, the Cooper pairs are correlated so that they all act together. In order for the electrons in a superconducting state to absorb or emit energy, the binding of the Cooper pairs must be broken. The energy needed to break up a Cooper pair is similar to that needed to break up a molecule into its constituent atoms. This energy is called the **superconducting energy gap** E_g. In the BCS theory, this energy at absolute zero is predicted to be

$$E_g = 3.5kT_c \qquad 6\text{-}40$$

A small, cubicle permanent magnet levitates above a disk of the superconductor yttrium-barium-copper oxide, cooled by liquid nitrogen to 77 K. At temperatures below 92 K, the disk becomes superconducting. The magnetic field of the cube sets up circulating electric currents in the superconducting disk, such that the resultant magnetic field in the superconductor is zero. These currents produce a magnetic field that repels the cube.

Example 6-9

Calculate the superconducting energy gap for mercury predicted by the BCS theory, and compare your result with the measured value of 1.65×10^{-3} eV.

From Table 6-3 on page 227, we have $T_c = 4.15$ K for mercury. The BCS prediction for the energy gap is then

$$E_g = 3.5kT_c = 3.5(1.38 \times 10^{-23} \text{ J/K})(4.15 \text{ K}) \frac{1 \text{ eV}}{1.6 \times 10^{-19} \text{ J}}$$

$$= 1.25 \times 10^{-3} \text{ eV}$$

This differs from the measured value of 1.65×10^{-3} eV by 24 percent.

Note that the energy gap for a typical superconductor is much smaller than the energy gap for a typical semiconductor, which is of the order of 1 eV. As the temperature is increased from $T = 0$, some of the Cooper pairs are broken. The resulting individual (unpaired) electrons interact with the Cooper pairs, reducing the energy gap until at $T = T_c$ the energy gap is zero (Figure 6-34).

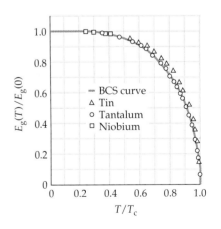

Figure 6-34 Ratio of the energy gap at temperature T to that at $T = 0$ as a function of the relative temperature T/T_c. The solid curve is that predicted by the BCS theory.

The Cooper pairs that we have discussed so far have zero momentum, so there are as many electrons traveling in one direction as the other and there is no current. Cooper pairs can also be formed with a net momentum P rather than zero momentum, but all the pairs have the same momentum. In this state, current is carried by the Cooper pairs. In ordinary conductors, resistance is present because the current carriers can be scattered with a change in momentum. As we have discussed, this scattering may be due to impurity atoms or thermal vibrations of the lattice ions. In a superconductor, the Cooper pairs are constantly scattering each other, but since the total momentum remains constant in this process, there is no change in the current. A Cooper pair cannot be scattered by a lattice ion because all the pairs act together. The only way that the current can be decreased by scattering is if a pair is broken up, which requires energy greater or equal to the energy gap E_g. At reasonably low currents, scattering events in which the total momentum of a Cooper pair is changed are completely prohibited, so there is no resistance.

Flux Quantization

Consider a superconducting ring of area A carrying a current. There can be a magnetic flux $\phi_m = B_n A$ through the ring due to the current in the ring and perhaps also due to other currents external to the ring. According to Faraday's law, if the flux changes, an emf will be induced in the ring that is proportional to the rate of change of the flux. But there can be no emf in the ring because it is resistanceless. The flux through the ring is thus frozen and cannot change. Another effect that results from the quantum-mechanical treatment of superconductivity is that the total flux through the loop is quantized and is given by

$$\phi_m = n \frac{h}{2e} \qquad n = 1, 2, 3, \ldots \qquad 6\text{-}41$$

The smallest unit of flux, called a **fluxon,** is

$$\phi_0 = \frac{h}{2e} = 2.0678 \times 10^{-15} \text{ T·m}^2 \qquad 6\text{-}42$$

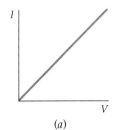

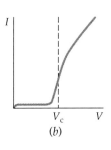

Figure 6-35 Tunneling current versus voltage for a junction of two metals separated by a thin oxide layer. (a) When both metals are normal metals, the current is proportional to the voltage as predicted by Ohm's law. (b) When one metal is a normal metal and one is a superconductor, the current is approximately zero until the applied voltage exceeds the critical voltage $V_c = E_g/2e$.

Tunneling

In Section 3-8, we discussed barrier penetration—the tunneling of a single particle through a potential barrier. The tunneling of electrons from one metal to another can be observed by separating the two metals with a thin layer only a few nanometers thick of an insulating material such as aluminum oxide. When both metals are normal metals (not superconductors), the current resulting from the tunneling of electrons through the insulating layer obeys Ohm's law for low applied voltages (Figure 6-35a). When one of the metals is a normal metal and the other is a superconductor, there is no current (at absolute zero) unless the applied voltage V is greater than a critical voltage $V_c = E_g/2e$, where E_g is the superconductor energy gap. Figure 6-35b shows the plot of current versus voltage for this situation. The current jumps abruptly when the energy $2eV$ absorbed by a Cooper pair is great enough to break up the pair. (At temperatures above absolute zero, there is a small current because some of the electrons in the superconductor are thermally excited above the energy gap and are therefore not paired.) The superconducting energy gap can thus be accurately measured by measuring the critical voltage V_c.

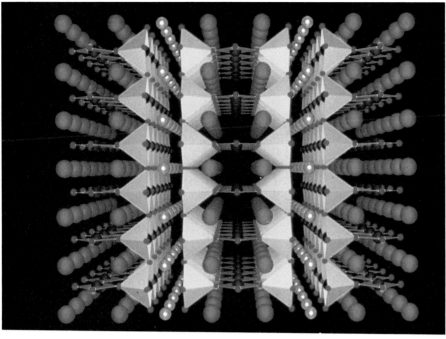

(a)

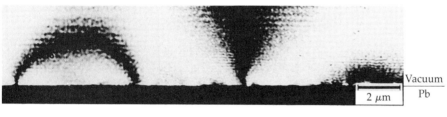

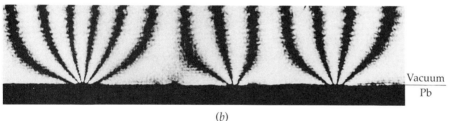

(b)

(c)

(*a*) A computer-generated image of the crystal structure of the high-temperature superconducting material yttrium-barium-copper oxide. (*b*) Fluxons penetrating a superconducting film. The image has been formed by a new technique—electron holography—in which coherent electron beams are used in place of coherent light beams to create a hologram. Electrons passing by a magnetic field are phase shifted; that is, the phase term in their wave function changes. (The shift arises from a phenomenon known as the Aharonov–Bohm effect.) By superposing such a phase-shifted beam with an unshifted reference beam, an interference pattern is created that can be interpreted as an image of the magnetic field. For the upper image, a magnetic field was applied perpendicular to a thin superconducting lead film. When the field was weak it was expelled by the Meissner effect. A stronger field, however, penetrated the film. The fluxons shown arose from vortexes of current set up in the superconductor—not from the applied field directly. In the upper right is an isolated fluxon; in the upper left is an antiparallel pair of fluxons. The lower micrograph, in which the lead film is thicker, shows penetration by bundles of fluxons. (*c*) A magnetohydrodynamic-powered ship, whose silent engine has no moving parts, is an application of superconductors. The propulsion system consists of a duct encased by an electromagnet. The windings around the magnet are superconductors. An electric current is passed through seawater in the duct and a resulting $I\ell \times \mathbf{B}$ force drives water out the back, propelling the ship forward. Reversing the current reverses the thrust.

In 1962, Brian Josephson proposed that when two superconductors form a junction, now called a **Josephson junction,** Cooper pairs could tunnel from one superconductor to the other with no resistance. The current is observed with no voltage applied across the junction and is given by

$$I = I_{max} \sin(\phi_2 - \phi_1) \qquad 6\text{-}43$$

where I_{max} is the maximum current, which depends on the thickness of the barrier, ϕ_1 is the phase of the wave function for the Cooper pairs in one of the superconductors, and ϕ_2 is the phase of the corresponding wave function in the other superconductor. This result has been observed experimentally and is known as the **dc Josephson effect.**

Josephson also predicted that if a dc voltage were applied across a Josephson junction, there would be a current that alternates with frequency f given by

$$f = \frac{2eV}{h} \qquad 6\text{-}44$$

This result, known as the **ac Josephson effect,** has been observed experimentally, and careful measurement of the frequency allows a precise determination of the ratio e/h. Because frequency can be measured so accurately, the ac Josephson effect is also used to establish precise voltage standards. The inverse effect, in which the application of an alternating voltage across a Josephson junction results in a dc current, has also been observed.

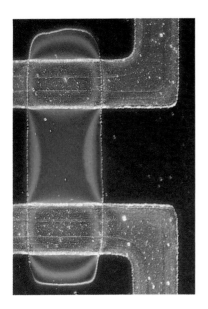

Two Josephson junctions.

Example 6-10

Using $e = 1.602 \times 10^{-19}$ C and $h = 6.626 \times 10^{-34}$ J·s, calculate the frequency of the Josephson current if the applied voltage is 1 μV.

From Equation 6-44, we obtain

$$f = \frac{2eV}{h} = \frac{2(1.602 \times 10^{-19} \text{ C})(10^{-6} \text{ V})}{6.626 \times 10^{-34} \text{ J·s}} = 4.836 \times 10^8 \text{ Hz}$$

$$= 483.6 \text{ MHz}$$

There is a third effect observed with Josephson junctions. When a dc magnetic field is applied through a superconducting ring containing two Josephson junctions, the maximum supercurrent shows interference effects that depend on the intensity of the magnetic field (Figure 6-36). This effect can be used to measure very weak magnetic fields. It is the basis for a device called a **SQUID** (for *Superconducting Quantum Interference Device*) that can detect magnetic fields as low as 10^{-14} T. Such a device can detect the magnetic fields produced by the tiny currents flowing in the heart and brain. (See the essay at the end of this chapter.)

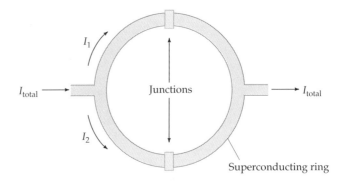

Figure 6-36 A superconducting ring with two Josephson junctions. When there is no applied magnetic field through the ring, the currents I_1 and I_2 are in phase. A very small applied magnetic field produces a phase difference in the two currents that produces interference in the total current exiting the ring.

High-Temperature Superconductivity

Until recently, the highest known critical temperature for a superconductor was 23.2 K for the alloy Nb$_3$Ge. In 1986, Bednorz and Muller found that an oxide of lanthanum, barium, and copper became superconducting at 30 K. Soon afterwards, in 1987, superconductivity with a critical temperature of 92 K was found in a copper oxide containing yttrium and barium (YBa$_2$Cu$_3$O$_7$). Since then, several copper oxides have been found with critical temperatures as high as 125 K. Table 6-3 includes some of the new, high-temperature superconductors along with their critical temperatures. These discoveries have revolutionized the study of superconductivity because relatively inexpensive liquid nitrogen, which boils at 77 K, can be used for a coolant. However, there are many problems, such as the brittleness of ceramics, that make these new superconductors difficult to use.

The new, high-temperature superconductors are all type II superconductors with very high upper critical fields. For some, B_{c2} is estimated to be as high as 100 T. Although the BCS theory appears to be the correct starting place for understanding these new superconductors, they have many features that are not clearly understood. Thus, there is much work, both experimental and theoretical, to be done.

Table 6-3 **Critical Temperatures for Some Superconducting Materials**

Material	T_c, K
Elements	
Al aluminum	1.14
Hg mercury	4.15
In indium	3.40
Pb lead	7.19
Sn tin	3.72
Ta tantalum	4.48
Compounds	
Nb$_3$Sn	18.05
Nb$_3$Ge	23.2
NbN	16.0
V$_3$Ga	16.5
V$_3$Si	17.1
La$_3$In	10.4
LaBaCuO	30
La$_2$CuO$_4$	40
YBa$_2$Cu$_3$O$_7$	92
DyBa$_2$Cu$_3$O$_7$	92.5
BiSrCaCuO	120
TlBaCaCuO	125

Summary

1. Solids are often found in crystalline form in which a small structure called the unit cell is repeated over and over. A crystal may have a face-centered-cubic, body-centered-cubic, hexagonal close-packed, or other structure depending on the type of bonding between the atoms, ions, or molecules in the crystal and on the relative sizes of the atoms if there are more than one kind as in NaCl.

2. The attractive part of the potential energy of an ion in an ionic crystal can be written

$$U_{att} = -\frac{\alpha k e^2}{r}$$

where r is the separation distance between neighboring ions and α is the Madelung constant, which depends on the geometry of the crystal and is of the order of 1.8. The repulsive part of the potential energy is due to the exclusion principle. An empirical expression that works fairly well is

$$U_{rep} = \frac{A}{r^n}$$

where n is about 9.

3. In the classical free-electron theory of metals, the electrical resistivity is given by

$$\rho = \frac{m_e v_{av}}{n e^2 \lambda}$$

where v_{av} is the mean speed of the electrons in the electron gas and λ is the mean free path between collisions. In the classical theory, v_{av} is given by the Maxwell–Boltzmann distribution and λ is related to the radius of a lattice ion r and the number density of ions n by

$$\lambda = \frac{1}{n \pi r^2}$$

The classical free-electron theory accounts for Ohm's law, but it gives the wrong temperature dependence of the resistivity and it gives magnitudes for the resistivity that do not agree with experiment. Also, the classical theory predicts that the heat capacity for metals should be $\frac{3}{2}R$ higher than that for other solids, which is not observed.

4. In the quantum-mechanical free-electron theory of metals, the Maxwell–Boltzmann distribution is replaced by the Fermi–Dirac distribution and the wave nature of electron scattering is taken into account. In the Fermi–Dirac distribution at $T = 0$, all the energy states below a certain energy called the Fermi energy E_F are filled and all those above this energy are empty. At higher temperatures, some electrons with energies of the order of kT below the Fermi energy are excited to energy states of the order of kT above that level. The Fermi energy at $T = 0$ depends on the electron number density $n = N/V$ and is given by

$$E_F = \frac{(hc)^2}{8m_ec^2}\left(\frac{3N}{\pi V}\right)^{2/3} = (0.365 \text{ eV·nm}^2)\left(\frac{N}{V}\right)^{2/3}$$

The Fermi energy for copper is about 7 eV, which is much greater than kT at ordinary temperatures. At a temperature T, the Fermi energy is defined to be that energy for which the probability of being occupied by an electron is $\frac{1}{2}$. The difference between the Fermi energy at a temperature T and that at $T = 0$ is usually negligible.

5. When two different metals are placed in contact, electrons flow from the metal with the higher Fermi energy to the metal with a lower Fermi energy until the Fermi energies of the two metals are equal. In equilibrium, there is a potential difference between the metals called the contact potential, which is equal to the difference in the work functions of the two metals divided by the electronic charge e:

$$V_{\text{contact}} = \frac{\phi_1 - \phi_2}{e}$$

6. In the quantum-mechanical theory of conduction, the quantity v_{av} in the expression for ρ is replaced by the Fermi speed u_F, which is essentially independent of temperature, and λ is interpreted as the mean free path for electrons in a lattice in which the ions are vibrating points. In the quantum theory, the contribution of the electron gas to the heat capacity is small because only those electrons within kT of the Fermi energy can be randomly excited.

7. When many atoms are brought together to form a solid, the individual energy levels are split into bands of allowed energies. The splitting depends on the type of bonding and the lattice separation. In a conductor, the uppermost band containing electrons is only partially full, so there are many available states for excited electrons. In an insulator, the uppermost band containing electrons, the valence band, is completely full and there is a large energy gap between it and the next allowed band, the conduction band. In a semiconductor, the energy gap between the filled valence band and the empty conduction band is small; so at ordinary temperatures, an appreciable number of electrons are thermally excited into the conduction band.

8. The conductivity of a semiconductor can be greatly increased by doping.

In an *n*-type semiconductor, the doping adds electrons just below the conduction band. In a *p*-type semiconductor, holes are added just above the valence band. A junction between an *n*-type and *p*-type semiconductor has applications in many devices, such as diodes, solar cells, and light emitting diodes.

9. A transistor consists of a very thin semiconductor of one type sandwiched between two semiconductors of the opposite type. Transistors are used in amplifiers because a small variation in the base current results in a large variation in the collector current.

10. In a superconductor, the resistance drops suddenly to zero below a critical temperature T_c. In the presence of an external magnetic field, the critical temperature is lowered, and above a critical field B_c, there is no superconductivity. Magnetic-field lines are completely expelled and **B** = 0 in type I superconductors. This is known as the Meissner effect. Type II superconductors have two critical fields. Below the first, B_{c1}, type II superconductors behave like type I superconductors. Above the second critical field, B_{c2}, superconductivity does not exist. The region between the two critical fields is called the vortex state. Superconductivity exists in this region but the Meissner effect is incomplete.

11. Superconductivity is described by a theory of quantum mechanics called the BCS theory in which the free electrons form Cooper pairs. The energy needed to break up a Cooper pair is called the energy gap E_g. When all the electrons are paired, individual electrons cannot be scattered by a lattice ion, so the resistance is zero.

12. The magnetic flux through a superconducting ring is quantized and takes on the values

$$\phi_m = n \frac{h}{2e} \quad n = 1, 2, 3, \ldots$$

The smallest unit of flux $\phi_0 = h/2e = 2.0678 \times 10^{-15}$ T·m² is called a fluxon.

13. When a normal conductor is separated from a superconductor by a thin layer of oxide, electrons can tunnel through the energy barrier if the bias voltage across the layer is $E_g/2e$, where E_g is the energy needed to break up a Cooper pair. The energy gap E_g can be determined by a measurement of the tunneling current versus the bias voltage. A system of two superconductors separated by a thin oxide layer is called a Josephson junction. A dc current is observed to tunnel through such a junction where there is no bias voltage. This is called the dc Josephson effect. When a dc voltage V is applied across a Josephson junction, an ac current is observed with a frequency $f = 2eV/h$. This is called the ac Josephson effect. Measurement of the frequency of this current allows a precise determination of the ratio e/h.

14. Recently a new type of superconductor called a high-temperature superconductor has been discovered. Such superconductors have critical temperatures as high as 125 K and can support magnetic fields as large as 100 T. These new superconductors are only partially understood in terms of the BCS theory.

SQUIDs

Samuel J. Williamson
New York University

Who would predict that the wave nature of an electron lets us study the extremely weak magnetic field of the human brain! This curious relationship is possible because of a device that relies on the quantum mechanical properties of electron waves for such exquisite sensitivity. It is known as the "SQUID", for Superconducting QUantum Interference Device.

Foundations

The SQUID's birth took place in the laboratories of low temperature physicists who were fascinated by superconducting phenomena. But the understanding of its principles rests on the theory developed by Brian Josephson in 1962 of how a pair of wave functions are coupled when two superconductors are separated by only a thin barrier of nonsuperconducting material. A practical example would be two films of niobium, one deposited on top of the other but separated by a thin layer of highly resistive niobium oxide (a "Nb-NbO-Nb" junction). If a source feeds electrical current into one film and takes it from the other at a temperature exceeding T_c for niobium, the corresponding electric field causes a multitude of individual electrons to diffuse across the barrier. The resistance of a typical-size example above the superconducting critical temperature T_c—which is 9.2 K for niobium—would be about 1 Ω.

This junction produces intriguing effects when cooled below T_c. Cooper pairs in each Nb film are described by a wave function, and these two states penetrate the barrier from opposite sides, with amplitudes decreasing exponentially with depth. If the barrier is sufficiently thin, the overlap of the two states near the center couples one to the other. The thinner the insulator, the stronger the coupling. This is called a tunnel junction, because Cooper pairs "tunnel" through the potential barrier where classically they would be prohibited.

The equation for the dc Josephson effect (6-43) shows that the current through a tunnel junction depends on the maximum superconducting current I_{max} and the phase difference between the wave functions on opposite sides. In practical devices such as the Nb-NbO-Nb tunnel junction, the value of I_{max} lies between 1 μA and 1 mA. When current provided to a junction just exceeds I_{max}, the voltage appearing across the barrier is much less than expected for a normal resistive barrier. The resistance is associated with the ac Josephson effect: Cooper pairs radiate energy at the Josephson frequency (Equation 6-44) as their potential energy decreases on crossing the junction.

Samuel J. Williamson is University Professor at New York University with appointments as Professor of Physics and Neural Science in the School of Arts and Science, and Adjunct Professor of Physiology and Biophysics in the School of Medicine at New York University. He received his SB and ScD from Massachusetts Institute of Technology. With a background in low-temperature physics, during the past 15 years he has developed techniques for applying SQUID magnetic-field sensors to studies of neuronal activity in the human brain. He is co-director of the Neuromagnetism Laboratory of NYU, where research focuses on the relationships between human physiology, performance, and cognition. He has written and edited several books in the physical sciences, and been actively involved in organizing international conferences on biomagnetism.

Basic Principles

How a SQUID works can be seen by considering a common type comprised of two junctions connected in parallel, as illustrated in Figure 6-36. (In practice, a resistor is also connected across each junction to minimize hysteresis effects associated with the small capacitance of the junction, but we shall disregard this technical point.) This is known as a dc SQUID, because dc current is provided at one end and taken from the other by attaching a current source. The strength of this "bias" current is chosen to exceed the maximum current I_{max} in both junctions, so that a voltage V appears across them.

When a magnetic field from an external source is applied to the area within the SQUID, the wavelength of the Cooper pairs taking one path around the area is increased, while the wavelength of those taking the other path is decreased. This is a remarkable quantum-mechanical property of a magnetic field—that its presence within the loop has a different effect on moving charges that pass around it in a clockwise sense from those that pass around it in a counterclockwise sense. The net result is that when the two currents rejoin they will generally have different phases. An interference phenomenon takes place that is very much like that of two-slit interference in optics. Constructive and destructive interference take place in succession if the strength of the external magnetic field is increased steadily. Correspondingly, the voltage appearing across the junctions increases and decreases periodically with increasing field.

Sensitivity

It should be no surprise that the period for one complete cycle in the SQUID's voltage corresponds to a change of flux within the SQUID equal to the fluxon $\phi_0 = h/2e$. Therefore, to go from one peak in the voltage to the next requires a flux change of only 2.07×10^{-15} T·m². If the area within the SQUID is 1 mm × 1 mm, this corresponds to a change in the external field of about 2×10^{-9} T. The value is impressively small compared with the earth's steady field of 60×10^{-6} T. Nevertheless, many orders of magnitude of greater sensitivity can be achieved by adding one more feature to the circuit.

A coil consisting of a few turns of wire placed near the SQUID is provided a current by a separate electronic circuit that can monitor the SQUID's voltage. If the SQUID's voltage begins to depart from, say, a minimum value because of a change in the external field, the circuit senses this and adjusts the current so that the coil's magnetic field counters the change in the external field. In this way, the total flux within the SQUID is held nearly constant. If a resistor is connected in series with the coil, the voltage across the resistor provides a continuous measurement of the external field. Such a circuit is so sensitive that a modern SQUID system detects changes of field amounting to only $10^{-6}\phi_0$. An example is shown in Figure 1. These SQUID sensors are being applied to a variety of magnetic studies where high sensitivity is essential.

Biomagnetic Research

Now we arrive at the point where we can relate the SQUID's capabilities to intriguing applications in brain research. Nerve cells ("neurons") in the surface layer, or cortex, of the brain respond to excitations from their neighbors by admitting ions through their membranes, a process that over a period of 10 to 100 ms produces an electrical current flowing along the length of the cell. There is much interest in studying this activity because that is how information from the sense organs such as the eyes, ears, and fingers is processed. The magnetic field of a single neuron is too weak to be detected outside the scalp, but coherent activity of populations of

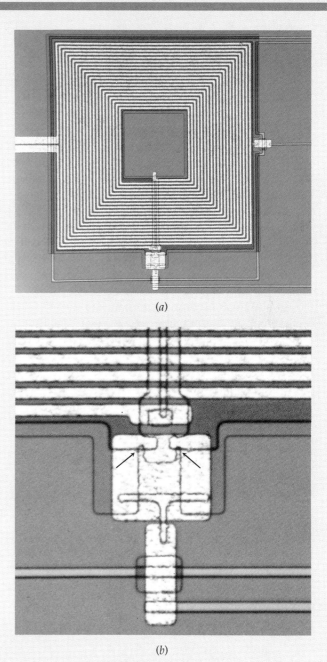

Figure 1 (a) Photomicrograph of a SQUID fabricated by depositing thin layers of appropriate materials on an insulating base. Because the layers are so thin, it is possible to see through them to layers beneath. The dominant feature is a 25-turn spiral input coil, only 0.3 mm on a side. Connections to the detection coil that responds to the field of interest will be attached to the two broad pads halfway up on the left side. The dc bias current for the two Josephson junctions comes in via the lead at the bottom right and leaves by the lead halfway up at the right. The bottom lead, shown magnified in (b), runs to the junctions, each only 2 μm long, which are the small tips of the U-shaped pad (indicated by arrows) just below the bottom of the input coil.

Continued

cells can be detected with a SQUID sensor, as shown in Figure 2. The sensing elements may be immersed in a bath of liquid helium (at 4.2 K) kept in a vacuum-insulated container called a dewar.

The principal challenge in biomagnetic measurements is not sensitivity for detecting the very weak signals, for field sensitivities as impressive as 3×10^{-15} T have been achieved (in comparison with neuronal signals that range up to 500×10^{-15} T). The major problem is environmental noise that can be 10^5 times greater than the fields of interest. One way to overcome much of this is by doing measurements within a magnetically shielded room. Another is by employing a "detection coil" having a geometry that rejects much of the environmental noise. The detection coil conveys field information to the SQUID by a "flux transporter", a closed superconducting circuit consisting of the detection coil and a smaller "input coil" positioned near the SQUID. Because of flux quantization, when the brain applies magnetic flux to the detection coil, current flows around the entire transporter to cancel the applied flux and maintain the total flux within it invariant. The SQUID experiences the correspondingly proportional field from the input coil. A detection coil wound as a gradiometer discriminates against distant noise sources, which produce relatively uniform fields, while retaining sensitivity to sources close by. A "first-order" gradiometer (one clockwise loop placed near a counter-clockwise loop) will not respond to a uniform field, for the positive flux imposed on one loop is cancelled by the negative flux imposed on the other. Yet it retains sensitivity to a field source that is placed much closer to one loop than the other. Figure 2 shows a second-order gradiometer, which provides even better discrimination against distant sources, for it may be viewed as two first-order gradiometers mounted back to back. Neither first-order gradiometer responds to a uniform field, and the combination of two first-order gradiometers will not respond to a uniform field gradient.

Measurements of magnetic field patterns across the scalp provide information from which the locations and strengths of underlying brain activity may be deduced (Figure 3). This is possible because the head is transparent to magnetic fields, so they emerge without distortion. As a result, the site of an active area in the cortex can be located with an accuracy of better than 3 mm. For instance, it was dis-

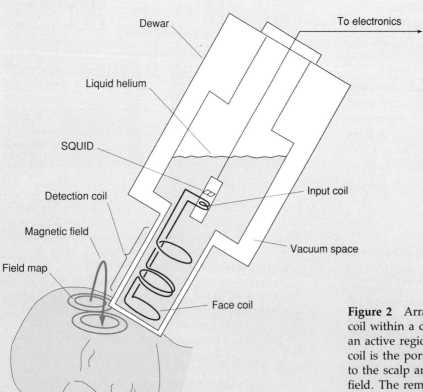

Figure 2 Arrangement of a SQUID and detection coil within a dewar, by which the magnetic field of an active region of the brain can be detected. The face coil is the portion of the detection coil that is closest to the scalp and receives the strongest biomagnetic field. The remaining turns of wire of the detection coil are arranged so that the net magnetic flux of a uniform field, or uniform field gradient, imposed by environmental field noise is made zero.

Figure 3 Three SQUID systems positioned to monitor the sequence of brain activity of a volunteer subject who responds by pressing an electrical switch when a visual display is projected onto the screen in front. The dewar suspended from the ceiling contains five SQUID sensors to monitor the response of visual cortex at five closely-spaced positions on the scalp. Each of the two dewars directed from the sides contains a single sensor to monitor the motor cortex controlling left and right finger motions. The SQUIDs in these smaller dewars are cooled by refrigerators that depend on the principle that high-pressure helium gas cools when it is allowed to expand rapidly at these low temperatures.

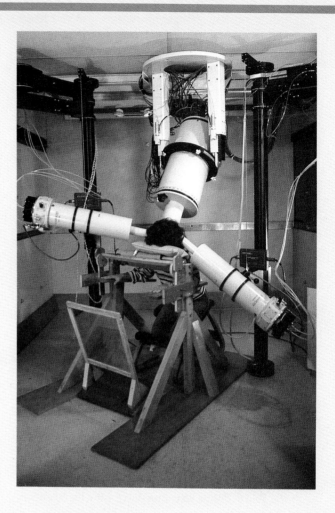

covered magnetically that humans have a "tone map" across the cortex concerned with auditory functions, located a few centimeters above the ear. Tones of different frequencies evoke activity at different locations. The cumulative distance across the cortex from one site of activity to another increases with the logarithm of the frequency (just as the piano keyboard is arranged). It was also discovered that tones of different intensity have relatively little effect on the strength of the neuronal response. Instead, the site of activity shifts across the cortex with increasing intensity, following a track that is almost at a right angle to the tone sequence. Again this is a logarithmic map, where the distance increases as the logarithm of the intensity of the tone. The strength of neuronal response is always the same, for easily heard tones. What affects the strength of activity is attention: neuronal response is more intense when a person pays attention to a sound than when it is disregarded. Many research groups are using noninvasive SQUID detectors to elucidate sensory and higher-level functions of the human brain.

Additional Applications

SQUIDs have a variety of applications in which extremely weak magnetic fields or currents must be measured. For instance, they are used in geophysical studies to monitor the low-frequency magnetic fields of electromagnetic waves that are reflected between the earth and ionosphere when solar activity creates disturbances in the upper atmosphere. Because these waves penetrate to considerable depth in the ground, measurements of the magnetic field and electric field (the latter obtained with electrodes placed in the soil) provide information that is useful in locating deposits of oil. It is also possible to locate bodies of hot water that may provide a geothermal source of energy.

Another application helps in the search for gravity waves. Here the goal is to detect minute changes in the shape of a massive bar of aluminum that would indicate the fleeting presence of such a wave. One type of detector would depend on a SQUID to indicate the movement of a nearby source of magnetic field, such as a small electromagnet mounted on the bar. Other uses in physics include measurements of extremely small changes in the resistance of a conductor, where a SQUID monitors the imbalance of current in an electrical bridge circuit that is produced by a change in resistance when some property such as temperature is altered.

Thus we see how the wave nature of the electron, exploited in the design of the SQUID sensor, provides opportunities for intriguing studies of very weak magnetic fields. One of the least expected when SQUIDs were first developed was the activity of neurons of the human brain!

Suggestions for Further Reading

Bentley, W. A., and W. J. Humphreys: *Snow Crystals*, Dover, New York, 1962.

A compilation of over 2000 photographs of snowflakes.

Feynman, Richard: *Lectures on Physics*, Vol. III, Addison-Wesley, Reading, Massachusetts, 1964.

The final volume of the three-volume set is devoted to quantum mechanics and its applications to matter. Based on lectures for freshmen and sophomores, this classic work has been enjoyed by students and faculty at all levels.

Hecht, Jeff: *The Laser Guidebook*, McGraw-Hill, New York, 1986.

A nonmathematical account of state-of-the-art laser technology.

Horowitz, Paul, and Winfield Hill: *The Art of Electronics*, 2nd ed., Cambridge University Press, Cambridge, 1989.

This text on electronic circuit design emphasizes intuitive and nonmathematical techniques. No previous exposure to electronics is assumed.

Kittel, Charles: *Introduction to Solid State Physics*, 6th ed., J. Wiley & Sons, New York, 1986.

The standard senior or beginning graduate text in solid state physics. Sections of the book are, nonetheless, accessible to less-advanced readers.

Rose-Innes, A. C., and E. H. Rhoderick: *Introduction to Superconductivity*, 2nd ed., Pergamon Press, Oxford, 1978.

Written for undergraduates or beginning graduate students. No previous knowledge of superconductivity is assumed.

Review

A. Objectives: After studying this chapter, you should:

1. Know what the Madelung constant is and why it differs from 1.

2. Be able to discuss the successes and failures of the classical free-electron theory of metals.

3. Be able to derive an expression for the Fermi energy at $T = 0$ in one dimension.

4. Be able to discuss qualitatively the band theory of solids.

5. Be able to discuss the effects of adding impurities to intrinsic semiconductors.

6. Know the general features of the current-versus-voltage curve for a typical *pn*-semiconductor junction.

7. Be able to discuss the characteristics and operation of a transistor.

8. Be able to discuss qualitatively the BCS theory of superconductivity.

B. Define, explain, or otherwise identify:

Amorphous solid
Unit cell
Madelung constant
Classical model of electric conduction
Drift velocity
Collision time
Mean free path
Dissociation energy
Fermi energy at $T = 0$
Fermi factor
Density of states
Fermi temperature
Fermi–Dirac distribution
Contact potential
Valence band
Conduction band
Forbidden energy band
Intrinsic semiconductor
Impurity semiconductor
Doping
Donor levels
n-type semiconductor
Acceptor levels
p-type semiconductor
Junction
Depletion region
Forward bias
Reverse bias
Avalanche breakdown
Zener diode
Tunnel diode
Solar cell
Surface-barrier detector
Light-emitting diode
Emitter
Base
Collector
Current gain
Voltage gain
Critical temperature
Superconductivity
Meissner effect
Type I superconductors
Type II superconductors
BCS Theory
Cooper pair
Superconducting energy gap
Fluxon
Josephson junction
dc Josephson effect
ac Josephson effect
SQUID

C. True or false: If the statement is true, explain why it is true. If it is false, give a counterexample.

1. Solids that are good electrical conductors are usually good heat conductors.

2. The classical free-electron theory adequately explains the heat capacity of metals.

3. At $T = 0$, the Fermi factor is either 1 or 0.

4. The Fermi energy is the average energy of an electron in a solid.

5. The contact potential between two metals is proportional to the difference in the work functions of the two metals.

6. At $T = 0$, an intrinsic semiconductor is an insulator.

7. Semiconductors conduct current in one direction only.

Problems

Level 1

6-1 The Structure of Solids

1. Suppose that hard spheres of radius R are located at the corners of a unit cell with a simple cubic structure. (*a*) If the hard spheres touch so as to take up the minimum volume possible, what is the size of the unit cell? (*b*) What fraction of the volume of the cubic structure is occupied by the hard spheres?

2. Calculate the distance r_0 between the K^+ and the Cl^- ions in KCl, assuming that each ion occupies a cubic volume of side r_0. The molar mass of KCl is 74.55 g/mol and its density is 1.984 g/cm^3.

3. The distance between the Li^+ and Cl^- ions in LiCl is 0.257 nm. Use this and the molecular mass of LiCl (42.4 g/mol) to compute the density of LiCl.

4. Find the value of n in Equation 6-6 that gives the measured dissociation energy of 741 kJ/mol for LiCl, which has the same structure as NaCl and for which $r_0 = 0.257$ nm.

6-2 The Classical Free-Electron Theory of Metals

5. A measure of the density of the free-electron gas in a metal is the distance r_s, which is defined as the radius of the sphere whose volume equals the volume per conduction electron. (*a*) Show that $r_s = (3/4\pi n)^{1/3}$, where n is the free-electron number density. (*b*) Calculate r_s for copper in nanometers.

6. (*a*) Given a mean free path $\lambda = 0.4$ nm and a mean speed $v_{av} = 1.17 \times 10^5$ m/s for the current flow in copper at a temperature of 300 K, calculate the classical value for the resistivity ρ of copper. (*b*) The classical model suggests that the mean free path is temperature independent and that v_{av} depends on temperature. From this model, what would ρ be at 100 K?

6-3 The Fermi Electron Gas

7. Calculate the number density of free electrons in (*a*) Ag ($\rho = 10.5$ g/cm^3) and (*b*) Au ($\rho = 19.3$ g/cm^3), assuming one free electron per atom, and compare your results with the values listed in Table 6-1.

8. Calculate the number density of free electrons for (*a*) Mg ($\rho = 1.74$ g/cm^3) and (*b*) Zn ($\rho = 7.1$ g/cm^3), assuming two free electrons per atom, and compare your results with the values listed in Table 6-1.

9. The density of aluminum is 2.7 g/cm^3. How many free electrons are there per aluminum atom? (Use Table 6-1 for the number density.)

10. The density of potassium is 0.851 g/cm^3. How many free electrons are there per potassium atom? (Use Table 6-1 for the number density.)

11. The density of tin is 7.3 g/cm^3. How many free electrons are there per tin atom? (Use Table 6-1 for the number density.)

12. Calculate the Fermi energy for (*a*) Al, (*b*) K, and (*c*) Sn using the number densities given in Table 6-1.

13. Calculate the Fermi temperature for (*a*) Al, (*b*) K, and (*c*) Sn.

14. Find the average energy of the conduction electrons at $T = 0$ in (*a*) copper and (*b*) lithium.

15. Calculate the (*a*) Fermi energy and (*b*) Fermi temperature for iron at $T = 0$.

16. (*a*) Which two metals in Table 6-2 develop the greatest potential when placed in contact? (*b*) What is the value of that contact potential?

17. (*a*) Which two metals in Table 6-2 develop the least potential when placed in contact? (*b*) What is the value of that contact potential?

18. Calculate the contact potential between (*a*) Ag and Cu, (*b*) Ag and Ni, and (*c*) Ca and Cu.

6-4 Quantum Theory of Electrical Conduction

19. Use Equation 6-33 with $\alpha = \pi^2/4$ to calculate the average energy of an electron in copper at $T = 300$ K. Compare your result with the average energy at $T = 0$ and the classical result of $\frac{3}{2}kT$.

20. What is the speed of a conduction electron whose energy is equal to the Fermi energy E_F for (*a*) Na, (*b*) Au, and (*c*) Sn?

21. The resistivities of Na, Au, and Sn at $T = 273$ K are 4.2 $\mu\Omega\cdot$cm, 2.04 $\mu\Omega\cdot$cm, and 10.6 $\mu\Omega\cdot$cm, respectively. Use these values and the Fermi speeds calculated in Problem 20 to find the mean free paths λ for the conduction electrons in these elements.

22. At what temperature is the heat capacity due to the electron gas in copper equal to 10 percent of that due to lattice vibrations?

6-5 Band Theory of Solids

23. The energy gap between the valence band and the conduction band in silicon is 1.14 eV at room temperature. What is the maximum wavelength of a photon that will excite an electron from the top of the valence band to the bottom of the conduction band?

24. Work Problem 23 for germanium, for which the energy gap is 0.74 eV.

25. Work Problem 23 for diamond, for which the energy gap is 7.0 eV.

26. A photon of wavelength 3.35 μm has just enough energy to raise an electron from the valence band to the conduction band in a lead sulfide crystal. (*a*) Find the energy gap between these bands in lead sulfide. (*b*) Find the temperature T for which kT equals this energy gap.

6-6 Impurity Semiconductors

27. What type of semiconductor is obtained if silicon is doped with (*a*) aluminum and (*b*) phosphorus? (See Table 4-1 for the electron configurations of these elements.)

28. What type of semiconductor is obtained if silicon is doped with (a) indium and (b) antimony? (See Table 4-1 for the electron configurations of these elements.)

29. The donor energy levels in an n-type semiconductor are 0.01 eV below the conduction band. Find the temperature for which $kT = 0.01$ eV.

6-7 Semiconductor Junctions and Devices

30. Simple theory for the current versus the bias voltage across a pn junction yields the equation

$$I = I_0(e^{eV_b/kT} - 1)$$

Sketch I versus V_b for both positive and negative values of V_b using this equation.

31. For a temperature of 300 K, use the equation in Problem 30 to find the bias voltage V_b for which the exponential term has the value (a) 10 and (b) 0.1.

6-8 Superconductivity

32. (a) Use Equation 6-40 to calculate the superconducting energy gap for tin and compare your result with the measured value of 6×10^{-4} eV. (b) Use the measured value to calculate the wavelength of a photon having sufficient energy to break up Cooper pairs in tin at $T = 0$.

33. Repeat Problem 32 for lead, which has a measured energy gap of 2.73×10^{-3} eV.

Level II

34. Suppose identical bowling balls of radius R are packed into a hexagonal close-packed structure. What fraction of the available volume of the unit cell is filled by the bowling balls?

35. Estimate the fraction of free electrons in copper that are in excited states above the Fermi energy at (a) room temperature of 300 K and (b) 1000 K.

36. A one-dimensional model of an ionic crystal consists of a line of alternating positive and negative ions with distance r_0 between each ion. (a) Show that the potential energy of attraction of one ion in the line is

$$V = -\frac{2ke^2}{r_0}\left(1 - \frac{1}{2} + \frac{1}{3} - \frac{1}{4} + \frac{1}{5} - \cdots\right)$$

(b) Using the result that

$$\ln(1 + x) = x - \frac{x^2}{2} + \frac{x^3}{3} - \frac{x^4}{4} + \cdots$$

show that the Madelung constant for this one-dimensional model is $\alpha = 2\ln 2 = 1.386$.

37. Estimate the Fermi energy of zinc from its electronic molar heat capacity of $(3.74 \times 10^{-4} \text{ J/mol·K}^2)T$.

38. Derive Equation 6-26 for the average energy of a Fermi electron gas at $T = 0$.

39. (a) How many atoms are in a 10-kg block of copper? (b) If there are eight conduction states per atom, find the number of states. (c) If these states are filled to the Fermi energy level, find the average separation in energy between the states. (d) Compare this average spacing to thermal energy kT at $T = 300$ K.

40. The density of the electron states in a metal can be written $g(E) = AE^{1/2}$, where A is a constant and E is measured from the bottom of the conduction band. (a) Show that the total number of states is $\frac{2}{3}AE_F^{3/2}$. (b) About what fraction of the conduction electrons are within kT of the Fermi energy? (c) Evaluate this fraction for copper at $T = 300$ K.

41. What energy is required to remove one ion pair from NaCl and convert the ions into an NaCl molecule? The energy is called the *cohesive energy per ion pair* for NaCl.

42. (a) Calculate the number of atoms in 10 g of KBr and from this the number of energy states in the conduction band, assuming eight states per atom. (b) The width of the KBr valence band is measured using x-ray spectroscopy to be about 1.5 eV. Estimate the average density of electron states in the valence band for a 10-g crystal. (c) How does the number of electron states per unit of energy change if the mass of the crystal is increased to 1 kg?

43. In Figure 6-30 for the pnp-transistor amplifier, suppose $R_b = 2$ kΩ and $R_L = 10$ kΩ. Suppose further that 10-μA ac base current generates a 0.5-mA ac collector current. What is the voltage gain of the amplifier?

44. Germanium can be used to measure the energy of incident particles. Consider a 660-keV gamma ray emitted from ^{137}Cs. (a) Given that the band gap in germanium is 0.72 eV, how many electron–hole pairs can be generated by this incident gamma ray? (b) The number of pairs N in part (a) will have statistical fluctuations given by $\pm\sqrt{N}$. What then is the energy resolution of this detector in this photon energy region?

45. A "good" silicon diode has a current–voltage characteristic given by

$$I = I_0(e^{eV_b/kT} - 1)$$

Let $kT = 0.025$ eV (room temperature) and the saturation current $I_0 = 1$ nA. (a) Show that for small reverse-bias voltages, the resistance is 25 MΩ. *Hint:* Do a Taylor expansion of the exponential function, or use your calculator and enter small values for V_b. (b) Find the dc resistance for a reverse bias of 0.5 V. (c) Find the dc resistance for a 0.5-V forward bias. What is the current in this case? (d) Calculate the ac resistance dV/dI for a 0.5-V forward bias.

46. The relative binding of the extra electron in the arsenic atom that replaces an atom in silicon or germanium can be understood from a calculation of the first Bohr orbit of this electron in these materials. Four of arsenic's outer electrons form covalent bonds, so the fifth electron sees a singularly charged center of attraction. This model is a modified hydrogen atom. In the Bohr model of the hydrogen atom, the electron moves in free space at a radius a_0 given by

$$a_0 = \frac{\epsilon_0 h^2}{\pi m_e e^2}$$

When an electron moves in a crystal, we can approximate the effect of the other atoms by replacing ϵ_0 with $\kappa\epsilon_0$ and m_e with an effective mass for the electron. For silicon κ is 12 and the effective mass is about $0.2m_e$, and for germanium κ is 16 and the effective mass is about $0.1m_e$. Estimate the Bohr radii for the outer electron as it orbits the impurity arsenic atom in silicon and germanium.

47. Modify Equation 4-9 for the ground-state energy of

the hydrogen atom in the spirit of the preceding problem by replacing ϵ_0 by $\kappa\epsilon_0$ and m_e by an effective mass for the electron to estimate the binding energy of the extra electron of an impurity arsenic atom in (a) silicon and (b) germanium. (c) Suppose you dope silicon with one part per million of arsenic. What fraction of these arsenic atoms would supply an electron to the conduction band at room temperature?

48. The number of electrons in the conduction band of an insulator or intrinsic semiconductor is governed chiefly by the Fermi factor, which in these cases is $e^{-(E-E_F)/kT}$, where E_F is the Fermi energy, which is approximately midway between the nearly filled valence band the nearly empty conduction band. If E_g is the energy gap between the valence and conduction bands, and E is measured from the top of the valence band, $E_F = \frac{1}{2}E_g$. The Fermi factor at the bottom of the conduction band $E = E_g$ is then $e^{-E_g/2kT}$. Calculate this Fermi factor at $T = 300$ K for a typical gap energy of (a) 6.0 eV for an insulator and (b) 1.0 eV for a semiconductor. Discuss the significance of these results if there are 10^{22} valence electrons per cubic centimeter and $e^{-(E-E_F)/kT}$ is the probability of one electron having an energy E in the conduction band.

49. The pressure of an ideal gas is related to the average energy of the gas particles by $PV = \frac{2}{3}NE_{av}$, where N is the number of particles and E_{av} is the average energy. Use this to calculate the pressure of the Fermi electron gas in copper in newtons per square meter, and compare your result with atmospheric pressure, which is about 10^5 N/m². (*Note:* The units are most easily handled by using the conversion factors 1 N/m² = 1 J/m³ and 1 eV = 1.6×10^{-19} J.)

50. The bulk modulus B of a material can be defined by

$$B = -V\frac{\partial P}{\partial V}$$

(a) Use the ideal-gas relation $PV = \frac{2}{3}NE$ and Equations 6-24 and 6-26 to show that

$$P = \frac{2NE_F}{5V} = CV^{-5/3}$$

where C is a constant independent of V. (b) Show that the bulk modulus of the Fermi electron gas is therefore

$$B = \frac{5}{3}P = \frac{2NE_F}{3V}$$

(c) Compute the bulk modulus in newtons per square meter for the Fermi electron gas in copper and compare your result with the measured value of 134×10^9 N/m².

51. The resistivity of pure copper is increased by about 1×10^{-8} Ω·m by the addition of 1 percent (by number of atoms) of an impurity throughout the metal. The mean free path depends on both the impurity and the oscillations of the lattice ions according to the equation

$$\frac{1}{\lambda} = \frac{1}{\lambda_t} + \frac{1}{\lambda_i}$$

(a) Estimate λ_i from the given data. (b) If d is the effective diameter of an impurity lattice ion seen by an electron, the scattering cross section is d^2. Estimate d^2 from $d = 2r$, where r is related to λ_i by Equation 6-14.

Level III

52. (a) Calculate the force $F = -dU/dr$ from Equation 6-5 and show that

$$F = \alpha\frac{ke^2}{r_0^2}\left(\frac{r_0^{n+1}}{r^{n+1}} - \frac{r_0^2}{r^2}\right)$$

(b) Note that $F = 0$ when $r = r_0$. Write $r = r_0 + \Delta r = r_0(1 + \epsilon)$, where $\epsilon = \Delta r/r_0$, and use the binomial expansion $(1 + \epsilon)^n = 1 + n\epsilon + n(n - 1)\epsilon^2/2$ to write F as a power series in Δr and show that, when $r = r_0 + \Delta r$, F is given by

$$F = -C\,\Delta r + B(\Delta r)^2 + \cdots$$

where

$$C = \alpha\,\frac{(n-1)ke^2}{r_0^3}$$

and

$$B = \alpha\,\frac{(n^2 + 3n - 4)ke^2}{2r_0^4}$$

(Note that the k in ke^2 in the expressions for C and B is the Coulomb constant.)

53. The quantity C in Problem 52 is the force constant for a "spring" consisting of a line of alternating positive and negative ions. If these ions are displaced slightly from their equilibrium separation r_0, they will vibrate with a frequency

$$f = \frac{1}{2\pi}\sqrt{\frac{C}{m}}$$

(a) Use the values of α, n, and r_0 for NaCl and the reduced mass for the NaCl molecule to calculate this frequency. (b) Calculate the wavelength of electromagnetic radiation corresponding to this frequency, and compare your result with the characteristic strong infrared absorption bands in the region of about $\lambda = 61$ μm that are observed for NaCl.

54. The term $B(\Delta r)^2$ in Problem 52 is related to the coefficient of thermal expansion of a solid. The time average of the force must be zero since there is no net acceleration averaged over time. Therefore, we have

$$\overline{\Delta r} = \frac{B}{C}\overline{(\Delta r)^2}$$

(a) Use the equipartition result that

$$\tfrac{1}{2}C\,\overline{(\Delta r)^2} = \tfrac{1}{2}kT$$

to show that $\overline{\Delta r} = (kB/C^2)T$, where k is Boltzmann's constant. (b) Evaluate the coefficient of thermal expansion kB/C^2r_0 for NaCl, and compare your result with the measured value of about 4×10^{-6}/K.

55. Consider a model for a metal in which the lattice of positive ions forms a container for a classical electron gas with n electrons per unit volume. In equilibrium, the average electron velocity is zero, but the application of an electric field produces an acceleration of the electrons. If we use a relaxation time τ to account for the electron–lattice collisions, then we have the equation

$$m\frac{dv}{dt} + \frac{m}{\tau}v = -eE$$

(a) Solve the equation for the drift velocity in the direction of the applied electric field. (b) Verify that Ohm's law is valid, and find the resistivity as a function of n, e, m, and the relaxation time τ.

Chapter 7

Nuclei

A nuclear power plant in West Germany. The fission reactor core is housed in the hemispherical containment structure at the center of the picture. To its left are two large cooling towers. A schematic of such a facility is shown on page 258.

The first information about the atomic nucleus came from the discovery of radioactivity by A. H. Becquerel in 1896. The rays emitted by radioactive nuclei were studied by many physicists in the early decades of the twentieth century. They were first classified by E. Rutherford as alpha, beta, and gamma rays according to their ability to penetrate matter and ionize air: Alpha radiation penetrates the least and produces the most ionization and gamma radiation penetrates the most and produces the least ionization. It was later found that alpha rays are helium nuclei, beta rays are electrons or positrons, and gamma rays are high-energy photons, that is, electromagnetic radiation of very short wavelength. (A positron is an antielectron, a particle identical to the electron except that its charge is positive.) The alpha-particle scattering experiments of H. W. Geiger and E. Marsden in 1911 and the successes of the Bohr model of the atom led to the modern picture of an atom as consisting of a tiny, massive nucleus with a radius of the order of 1 to 10 fm (1 fm = 10^{-15} m) surrounded by a cloud of electrons at a relatively great distance, of the order of 0.1 nm = 100,000 fm, from the nucleus.

Artificial nuclear disintegration was first observed in 1919 when Rutherford detected protons that were emitted when he bombarded nitrogen with alpha particles. Such experiments were extended to many other elements in the next few years.

In 1932, the neutron was discovered by J. Chadwick, the positron was discovered by C. Anderson, and the first nuclear reaction using artificially accelerated particles was observed by J. D. Cockcroft and E. T. S. Walton. It is therefore reasonable to mark that year as the beginning of modern nuclear physics. With the discovery of the neutron, it became possible to understand some of the properties of nuclear structure. The advent of nuclear accelerators removed the severe limitations on particle type and energy imposed by the need to use naturally occurring radioactive sources, and made many additional experimental studies possible.

In this chapter, we will first discuss some of the general properties of the atomic nucleus and the important features of radioactivity. We will then look at some nuclear reactions, including the important reactions of fission and fusion. Finally, we will look at the interactions of nuclear particles with matter, a subject important for understanding the detection of nuclear particles, the shielding of reactors, and the effects of radiation on the human body. Our discussions will be descriptive and phenomenological, with the aim of presenting general information rather than a theoretical understanding of nuclear physics.

7-1 Properties of Nuclei

The nucleus of an atom contains just two kinds of particles: protons and neutrons. (The normal hydrogen nucleus consists of a single proton.) These particles have approximately the same mass. (The neutron is about 0.2 percent more massive.) The proton has a charge of $+e$ and the neutron is uncharged. The number of protons Z is the **atomic number** of the atom, which also equals the number of electrons in the atom. The number of neutrons N is approximately equal to Z for light nuclei and is somewhat greater than Z for heavier nuclei. The total number of nucleons $A = N + Z$ is called the **mass number** of the nucleus. (The term **nucleon** refers to either a neutron or a proton.) A particular nuclear species, called a **nuclide,** is designated by its atomic symbol (H for hydrogen, He for helium, and so forth) with the mass number A as a presuperscript. Two or more nuclides with the same atomic number Z but different N and A numbers are called **isotopes.** The lightest element, hydrogen, has three isotopes: ordinary hydrogen (^{1}H), whose nucleus is just a single proton; deuterium (^{2}H), whose nucleus contains one proton and one neutron; and tritium (^{3}H), whose nucleus contains one proton and two neutrons. Although the mass of the deuterium atom is about twice that of the ordinary hydrogen atom and that of the tritium atom is three times as great, these three atoms have nearly identical chemical properties because they each have one electron. On the average, there are about 2.6 stable isotopes for each element, although some have only one and others have five or six. The most common isotope of the second lightest atom, helium, is ^{4}He. The nucleus of the ^{4}He atom is also known as an alpha particle. Another isotope of helium is ^{3}He.

Inside the nucleus, the nucleons exert strong attractive forces on their nearby neighbors. This force, called the **strong nuclear force** or the **hadronic force,** is much stronger than the electrostatic force of repulsion between the protons and is very much stronger than the gravitational forces between the nucleons. (Gravity is so weak that it can always be neglected in nuclear physics.) The strong nuclear force is roughly the same between two neutrons, two protons, or a neutron and a proton. Two protons, of course, also exert a repulsive electrostatic force on each other due to their charges, which tends to weaken the attraction between them somewhat. The strong nuclear force decreases rapidly with distance, and it is negligible when two nucleons are more than a few femtometers apart.

Nuclear Size and Shape

The size and shape of the nucleus can be determined by bombarding it with high-energy particles and observing the scattering (similar to the experiments of Geiger and Marsden) or, in some cases, from measurements of its radioactivity. The results depend somewhat on the kind of experiment. For example, a scattering experiment using electrons measures the charge distribution of the nucleus, whereas one using neutrons determines the region of influence of the strong nuclear force. Despite these differences, a wide variety of experiments suggest that most nuclei are approximately spherical, with radii given approximately by

Nuclear radius
$$R = R_0 A^{1/3} \qquad 7\text{-}1$$

where R_0 is about 1.5 fm. The fact that the radius of a spherical nucleus is proportional to $A^{1/3}$ implies that the volume of the nucleus is proportional to A. Since the mass of the nucleus is also approximately proportional to A, the densities of all nuclei are approximately the same. The fact that a drop of liquid also has a constant density independent of its size has led to a model in which the nucleus is viewed as analogous to a drop of liquid. This model has proved quite successful in understanding nuclear behavior, especially the fission of heavy nuclei.

N and Z Numbers

For light nuclei, the greatest stability is achieved when the numbers of protons and neutrons are approximately equal, that is, when $N \approx Z$. For heavier nuclei, electrostatic repulsion between the protons leads to greater stability when there are more neutrons than protons. We can see this by looking at the N and Z numbers for the most abundant isotopes of some representative elements: for $^{16}_{8}O$, $N = 8$ and $Z = 8$; for $^{40}_{20}Ca$, $N = 20$ and $Z = 20$; for $^{56}_{26}Fe$, $N = 30$ and $Z = 26$; for $^{207}_{82}Pb$, $N = 125$ and $Z = 82$; and for $^{238}_{92}U$, $N = 146$ and $Z = 92$. (The atomic number Z has been included here as a presubscript of the atomic symbol for emphasis. It is not actually needed because the atomic number is implied by the atomic symbol.)

Figure 7-1 shows a plot of N versus Z for the known stable nuclei. The straight line $N = Z$ is followed for small values of N and Z. We can understand this tendency for N and Z to be equal by considering the total energy of A particles in a one-dimensional box (see Section 3-6). Figure 7-2 shows the energy levels for eight neutrons and for four neutrons and four protons. Because of the exclusion principle, only two identical particles (with opposite spins) can be in the same space state. Since protons and neutrons are not identical, we can put two each in a state as in Figure 7-2b. Thus, the total energy for four protons and four neutrons is less than that for eight neutrons (or eight protons) as in Figure 7-2a. When the Coulomb energy of repulsion is included, this result changes somewhat. This potential energy is proportional to Z^2. For large values of A (and therefore for large values of Z), the

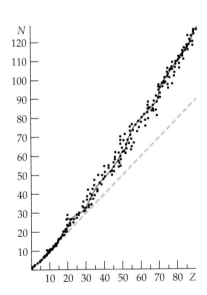

Figure 7-1 Plot of number of neutrons N versus number of protons Z for the stable nuclides. The dashed line is $N = Z$.

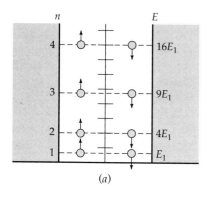
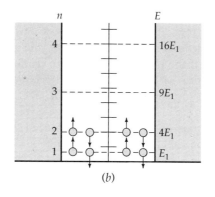

Figure 7-2 (a) Eight neutrons in a one-dimensional box. In accordance with the exclusion principle, only two neutrons (with opposite spins) can be in a given energy level. (b) Four neutrons and four protons in a one-dimensional box. Because protons and neutrons are not identical particles, two of each can be in each energy level. The total energy is much less for this case than for that in (a).

total energy may be increased less by adding two neutrons than by adding one neutron and one proton because of the electrostatic repulsion. This explains why $N > Z$ for the heavier nuclei.

Mass and Binding Energy

In our study of relativistic energy (Section 1-10), we saw that the mass of a nucleus is not equal to the sum of the masses of the individual nucleons that make up the nucleus. When two or more nucleons fuse together to form a nucleus, the total rest mass decreases and energy is given off. Conversely, to break up a nucleus into its parts, energy must be put into the system to increase the rest mass. The energy involved is c^2 times the change in mass, where c is the speed of light in vacuum. The difference between the rest energy of the parts of a nucleus and the rest energy of the nucleus is the total **binding energy** of the nucleus.

As we saw in Section 5-3, atomic and nuclear masses are often given in unified mass units (u), which are defined as one-twelfth of the mass of the neutral carbon-12 atom. The rest energy of one unified mass unit is

$$(1 \text{ u})c^2 = 931.5 \text{ MeV} \qquad 7\text{-}2$$

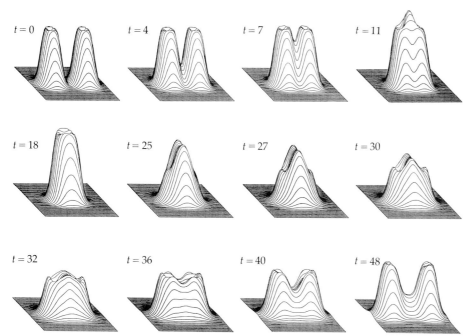

A computer simulation of a collision between two carbon-12 nuclei. The incoming carbon nucleus is considered to have an energy of 768 MeV, distributed equally among the 12 nucleons. The base of the plot corresponds to an area 18 fm on a side. The vertical dimension represents the density of protons and neutrons on a scale where the original height of each nucleus corresponds to 1.5×10^{11} nucleons/fm^3. Increments of time t are in units of 3.3×10^{-24} s. At certain energies, colliding nuclei bind together for a brief period before separating. Afterward, some of the initial kinetic energy appears in the form of internal excitations of the nuclei.

Consider ^{4}He, for example, which consists of two protons and two neutrons. The mass of an atom can be accurately measured using a mass spectrometer. The mass of the ^{4}He atom is 4.002603 u. This includes the masses of the two electrons in the atom. The mass of the ^{1}H atom is 1.007825 u, and that of the neutron is 1.008665 u. The sum of the masses of two ^{1}H atoms plus two neutrons is 2(1.007825 u) + 2(1.008665 u) = 4.032980 u, which is greater than the mass of the ^{4}He atom by 0.030377 u. Note that by using the masses of two ^{1}H atoms rather than two protons, the masses of the electrons in the atoms are accounted for. We do this because it is atomic masses that are measured directly and listed in mass tables.

We can find the binding energy of the ^{4}He nucleus from the mass difference of 0.030377 u by using the conversion factor 1 uc^2 = 931.5 MeV from Equation 7-2:

$$(0.030377 \text{ u})c^2 = 0.030377 \text{ u}c^2 \times \frac{931.5 \text{ MeV}}{1 \text{ u}c^2} = 28.30 \text{ MeV}$$

The total binding energy of ^{4}He is thus 28.3 MeV. In general, the binding energy E_b of the nucleus of an atom of atomic mass M_A containing Z protons and N neutrons is found by calculating the difference between the mass of the parts and the mass of the nucleus and then multiplying by c^2:

Total nuclear binding energy

$$E_b = (ZM_H + Nm_n - M_A)c^2 \qquad 7\text{-}3$$

where M_H is the mass of the ^{1}H atom and m_n is the mass of the neutron. Note again that this formula is written in terms of atomic masses rather than nuclear masses. The mass of the Z electrons in the term ZM_H is canceled by the mass of the Z electrons in the term M_A. The atomic masses of the neutron and some selected isotopes are listed in Table 7-1.

Table 7-1 Atomic Masses of the Neutron and Selected Isotopes

Element	Symbol	Z	Atomic mass, u
Neutron	n	0	1.008 665
Hydrogen	^{1}H	1	1.007 825
(deuterium)	^{2}H or D	1	2.014 102
(tritium)	^{3}H or T	1	3.016 050
Helium	^{3}He	2	3.016 030
	^{4}He	2	4.002 603
Lithium	^{6}Li	3	6.015 125
Boron	^{10}B	5	10.012 939
Carbon	^{12}C	6	12.000 000
	^{14}C	6	14.003 242
Oxygen	^{16}O	8	15.994 915
Sodium	^{23}Na	11	22.989 771
Potassium	^{39}K	19	38.963 710
Iron	^{56}Fe	26	55.939 395
Copper	^{63}Cu	29	62.929 592
Silver	^{107}Ag	47	106.905 094
Gold	^{197}Au	79	196.966 541
Lead	^{208}Pb	82	207.976 650
Polonium	^{212}Po	84	211.989 629
Radon	^{222}Rn	86	222.017 531
Radium	^{226}Ra	88	226.025 360
Uranium	^{238}U	92	238.048 608
Plutonium	^{242}Pu	94	242.058 725

Example 7-1

Find the binding energy of the last neutron in ^{4}He.

From Table 7-1, the rest mass of ^{4}He is 4.00260 u and that of ^{3}He is 3.01603 u. The rest mass of ^{3}He plus that of the neutron is 3.01603 u + 1.00866 u = 4.02469 u. This is greater than the rest mass of ^{4}He by 4.02469 u − 4.00260 u = 0.02209 u. The binding energy of the last neutron is thus

$$(\Delta m)c^2 = (0.02209 \text{ u})c^2 \times \frac{931.5 \text{ MeV}}{1 \text{ u}c^2} = 20.58 \text{ MeV}$$

Once the atomic mass of a nuclide has been determined, its binding energy can be computed from Equation 7-3. Figure 7-3 shows the binding energy per nucleon E_b/A versus A. The mean value of E_b/A is about 8.3 MeV. The flatness of this curve for $A > 50$ shows that E_b is approximately proportional to the number of nucleons A in the nucleus. This indicates that there is saturation of nuclear bonds in the nucleus. That is, each nucleon bonds to only a certain maximum number of other nucleons independent of the total number of nucleons in the nucleus. If there were no saturation and each nucleon bonded to every other nucleon, there would be $A − 1$ bonds for each nucleon and a total of $A(A − 1)$ bonds altogether. The

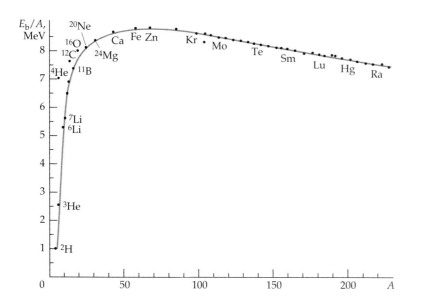

Figure 7-3 The binding energy per nucleon versus the mass number A. For nuclei with values of A greater than about 50, the curve is approximately flat, indicating that the total binding energy is approximately proportional to A.

total binding energy, which is a measure of the energy needed to break all the bonds between nucleons, would then be proportional to $A(A − 1)$, and E_b/A would not be approximately constant. Figure 7-3 indicates that, above a certain value of A, there is a fixed number of bonds per nucleon, as would be the case if each nucleon were attracted only to its nearest neighbors. Such a situation also leads to a constant nuclear density, which is consistent with measurements of nuclear radii. The steep rise in the curve for low values of A is due to the increase in the number of nearest neighbors and therefore to the increase in the number of bonds per nucleon. The gradual decrease in E_b/A at high values of A is due to the Coulomb repulsion between the protons, which increases as Z^2 and decreases the binding energy. For values of A greater than about 260, the Coulomb repulsion becomes so great that the nucleus is unstable and undergoes spontaneous fission.

Spin and Magnetic Moment

The spin quantum number of the neutron and the proton is $\frac{1}{2}$. The angular momentum of the nucleus is a combination of the spin angular momenta of the nucleons plus any orbital angular momenta due to the motion of the nucleons. The resultant angular momentum is called the **nuclear spin I**. The spin of all nuclei with an even number of protons and an even number of neutrons is zero. Evidently, the nucleons couple together in pairs in such a way that their angular momenta add to zero, as is often the case for electrons in atoms.

In Section 4-3, we found that the magnetic moment associated with the angular momentum of an electron in an atom was of the order of a Bohr magneton $\mu_B = e\hbar/2m_e$ and that the magnetic moment associated with electron spin was 1 μ_B in agreement with the Dirac relativistic wave equation rather than the $\frac{1}{2}$ μ_B predicted by classical physics. The magnetic moment of the nucleus is of the order of the **nuclear magneton**

$$\mu_N = \frac{e\hbar}{2m_p} = 5.05 \times 10^{-27} \text{ J/T} = 3.15 \times 10^{-8} \text{ eV/T} \qquad 7\text{-}4$$

where m_p is the mass of the proton. If the proton and neutron obeyed the Dirac relativistic wave equation as the electron does, the (z components of the) magnetic moments associated with their spins would be 1 μ_N for the proton and 0 for the neutron because it has no charge. The experimentally determined moment of the proton is

$$(\mu_z)_{\text{proton}} = +2.7928 \ \mu_N \qquad 7\text{-}5$$

and that of the neutron is

$$(\mu_z)_{\text{neutron}} = -1.9135 \ \mu_N \qquad 7\text{-}6$$

The negative sign for the z component of the magnetic moment of the neutron indicates that its direction is opposite that of the neutron's spin. It is interesting to note that the deviations of these moments from those predicted by the Dirac equation are of about the same magnitude—1.91 μ_N for the neutron and 1.79 μ_N for the proton. These deviations are due to the fact that the proton and neutron are not point particles like the electron but are instead composite particles made up of elementary particles called quarks, which we will discuss in Chapter 8.

Questions

1. How does the strong nuclear force differ from the electromagnetic force?
2. What property of the strong nuclear force is indicated by the fact that all nuclei have about the same density?
3. The mass of ^{12}C, which contains six protons and six neutrons, is exactly 12.000 u by the definition of the unified mass unit. Why isn't the mass of ^{16}O, which contains eight protons and eight neutrons, exactly 16.000 u?

7-2 Nuclear Magnetic Resonance

In Section 4-5, we saw that the energy levels of the atom were split in the presence of an external magnetic field (the Zeeman effect) because of the interaction of the atomic magnetic moment and the field. Since nuclei also have magnetic moments, the energy of a nucleus is also split in the presence of a magnetic field. We will consider only the simplest case, the hydrogen atom for which the nucleus is a single proton.

The potential energy of a magnetic moment $\boldsymbol{\mu}$ in an external magnetic field $\mathbf{B}$ is given by

$$U = -\boldsymbol{\mu} \cdot \mathbf{B} \qquad 7\text{-}7$$

The potential energy is lowest when the magnetic moment is aligned with the field and highest when it is in the opposite direction. Since the spin quantum number of the proton is $\frac{1}{2}$, the proton's magnetic moment has two possible orientations in an external magnetic field: parallel to the field (spin up) or antiparallel to the field (spin down). The difference in energy of these two orientations (Figure 7-4) is

$$\Delta E = 2(\mu_z)_\mathrm{P} B \qquad 7\text{-}8$$

When hydrogen atoms are irradiated with photons of energy ΔE, some of the nuclei are induced to make transitions from the lower state to the upper state by resonance absorption. These nuclei then decay back to the lower state, emitting photons of energy ΔE. The frequency of the photons absorbed and emitted is found from

$$hf = \Delta E = 2(\mu_z)_\mathrm{P} B$$

In a magnetic field of 1 T, this energy is

$$\Delta E = 2(\mu_z)_\mathrm{P} B$$

$$= 2(2.79 \ \mu_\mathrm{N}) \left(\frac{3.15 \times 10^{-8} \ \mathrm{eV/T}}{1 \ \mu_\mathrm{N}} \right) (1 \ \mathrm{T})$$

$$= 1.76 \times 10^{-7} \ \mathrm{eV}$$

and the frequency of the photons is

$$f = \frac{\Delta E}{h} = \frac{1.76 \times 10^{-7} \ \mathrm{eV}}{4.14 \times 10^{-15} \ \mathrm{eV \cdot s}}$$

$$= 4.25 \times 10^7 \ \mathrm{Hz} = 42.5 \ \mathrm{MHz}$$

This frequency is in the radio band of the electromagnetic spectrum, and the radiation is called **RF (radiofrequency) radiation.** The measurement of this resonance frequency for free protons can be used to determine the magnetic moment of the proton.

When a hydrogen atom is in a molecule, the magnetic field at the nucleus is the sum of the external magnetic field and the local magnetic field due to the atoms and nuclei of the surrounding material. Since the resonance frequency is proportional to the total magnetic field seen by the proton, a measurement of this frequency can give information about the internal magnetic field seen in the molecule. This is called **nuclear magnetic resonance.**

Nuclear magnetic resonance is used as an alternative to x rays or ultrasound for imaging. A patient can be placed in a magnetic field that is constant in time but not in space. When the patient is irradiated by a broadband RF source, the resonance frequency of the absorbed and emitted RF photons is then dependent on the value of the magnetic field, which can be related to specific positions in the body of the patient. Since the energy of the photons is much less than the energy of molecular bonds and the intensity used is low enough so that it produces negligible heating, the RF photons produce little (if any) biological damage.

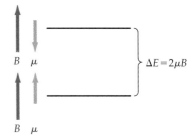

Figure 7-4 A proton has two energy states in the presence of a magnetic field corresponding to whether the magnetic moment of the proton is aligned parallel or antiparallel to the field.

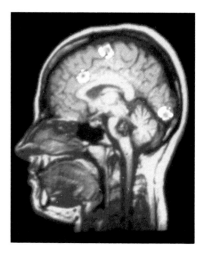

Positron emission tomography images three areas (small pink dots surrounded by white) in the left brain that are active during language tasks. The area in the back is active during reading, the middle area during speech, and the front area while thinking about the meaning of a word. A PET image, which is here superposed over an NMR scan (blue and purple), is obtained by injecting a radioactive substance, for instance an isotope of oxygen, into the body. As the tracer circulates, it emits positrons. The annihilation of the positrons with electrons produces pairs of γ rays from which the location of the reaction can be inferred. An increase in the reaction rate can be interpreted as an increase in blood flow, corresponding to an increase in physiological activity.

7-3 Radioactivity

Nuclei that are not stable are radioactive; that is, they decay into other nuclei by the emission of radiation. The term radiation refers to the emission of particles, such as electrons, neutrons, or alpha particles, as well as to electromagnetic radiation. The three kinds of radioactivity—**alpha (α) decay, beta (β) decay,** and **gamma (γ) decay**—were named before it was known that alpha particles are ^{4}He nuclei, beta particles are electrons (e^-) or positrons (e^+), and gamma rays are photons.

In 1900, Rutherford discovered that the rate of emission of radioactive particles from a substance is not constant over time but decreases exponentially. *This exponential time dependence is characteristic of all radioactivity and indicates that radioactive decay is a statistical process.* Because each nucleus is well shielded from others by the atomic electrons, pressure and temperature changes have little or no effect on the rate of radioactive decay or other nuclear properties.

Let N be the number of radioactive nuclei at some time t. For a statistical decay process in which the decay of any individual nucleus is a random event, we expect the number of nuclei that decay in some time interval dt to be proportional to N and to dt. Because of these decays, the number N will decrease. The change in N is given by

$$dN = -\lambda N\, dt \qquad 7\text{-}9$$

where λ is a constant of proportionality called the **decay constant.** The rate of change of N, dN/dt, is proportional to N. This is characteristic of exponential decay. To solve Equation 7-9 for N, we first divide each side by N, thus separating the variables N and t:

$$\frac{dN}{N} = -\lambda\, dt$$

Integrating, we obtain

$$\ln N = -\lambda t + C \qquad 7\text{-}10$$

where C is some constant of integration. Taking the exponential of each side of Equation 7-10, we obtain

$$N = e^{-\lambda t + C} = e^C e^{-\lambda t}$$

or

$$N = N_0 e^{-\lambda t} \qquad 7\text{-}11$$

where $N_0 = e^C$ is the number of nuclei at $t = 0$. The number of radioactive decays per second is called the decay rate R:

Decay rate

$$R = -\frac{dN}{dt} = \lambda N = \lambda N_0 e^{-\lambda t} = R_0 e^{-\lambda t} \qquad 7\text{-}12$$

where

$$R_0 = \lambda N_0 \qquad 7\text{-}13$$

is the rate of decay at time $t = 0$. The decrease in the decay rate R is the quantity that is determined experimentally.

The average or **mean lifetime** τ is the reciprocal of the decay constant:

$$\tau = \frac{1}{\lambda} \qquad 7\text{-}14$$

(See Problem 54.) The mean lifetime is analogous to the time constant in the exponential decrease in the charge on a capacitor in an *RC* circuit. After a time equal to the mean lifetime the number of radioactive nuclei and the decay rate have each decreased to 37 percent of their original values. The **half-life** $t_{1/2}$ is defined as the time it takes for the number of nuclei and the decay rate to decrease by half. Setting $t = t_{1/2}$ and $N = N_0/2$ in Equation 7-11, we obtain

$$\frac{N_0}{2} = N_0 e^{-\lambda t_{1/2}} \qquad 7\text{-}15$$

Solving for $t_{1/2}$, we obtain

$$t_{1/2} = \frac{\ln 2}{\lambda} = \frac{0.693}{\lambda} = 0.693\tau \qquad 7\text{-}16$$

Exercise

Show the mathematical steps needed to obtain Equation 7-16 from Equation 7-15.

Figure 7-5 shows a plot of N versus t. If we multiply the numbers on the N axis by λ, this graph becomes a plot of R versus t. After each time interval of one half-life, the number of nuclei left and the decay rate have decreased to half of their previous values. For example, if the decay rate is R_0 initially, it will be $\frac{1}{2}R_0$ after one half-life, $(\frac{1}{2})(\frac{1}{2})R_0$ after two half-lives, and so forth. After n half-lives, the decay rate will be

$$R = \left(\frac{1}{2}\right)^n R_0 \qquad 7\text{-}17$$

The half-lives of radioactive nuclei vary from very small times (less than 1 μs) to very large times (up to 10^{16} y).

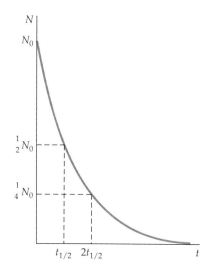

Figure 7-5 Exponential radioactive decay. After each half-life $t_{1/2}$, the number of nuclei remaining has decreased by one-half. The decay rate $R = \lambda N$ has the same time dependence.

Example 7-2

A radioactive source has a half-life of 1 min. At time $t = 0$, it is placed near a detector, and the counting rate (the number of decay particles detected per unit time) is observed to be 2000 counts/s. Find the counting rate at times $t = 1, 2, 3,$ and 10 min.

Since the half-life is 1 min, the counting rate will be half as great at $t = 1$ min as at $t = 0$, so $r_1 = 1000$ counts/s at 1 min, $r_2 = 500$ counts/s at $t = 2$ min, and $r_3 = 250$ counts/s at 3 min. At $t = 10$ min $= 10 t_{1/2}$, the rate will be $r_{10} = (\frac{1}{2})^{10}(2000) = 1.95$ counts/s ≈ 2 counts/s.

Example 7-3

If the detection efficiency in Example 7-2 is 20 percent, how many radioactive nuclei are there at time $t = 0$? At time $t = 1$ min? How many nuclei decay in the first minute?

The detection efficiency depends on the distance from the source to the detector and the probability that a radioactive decay particle entering the detector will produce a count. If the source is smaller than the detector and is placed very close to the detector, about half the emitted particles will enter the detector. If the counting rate at $t = 0$ is 2000 counts/s and the efficiency is 20 percent, the decay rate at $t = 0$ must be 10,000 s^{-1}. The number of radioactive nuclei can be found from Equation 7-12:

$$R = \lambda N$$

The decay constant is related to the half-life by Equation 7-16:

$$\lambda = \frac{0.693}{t_{1/2}} = \frac{0.693}{1 \text{ min}}$$

The number of nuclei at $t = 0$ is therefore

$$N = \frac{R}{\lambda} = \frac{10,000 \text{ s}^{-1}}{0.693 \text{ min}^{-1}} \times \frac{60 \text{ s}}{1 \text{ min}} = 8.66 \times 10^5$$

At time $t = 1 \text{ min} = t_{1/2}$, there are half as many nuclei as at $t = 0$, so $N_1 = \frac{1}{2}(8.66 \times 10^5) = 4.33 \times 10^5$. The number of nuclei that decay in the first minute is therefore 4.33×10^5.

The SI unit of radioactive decay is the **becquerel** (Bq), which is defined as one decay per second:

$$1 \text{ Bq} = 1 \text{ decay/s} \quad\quad\quad 7\text{-}18$$

A historical unit that applies to all types of radioactivity is the **curie** (Ci), which is defined as

$$1 \text{ Ci} = 3.7 \times 10^{10} \text{ decays/s} = 3.7 \times 10^{10} \text{ Bq} \quad\quad\quad 7\text{-}19$$

The curie is the rate at which radiation is emitted by 1 g of radium. Since this is a very large unit, the millicurie (mCi) or microcurie (μCi) is often used.

Beta Decay

Beta decay occurs in nuclei that have too many or too few neutrons for stability. The energy released in β decay can be determined by computing the difference between the rest mass of the original nucleus and that of the decay products. In β decay, A remains the same while Z either increases by 1 (β^- decay) or decreases by 1 (β^+ decay).*

The simplest example of β decay is the decay of the free neutron into a proton plus an electron. (The half-life of a free neutron is about 10.8 min.) The energy of decay is 0.782 MeV, which is the difference between the rest energy of the neutron and that of the proton plus electron. More generally, in β^- decay, a nucleus of mass number A and atomic number Z decays into a nucleus, referred to as the **daughter nucleus,** of mass number A and atomic number $Z' = Z + 1$ with the emission of an electron. If the decay energy is shared by the daughter nucleus and the emitted electron, the energy of the electron is uniquely determined by the conservation of energy and momentum. Experimentally, however, the energies of the electrons emitted in β^- decay are observed to vary from zero to the maximum energy available. A typical energy spectrum for these electrons is shown in Figure 7-6.

To explain the apparent nonconservation of energy in beta decay, W. Pauli in 1930 suggested that a third particle, which he called the **neutrino,** is also emitted. The mass of the neutrino was originally assumed to be

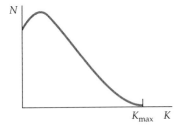

Figure 7-6 Number of electrons emitted in β^- decay versus kinetic energy. The fact that all the electrons do not have the same energy K_{max} suggests that another particle that shares the energy available for decay is emitted.

*The notations β^- and e^- are used interchangeably for the electron. Similarly, β^+ and e^+ are used interchangeably for the positron.

zero because the maximum energy of the emitted electrons is equal to the total available for the decay. In 1948, measurements of the momenta of the emitted electron and the recoiling nucleus showed that the neutrino was also needed for the conservation of linear momentum in β decay. The neutrino was first observed experimentally in 1957. It is now known that there are at least three kinds of neutrinos, one (ν_e) associated with electrons, one (ν_μ) associated with muons, and one (ν_τ) not yet observed experimentally, associated with the newly discovered tau particle τ. Moreover, each neutrino has an antiparticle, written $\bar{\nu}_e$, $\bar{\nu}_\mu$, and $\bar{\nu}_\tau$. It is the electron antineutrino that is emitted in the decay of a neutron, which is written

$$n \rightarrow p + \beta^- + \bar{\nu}_e \qquad 7\text{-}20$$

The mass of the electron neutrino or electron antineutrino is known to be less than 4×10^{-5} times the mass of the electron.

In β^+ decay, a proton changes into a neutron with the emission of a positron (and a neutrino). A free proton cannot decay by positron emission because of conservation of energy (the rest mass of the neutron plus the positron is greater than that of the proton), but because of binding-energy effects, a proton inside a nucleus can decay. A typical β^+ decay is

$$^{13}_{7}N \rightarrow {}^{13}_{6}C + \beta^+ + \nu_e \qquad 7\text{-}21$$

The electrons or positrons emitted in β decay do not exist inside the nucleus. They are created in the process of decay, just as photons are created when an atom makes a transition from a higher to a lower energy state.

An important example of β decay is that of ^{14}C, which is used in **radioactive carbon dating:**

$$^{14}C \rightarrow {}^{14}N + \beta^- + \bar{\nu}_e \qquad 7\text{-}22$$

The half-life for this decay is 5730 y. The radioactive isotope ^{14}C is produced in the upper atmosphere in nuclear reactions caused by cosmic rays. The chemical behavior of carbon atoms with ^{14}C nuclei is the same as those with ordinary ^{12}C nuclei. For example, atoms with these nuclei combine with oxygen to form CO_2 molecules. Since living organisms continually exchange CO_2 with the atmosphere, the ratio of ^{14}C to ^{12}C in a living organism is the same as the equilibrium ratio in the atmosphere, which is about 1.3×10^{-12}. After an organism dies, it no longer absorbs ^{14}C from the atmosphere, so the ratio of ^{14}C to ^{12}C continually decreases due to the radioactive decay of ^{14}C. The number of ^{14}C decays per minute per gram of carbon in a living organism can be calculated from the known half-life of ^{14}C and the number of ^{14}C nuclei in a gram of carbon. The result is that there are about 15.0 decays per minute per gram of carbon in a living organism. Using this result and the measured number of decays per minute per gram of carbon in a nonliving sample of bone, wood, or other object containing carbon, we can determine the age of the sample. For example, if the measured rate were 7.5 decays per minute per gram, the sample would be one half-life = 5730 years old.

Example 7-4

A bone containing 200 g of carbon has a β-decay rate of 400 decays/min. How old is the bone?

We first obtain a rough estimate of the age of the bone. If the bone were from a living organism, we would expect the decay rate to be (15 decays/min·g)(200 g) = 3000 decays/min. Since 400/3000 is roughly 1/8 (actually 1/7.5), the sample must be about three half-lives old which is about 3(5730) y = 17,190 y.

To find the age of the bone more accurately, we note that after n half-lives, the decay rate will have decreased by a factor of $(\frac{1}{2})^n$. We can find n from

$$\left(\frac{1}{2}\right)^n = \frac{400}{3000}$$

or

$$2^n = \frac{3000}{400} = 7.5$$

We solve for n by taking the logarithm of each side:

$$n \ln 2 = \ln 7.5$$

$$n = \frac{\ln 7.5}{\ln 2} = 2.91$$

The age of the bone is therefore

$$t = nt_{1/2} = 2.91(5730 \text{ y}) = 16{,}700 \text{ y}$$

Gamma Decay

In gamma decay a nucleus in an excited state decays to a lower-energy state by the emission of a photon. This is the nuclear counterpart of spontaneous emission of photons by atoms and molecules. Unlike β or α decay, the radioactive nucleus remains the same nucleus as it decays via γ decay. Since the spacing of the nuclear energy levels is of the order of 1 MeV (as compared with spacing of the order of 1 eV in atoms), the wavelengths of the emitted photons are of the order of 1 pm:

$$\lambda = \frac{hc}{E} \approx \frac{1240 \text{ eV·nm}}{1 \text{ MeV}} = 0.00124 \text{ nm} = 1.24 \text{ pm}$$

The emission of gamma rays normally happens very quickly and is observed only because it usually follows either α or β decay. For example, if a radioactive parent nucleus decays by β decay to an excited state of the daughter nucleus, the daughter nucleus then decays to its ground state by γ emission. The mean lifetime for γ decay is often very short. Direct measurements of mean lifetimes as short as about 10^{-11} s are possible. Measurements of mean lifetimes shorter than 10^{-11} s are difficult, but they can sometimes be made by indirect methods.

A few γ emitters have very long lifetimes, of the order of hours. Nuclear energy states that have such long lifetimes are called **metastable states.**

Alpha Decay

All very heavy nuclei ($Z > 83$) are theoretically unstable to alpha decay because the mass of the original radioactive nucleus is greater than the sum of the masses of the decay products—an α particle and the daughter nucleus. Consider the decay of ^{232}Th ($Z = 90$) into ^{228}Ra ($Z = 88$) plus an α particle. This is written as

$$^{232}\text{Th} \rightarrow {}^{228}\text{Ra} + \alpha = {}^{228}\text{Ra} + {}^{4}\text{He} \qquad 7\text{-}23$$

The mass of the ^{232}Th atom is 232.038124 u. The mass of the daughter atom ^{228}Ra is 228.031139 u. Adding to this 4.002603 u for the mass of ^{4}He, we get 232.033742 u for the total mass of the decay products. This is less than the mass of ^{232}Th by 0.004382 u, which multiplied by 931.5 MeV/c^2 gives

4.08 MeV/c^2 for the excess rest mass of ^{232}Th over that of the decay products. The isotope ^{232}Th is therefore theoretically unstable to α decay. This decay does in fact occur in nature with the emission of an α particle of kinetic energy 4.08 MeV. (The kinetic energy of the α particle is actually somewhat less than 4.08 MeV because some of the decay energy is shared by the recoiling ^{228}Ra nucleus.)

In general, when a nucleus emits an α particle, both N and Z decrease by 2 and A decreases by 4. The daughter of a radioactive nucleus is often itself radioactive and decays by either α or β decay or both. If the original nucleus has a mass number A that is 4 times an integer, the daughter nucleus and all those in the chain will also have mass numbers equal to 4 times an integer. Similarly, if the mass number of the original nucleus is $4n + 1$, where n is an integer, all the nuclei in the decay chain will have mass numbers given by $4n + 1$, with n decreasing by one at each decay. We can see, therefore, that there are four possible α-decay chains, depending on whether A equals $4n$, $4n + 1$, $4n + 2$, or $4n + 3$, where n is an integer. All but one of these decay chains are found in nature. The $4n + 1$ series is not found because its longest-lived member (other than the stable end product ^{209}Bi) is ^{237}Np, which has a half-life of only 2×10^6 y. As this is much less than the age of the earth, this series has disappeared.

Figure 7-7 shows the thorium series, for which $A = 4n$. It begins with an α decay from ^{232}Th to ^{228}Ra. The daughter nuclide of an α decay is on the left or neutron-rich side of the stability curve (the dashed line in the figure), so it often decays by β^- decay. In the thorium series, ^{228}Ra decays by β^- decay to ^{228}Ac, which in turn decays by β^- decay to ^{228}Th. There are then four α decays to ^{212}Pb, which decays by β^- decay to ^{212}Bi. The series branches at ^{212}Bi, which decays either by α decay to ^{208}Tl or by β^- decay to ^{212}Po. The branches meet at the stable lead isotope ^{208}Pb.

The energies of α particles from natural radioactive sources range from about 4 to 7 MeV, and the half-lives range from about 10^{-5} s to 10^{10} y. In general, the smaller the energy of the emitted α particle, the longer the half-life. As we discussed in Section 3-9, the enormous variation in half-lives was explained by George Gamow in 1928. He considered α decay to be

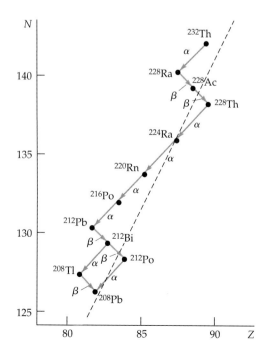

Figure 7-7 The thorium ($4n$) α-decay series. The dashed line is the curve of stability.

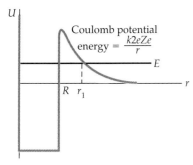

Figure 7-8 A model of the potential energy for an α particle and a nucleus. The strong attractive nuclear force that exists for values of r less than the nuclear radius R is indicated by the potential well. Outside the nucleus, the nuclear force is negligible, and the potential energy is given by Coulomb's law $U = +k2eZe/r$, where Ze is the nuclear charge and $2e$ is the charge of the α particle. The energy E is the kinetic energy of the α particle when it is far away from the nucleus. A small increase in E reduces the relative height of the barrier and also its thickness, leading to a much greater chance of penetration. An increase in the energy of the emitted α particles by a factor of 2 results in a reduction of the half-life by a factor of more than 10^{20}.

a process in which an α particle is first formed inside a nucleus and then tunnels through the Coulomb barrier (Figure 7-8). A slight increase in the energy of the α particle reduces the relative height $U - E$ of the barrier and also the thickness. Because the probability of penetration is so sensitive to the relative height and thickness of the barrier, a small increase in E leads to a large increase in the probability of barrier penetration and therefore to a shorter lifetime. Gamow was able to derive an expression for the half-life as a function of E that is in excellent agreement with experimental results.

Questions

4. Why do extreme changes in the temperature or pressure of a radioactive sample have little or no effect on the radioactivity?
5. Why is the decay series $A = 4n + 1$ not found in nature?
6. A decay by α emission is often followed by β decay. When this occurs, it is by β^- and not β^+ decay. Why?
7. The half-life of ^{14}C is much less than the age of the universe, yet ^{14}C is found in nature. Why?
8. What effect would a long-term variation in cosmic-ray activity have on the accuracy of ^{14}C dating?

7-4 Nuclear Reactions

Information about nuclei is typically obtained by bombarding them with various particles and observing the results. Although the first experiments of this type were limited by the need to use naturally occurring radiation, they produced many important discoveries. In 1932 J. D. Cockcroft and E. T. S. Walton succeeded in producing the reaction

$$p + {}^7\text{Li} \rightarrow {}^8\text{Be} \rightarrow {}^4\text{He} + {}^4\text{He}$$

using artificially accelerated protons. At about the same time, the Van de Graaff electrostatic generator was built (by R. Van de Graaff in 1931), as was the first cyclotron (by E. O. Lawrence and M. S. Livingston in 1932). Since then, enormous advances in the technology for accelerating and detecting particles have been made, and many nuclear reactions have been studied.

When a particle is incident on a nucleus, several different things can happen. The incident particle may be scattered elastically or inelastically, in which case the nucleus is left in an excited state and decays by emitting photons (or other particles); or the incident particle may be absorbed by the nucleus, and another particle or particles may be emitted.

When energy is released by a nuclear reaction, the reaction is said to be an **exothermic reaction.** The amount of energy released is called the **Q value** of the reaction. In an exothermic reaction, the total mass of the incoming particles is greater than that of the outgoing particles. The Q value equals c^2 times this mass difference. If the total mass of the incoming particles is less than that of the outgoing particles, energy is required for the reaction to take place, and the reaction is said to be an **endothermic reaction.** The Q value of an endothermic reaction is also equal to c^2 times the initial mass minus the final mass, but in this case the Q value is negative. An endothermic reaction cannot take place below a certain threshold energy. The threshold energy is usually somewhat greater than $|Q|$ because the outgoing particles must have some kinetic energy to conserve momentum.

A measure of the effective size of a nucleus for a particular nuclear reaction is the **cross section** σ. If I is the number of incident particles per unit time per unit area (the incident intensity) and R is the number of reactions per unit time per nucleus, the cross section is

$$\sigma = \frac{R}{I} \qquad 7\text{-}24$$

The cross section σ has the dimensions of area. Since nuclear cross sections are of the order of the square of the nuclear radius, a convenient unit for them is the **barn,** which is defined as

$$1 \text{ barn} = 10^{-28} \text{ m}^2 \qquad 7\text{-}25$$

The cross section for a particular reaction is a function of energy. For an endothermic reaction, it is zero for energies below the threshold energy.

Example 7-5

Find the Q value of the reaction

$$p + {}^7\text{Li} \rightarrow {}^4\text{He} + {}^4\text{He}$$

and state whether the reaction is exothermic or endothermic. The atomic mass of ${}^7\text{Li}$ is 7.016004 u.

Using 1.007825 u for the mass of ${}^1\text{H}$ and 4.002603 u for the mass of ${}^4\text{He}$ from Table 7-1, we have for the total mass of the original particles

$$m_i = 1.007825 \text{ u} + 7.016004 \text{ u} = 8.023829 \text{ u}$$

and for the total mass of the final particles

$$m_f = 2(4.002603 \text{ u}) = 8.005206 \text{ u}$$

Since the initial mass is greater than the final mass by

$$\Delta m = m_i - m_f = 8.023829 \text{ u} - 8.005206 \text{ u} = 0.018623 \text{ u}$$

mass is converted into energy and the reaction is exothermic. The Q value is positive and is given by

$$Q = (\Delta m)c^2 = (0.018623 \text{ u})c^2(931.5 \text{ MeV}/uc^2) = 17.35 \text{ MeV}$$

Note that we used the mass of atomic hydrogen rather than that of the proton and the atomic masses of the ${}^7\text{Li}$ and ${}^4\text{He}$ atoms rather than the masses of the individual nuclei so that the masses of the eight electrons on each side of the reaction cancel.

Reactions with Neutrons

Nuclear reactions involving neutrons are important for understanding nuclear reactors. The most likely reaction with a nucleus for a neutron of energy of more than about 1 MeV is scattering. However, even if the scattering is elastic, the neutron loses some energy to the nucleus because the nucleus recoils. If a neutron is scattered many times in a material, its energy decreases until it is of the order of the energy of thermal motion kT, where k is Boltzmann's constant and T is the absolute temperature. (At ordinary room temperatures, kT is about 0.025 eV.) The neutron is then equally likely to gain or lose energy from a nucleus when it is elastically scattered. A neutron with energy of the order of kT is called a **thermal neutron.**

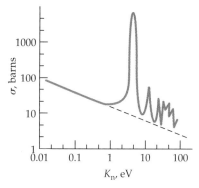

Figure 7-9 Neutron-capture cross section for silver versus energy of the neutron. The straight line indicates the $1/v$ dependence of the cross section, which is proportional to the time spent by the neutron near the silver nucleus. Superimposed on this dependence are a large resonance and several smaller resonances.

At low energies, a neutron is likely to be captured, with the emission of a γ ray from the excited nucleus. Figure 7-9 shows the neutron-capture cross section for silver as a function of the energy of the neutron. The large peak in this curve is called a **resonance**. Except for the resonance, the cross section varies fairly smoothly with energy, decreasing with increasing energy roughly as $1/v$, where v is the speed of the neutron. We can understand this energy dependence as follows: Consider a neutron moving with speed v near a nucleus of diameter $2R$. The time it takes the neutron to pass the nucleus is $2R/v$. Thus, the neutron-capture cross section is proportional to the time spent by the neutron in the vicinity of the nucleus. The dashed line in Figure 7-9 indicates this $1/v$ dependence. At the maximum of the resonance, the value of the cross section is very large ($\sigma > 5000$ barns) compared with a value of only about 10 barns just past the resonance. Many elements show similar resonances in their neutron-capture cross sections. For example, the maximum cross section for ^{113}Cd is about 57,000 barns. This material is thus very useful for shielding against low-energy neutrons.

An important nuclear reaction that involves neutrons is fission, which will be discussed in the next section.

Questions

9. What is meant by the cross section for a nuclear reaction?
10. Why is the neutron-capture cross section (excluding resonances) proportional to $1/v$?
11. What is meant by the Q value of a reaction?

(a)

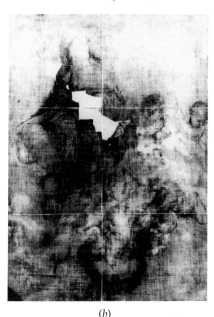

(b)

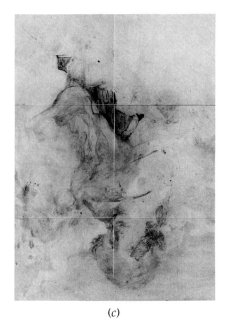

(c)

Hidden layers in paintings are analyzed by bombarding the painting with neutrons and observing the radiative emissions from nuclei that have captured a neutron. Different elements used in the painting have different half-lives. (a) Van Dyck's painting "Saint Rosalie Interceding for the Plague-Stricken of Palermo." The black and white images in (b) and (c) were formed using a special film sensitive to electrons emitted by the radioactively decaying elements. Image (b), taken a few hours after the neutron irradiation, reveals the presence of manganese, found in umber, a dark earth-pigment used for the painting's base layer. (Blank areas show where modern repairs, free of manganese, have been made.) The image in (c) was taken four days later, after the umber emissions had died away and when phosphorus, found in charcoal and boneblack, was the main radiating element. Upside down, is revealed a sketch of Van Dyck himself. The self-portrait, executed in charcoal, had been overpainted by the artist.

7-5 Fission, Fusion, and Nuclear Reactors

Figure 7-10 shows a plot of the nuclear mass difference per nucleon ($M - Zm_p - Nm_n$)/A in units of MeV/c^2 versus A. This is just the negative of the binding-energy curve shown in Figure 7-3. From Figure 7-10, we can see that the rest mass per nucleon for both very heavy ($A \approx 200$) and very light ($A \lesssim 20$) nuclides is more than that for nuclides of intermediate mass. Thus, energy is released when a very heavy nucleus, such as ^{235}U, breaks up into two lighter nuclei—a process called **fission**—or when two very light nuclei, such as ^{2}H and ^{3}H, fuse together to form a nucleus of greater mass—a process called **fusion**.

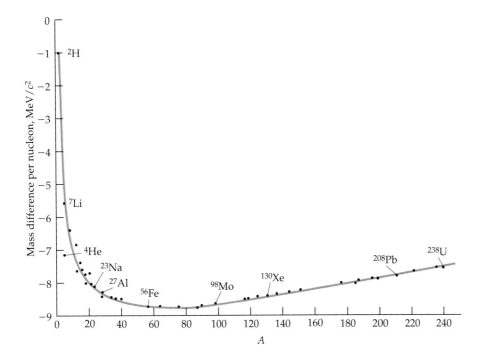

Figure 7-10 Plot of mass difference per nucleon ($M - Zm_p - Nm_n$)/A in units of MeV/c^2 versus A. The rest mass per nucleon is less for intermediate mass nuclei than for either very light or very heavy nuclei.

The application of both fission and fusion to the development of nuclear weapons has had a profound effect on our lives during the past 45 years. The peaceful application of these reactions to the development of energy resources may have an even greater effect in the future. In this section, we will look at some of the features of fission and fusion that are important for their application in reactors to generate power.

Fission

Very heavy nuclei ($Z > 92$) are subject to spontaneous fission. They break apart into two nuclei even if left to themselves with no outside disturbance. We can understand this by considering the analogy of a charged liquid drop. If the drop is not too large, surface tension can overcome the repulsive forces of the charges and hold the drop together. There is, however, a certain maximum size beyond which the drop will be unstable and will spontaneously break apart. Spontaneous fission puts an upper limit on the size of a nucleus and therefore on the number of elements that are possible.

Some heavy nuclei, uranium and plutonium in particular, can be induced to fission by the capture of a neutron. In the fission of ^{235}U, for example, the uranium nucleus is excited by the capture of a neutron, causing it to split into two nuclei and emit several neutrons. The Coulomb force of repulsion drives the fission fragments apart, with the energy eventually showing up as thermal energy. Consider, for example, the fission of a nucleus of mass number $A = 200$ into two nuclei of mass number $A = 100$. Since the rest energy for $A = 200$ is about 1 MeV per nucleon greater than that for $A = 100$, about 200 MeV per nucleus is released in such a fission. This is a large amount of energy. By contrast, in the chemical reaction of combustion, only about 4 eV of energy is released per molecule of oxygen consumed.

Example 7-6

Calculate the total energy in kilowatt-hours released in the fission of 1 g of ^{235}U, assuming that 200 MeV is released per fission.

Since 1 mol of ^{235}U has a mass of 235 g and contains $N_A = 6.02 \times 10^{23}$ nuclei, the number of ^{235}U nuclei in 1 g is

$$N = \frac{6.02 \times 10^{23} \text{ nuclei/mol}}{235 \text{ g/mol}} = 2.56 \times 10^{21} \text{ nuclei/g}$$

The energy released per gram is then

$$\frac{200 \text{ MeV}}{\text{nucleus}} \times \frac{2.56 \times 10^{21} \text{ nuclei}}{1 \text{ g}} \times \frac{1.6 \times 10^{-19} \text{ J}}{1 \text{ eV}} \times \frac{1 \text{ h}}{3600 \text{ s}} \times \frac{1 \text{ kW}}{1000 \text{ J/s}}$$
$$= 2.28 \times 10^4 \text{ kW·h/g}$$

The fission of uranium was discovered in 1939 by Hahn and Strassmann, who found, by careful chemical analysis, that medium-mass elements (such as barium and lanthanum) were produced in the bombardment of uranium with neutrons. The discovery that several neutrons are emitted in the fission process led to speculation concerning the possibility of using these neutrons to cause further fissions, thereby producing a chain reaction. When ^{235}U captures a neutron, the resulting ^{236}U nucleus emits gamma rays as it deexcites to the ground state about 15 percent of the time and undergoes fission about 85 percent of the time. The fission process is somewhat analogous to the oscillation of a liquid drop, as shown in Figure 7-11. If the oscillations are violent enough, the drop splits in two. Using the liquid-drop model, Bohr and Wheeler calculated the critical energy E_c needed by the ^{236}U nucleus to undergo fission. (^{236}U is the nucleus formed momentarily by the capture of a neutron by ^{235}U.) For this nucleus, the critical energy is 5.3 MeV, which is less than the 6.4 MeV of excitation energy produced when ^{235}U captures a neutron. The capture of a neutron by ^{235}U therefore produces an excited state of the ^{236}U nucleus that has more than enough energy to break apart. On the other hand, the critical energy for fission of the ^{239}U nucleus is 5.9 MeV. The capture of a neutron by a ^{238}U nucleus produces an excitation energy of only 5.2 MeV. Therefore, when a neutron is captured by ^{238}U to form ^{239}U, the excitation energy is not great enough for fission to occur. In this case, the excited ^{239}U nucleus deexcites by γ or α emission.

A fissioning nucleus can break into two medium-mass fragments in many different ways, as shown in Figure 7-12. Depending on the particular

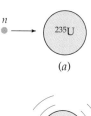

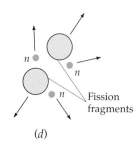

Figure 7-11 Schematic illustration of nuclear fission. (a) The absorption of a neutron by ^{235}U leads to (b) ^{236}U in an excited state. In (c), the oscillation of ^{236}U has become unstable. (d) The nucleus splits apart into two nuclei of medium mass and emits several neutrons that can produce fission in other nuclei.

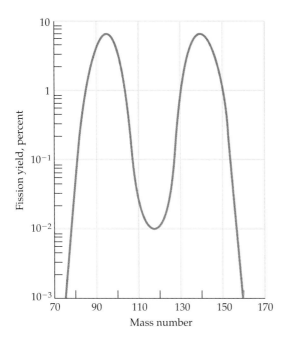

Figure 7-12 Distribution of the possible fission fragments of ^{235}U. The splitting of ^{235}U into two fragments of unequal mass is more likely than its splitting into fragments of equal mass.

reaction, 1, 2, or 3 neutrons may be emitted. The average number of neutrons emitted in the fission of ^{235}U is about 2.5. A typical fission reaction is

$$n + {}^{235}\text{U} \rightarrow {}^{141}\text{Ba} + {}^{92}\text{Kr} + 3n$$

Nuclear Fission Reactors

To sustain a chain reaction in a fission reactor, one of the neutrons (on the average) emitted in the fission of ^{235}U must be captured by another ^{235}U nucleus and cause it to fission. The **reproduction constant** k of a reactor is defined as the average number of neutrons from each fission that cause a subsequent fission. The maximum possible value of k is 2.5, but it is normally less than this for two important reasons: (1) Some of the neutrons may escape from the region containing fissionable nuclei, and (2) some of the neutrons may be captured by nonfissioning nuclei in the reactor. If k is exactly 1, the reaction will be self-sustaining. If it is less than 1, the reaction will die out. If k is significantly greater than 1, the reaction rate will increase rapidly and "run away." In the design of nuclear bombs, such a runaway reaction is desired. In power reactors, the value of k must be kept very nearly equal to 1.

Since the neutrons emitted in fission have energies of the order of 1 MeV, whereas the cross section for neutron capture leading to fission in ^{235}U is largest at small energies, the chain reaction can be sustained only if the neutrons are slowed down before they escape from the reactor. At high energies (1 to 2 MeV), neutrons lose energy rapidly by inelastic scattering from ^{238}U, the principal constituent of natural uranium. (Natural uranium contains 99.3 percent ^{238}U and only 0.7 percent fissionable ^{235}U.) Once the neutron energy is below the excitation energies of the nuclei in the reactor (about 1 MeV), the main process of energy loss is by elastic scattering, in which a fast neutron collides with a nucleus at rest and transfers some of its kinetic energy to that nucleus. Such energy transfers are efficient only if the masses of the two bodies are comparable. A neutron will not transfer much energy in an elastic collision with a heavy uranium nucleus. Such a collision is like one between a marble and a billiard ball. The marble will be deflected by the much more massive billiard ball, but very little of its kinetic energy will be transferred to the billiard ball. A **moderator** consisting of material such as water or carbon that contains light nuclei is therefore placed around the fissionable material in the core of the reactor to slow down the neutrons. The neutrons are slowed down by elastic collisions with the nuclei of the moderator until they are in thermal equilibrium with the moderator. Because of the relatively large neutron-capture cross section of the hydrogen nucleus in water, reactors using ordinary water as a moderator cannot easily achieve $k \approx 1$ unless they use enriched uranium, in which the ^{235}U content has been increased from 0.7 to between 1 and 4 percent. Natural uranium can be used if heavy water (D_2O) is used instead of ordinary (light) water (H_2O) as the moderator. Although heavy water is expensive, most Canadian reactors use it for a moderator to avoid the cost of constructing uranium-enrichment facilities.

Figure 7-13 shows some of the features of a pressurized-water reactor commonly used in the United States to generate electricity. Fission in the core heats the water in the primary loop, which is closed, to a high temperature. This water, which also serves as the moderator, is under high pressure to prevent it from boiling. The hot water is pumped to a heat exchanger, where it heats the water in the secondary loop and converts it to steam, which is then used to drive the turbines that produce electrical power. Note that the water in the secondary loop is isolated from that in the primary loop to prevent its contamination by the radioactive nuclei in the reactor core.

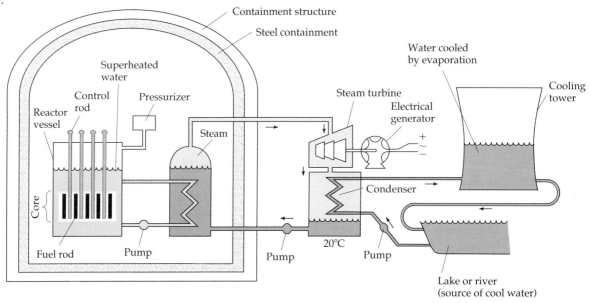

Figure 7-13 Simplified drawing of a pressurized-water reactor. The water in contact with the reactor core serves as both the moderator and the heat-transfer material. It is isolated from the water used to produce the steam that drives the turbines. Many features, such as the back-up cooling mechanisms, are not shown here.

The ability to control the reproduction factor k precisely is important if a power reactor is to be operated safely. Both natural negative-feedback mechanisms and mechanical methods of control are used. If k is greater than 1 and the reaction rate increases, the temperature of the reactor increases. If water is used as a moderator, its density decreases with increasing temperature and it becomes a less effective moderator. A second important control method is the use of control rods made of a material, such as cadmium, that has a very large neutron-capture cross section. When the reactor is started up, the control rods are inserted so that k is less than 1. As the rods are gradually withdrawn from the reactor, fewer neutrons are captured by them and k increases to 1. If k becomes greater than 1, the rods are inserted again.

Mechanical control of the reaction rate of a nuclear reactor using control rods is possible only because some of the neutrons emitted in the fission process are **delayed neutrons**. The time needed to slow down a neutron from 1 or 2 MeV to the thermal-energy level is only of the order of a millisecond. If all the neutrons emitted in fission were prompt neutrons, that is, emitted immediately in the fission process, mechanical control would not be possible because the reactor would run away before the rods could be inserted. However, about 0.65 percent of the neutrons emitted are delayed by an average time of about 14 s. These neutrons are emitted not in the fission process itself but in the decay of the fission fragments. The effect of the delayed neutrons can be seen in the following examples.

Example 7-7

If the average time between fission generations (the time it takes for a neutron emitted in one fission to cause another) is 1 ms = 0.001 s and the reproduction constant is 1.001, how long will it take for the reaction rate to double?

If $k = 1.001$, the reaction rate after N generations is 1.001^N. Setting this rate equal to 2 and solving for N, we obtain

$$(1.001)^N = 2$$

$$N \ln 1.001 = \ln 2$$

$$N = \frac{\ln 2}{\ln 1.001} = 693 \approx 700$$

It thus takes about 700 generations for the reaction rate to double. The time for 700 generations is $700(0.001 \text{ s}) = 0.70$ s. This is not enough time for mechanical control by the insertion of control rods.

Example 7-8

Assuming that 0.65 percent of the neutrons emitted are delayed by 14 s, find the average generation time and the doubling time if $k = 1.001$.

Since 99.35 percent of the generation times are 0.001 s and 0.65 percent are 14 s, the average generation time is

$$t_{av} = 0.9935(0.001 \text{ s}) + 0.0065(14 \text{ s}) = 0.092 \text{ s}$$

Note that these few delayed neutrons increase the generation time by nearly a hundredfold. The time for 700 generations is

$$700(0.092 \text{ s}) = 64.4 \text{ s}$$

This is plenty of time for the mechanical insertion of control rods.

Because of the limited supply of natural uranium, the small fraction of ^{235}U in natural uranium, and the limited capacity of enrichment facilities, reactors based on the fission of ^{235}U cannot meet our energy needs for very long. A promising alternative is the **breeder reactor**. When the relatively plentiful but nonfissionable ^{238}U nucleus captures a neutron, it decays by β decay (with a half-life of 20 min) to ^{239}Np, which in turn decays by β decay (with a half-life of 2 days) to the fissionable nuclide ^{239}Pu. Since ^{239}Pu fissions with fast neutrons, no moderator is needed. A reactor initially fueled with a mixture of ^{238}U and ^{239}Pu will breed as much fuel as it uses or more if one or more of the neutrons emitted in the fission of ^{239}Pu is captured by ^{238}U. Practical studies indicate that a typical breeder reactor can be expected to double its fuel supply in 7 to 10 years.

The inside of a nuclear power plant in Kent, England. A technician is standing on the reactor charge transfer plate, into which uranium fuel rods fit.

There are two major safety problems inherent with breeder reactors. The fraction of delayed neutrons is only 0.3 percent for the fission of ^{239}Pu, so the time between generations is much less than that for ordinary reactors. Mechanical control is therefore much more difficult. Also, since the operating temperature of a breeder reactor is relatively high and a moderator is not desired, a heat-transfer material such as liquid sodium metal is used rather than water (which is the moderator as well as the heat-transfer material in an ordinary reactor). If the temperature of the reactor increases, the resulting decrease in the density of the heat-transfer material leads to positive feedback since it will absorb fewer neutrons than before. Because of these safety considerations, breeder reactors are not yet in use in the United States. There are, however, several in operation in France, Great Britain, and the Soviet Union.

Safety of Fission Reactors

There has been heated debate about the safety of fission reactors, particularly since the disastrous accident at Chernobyl in the U.S.S.R. in 1986 and the much less serious accident at the Three Mile Island plant in the United States in 1979. A common fear is that a reactor might blow up as a uranium bomb. This is virtually impossible. Even in light-water reactors, the enriched uranium used contains only 1 to 4 percent ^{235}U, whereas a uranium bomb typically requires enriched uranium with 90 percent ^{235}U. Another fear is **meltdown,** the melting of the fuel core because of the heat produced by the radioactive decay of the fission fragments that occurs even after the reactor is shut down. If the cooling system fails, it is possible that the core would melt and, in a worst-case scenario, plunge through the containment building into the ground. Meltdown did not occur at either Chernobyl or Three Mile Island.

A much more serious fear is that radioactive material may be released into the atmosphere, as did occur at Chernobyl. The reactor at Chernobyl was a graphite-moderated reactor designed to produce plutonium for weapons as well as electrical power. It was run at a very high power sufficient to ignite the graphite if the cooling system failed. There are no comparable dual-purpose reactors outside the Soviet Union that are operated in this way. A similar accident with water-cooled reactors is probably not possible. Furthermore, a common safety feature not used at Chernobyl is a containment building with walls of concrete and steel at least 1 m thick.

With any type of fission reactor, there is the problem of the storage of the long-lived radioactive waste products produced. Despite the fact that elaborate storage methods are used, their long-term efficacy is always open to question. In assessing the danger of nuclear reactors, however, we should compare them with the dangers of ordinary power plants, such as coal-burning plants. The deaths in the United States due to lung disease caused by coal-burning plants number in the tens of thousands in a single year.

Fusion

In fusion, two light nuclei such as deuterium (^{2}H) and tritium (^{3}H) fuse together to form a heavier nucleus. A typical fusion reaction is

$$^2H + {}^3H \rightarrow {}^4He + n + 17.6 \text{ MeV} \qquad 7\text{-}26$$

The energy released in fusion depends on the particular reaction. For the ^{2}H + ^{3}H reaction, it is 17.6 MeV. Although this is less than the energy released in a fission reaction, it is a greater amount of energy per unit mass. The energy released in this fusion reaction is (17.6 MeV)/(5 nucleons) = 3.52 MeV per nucleon. This is about 3.5 times as great as the 1 MeV per nucleon released in fission.

The production of power from the fusion of light nuclei holds great promise because of the relative abundance of the fuel and the absence of some of the dangers inherent in fission reactors. Unfortunately, the technology necessary to make fusion a practical source of energy has not yet been developed. We will consider the ^{2}H + ^{3}H reaction; other reactions present similar problems.

Because of the Coulomb repulsion between the ^{2}H and ^{3}H nuclei, very large kinetic energies, of the order of 1 MeV, are needed to get the nuclei close enough together for the attractive nuclear forces to become effective and cause fusion. Such energies can be obtained in an accelerator, but since the scattering of one nucleus by the other is much more probable than fusion, the bombardment of one nucleus by another in an accelerator requires the input of more energy than is recovered. To obtain energy from fusion, the particles must be heated to a temperature great enough for the fusion reaction to occur as the result of random thermal collisions. Because a significant number of particles have kinetic energies greater than the mean kinetic energy $\frac{3}{2}kT$ and because some particles can tunnel through the Coulomb barrier, a temperature T corresponding to $kT \approx 10$ keV is adequate to ensure that a reasonable number of fusion reactions will occur if the density of particles is sufficiently high. The temperature corresponding to $kT = 10$ keV is of the order of 10^8 K. Such temperatures occur in the interiors of stars, where such reactions are common. At these temperatures, a gas consists of positive ions and negative electrons and is called a **plasma**. One of the problems arising in attempts to produce controlled fusion reactions is that of confining the plasma long enough for the reactions to take place. In the interior of the sun the plasma is confined by the enormous gravitational field of the sun. In a laboratory on earth, confinement is a difficult problem.

The energy required to heat a plasma is proportional to the density of its ions n, whereas the collision rate is proportional to n^2, the square of the density. If τ is the confinement time, the output energy is proportional to $n^2\tau$. If the output energy is to exceed the input energy, we must have

$$C_1 n^2 \tau > C_2 n$$

where C_1 and C_2 are constants. In 1957, the British physicist J. D. Lawson evaluated these constants from estimates of the efficiencies of various hypothetical fusion reactors and derived the following relation between density and confinement time, known as **Lawson's criterion:**

Lawson's criterion $n\tau > 10^{20}$ s·particles/m³ 7-27

If Lawson's criterion is met and the thermal energy of the ions is great enough ($kT \sim 10$ keV), the energy released by a fusion reactor will just equal the energy input; that is, the reactor will just break even. For the reactor to be practical, much more energy must be released.

Two schemes for achieving Lawson's criterion are currently under investigation. In one scheme, **magnetic confinement,** a magnetic field is used to confine the plasma. In the most common arrangement, first developed in the U.S.S.R. and called the Tokamak, the plasma is confined in a large toroid. The magnetic field is a combination of the doughnut-shaped magnetic field due to the windings of the toroid and the self-field due to the current of the circulating plasma. The break-even point has been achieved recently using magnetic confinement, but we are still a long way from building a practical fusion reactor.

(*a*) Schematic of the tokamak fusion test reactor. The toroidal coils, encircling the doughnut-shaped vacuum vessel, are designed to conduct current for 3-s pulses, separated by waiting times of 5 min. Pulses peak at 73,000 A, producing a magnetic field of 5.2 T. This field is the principal means of confining the deuterium–tritium plasma that circulates within the vacuum vessel. Current for the pulses is delivered by converting the rotational energy of two 600-ton flywheels. Sets of poloidal coils, perpendicular to the toroidal coils, carry an oscillating current that generates a current through the confined plasma itself, heating it ohmically. Additional poloidal fields help stabilize the confined plasma. Between four and six neutral-beam injection systems (only one of which is shown in the schematic) are used to inject high-energy deuterium atoms into the deuterium–tritium plasma, heating it beyond what could be obtained ohmically—ultimately to the point of fusion. (*b*) The TFTR itself. The diameter of the vacuum vessel is 7.7 m. (*c*) An 800-kA plasma, lasting 1.6 s, as it discharges within the vacuum vessel.

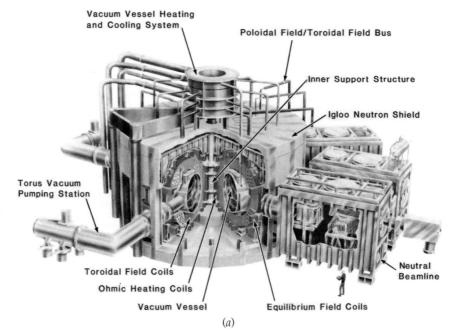

(*a*)

(*b*)

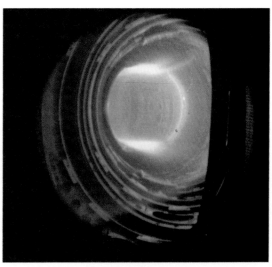

(*c*)

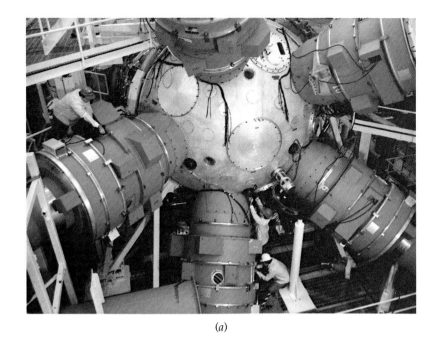
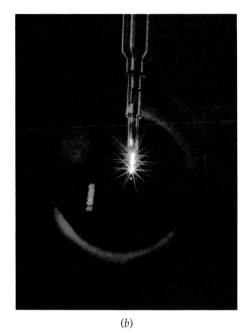

(a) (b)

In a second scheme, called **inertial confinement,** a pellet of frozen solid deuterium and tritium is bombarded from all sides by intense pulsed laser beams of energies of the order of 10^6 J lasting about 10^{-8} s. (Intense ion beams are also used.) Computer simulation studies indicate the pellet should be compressed to about 10^4 times its normal density and heated to a temperature greater than 10^8 K. This should produce about 10^6 J of fusion energy in 10^{-10} s, which is so brief that confinement is achieved by inertia alone.

Because the break-even point is just barely being achieved in magnetic-confinement fusion, and because the building of a fusion reactor involves many practical problems that have not yet been solved, the availability of fusion to meet our energy needs is not expected for at least several decades. However, fusion holds great promise as an energy source for the future.

(a) The Nova target chamber, an aluminum sphere approximately 5 m in diameter, inside which ten beams from the world's most powerful laser converge onto a hydrogen-containing pellet 0.5 mm in diameter. The resulting fusion reaction is visible as a tiny star (b), lasting 10^{-10} s, releasing 10^{13} neutrons.

Questions

12. Why isn't there an element with $Z = 130$?
13. Why is a moderator needed in an ordinary nuclear-fission reactor?
14. Explain why water is more effective than lead in slowing down fast neutrons.
15. What happens to the neutrons produced in fission that do not produce another fission?
16. What is the advantage of a breeder reactor over an ordinary one? What are the disadvantages?
17. Why does fusion occur spontaneously in the sun but not on earth?

7-6 The Interaction of Particles with Matter

In this section, we will discuss briefly the main interactions of charged particles, neutrons, and photons with matter. Understanding these interactions is important for study of nuclear detectors, shielding, and the effects of radiation on living organisms. In each case, we will examine the principal factors involved in stopping or attenuating a beam of particles.

Charged Particles

When a charged particle traverses matter, it loses energy mainly through collisions with electrons. This often leads to the ionization of the atoms in the matter, in which case the particle leaves a trail of ionized atoms in its path. If the energy of the particle is large compared with the ionization energies of the atoms, the energy loss in each encounter with an electron will be only a small fraction of the particle's energy. (A heavy particle cannot lose a large fraction of its energy to a free electron because of conservation of momentum. For example, when a billiard ball collides with a marble, only a very small fraction of the energy of the billiard ball can be lost.) Since the number of electrons in matter is so large, we can treat the loss of energy as continuous. After a fairly well-defined distance, called the **range**, the particle will have lost all its kinetic energy and will come to a stop. Near the end of the range, the view of energy loss as continuous is not valid because individual encounters are important. For electrons, this can lead to a significant statistical variation in path length, but for protons and other heavy particles with energies of several MeV or more, the path lengths vary by only a few percent or less.

Figure 7-14 shows the rate of energy loss per unit path length $-dK/dx$ versus the energy of the ionizing particle. We can see from this figure that the rate of energy loss $-dK/dx$ is maximum at low energies and that at high energies it is approximately independent of the energy. Particles with kinetic energies greater than their rest energies mc^2 are called **minimum ionizing particles**. Their energy loss per unit path length is approximately constant, and their range is roughly proportional to their energy. Figure 7-15 shows the range-versus-energy curve for protons in air.

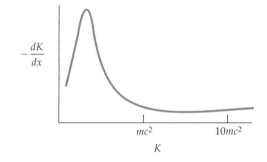

Figure 7-14 Energy loss per unit path length $-dK/dx$ versus kinetic energy for a charged particle. For particles with kinetic energies greater than their rest energies, the energy loss per unit path length is approximately constant. The distance traveled by such a particle is therefore approximately proportional to its energy.

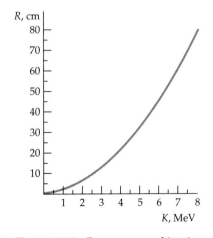

Figure 7-15 Range versus kinetic energy for protons in dry air. Except at low energies, the relationship between range and energy is approximately linear.

Since a charged particle loses energy through collisions with the electrons in a material, the greater the number of electrons, the greater the rate of energy loss. The energy loss rate $-dK/dx$ is approximately proportional to the density of the material. For example, the range of a 6-MeV proton is about 40 cm in air; but in water, which is about 800 times more dense than air, its range is only 0.5 mm.

If the energy of the charged particle is large compared with its rest energy, the particle also radiates some energy away in the form of bremsstrahlung.

The fact that the rate of energy loss for heavy charged particles is large at very low energies (as shown by the low-energy peak in Figure 7-14) has important applications in nuclear radiation therapy. Figure 7-16 shows a plot of energy loss versus penetration distance for charged particles in water. Most of the energy is lost near the end of the range. The peak in this curve is called the **Bragg peak**. A beam of heavy charged particles can be used to destroy cancer cells at a given depth in the body without destroying other, healthy cells if the energy is carefully chosen so that most of the energy loss occurs at the proper depth.

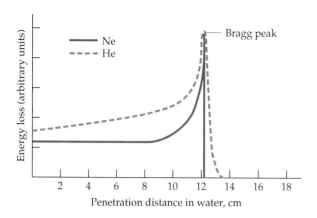

Figure 7-16 Energy loss of helium ions and neon ions in water versus penetration distance. Most of the energy loss occurs near the end of the path, as shown by the Bragg peak. The heavier the ion, the narrower the peak.

Neutrons

Since neutrons are uncharged, they do not interact with electrons in matter. Neutrons are removed from a beam by nuclear scattering or by capture. For energies that are large compared with thermal energies (kT), the most important processes are elastic and inelastic scattering. If we have a collimated neutron beam, any scattering or absorption will remove neutrons from the beam. This is very different from the case of a charged particle, which undergoes many collisions that decrease the energy of the particle but do not remove it from the beam until its energy is essentially zero. A neutron is removed from the beam at its first collision.

The chance of a neutron being removed from a beam within a given path distance is proportional to the number of neutrons in the beam and to the path distance. Let σ be the total cross section for the scattering plus the absorption of a neutron. If I is the incident intensity of the neutron beam (the number of particles per unit time per unit area), the number of neutrons removed from the beam per unit time will be $R = \sigma I$ per nucleus (Equation 7-24). If n is the number density of the nuclei (the nuclei per unit volume) and A is the area of the incident beam, the number of nuclei encountered in a distance dx is $nA\,dx$. The number of neutrons removed from the beam in a distance dx is thus

$$-dN = \sigma I(nA\,dx) = \sigma n N\,dx \qquad 7\text{-}28$$

where $N = IA$ is the total number of neutrons per unit time in the beam. Solving Equation 7-28 for N, we obtain

$$N = N_0 e^{-\sigma n x} \qquad 7\text{-}29$$

If we divide each side of Equation 7-29 by the area of the beam, we obtain a similar equation for the intensity of the beam:

$$I = I_0 e^{-\sigma n x} \qquad 7\text{-}30$$

We thus have an exponential decrease in the neutron intensity with penetration. After a certain characteristic distance, half the neutrons in a beam are removed. Then after an equal distance, half of the remaining neutrons are removed, and so on. Thus, there is no well-defined range.

At the half-penetration distance $x_{1/2}$, the number of neutrons will be $\tfrac{1}{2}N_0$. From Equation 7-29,

$$\tfrac{1}{2}N_0 = N_0 e^{-\sigma n x_{1/2}}$$

$$e^{\sigma n x_{1/2}} = 2$$

$$x_{1/2} = \frac{\ln 2}{\sigma n} \qquad 7\text{-}31$$

Example 7-9

The total cross section for the scattering and absorption of neutrons of a certain energy is 0.3 barn for copper. (*a*) Find the fraction of neutrons of that energy that penetrates 10 cm in copper. (*b*) To what distance will half the neutrons penetrate?

(*a*) Using $n = 8.45 \times 10^{28}$ nuclei/m^3 for copper, we have

$$\sigma n x = (0.3 \times 10^{-28} \text{ m}^2)(8.45 \times 10^{28}/\text{m}^3)(0.10 \text{ m}) = 0.254$$

According to Equation 7-29, if we have N_0 neutrons at $x = 0$, the number at $x = 0.10$ m is

$$N = N_0 e^{-\sigma n x} = N_0 e^{-0.254} = 0.776 N_0$$

The fraction that penetrates 10 cm is thus 0.776 or 77.6 percent.

(*b*) For $n = 8.45 \times 10^{28}$ nuclei/m^3 and $\sigma = 0.3 \times 10^{-28}$ m^2, we have from Equation 7-31

$$x_{1/2} = \frac{\ln 2}{(0.3 \times 10^{-28} \text{ m}^2)(8.45 \times 10^{28}/\text{m}^3)} = \frac{0.693}{2.54} \text{ m}$$

$$= 0.273 \text{ m} = 27.3 \text{ cm}$$

Photons

The intensity of a photon beam, like that of a neutron beam, decreases exponentially with distance in an absorbing material. The intensity versus penetration is given by Equation 7-30, where σ is the total cross section for absorption and scattering. The important processes that remove photons from a beam are the photoelectric effect, Compton scattering, and pair production. The total cross section for absorption and scattering is the sum of the partial cross sections for these three processes: σ_{pe}, σ_{cs}, and σ_{pp}. These partial cross sections and the total cross section are shown as functions of energy in Figure 7-17. The cross section for the photoelectric effect dominates at very low energies (less than 1 MeV, which are not shown in the figure), but it decreases rapidly with increasing energy. If the photon energy is large compared with the binding energy of the electrons (a few keV), the electrons can be considered to be free and Compton scattering is the principal mechanism for the removal of photons from the beam. If the photon energy is greater than $2m_e c^2 = 1.02$ MeV, the photon can disappear, with the creation of an electron–positron pair, a process called **pair production.** The cross section for pair production increases rapidly with the photon energy and is the dominant component of the total cross section at high ener-

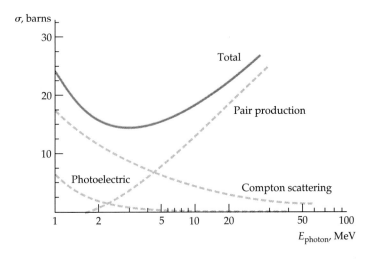

Figure 7-17 Photon interaction cross sections versus energy for lead. The total cross section is the sum of the cross sections for the photoelectric effect, Compton scattering, and pair production.

gies. Pair production cannot occur in free space. Momentum conservation requires that a nucleus be nearby to absorb momentum by recoil. The cross section for pair production is proportional to Z^2 of the absorbing material.

Dosage

The biological effects of radiation are principally due to the ionization it produces. Even a small amount of ionization can seriously disrupt the functioning of sensitive living cells or even kill them. Three different units are used to measure these effects: the roentgen, the rad, and the rem.

The **roentgen** (R) is defined as the amount of radiation that produces $\frac{1}{3} \times 10^{-9}$ C of electric charge (either positive ions or electrons) in 1 cm^3 of dry air at standard conditions. It is a measure of exposure to radiation. The roentgen has been largely replaced by the **rad** (*r*adiation *a*bsorbed *d*ose), a measure of the energy absorbed, which is defined as the amount of radiation that deposits 10^{-2} J/kg of energy in any material. The SI unit, joules per kilogram, is called a **gray** (Gy). Thus

$$1 \text{ rad} = 10^{-2} \text{ Gy} \qquad 7\text{-}32$$

Since 1 R is equivalent to the deposit of about 8.7×10^{-3} J/kg of energy, the rad and the roentgen are roughly equal.

The amount of biological damage depends not only on the energy absorbed, which is equivalent to the number of ion pairs formed, but also on the spacing of the ion pairs. If the ion pairs are closely spaced, as in the ionization caused by α particles, the biological effect is increased. The unit **rem** (*r*oentgen *e*quivalent in *m*an) is the dose that has the same biological effect as 1 rad of β or γ radiation:

$$1 \text{ rem} = 1 \text{ rad} \times \text{RBE} \qquad 7\text{-}33$$

where RBE is the *r*elative *b*iological *e*ffectiveness factor. Table 7-2 gives approximate values of the RBE factors for different types of radiation. The SI unit for dose equivalent is the seivert (Sv), which is defined as the product of the gray and the RBE:

$$1 \text{ Sv} = 1 \text{ Gy} \times \text{RBE} = 100 \text{ rem} \qquad 7\text{-}34$$

Table 7-3 compares the various radiation units we have discussed.

Our knowledge of the effects of large radiation doses comes mainly from studies of victims of atomic bomb explosions. Doses under 25 rem over the entire body seem to have no immediate effects. Doses over 100 rem damage the blood-forming tissues, and those over 500 rem usually lead to death in a short time.

The long-term effects of sublethal doses acquired over a period of time are more difficult to measure. The chances of dying of cancer are doubled by

Table 7-2 **Approximate RBE Factors**

Type of radiation	RBE factor
Photons < 4 MeV	1
Photons > 4 MeV	0.7
β particles < 30 keV	1.7
β particles > 30 keV	1
Slow neutrons	4 or 5
Fast neutrons	10
Protons	10
α particles	10
Heavy ions	20

Table 7-3 **Radiation and Dose Units***

	Customary unit		SI unit		
Quantity	Name	Symbol	Name	Symbol	Conversion
Energy	electron volt	eV	joule	J	1 MeV = 1.602×10^{-13} J
Exposure	roentgen	R	coulomb per kilogram	C/kg	1 R = 2.58×10^{-4} C/kg
Absorbed dose	rad	rad or rd	gray	Gy = J/kg	1 rad = 10^{-2} J/kg = 10^{-2} Gy
Dose equivalent	rem	rem	seivert	Sv	1 rem = 10^{-2} Sv
Activity	curie	Ci	becquerel	Bq = 1/s	1 Ci = 3.7×10^{10} decays/s = 3.7×10^{10} Bq

*Adapted from S. C. Bushong, *The Physics Teacher*, vol. 15, no. 3, p. 135, 1977.

a dose of somewhere between 100 and 500 rem. Not much is known about the effects of very low-level doses. It is possible that there is some threshold dose below which the damage done is repaired so that there is no resulting increase in the chance of cancer. But it is also possible (and is generally believed) that there is no such threshold and that the cancer-causing effects of radiation are proportional to the cumulative dose even at low levels.

Some typical human radiation exposures are listed in Table 7-4. The internal dose listed in this table is from radioactive nuclei, such as potassium-40 and uranium and its decay products, inside our bodies. Most of the radioactive fallout due to nuclear weapons testing is strontium-90 and cesium-137, both of which have half-lives of about 30 y. If there is no further testing, this source of radiation will eventually become negligible. We are shielded from most cosmic rays by the atmosphere. The dose we receive is about 40 mrem/y at sea level; it increases by about 1 mrem/y for every 30 m of altitude.

One of the decay products of uranium is radon-222 (^{222}Rn), which decays by α emission with a half-life of 3.82 days. This decay is followed by other α and β decays that result in the relatively stable lead-210, which has a half-life of 22 y. Since radon is an inert gas, it diffuses through materials without interacting with them chemically. It has recently been recognized as a health hazard because it seeps into homes from the ground, where it accumulates and can be breathed into our lungs.

The largest source of artificial exposure to radiation is currently medical diagnostic x rays. The dose received varies enormously, depending on the type of machine used, the sensitivity of the film, and so forth. For a chest x ray, some mobile units give doses of 1000 millirems, and the average dose is around 200 millirems. If the best procedures are used, however, the dose from a chest x ray can be limited to 6 millirems.

Because of our lack of knowledge of the risks of radiation, we should clearly limit our exposure to it as much as possible. Table 7-5 lists some of the dose limits recommended by the National Council on Radiation Protection.

Table 7-4 **Average Radiation Doses Received by the United States Population**

Radiation source	Average effective dose (mrem/y)
Cosmic rays	45
Internal radioactive nuclides	35
Building materials	40
Ground	11
Air	5
Diagnostic x rays	70
Global fallout	4
Television	1
Nuclear power	0.003

Table 7-5 **Recommended Dose Limits***

	Maximum permissible dose equivalent for occupational exposure
Combined whole-body occupational exposure	
Prospective annual limit	5 rems in any one year
Retrospective annual limit	10–15 rems in any one year
Long-term accumulation to age N years	$(N - 18) \times 5$ rems
Skin	15 rems in any one year
Hands	75 rems in any one year (25 per quarter)
Forearms	30 rems in any one year (10 per quarter)
Other organs, tissues, and organ systems	15 rems in any one year (5 per quarter)
Fertile women (with respect to fetus)	0.5 rem in gestation period
Dose limits for nonoccupationally exposed	
Population average	0.17 rem in any one year
An individual in the population	0.5 rem in any one year
Students	0.1 rem in any one year

*Adapted from *NCRP Rep.* 39, 1971, as given in S. C. Bushong, *The Physics Teacher*, vol. 15, no. 3, p. 135, 1977.

Summary

1. Nuclei have N neutrons, Z protons, and a mass number $A = N + Z$. Two or more nuclei having the same value of Z but different values of N and A are called isotopes.

2. For light nuclei, N and Z are approximately equal, whereas for heavy nuclei, N is greater than Z.

3. Most nuclei are approximately spherical in shape and have a volume that is proportional to A, implying that nuclear density is independent of A. The radius of a nucleus is given approximately by

$$R = R_0 A^{1/3}$$

   where R_0 is about 1.5 fm.

4. The mass of a stable nucleus is less than the sum of the masses of its nucleons. The mass difference times c^2 equals the binding energy of the nucleus. The binding energy is approximately proportional to the mass number A.

5. Nuclei have magnetic moments associated with their spin. The magnetic moment of the proton is $+2.7928\ \mu_N$ and that of the neutron is $-1.9135\ \mu_N$, where

$$\mu_N = \frac{e\hbar}{2m_p} = 5.05 \times 10^{-27}\ \text{J/T} = 3.15 \times 10^{-8}\ \text{eV/T}$$

   is the nuclear magneton.

6. A proton may have its magnetic moment aligned parallel or antiparallel to an external magnetic field. The difference in energy between these two states is given by

$$\Delta E = 2(\mu_z)_p B$$

   Transitions between these two states can be induced by RF photons. The resonance absorption of these RF photons is called nuclear magnetic resonance.

7. Unstable nuclei are radioactive and decay by emitting α particles (^{4}He nuclei), β particles (electrons or positrons), or γ rays (photons). All radioactivity is statistical in nature and follows an exponential decay law:

$$N = N_0 e^{-\lambda t}$$

   where λ is the decay constant. The rate of decay is given by

$$R = \lambda N = R_0 e^{-\lambda t}$$

   The time it takes for the number of nuclei or the decay rate to decrease by half is called the half-life:

$$t_{1/2} = \frac{0.693}{\lambda}$$

   Half-lives of alpha decay range from a fraction of a second to millions of years. For β decay they range up to hours or days and for γ decay half-lives are usually less than a microsecond. The number of decays per second of 1 g of radium is the curie, which equals 3.7×10^{10} decays/s $= 3.7 \times 10^{10}$ Bq.

8. The cross section σ is a measure of the effective size of a nucleus for a particular nuclear reaction. Cross sections are measured in barns, where 1 barn = 10^{-28} m^2. An important nuclear reaction is neutron capture, in which a neutron is captured by a nucleus with the emission of a photon. The neutron-capture cross section exhibits strong resonances superimposed on a $1/v$ dependence.

9. Fission occurs when some heavy elements, such as ^{235}U or ^{239}Pu, capture a neutron and split apart into two medium-mass nuclei. The two nuclei then fly apart because of electrostatic repulsion, releasing a large amount of energy. A chain reaction is possible because several neutrons are emitted by a nucleus when it undergoes fission. A chain reaction can be sustained if, on the average, one of the emitted neutrons is slowed down by scattering in the reactor and is then captured by another fissionable nucleus. Very heavy nuclei ($Z > 92$) are subject to spontaneous fission.

10. A large amount of energy is also released when two light nuclei, such as ^{2}H and ^{3}H, fuse together. Fusion takes place spontaneously inside the sun and other stars, where the temperature is great enough (about 10^8 K) for thermal motion to bring the charged hydrogen ions close enough together to fuse. Although controlled fusion holds great promise as a future energy source, practical difficulties have thus far prevented its development.

11. Charged particles lose energy nearly continuously when traversing matter because they interact with the electrons in the matter. They have fairly well-defined ranges in a particular material that are roughly proportional to the energy of the particle and vary inversely with the density of the material. Neutrons and photons do not have well-defined ranges. Instead, the intensity of a neutron or photon beam decreases exponentially with the penetration distance. Neutrons are removed from a beam by scattering or by capture. Photons are absorbed by the photoelectric effect at very low energies and by pair production at high energies. At intermediate energies they undergo Compton scattering from electrons in the material.

12. Radiation dosages are measured in rads, where 1 rad is the amount of radiation that deposits 10^{-2} J/kg of energy in any material. A rem is the dose that has the same biological effect as 1 rad of β or γ radiation. It equals the rad multiplied by the relative biological effectiveness factor RBE, which is approximately 1 for gamma and beta rays, about 4 or 5 for slow neutrons, about 10 for fast neutrons, and between 10 and 20 for alpha particles of energies of the order of 5 to 10 MeV.

Suggestions for Further Reading

Bulletin of the Atomic Scientists, published monthly except for July and August by the Educational Foundation for Nuclear Science, Chicago, Illinois.

Founded in 1945 by Western scientists concerned about the consequences of our new-found ability to induce nuclear fission, this magazine has numbered among its sponsors many eminent physicists including Albert Einstein. Its articles deal with current issues relating mainly to nuclear weapons and the policies of countries possessing them but also to nuclear reactors.

Černy, Joseph, and Arthur M. Poskanzer: "Exotic Light Nuclei," *Scientific American*, June 1978, p. 60.

Many isotopes of most of the light elements can be produced, but some decay so quickly that there is only just sufficient time to detect them before they are gone.

Engelman, Donald M., and Peter B. Moore: "Neutron-Scattering Studies of the Ribosome," *Scientific American*, October 1976, p. 44.

This article describes how a beam of neutrons can be used to obtain information about the relative positions of large molecules in biological structures such as the ribosome.

Furth, Harold P.: "Progress Toward a Tokamak Fusion Reactor," *Scientific American*, August 1979, p. 50.

Nuckolls, John H.: "The Feasibility of Inertial-Confinement Fusion," *Physics Today*, vol. 35, no. 9, 1982, p. 24.

These articles describe two possible methods of achieving the fusion of light nuclei under controlled conditions suitable for power generation.

Levi, Barbara G.: "Radionuclide Releases from Severe Accidents at Nuclear Power Plants," *Physics Today*, vol. 38, no. 5, 1985, p. 67.

Lewis, Harold W.: "The Safety of Fission Reactors," *Scientific American*, March 1980, p. 53.

Both of these articles deal with risk-assessment studies of American fission reactor technology undertaken after the Three Mile Island reactor accident.

Perrow, Charles: *Normal Accidents: Living with High-Risk Technologies*, Basic Books, New York, 1984.

This book is also the result of a study undertaken after the Three Mile Island reactor accident. It provides an alternative view to that of risk assessment in making decisions about such complex, high-risk systems as fission reactors.

Schramm, David N.: "The Age of the Elements," *Scientific American*, January 1974, p. 69.

The age of the universe and structures in it can be determined from abundance ratios of various radioactive elements and their daughter nuclei, based on certain assumptions about the processes by which they were formed.

Schroeer, Dietrich: *Science, Technology, and the Nuclear Arms Race*, John Wiley & Sons, New York, 1984.

This book describes in a clear and understandable way the principles of operation of modern fission and fusion bombs. The technology of bomb delivery systems, nuclear deterrence, alternatives to deterrence, and arms control and disarmament are discussed with reference to the specific technologies involved.

Review

A. Objectives: After studying this chapter, you should:

1. Be able to give the order of magnitude of the radius of an atom and of a nucleus.

2. Be able to sketch the *N*-versus-*Z* curve for stable nuclei.

3. Be able to sketch the binding energy per nucleon versus *A* and discuss the significance of this curve for fission and fusion.

4. Know the exponential law of radioactive decay and be able to work problems using it.

5. Be able to describe the nuclear-fission chain reaction and discuss the advantages and disadvantages of fission reactors.

6. Be able to state Lawson's criterion for nuclear-fusion reactors.

7. Be able to discuss the chief mechanisms by which particles lose energy in matter and explain why some particles have well-defined ranges and others do not.

8. Be able to discuss the three units of radiation dosage: the roentgen, the rad, and the rem.

B. Define, explain, or otherwise identify:

Atomic number
Mass number
Nucleon
Nuclide
Isotopes
Strong nuclear force
Hadronic force
Binding energy
Nuclear spin
Nuclear magneton
RF radiation
Nuclear magnetic resonance
Alpha decay
Beta decay
Gamma decay
Decay constant
Mean lifetime
Half-life
Becquerel
Curie
Daughter nucleus
Neutrino
Radioactive carbon dating
Metastable state
Exothermic reaction
Q value of reaction
Endothermic reaction
Cross section
Barn
Thermal neutron
Resonance
Fission
Fusion
Reproduction constant
Moderator
Delayed neutron
Breeder reactor
Meltdown
Plasma
Lawson's criterion
Magnetic confinement
Inertial confinement
Range
Minimum ionizing particles
Bragg peak
Pair production
Roentgen
Rad
Gray
Rem

C. True or false: If the statement is true, explain why it is true. If it is false, give a counterexample.

1. The atomic nucleus contains protons, neutrons, and electrons.

2. The mass of ^{2}H is less than the mass of a proton plus a neutron.

3. After two half-lives, all the radioactive nuclei in a given sample have decayed.

4. Exothermic reactions have no threshold energy.

5. In a breeder reactor, fuel can be produced as fast as it is consumed.

6. The rad is a measure of the energy deposited per unit volume in matter.

Problems

Level I

7-1 Properties of Nuclei

1. Give the symbols for two other isotopes of (a) ^{14}N, (b) ^{56}Fe, and (c) ^{118}Sn.

2. Calculate the binding energy and the binding energy per nucleon from the masses given in Table 7-1 for (a) ^{12}C, (b) ^{56}Fe, and (c) ^{238}U.

3. Repeat Problem 2 for (a) ^{6}Li, (b) ^{39}K, and (c) ^{208}Pb.

4. Use Equation 7-1 to compute the radii of the following nuclei: (a) ^{16}O, (b) ^{56}Fe, and (c) ^{197}Au.

5. Find the energy needed to remove a neutron from (a) ^{4}He and (b) ^{7}Li (mass, 7.016004 u).

6. (a) Given that the mass of a nucleus of mass number A is approximately $m = CA$, where C is a constant, find an expression for the nuclear density in terms of C and the constant R_0 in Equation 7-1. (b) Compute the value of this nuclear density in grams per cubic centimeter using the fact that C has the approximate value of 1 g per Avogadro's number of nucleons.

7. Derive Equation 7-2; that is, show that the rest energy of one unified mass unit is 931.5 MeV.

8. Use Equation 7-1 for the radius of a spherical nucleus and the approximation that the mass of a nucleus of mass number A is A u to calculate the density of nuclear matter in grams per cubic centimeter.

7-2 Nuclear Magnetic Resonance

There are no problems for this section.

7-3 Radioactivity

9. The counting rate from a radioactive source is 4000 counts/s at time $t = 0$. After 10 s, the counting rate is 1000 counts/s. (a) What is the half-life? (b) What is the counting rate after 20 s?

10. A certain source gives 2000 counts/s at time $t = 0$. Its half-life is 2 min. (a) What is the counting rate after 4 min? (b) After 6 min? (c) After 8 min?

11. The counting rate from a radioactive source is 6400 counts/s. The half-life of the source is 10 s. Make a plot of the counting rate as a function of time for times up to 1 min. What is the decay constant for this source?

12. The counting rate from a radioactive source is 8000 counts/s at time $t = 0$, and 10 min later the rate is 1000 counts/s. (a) What is the half-life? (b) What is the decay constant? (c) What is the counting rate after 20 min?

13. The half-life of radium is 1620 y. Calculate the number of disintegrations per second of 1 g of radium, and show that the disintegration rate is approximately 1 Ci.

14. A radioactive silver foil ($t_{1/2}$ = 2.4 min) is placed near a Geiger counter and 1000 counts/s are observed at time $t = 0$. (a) What is the counting rate at $t = 2.4$ min and at $t = 4.8$ min? (b) If the counting efficiency is 20 percent, how many radioactive nuclei are there at time $t = 0$? At time $t = 2.4$ min? (c) At what time will the counting rate be about 30 counts/s?

15. The stable isotope of sodium is ^{23}Na. What kind of radioactivity would you expect of (a) ^{22}Na and (b) ^{24}Na?

16. Use Table 7-1 to calculate the energy in MeV for the α decay of (a) ^{226}Ra and (b) ^{242}Pu.

17. A sample of wood contains 10 g of carbon and shows a ^{14}C decay rate of 100 counts/min. What is the age of the sample?

18. A bone claimed to be 10,000 years old contains 15 g of carbon. What should the decay rate of ^{14}C be for this bone?

19. A sample of animal bone unearthed at an archaeological site is found to contain 175 g of carbon, and the decay rate of ^{14}C in the sample is measured to be 8.1 Bq. How old is the bone?

20. A sample of a radioactive isotope is found to have an activity of 115.0 Bq immediately after it is pulled from the reactor that formed it. Its activity 2 h 15 min later is measured to be 85.2 Bq. (a) Calculate the decay constant and the half-life of the sample. (b) How many radioactive nuclei were there in the sample initially?

7-4 Nuclear Reactions

21. Using Table 7-1, find the Q values for the following reactions: (a) ^{1}H + ^{3}H → ^{3}He + n + Q and (b) ^{2}H + ^{2}H → ^{3}He + n + Q.

22. Using Table 7-1, find the Q values for the following reactions: (a) ^{2}H + ^{2}H → ^{3}H + ^{1}H + Q, (b) ^{2}H + ^{3}He → ^{4}He + ^{1}H + Q, and (c) ^{6}Li + n → ^{3}H + ^{4}He + Q.

7-5 Fission, Fusion, and Nuclear Reactors

23. Assuming an average energy of 200 MeV per fission, calculate the number of fissions per second needed for a 500-MW reactor.

24. If the reproduction factor in a reactor is $k = 1.1$, find the number of generations needed for the power level to (a) double, (b) increase by a factor of 10, and (c) increase by a factor of 100. Find the time needed in each case if (d) there are no delayed neutrons, so the time between generations is 1 ms, and (e) there are delayed neutrons that make the average time between generations 100 ms.

25. Compute the temperature T for which $kT = 10$ keV, where k is Boltzmann's constant.

7-6 The Interaction of Particles with Matter

26. The range of 4-MeV α particles in air (ρ = 1.29 mg/cm^3) is 2.5 cm. Assuming the range to be inversely proportional to the density of matter, find the range of 4-MeV α particles in (a) water and (b) lead (ρ = 11.2 g/cm^3).

27. The range of 6-MeV protons in air is approximately 45 cm. Find the approximate range of 6-MeV protons in (a) water and (b) lead (see Problem 26).

28. A neutron beam has its intensity reduced by a factor of 2 at a penetration distance of 3 cm in iron. How great a thickness of iron is needed to reduce the intensity by a factor of (*a*) 8 and (*b*) 128?

29. The number density of iron nuclei is $n = 8.50 \times 10^{28}$ nuclei/m^3. Find the total scattering and absorption cross section for the neutron beam of Problem 28.

30. A 1.0-cm-thick piece of lead shielding reduces the intensity of a beam of 15-MeV γ rays by a factor of 2. (*a*) By how much will 5 cm of lead reduce the intensity of this beam? (*b*) Approximately what thickness of lead is needed to reduce the intensity by a factor of 1000?

31. Use the data of Problem 30 and the number density of lead $n = 3.30 \times 10^{28}$ nuclei/m^3 to find the total cross section for the removal of 15-MeV photons in lead.

Level II

32. (*a*) Show that $ke^2 = 1.44$ MeV $\cdot$ fm, where k is the Coulomb constant and e is the electron charge. (*b*) Show that $hc = 1240$ MeV $\cdot$ fm.

33. Twelve nucleons are in a one-dimensional infinite square well of length $L = 3$ fm. (*a*) Using the approximation that the mass of a nucleon is 1 u, find the lowest energy of a nucleon in the well. Express your answer in MeV. What is the ground-state energy of the system of 12 nucleons in the well if (*b*) all the nucleons are neutrons so that there can be only 2 in each state and (*c*) 6 of the nucleons are neutrons and 6 are protons so that there can be 4 nucleons in each state? (Neglect the energy of Coulomb repulsion of the protons.)

34. Derive the result that the activity of 1 g of natural carbon due to the β decay of ^{14}C is 15 decays/min = 0.25 Bq.

35. A 0.05394-kg sample of ^{144}Nd (atomic mass 143.91 u) emits an average of 2.36 α particles each second. Find the decay constant in s^{-1} and the half-life in years.

36. (*a*) Show that if the decay rate is R_0 at time $t = 0$ and R_1 at some later time t_1, the decay constant is given by

$$\lambda = t_1^{-1} \ln \frac{R_0}{R_1}$$

and the half-life is given by

$$t_{1/2} = \frac{0.693 \, t_1}{\ln (R_0/R_1)}$$

(*b*) Use these results to find the decay constant and the half-life if the decay rate is 1200 Bq at $t = 0$ and 800 Bq at $t_1 = 60$ s.

37. A 1.00-mg sample of substance of atomic mass 59.934 u emits β particles with an activity of 1.131 Ci. Find the decay constant for this substance in s^{-1} and its half-life in years.

38. The counting rate from a radioactive source is measured every minute. The resulting counts per second are 1000, 820, 673, 552, 453, 371, 305, 250. Plot the counting rate versus time on semilog graph paper, and use your graph to find the half-life of the source.

39. A sample of radioactive material is initially found to have an activity of 115.0 decays/min. After 4 d 5 h, its activity is measured to be 73.5 decays/min. (*a*) Calculate the half-life of the material. (*b*) How long (from the initial time) will it take for the sample to reach an activity level of 10.0 decays/min?

40. (*a*) Use the atomic masses $m = 14.00324$ u for $^{14}_{6}$C and $m = 14.00307$ u for $^{14}_{7}$N to calculate the Q value (in MeV) for the beta decay

$$^{14}_{6}\text{C} \rightarrow {}^{14}_{7}\text{N} + \beta^- + \bar{\nu}_e$$

(*b*) Explain why you do not need to add the mass of the β^- to that of atomic $^{14}_{7}$N for this calculation.

41. (*a*) Use the atomic masses $m = 13.00574$ u for $^{13}_{7}$N and $m = 13.003354$ u for $^{13}_{6}$C to calculate the Q value (in MeV) for the beta decay

$$^{13}_{7}\text{N} \rightarrow {}^{13}_{6}\text{C} + \beta^+ + \nu_e$$

(*b*) Explain why you need to add two electron masses to the mass of $^{13}_{6}$C in the calculation of the Q value for this reaction.

42. The electrostatic potential energy of two charges q_1 and q_2 separated by a distance r is $U = kq_1q_2/r$, where k is the Coulomb constant. (*a*) Use Equation 7-1 to calculate the radii of ^{2}H and ^{3}H. (*b*) Find the electrostatic potential energy when these two nuclei are just touching, that is, when their centers are separated by the sum of their radii.

43. (*a*) Calculate the radii of $^{141}_{56}$Ba and $^{92}_{36}$Kr from Equation 7-1. (*b*) Assume that after the fission of ^{235}U into ^{141}Ba and ^{92}Kr, the two nuclei are momentarily separated by a distance r equal to the sum of the radii found in (*a*), and calculate the electrostatic potential energy for these two nuclei at this separation. (See Problem 42.) Compare your result with the measured fission energy of 175 MeV.

44. (*a*) Find the wavelength of a particle in the ground state of a one-dimensional infinite square well of length $L = 2$ fm. (*b*) Find the momentum in units of MeV/c for a particle with this wavelength. (*c*) Show that the total energy of an electron with this wavelength is approximately $E \approx pc$. (*d*) What is the kinetic energy of an electron in the ground state of this well? This calculation shows that if an electron were confined in a region of space as small as a nucleus, it would have a very large kinetic energy.

45. The intensity of a neutrino beam decreases exponentially with distance according to Equation 7-30b, just like that of a beam of neutrons or photons. The absorption cross section for neutrinos is of the order of 10^{-20} barn. Find the thickness of iron ($n = 8.50 \times 10^{28}$ nuclei/m^3) needed to reduce the intensity of a neutrino beam by a factor of e. Compare this thickness with the distance from the earth to the sun (about 150 Gm).

46. A shielded γ source yields a dose rate of 0.0500 rad/h at a distance of 1.00 m for an average-sized person. If workers are allowed a maximum dose rate of 5.00 rem/y, how close to the source may they work, assuming that they work 2000 hours per year? Assume that the radiation intensity follows the inverse-square law. (It will actually fall off more rapidly than $1/r^2$ because of the absorption of photons by the air, so the result for this problem will give a better-than-permissible value.)

47. In 1989, researchers claimed to have achieved fusion in an electrochemical cell at room temperature. They claimed a power output of 4 W from deuterium fusion reactions in the palladium electrode of their apparatus. (a) If the two most likely reactions are

$$^2H + {}^2H \rightarrow {}^3He + n + 3.27 \text{ MeV}$$

and

$$^2H + {}^2H \rightarrow {}^3H + {}^1H + 4.03 \text{ MeV}$$

with 50 percent of the reactions going by each branch, how many neutrons per second would we expect to be emitted in the generation of 4 W of power? (b) If one-tenth of these neutrons were absorbed by the body of an 80.0-kg worker near the device, and if each absorbed neutron carries an average energy of 0.5 MeV with an RBE of 4, to what radiation dose rate in rems per hour would this correspond? (c) How long would it take for a person to receive a total dose of 500 rems? (This is the dose that is usually lethal to half of those receiving it.)

48. The total energy consumed in the United States in 1 y is about 7.0×10^{19} J. How many kilograms of ^{235}U would be needed to provide this amount of energy if we assume that 200 MeV of energy is released by each fissioning uranium nucleus, that all of the uranium atoms undergo fission, and that all of the energy-conversion mechanisms used are 100 percent efficient?

49. The rubidium isotope ^{87}Rb is a β emitter with a half-life of 4.9×10^{10} y that decays into ^{87}Sr. It is used to determine the age of rocks and fossils. Rocks containing the fossils of early animals contain a ratio of ^{87}Sr to ^{87}Rb of 0.0100. Assuming that there were no ^{87}Sr present when the rocks were formed, calculate the age of these fossils.

50. (a) How many α decays and how many β decays must a ^{222}Rn nucleus undergo before it becomes a ^{210}Pb nucleus? (b) Calculate the total energy released in the decay of one ^{222}Rn nucleus to ^{210}Pb. (The mass of ^{210}Pb is 209.984187 u.)

51. The EPA standard for the maximum indoor radon exposure is 4 pCi per liter of air. (a) If the lung capacity of a person is 3.5 L, how many atoms of ^{222}Rn are in the lungs of a person in a room that has the maximum allowed amount of radon? (b) If the total energy absorbed in the lungs from the decay of one ^{222}Rn nucleus to ^{210}Pb is 20.3 MeV and the lungs have a mass of 2.0 kg, what dose in rems does a person who breathes nothing but contaminated air for 1 y receive? (c) Assuming that the risk of lung cancer in nonsmokers is proportional to the dose in rems of the radiation received by the lungs, by what factor is the probability of lung cancer increased for this person, assuming a background dose of 150 mrem per year?

52. A fusion reactor using only deuterium for fuel would have the following two reactions taking place in it:

$$^2H + {}^2H \rightarrow {}^3He + n + 3.27 \text{ MeV}$$

and

$$^2H + {}^2H \rightarrow {}^3H + {}^1H + 4.03 \text{ MeV}$$

The ^3H produced in the second reaction reacts immediately with another ^2H to produce

$$^3H + {}^2H \rightarrow {}^4He + n + 17.7 \text{ MeV}$$

The ratio of ^2H to ^1H atoms in naturally occurring hydrogen is 1.5×10^{-4}. How much energy would be produced from 4 L of water if all of the ^2H nuclei undergo fusion?

Level III

53. The fusion reaction between ^2H and ^3H is

$$^3H + {}^2H \rightarrow {}^4He + n + 17.7 \text{ MeV}$$

Using the conservation of momentum and the given Q value, find the final energies of both the ^4He nucleus and the neutron, assuming that the initial momentum of the system is zero.

54. If there are N_0 radioactive nuclei at time $t = 0$, the number that decay in some time interval dt at time t is $-dN = \lambda N_0 e^{-\lambda t} dt$. If we multiply this number by the lifetime t of these nuclei, sum over all the possible lifetimes from $t = 0$ to $t = \infty$, and divide by the total number of nuclei, we get the mean lifetime τ:

$$\tau = \frac{1}{N_0} \int_0^\infty t \, |dN| = \int_0^\infty t \lambda e^{-\lambda t} \, dt$$

Show that $\tau = 1/\lambda$.

55. Assume that a neutron decays into a proton plus an electron without the emission of a neutrino. The energy shared by the proton and electron is then 0.782 MeV. In the rest frame of the neutron, the total momentum is zero, so the momentum of the proton must be equal and opposite that of the electron. This determines the relative energies of the two particles, but because the electron is relativistic, the exact calculation of these relative energies is somewhat difficult. (a) Assume that the kinetic energy of the electron is 0.782 MeV and calculate the momentum p of the electron in units of MeV/c. (Hint: Use Equation 1-34.) (b) From your result for (a), calculate the kinetic energy $p^2/2m_p$ of the proton. (c) Since the total energy of the electron plus proton is 0.782 MeV, the calculation in (b) gives a correction to the assumption that the energy of the electron is 0.782 MeV. What percentage of 0.782 MeV is this correction?

56. Consider a neutron of mass m moving with speed v_L and making an elastic head-on collision with a nucleus of mass M that is at rest in the laboratory frame of reference. (a) Show that the speed of the center of mass in the lab frame is $V = mv_L/(m + M)$. (b) What is the speed of the nucleus in the center-of-mass frame before the collision? After the collision? (c) What is the speed of the nucleus in the lab frame after the collision? (d) Show that the energy of the nucleus after the collision in the lab frame is

$$\frac{1}{2}M(2V)^2 = \frac{4mM}{(m+M)^2}\left(\frac{1}{2}mv_L^2\right)$$

(e) Show that the fraction of the energy lost by the neutron in this elastic collision is

$$\frac{-\Delta E}{E} = \frac{4mM}{(m+M)^2} = \frac{4(m/M)}{(1+m/M)^2} \qquad 7\text{-}35$$

57. (a) Use the result of part (e) of Problem 56 (Equation 7-35) to show that after N head-on collisions of a neutron with carbon nuclei at rest, the energy of the neutron is approximately $0.714^N E_0$, where E_0 is its original energy. (b) How many head-on collisions are required to reduce the energy of the neutron from 2 MeV to 0.02 eV, assuming stationary carbon nuclei?

58. On the average, a neutron loses 63 percent of its energy in a collision with a hydrogen atom and 11 percent of its energy in a collision with a carbon atom. Calculate the number of collisions needed to reduce the energy of a neutron from 2 MeV to 0.02 eV if the neutron collides with (a) hydrogen atoms and (b) carbon atoms. (See Problem 57.)

59. Energy is generated in the sun and other stars by fusion. One of the fusion cycles, the proton–proton cycle, consists of the following reactions:

$$^1H + {}^1H \rightarrow {}^2H + \beta^+ + \nu_e$$
$$^1H + {}^2H \rightarrow {}^3He + \gamma$$

followed by

$$^1H + {}^3He \rightarrow {}^4He + \beta^+ + \nu_e$$

(a) Show that the net effect of these reactions is

$$4{}^1H \rightarrow {}^4He + 2\beta^+ + 2\nu_e + \gamma$$

(b) Show that rest energy of 24.7 MeV is released in this cycle (not counting the energy of 1.02 MeV released when each positron meets an electron and the two annihilate). (c) The sun radiates energy at the rate of about 4×10^{26} W. Assuming this is due to the conversion of four protons into helium plus γ rays and neutrinos, which releases 26.7 MeV, what is the rate of proton consumption in the sun? How long will the sun last if it continues to radiate at its present level? [Assume that protons constitute about half of the total mass (2×10^{30} kg) of the sun.]

60. Radioactive nuclei with a decay constant of λ are produced in an accelerator at a constant rate R_p. The number of radioactive nuclei N then obeys the equation $dN/dt = R_p - \lambda N$. (a) If N is zero at $t = 0$, sketch N versus t for this situation. (b) The isotope ^{62}Cu is produced at a rate of 100 per second by placing ordinary copper (^{63}Cu) in a beam of high-energy photons. The reaction is

$$\gamma + {}^{63}Cu \rightarrow {}^{62}Cu + n$$

^{62}Cu decays by β decay with a half-life of 10 min. After a time long enough so that $dN/dt \approx 0$, how many ^{62}Cu nuclei are there?

61. A common medical imaging procedure requires the injection of 0.3 mCi of technetium-99, which is in an excited metastable state and emits γ rays with half-life of 6.0 h. The energy of the emitted photons is 0.143 MeV. If the body is able to excrete 60 percent of the technetium in it every hour, what is the total radiation dose that would be given to a 50-kg woman?

Chapter 8

Elementary Particles

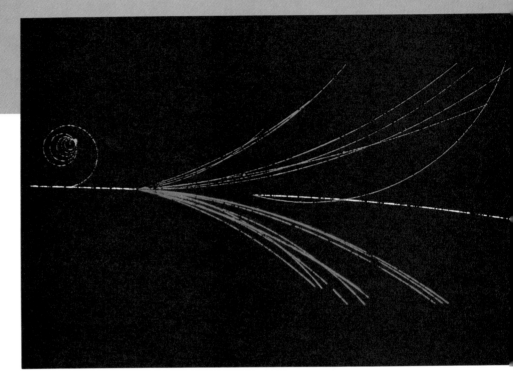

Tracks in a bubble chamber produced by an incoming high-energy proton (yellow) incident from the left, colliding with a proton at rest. The small green spiral is an electron knocked out of an atom. It curves to the left because of an external magnetic field in the chamber. The collision produces seven negative particles (π^-)(blue), a neutral particle Λ^0 that leaves no track, nine positive particles (red) that include seven π^+, a K^+, and a proton. The Λ^0 travels in the original direction of the incoming proton before decaying into a proton (yellow) and a π^- (purple).

In Dalton's atomic theory of matter (1808), the atom was considered to be the smallest indivisible constituent of matter, that is, an elementary particle. Then, with the discovery of the electron by Thomson (1897), the Bohr theory of the nuclear atom (1913), and the discovery of the neutron (1932), it became clear that atoms and even nuclei have considerable structure. For a time, it was thought there were just four "elementary" particles: the proton, neutron, electron, and photon. However, the positron or antielectron was discovered in 1932, and shortly thereafter, the muon, the pion, and many other particles were predicted and discovered.

Since the 1950s, enormous sums of money have been spent constructing particle accelerators of greater and greater energies in hopes of finding particles predicted by various theories. At present, we know of several hundred particles that at one time or another have been considered to be elementary, and research teams at the giant accelerator laboratories around the world are searching for and finding new particles. Some of these have such short lifetimes (of the order of 10^{-23} s) that they can be detected only indirectly. Many are observed only in nuclear reactions with high-energy accelerators. In addition to the usual particle properties of mass, charge, and spin, new properties have been found and given whimsical names such as strangeness, charm, color, topness, and bottomness.

In this chapter, we will first look at the various ways of classifying the multitude of particles that have been found. We will then describe the current theory of elementary particles, called the *standard model*, in which all matter in nature—from the exotic particles produced in the giant accelerator laboratories to ordinary grains of sand—is considered to be constructed from just two families of elementary particles, leptons and quarks.

8-1 Hadrons and Leptons

All the different forces observed in nature, from ordinary friction to the tremendous forces involved in supernova explosions, can be understood in terms of the four basic interactions. In order of decreasing strength, these are

1. The strong nuclear interaction (also called the hadronic interaction)
2. The electromagnetic interaction
3. The weak (nuclear) interaction
4. The gravitational interaction

Molecular forces and most of the everyday forces we observe between macroscopic objects (for example, friction, contact forces, and forces exerted by springs and strings) are complex manifestations of the electromagnetic interaction. Although gravity plays an important role in our lives, it is so weak compared with the other forces that its role in the interactions between elementary particles is essentially negligible. The weak interaction describes the interaction between electrons or positrons and nucleons that results in beta decay, which we discussed in Chapter 7. The strong interaction describes the forces between nucleons (neutrons and protons) that hold nuclei together. The four basic interactions provide a convenient structure for the classification of particles. Some particles participate in all four interactions, whereas others participate in only some of them.

Particles that interact via the strong interaction are called **hadrons**. There are two kinds of hadrons: **baryons**, which have spin $\frac{1}{2}$ (or $\frac{3}{2}$, $\frac{5}{2}$, and so on); and **mesons**, which have zero or integral spin. Baryons, which include nucleons, are the most massive of the elementary particles. Mesons have intermediate masses between the mass of the electron and the mass of the proton.

The existence of mesons was predicted in 1935 by the Japanese physicist H. Yukawa in a theory of nuclear forces that involved the exchange of a particle whose mass is related to the range of the strong nuclear force. Yukawa's theory was analogous to the quantum theory of the electrodynamic interaction. In classical electrodynamics, we avoid action at a distance by describing the force between two electric charges as being carried by the electromagnetic field. In an alternative description, called **quantum electrodynamics (QED),** the electromagnetic field is described in terms of photons, or field quanta, that are emitted by one charged particle and absorbed by another. Because these photons are not observed directly, they are called **virtual photons.** The emission of a virtual photon by a particle such as an electron violates the conservation of energy. However, in quantum mechanics, energy conservation can be violated if it occurs for such a short time that it cannot be observed in accordance with the uncertainty principle. If ΔE is the energy needed to create a virtual photon, such a process is allowed if the photon is absorbed by another particle (or is reabsorbed by the emitting particle) in time Δt given by $\Delta t \approx \hbar/\Delta E$.

Yukawa described the strong force between two nucleons in terms of the emission and absorption of a virtual particle of mass m_π. To avoid violating the conservation of energy, this particle must be absorbed in time Δt given by the uncertainty relation $\Delta t \approx \hbar/\Delta E$ with $\Delta E = m_\pi c^2$:

$$\Delta t \approx \frac{\hbar}{\Delta E} \approx \frac{\hbar}{m_\pi c^2} \qquad 8\text{-}1$$

Since no particle can travel faster than the speed of light c, the maximum distance this particle can travel is

$$d = c\,\Delta t = \frac{\hbar}{m_\pi c} \qquad 8\text{-}2$$

Setting this distance equal to the range of the nuclear force, which is about 1.5×10^{-15} m, Yukawa obtained an estimate of the mass m_π of his new particle. Solving Equation 8-2 for m_π and multiplying by c^2, we obtain for the rest energy of the particle

$$m_\pi c^2 \approx \frac{\hbar c}{d} = \frac{(6.58 \times 10^{-16}\text{ eV·s})(3 \times 10^8\text{ m/s})}{1.5 \times 10^{-15}\text{ m}}$$

$$\approx 1.30 \times 10^8 \text{ eV} = 130 \text{ MeV}$$

The π meson or **pion** discovered in 1947 has just the properties described by Yukawa.

Particles that decay via the strong interaction have very short lifetimes of the order of 10^{-23} s, which is about the time it takes light to travel a distance equal to the diameter of a nucleus. On the other hand, particles that decay via the weak interaction have much longer lifetimes of the order of 10^{-10} s.

Hadrons are rather complicated entities with complex structures. If we use the term "elementary particle" to mean a point particle without structure that is not constructed from some more elementary entities, hadrons do not fit the bill. It is now believed that all hadrons are composed of more fundamental entities called *quarks*, which are truly elementary particles. We will discuss quarks in Section 8.4. Table 8-1 lists some of the properties of the hadrons that are stable against decay via the strong interaction.

Table 8-1 Hadrons That Are Stable Against Decay Via the Strong Nuclear Interaction

Name	Symbol	Mass, MeV/c^2	Spin, $\hbar$	Charge, e	Antiparticle	Mean lifetime, s	Typical decay products[†]
Baryons							
Nucleon	p (proton)	938.3	$\frac{1}{2}$	+1	$\bar{p}$	infinite	
	n (neutron)	939.6	$\frac{1}{2}$	0	$\bar{n}$	930	$p + e^- + \bar{\nu}_e$
Lambda	Λ^0	1116	$\frac{1}{2}$	0	$\bar{\Lambda}^0$	2.5×10^{-10}	$p + \pi^-$
Sigma	Σ^+	1189	$\frac{1}{2}$	+1	$\bar{\Sigma}^-$	0.8×10^{-10}	$n + \pi^+$
	Σ^0	1193	$\frac{1}{2}$	0	$\bar{\Sigma}^0$	10^{-20}	$\Lambda^0 + \gamma$
	Σ^-	1197	$\frac{1}{2}$	−1	$\bar{\Sigma}^+$	1.7×10^{-10}	$n + \pi^-$
Xi	Ξ^0	1315	$\frac{1}{2}$	0	$\bar{\Xi}^0$	3.0×10^{-10}	$\Lambda^0 + \pi^0$
	Ξ^-	1321	$\frac{1}{2}$	−1	$\bar{\Xi}^+$	1.7×10^{-10}	$\Lambda^0 + \pi^-$
Omega	Ω^-	1672	$\frac{3}{2}$	−1	Ω^+	1.3×10^{-10}	$\Xi^0 + \pi^-$
Mesons							
Pion	π^+	139.6	0	+1	π^-	2.6×10^{-8}	$\mu^+ + \nu_\mu$
	π^0	135	0	0	π^0	0.8×10^{-16}	$\gamma + \gamma$
	π^-	139.6	0	−1	π^+	2.6×10^{-8}	$\mu^- + \bar{\nu}_\mu$
Kaon	K^+	493.7	0	+1	K^-	1.24×10^{-8}	$\pi^+ + \pi^0$
	K^0	497.7	0	0	$\bar{K}^0$	0.88×10^{-10} and 5.2×10^{-8}[‡]	$\pi^+ + \pi^-$ $\pi^+ + e^- + \bar{\nu}_e$
Eta	η^0	549	0	0		2×10^{-19}	$\gamma + \gamma$

[†]Other decay modes also occur for most particles.
[‡]The K^0 has two distinct lifetimes, sometimes referred to as K^0_{short} and K^0_{long}. All other particles have a unique lifetime.

Particles that participate in the weak interaction but not in the strong interaction are called **leptons**. These include electrons, muons, and neutrinos, which are all less massive than the lightest hadron. The word *lepton*, meaning "light particle," was chosen to reflect the relatively small mass of these particles. However, the most recently discovered lepton, the *tau*, found by Perl in 1975, has a mass of about 1780 MeV/c^2, nearly twice that of the proton (938 MeV/c^2), so we now have a "heavy lepton." As far as we know, leptons are point particles with no structure and can be considered to be truly elementary in the sense that they are not composed of other particles.

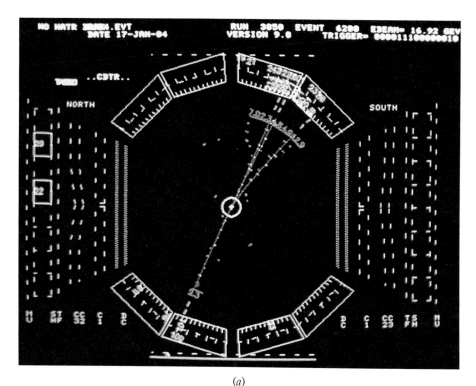

(a)

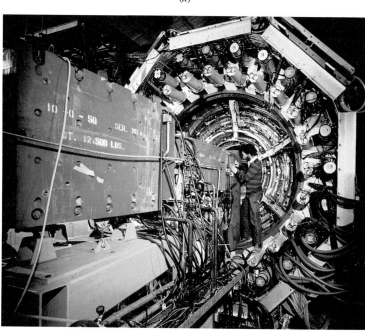

(b)

(a) A computer display of the production and decay of a τ^+ and τ^- pair. An electron and positron annihilate at the center marked by the yellow cross—producing a τ^+ and τ^- pair, which travel in opposite directions, but quickly decay while still inside the beam pipe (yellow circle). The τ^+ decays into two invisible neutrinos and a μ^+, which travels toward the bottom left. Its track in the drift chamber is calculated by a computer and indicated in red. It penetrates the lead-argon counters outlined in purple and is detected at the blue dot near the bottom blue line that marks the end of a muon detector. The τ^- decays into three charged pions (red tracks moving upward) plus invisible neutrinos. (b) The Mark I detector, built by a team from the Stanford Linear Accelerator Center (SLAC) and the Lawrence Berkeley Laboratory, became famous for many discoveries, including the ψ/J meson and the τ lepton. Tracks of particles are recorded by wire spark chambers wrapped in concentric cylinders around the beam pipe extending out to the ring where physicist Carl Friedberg has his right foot. Beyond this are two rings of protruding tubes, housing photomultipliers that view various scintillation counters. The rectangular magnets at the left guide the counterrotating beams that collide in the center of the detector.

It is currently thought that there are six leptons, each of which has an antiparticle. They are the electron, the muon, and the tau and a distinct neutrino associated with each of these three particles. (The neutrino associated with the tau has not yet been observed experimentally.) The masses of these particles are quite different. The mass of the electron is 0.511 MeV/c^2, the mass of the muon is 105 MeV/c^2, and that of the tau is 1780 MeV/c^2. The neutrinos are thought to be massless, but there is considerable debate as to the possibility that they may have a very small but nonzero mass, perhaps of the order of a few eV/c^2. Experiments designed to detect neutrinos emitted from the sun have found a much smaller number than expected, which could be explained if the mass of the neutrino were not zero. In addition, a mass as small as 40 eV/c^2 for the neutrino would have great cosmological significance. The answer to the question of whether the universe will continue to expand indefinitely or will reach a maximum size and begin to contract depends on the total mass in the universe. Thus, the answer could depend on whether the rest mass of the neutrino is merely small rather than zero. The observation of electron neutrinos from the supernova 1987A puts an upper limit on the mass of these neutrinos. Since the velocity of a particle with mass depends on its energy, the arrival time of a burst of neutrinos with mass from a supernova would be spread out in time. The fact that the electron neutrinos from the 1987 supernova all arrived at the earth within 13 seconds of one another results in an upper limit of about 16 eV/c^2 for their mass. This upper limit does not rule out the possibility of zero mass for these neutrinos.

Questions

1. How are baryons and mesons similar? How are they different?
2. The muon and the pion have nearly the same mass. How do these particles differ?

8-2 Spin and Antiparticles

One important characteristic of a particle is its intrinsic spin angular momentum. We have already discussed the fact that the electron has a quantum number m_s that corresponds to the z component of its intrinsic spin characterized by the quantum number $s = \frac{1}{2}$. Protons, neutrons, neutrinos, and the various other particles that also have an intrinsic spin characterized by the quantum number $s = \frac{1}{2}$ are called **spin-$\frac{1}{2}$ particles.** Particles that have spin $\frac{1}{2}$ (or $\frac{3}{2}, \frac{5}{2}, \ldots$) are called **fermions** and obey the Pauli exclusion principle. Particles such as pions and other mesons have zero spin or integral spin ($s = 0, 1, 2, \ldots$). These particles are called **bosons** and do not obey the Pauli exclusion principle. Any number of these particles can be in the same quantum state.

Spin-$\frac{1}{2}$ particles are described by the Dirac equation, an extension of the Schrödinger equation to include special relativity. One feature of Dirac's theory proposed in 1927 is the prediction of the existence of antiparticles. In special relativity, the energy of a particle is related to the mass and momentum of the particle by $E = \pm\sqrt{p^2c^2 + m^2c^4}$. We usually choose the positive solution and dismiss the negative-energy solution with a physical argument. However, the Dirac equation requires the existence of wave functions that correspond to the negative-energy states. Dirac got around this difficulty by postulating that all the negative-energy states were filled and would therefore not be observable. Only holes in the "infinite sea" of negative-energy states would be observed. This interpretation received little attention until the positron was discovered in 1932 by Carl Anderson.

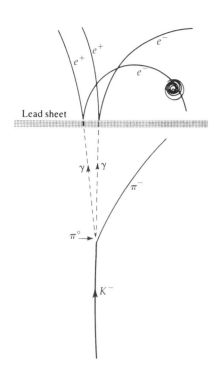

A negative kaon (K^-) enters a bubble chamber from the bottom and decays into a π^-, which moves off to the right, and a π^0, which immediately decays into two photons whose paths are indicated by the dashed lines in the drawing. Each photon interacts in the lead sheet, producing an electron-positron pair. The spiral at the right is another electron that has been knocked out of an atom in the chamber. (Other extraneous tracks have been removed from the photograph.)

Antiparticles are never created alone but always in particle–antiparticle pairs. In the creation of an electron–positron pair by a photon, the energy of the photon must be greater than the rest energy of the electron plus the positron, which is $2m_e c^2 \approx 1.02$ MeV, where m_e is the mass of the electron. Although the positron is stable, it has only a short-term existence in our universe because of the large supply of electrons in matter. The fate of a positron is annihilation according to the reaction

$$e^+ + e^- \rightarrow \gamma + \gamma \qquad 8\text{-}3$$

The probability of this reaction is large only if the positron is at rest or nearly at rest. Two photons moving in opposite directions are needed to conserve linear momentum.

The fact that we call electrons *particles* and positrons *antiparticles* does not imply that positrons are less fundamental than electrons. It merely reflects the nature of our part of the universe. If our matter were made up of negative protons and positive electrons, then positive protons and negative electrons would suffer quick annihilation and would be called antiparticles.

The antiproton (p^-) was discovered in 1955 by E. Segré and O. Chamberlain using a beam of protons in the Bevatron at Berkeley to produce the reaction*

$$p^+ + p^+ \rightarrow p^+ + p^+ + p^+ + p^- \qquad 8\text{-}4$$

The creation of a proton–antiproton pair (Figure 8-1) requires kinetic energy of at least $2m_p c^2 = 1877$ MeV $= 1.877$ GeV in the zero-momentum reference

Figure 8-1 Bubble-chamber tracks showing the creation of a proton–antiproton pair in the collision of an incident 25-GeV proton with a stationary proton in liquid hydrogen.

---

*The antiproton is sometimes denoted by $\bar{p}$ rather than p^-. For neutral particles, such as the neutron, the bar must be used to denote the antiparticle. Thus the antineutron is denoted by $\bar{n}$. The normal electron and proton are often denoted by e and p without the minus or plus superscripts.

frame in which the two protons approach each other with equal and opposite momenta. In the laboratory frame in which one of the protons is initially at rest, the kinetic energy of the incoming proton must be at least $6m_pc^2 = 5.63$ GeV (see Problem 76 of Chapter 1). This energy was not available in laboratories before the development of high-energy accelerators in the 1950s. Antiprotons annihilate with protons to produce two gamma rays in a reaction similar to that in Equation 8-3.

Example 8-1

A proton and an antiproton at rest annihilate according to the reaction

$$p^+ + p^- \rightarrow \gamma + \gamma$$

Find the energies and wavelengths of the photons.

Since the proton and the antiproton are at rest, conservation of momentum requires that the two photons created in their annihilation have equal and opposite momenta and therefore equal energies. Since the total energy on the left side of the reaction is $2m_pc^2$, the energy of each photon is

$$E_\gamma = m_pc^2 = 938 \text{ MeV}$$

The wavelength is

$$\lambda = \frac{c}{f} = \frac{hc}{hf} = \frac{hc}{E_\gamma} = \frac{1240 \text{ eV·nm}}{9.38 \times 10^8 \text{ eV}} = 1.32 \times 10^{-15} \text{ m} = 1.32 \text{ fm}$$

Air view of the European Laboratory for Particle Physics (CERN) just outside of Geneva, Switzerland. The large circle shows the Large Electron-Positron collider (LEP) tunnel that is 27 kilometers in circumference. The irregular dashed line is the border between France and Switzerland.

The tunnel of the proton–antiproton collider at CERN. The same bending magnets and focusing magnets can be used for protons or antiprotons moving in opposite directions. The rectangular box in the foreground is a focusing magnet; the next four boxes are bending magnets.

8-3 The Conservation Laws

One of the maxims of nature is "anything that can happen does." If a conceivable decay or reaction does not occur, there must be a reason. The reason is usually expressed in terms of a conservation law. The conservation of energy rules out the decay of any particle for which the total rest mass of the decay products would be greater than the initial rest mass of the particle before decay. The conservation of linear momentum requires that when an electron and positron at rest annihilate, two photons must be emitted. Angular momentum must also be conserved in a reaction or decay. A fourth conservation law that restricts the possible particle decays and reactions is that of electric charge. The net electric charge before a decay or reaction must equal the net charge after the decay or reaction.

There are two additional conservation laws that are important in the reactions and decays of elementary particles: the conservation of baryon number and the conservation of lepton number. Consider the possible decay

$$p \rightarrow \pi^0 + e^+$$

This decay would conserve charge, energy, angular momentum, and linear momentum, but it does not occur. It does not conserve either lepton number or baryon number. The conservation of lepton number and baryon number implies that whenever a lepton or baryon particle is created, an antiparticle of the same type is also created. We assign the **lepton number** $L = +1$ to all leptons, $L = -1$ to all antileptons, and $L = 0$ to all other particles. Similarly, the **baryon number** $B = +1$ is assigned to all baryons, $B = -1$ to all antibaryons, and $B = 0$ to all other particles. The baryon and lepton numbers cannot change in a reaction or decay. The conservation of baryon number along with the conservation of energy implies that the least massive baryon, the proton, must be stable.

The conservation of lepton number implies that the neutrino emitted in the beta decay of the free neutron is an antineutrino:

$$n \rightarrow p^+ + e^- + \bar{\nu}_e \qquad 8\text{-}5$$

The fact that neutrinos and antineutrinos are different is illustrated by an experiment in which ^{37}Cl is bombarded with an intense antineutrino beam from the decay of reactor neutrons. If neutrinos and antineutrinos were the same, we would expect the following reaction:

$$^{37}\text{Cl} + \bar{\nu}_e \rightarrow {}^{37}\text{Ar} + e^- \qquad 8\text{-}6$$

This reaction is not observed. However, if protons are bombarded with antineutrinos, the reaction

$$p + \bar{\nu}_e \rightarrow n + e^+ \qquad 8\text{-}7$$

is observed. Note that the lepton number is -1 on the left side of reaction 8-6 and $+1$ on the right side. But the lepton number is -1 on both sides of reaction 8-7.

Not only are neutrinos and antineutrinos distinct particles, but the neutrinos associated with electrons are distinct from the neutrinos associated with muons. It is also believed that the recently discovered heavy lepton, the tau, has a neutrino associated with it. Electronlike leptons (e and ν_e), muonlike leptons (μ and ν_μ), and presumably taulike leptons (τ and ν_τ) are each separately conserved. This is easily handled by assigning separate lepton numbers L_e, L_μ, and L_τ to the particles. For e and ν_e, $L_e = +1$; for their antiparticles, $L_e = -1$; and for all other particles, $L_e = 0$. The lepton numbers L_μ and L_τ are similarly assigned.

Example 8-2

What conservation laws (if any) are violated by the following decays?
(a) $n \rightarrow p + \pi^-$ (b) $\Lambda^0 \rightarrow p^- + \pi^+$ (c) $\mu^- \rightarrow e^- + \gamma$

(a) There are no leptons in this decay, so there is no problem with the conservation of lepton number. The net charge is zero before and after the decay, so charge is conserved. Also, the baryon number is $+1$ before and after the decay. However, the rest energy of the proton (938.3 MeV) plus that of the pion (139.6 MeV) is greater than the rest energy of the neutron (939.6 MeV). Thus, this decay violates the conservation of energy.

(b) Again, there are no leptons involved, and the net charge is zero before and after the decay. Also, the rest energy of the Λ^0 (1116 MeV) is greater than the rest energy of the antiproton (938.3 MeV) plus that of the pion (139.6 MeV), so energy is conserved with the loss in rest energy equaling the gain in kinetic energy of the decay products. However, this decay does not conserve baryon number, which is $+1$ for the Λ^0 and -1 for the antiproton.

(c) This reaction does not conserve muon lepton number or electron lepton number. The muon does decay via

$$\mu^- \rightarrow e^- + \bar{\nu}_e + \nu_\mu$$

which does conserve both muon and electron lepton numbers.

There are some conservation laws that are not universal but apply only to certain kinds of interactions. In particular, there are quantities that are conserved in decays and reactions that occur via the strong interaction but not in decays or reactions that occur via the weak interaction. One of these quantities that is particularly important is **strangeness** introduced by M. Gell-Mann and K. Nishijima in 1952 to explain the strange behavior of the heavy baryons and mesons. Consider the reaction

$$p + \pi^- \rightarrow \Lambda^0 + K^0 \qquad 8\text{-}8$$

The cross section for this reaction is large, as would be expected if it takes place via the strong interaction. However, the decay times for both the Λ^0 and K^0 are of the order of 10^{-10} s, which is characteristic of the weak interaction, rather than 10^{-23} s, which would be expected for the strong interaction. Other particles showing similar behavior were called **strange particles.**

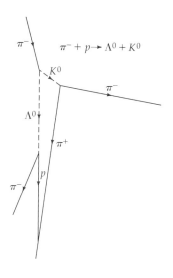
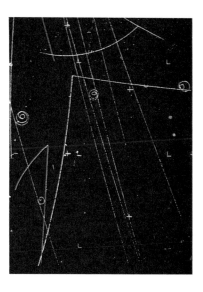

An early photograph of bubble-chamber tracks at the Lawrence Berkeley Laboratory, showing the production and decay of two strange particles, the K^0 and the Λ^0. These neutral particles are identified by the tracks of their decay particles. The lambda particle was named because of the similarity of the tracks of its decay particles and the Greek letter Λ. (The blue tracks are particles not involved in the reaction of Equation 8-8.)

These particles are always produced in pairs and never singly, even when all other conservation laws are met. This behavior is described by assigning a new property called strangeness to these particles. The strangeness of the ordinary hadrons—the nucleons and pions—was arbitrarily taken to be zero. The strangeness of the K^0 was arbitrarily chosen to be +1. Therefore, the strangeness of the Λ^0 particle must be −1 so that strangeness is conserved in the reaction of Equation 8-8. The strangeness of other particles could then be assigned by looking at their various reactions and decays. In reactions and decays that occur via the strong interaction, strangeness is conserved. In those that occur via the weak interaction, the strangeness can change by ±1.

Figure 8-2 shows the masses of the baryons and mesons that are stable against decay via the strong interaction versus strangeness. We can see from this figure that these particles cluster in multiplets of one, two, or three particles of approximately equal mass, and that the strangeness of a multiplet of particles is related to the "center of charge" of the multiplet.

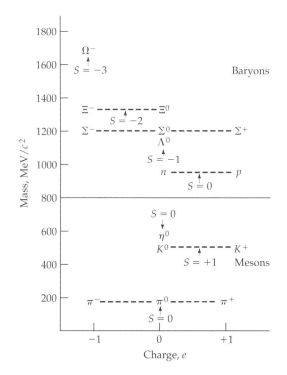

Figure 8-2 The strangeness of hadrons shown on a plot of rest mass versus charge. The strangeness of a baryon-charge multiplet is related to the number of places the center of charge of the multiplet is displaced from that of the nucleon doublet. For each displacement of $\frac{1}{2}e$, the strangeness changes by ±1. For mesons, the strangeness is related to the number of places the center of charge is displaced from that of the pion triplet. Because of the unfortunate original assignment of +1 for the strangeness of kaons, all of the baryons that are stable against decay via the strong interaction have negative or zero strangeness.

Example 8-3

State whether the following decays can occur via the strong interaction, via the weak interaction, or not at all:

(a) $\Sigma^+ \rightarrow p + \pi^0$ (b) $\Sigma^0 \rightarrow \Lambda^0 + \gamma$ (c) $\Xi^0 \rightarrow n + \pi^0$

We first note the mass of each decaying particle is greater than that of the decay products, so there is no problem with energy conservation in any of the decays. In addition, there are no leptons involved in any of the decays, and charge and baryon number are both conserved in all the decays.

(a) From Figure 8-2, we can see that the strangeness of the Σ^+ is -1 whereas the strangeness of both the proton and the pion is zero. This decay is possible via the weak interaction but not the strong interaction. It is, in fact, one of the decay modes of the Σ^+ particle with a lifetime of the order of 10^{-10} s.

(b) Since the strangeness of both the Σ^0 and Λ^0 is -1, this decay can proceed via the strong interaction. It is, in fact, the dominant mode of decay of the Σ^0 particle with a lifetime of about 10^{-20} s.

(c) The strangeness of the Ξ^0 is -2 whereas the strangeness of both the neutron and pion is zero. Since strangeness cannot change by 2 in a decay or reaction, this decay cannot occur.

Question

3. How can you tell whether a decay proceeds via the strong interaction or the weak interaction?

8-4 The Quark Model

We have seen that leptons appear to be truly elementary particles in that they do not break down into smaller entities and they seem to have no measurable size or structure. Hadrons, on the other hand, are complex particles with size and structure, and they decay into other hadrons. Furthermore, at the present time, there are only six known leptons, whereas there are many more hadrons. Table 8-1 includes only hadrons that are stable against decay via the strong interaction. Hundreds of other hadrons have been discovered, and their properties, such as charge, spin, mass, strangeness, and decay schemes, have been measured.

The most important advance in our understanding of elementary particles was the quark model proposed by M. Gell-Mann and G. Zweig in 1963. According to this model, all hadrons are thought to consist of combinations of two or three truly elementary particles called **quarks.** (The name *quark* was chosen by Gell-Mann from a quotation from *Finnegans Wake* by James Joyce.) In the original model, quarks came in three types, called **flavors,** labeled *u*, *d*, and *s* (for *up*, *down*, and *strange*). An unusual property of quarks is that they carry fractional electron charges. The charge of the *u* quark is $+\frac{2}{3}e$ and that of the *d* and *s* quarks is $-\frac{1}{3}e$. Each quark has spin $\frac{1}{2}\hbar$ and a baryon number of $\frac{1}{3}$. The strangeness of the *u* and *d* quark is 0 and that of the *s* quark is -1. Each quark has an antiquark with the opposite electric charge, baryon number, and strangeness. These properties are listed in Table 8-2. Baryons consist of three quarks (or three antiquarks for antiparticles), whereas mesons consist of a quark and an antiquark, giving them a baryon number $B = 0$, as required. The proton consists of the combination *uud* and the neutron, *udd*. Baryons with a strangeness $S = -1$ contain one *s* quark. All the particles listed in Table 8-1 can be constructed from these

Table 8-2 Properties of Quarks and Antiquarks

			Quarks				
Flavor	Spin	Charge	Baryon number	Strangeness	Charm	Topness	Bottomness
u (up)	$\frac{1}{2}\hbar$	$+\frac{2}{3}e$	$+\frac{1}{3}$	0	0	0	0
d (down)	$\frac{1}{2}\hbar$	$-\frac{1}{3}e$	$+\frac{1}{3}$	0	0	0	0
s (strange)	$\frac{1}{2}\hbar$	$-\frac{1}{3}e$	$+\frac{1}{3}$	-1	0	0	0
c (charmed)	$\frac{1}{2}\hbar$	$+\frac{2}{3}e$	$+\frac{1}{3}$	0	$+1$	0	0
t (top)	$\frac{1}{2}\hbar$	$+\frac{2}{3}e$	$+\frac{1}{3}$	0	0	$+1$	0
b (bottom)	$\frac{1}{2}\hbar$	$-\frac{1}{3}e$	$+\frac{1}{3}$	0	0	0	$+1$
			Antiquarks				
$\bar{u}$	$\frac{1}{2}\hbar$	$-\frac{2}{3}e$	$-\frac{1}{3}$	0	0	0	0
$\bar{d}$	$\frac{1}{2}\hbar$	$+\frac{1}{3}e$	$-\frac{1}{3}$	0	0	0	0
$\bar{s}$	$\frac{1}{2}\hbar$	$+\frac{1}{3}e$	$-\frac{1}{3}$	$+1$	0	0	0
$\bar{c}$	$\frac{1}{2}\hbar$	$-\frac{2}{3}e$	$-\frac{1}{3}$	0	-1	0	0
$\bar{t}$	$\frac{1}{2}\hbar$	$-\frac{2}{3}e$	$-\frac{1}{3}$	0	0	-1	0
$\bar{b}$	$\frac{1}{2}\hbar$	$+\frac{1}{3}e$	$-\frac{1}{3}$	0	0	0	-1

three quarks and three antiquarks.* The great strength of the quark model is that all the allowed combinations of three quarks or quark–antiquark pairs result in known hadrons. Strong evidence for the existence of quarks inside a nucleon is provided by high-energy scattering experiments called *deep inelastic scattering*. In these experiments, a nucleon is bombarded with electrons, muons, or neutrinos of energies from 15 to 200 GeV. Analyses of particles scattered at large angles indicate the presence within the nucleon of spin-$\frac{1}{2}$ particles of sizes much smaller than that of the nucleon. These experiments are analogous to Rutherford's scattering of α particles by atoms in which the presence of a tiny nucleus in the atom was inferred from the large-angle scattering of the α particles.

Example 8-4

What are the properties of the particles made up of the following quarks: (a) $u\bar{d}$, (b) $\bar{u}d$, (c) dds, and (d) uss?

(a) Since $u\bar{d}$ is a quark–antiquark combination, it has baryon number 0 and is therefore a meson. There is no strange quark here, so the strangeness of the meson is zero. The charge of the up quark is $+\frac{2}{3}e$ and that of the anti-down quark is $+\frac{1}{3}e$, so the charge of the meson is $+1e$. This is the quark combination of the π^+ meson.

(b) The particle $\bar{u}d$ is also a meson with zero strangeness. Its electric charge is $-\frac{2}{3}e + (-\frac{1}{3}e) = -1e$. This is the quark combination of the π^- meson.

(c) The particle dds is a baryon with strangeness -1 since it contains one strange quark. Its electric charge is $-\frac{1}{3}e - \frac{1}{3}e - \frac{1}{3}e = -1e$. This is the quark combination for the Σ^- particle.

(d) The particle uss is a baryon with strangeness -2. Its electric charge is $+\frac{2}{3}e - \frac{1}{3}e - \frac{1}{3}e = 0$. This is the quark combination for the Ξ^0 particle.

---

*The correct quark combinations of hadrons are not always obvious because of the symmetry requirements on the total wave function. For example, the π^0 meson is represented by a linear combination of $u\bar{u}$ and $d\bar{d}$.

In 1967, a fourth quark was proposed to explain some discrepancies between experimental determinations of certain decay rates and calculations based on the quark model. The fourth quark is labeled c for a new property called **charm**. Like strangeness, charm is conserved in strong interactions but changes by ± 1 in weak interactions. In 1975, a new heavy meson called the ψ/J **particle** (or simply the ψ **particle**) was discovered that has the properties expected of a $c\bar{c}$ combination. Since then other mesons with combinations such as $c\bar{d}$ and $\bar{c}d$, as well as baryons containing the charmed quark, have been discovered. Two more quarks labeled t and b (for *t*op and *b*ottom or, as some prefer, *t*ruth and *b*eauty) have been proposed. In 1977, a massive new meson called the **Y meson** or **bottomonium**, which is considered to have the quark combination $b\bar{b}$, was discovered. At the present time there is no direct evidence of the existence of the top quark.

The six quarks and six leptons (and their antiparticles) are thought to be the fundamental, elementary particles of which all matter is composed. Table 8-3 lists some of the properties of quarks and leptons. In this table, the masses given for neutrinos are upper limits, and those given for quarks are educated guesses. There is experimental evidence for the existence of each of these particles except for the top quark. The existence of the top quark is strongly suspected because of symmetry between the number of quarks and the number of leptons.

Table 8-3 Fundamental Particles and Their Approximate Masses[†]

	Light	Medium	Heavy	Charge
Quarks	$u(\sim 400 \text{ MeV}/c^2)$	$c(\sim 1.5 \text{ GeV}/c^2)$	$t(>89 \text{ GeV}/c^2)$	$+\frac{2}{3}e$
	$d(\sim 700 \text{ MeV}/c^2)$	$s(\sim 0.15 \text{ GeV}/c^2)$	$b(\sim 4.7 \text{ GeV}/c^2)$	$-\frac{1}{3}e$
Leptons	$e(0.511 \text{ MeV}/c^2)$	$\mu(106 \text{ MeV}/c^2)$	$\tau(1.78 \text{ GeV}/c^2)$	$-1e$
	$\nu_e(<16 \text{ eV}/c^2)$[‡]	$\nu_\mu(<300 \text{ keV}/c^2)$[‡]	$\nu_\tau(<40 \text{ MeV}/c^2)$[‡]	0

[†]Because quarks are always bound in mesons or baryons, their masses are not well understood. The values given here are merely educated guesses.

[‡]These masses are upper limits. Neutrinos may be massless particles.

Quark Confinement

Despite considerable experimental effort, no isolated quark has ever been observed. It is now believed that it is impossible to obtain an isolated quark. This would be true, for example, if the force between two quarks remains constant regardless of separation distance, rather than decreasing with increasing separation distance as is the case for other fundamental forces such as the electric force between two charges, the gravitational force between two masses, and the strong nuclear force between two hadrons. As a result, the potential energy of two quarks increases with increasing separation distance so that an infinite amount of energy would be needed to separate the quarks completely.

When a large amount of energy is added to a quark system such as a nucleon, a quark–antiquark pair is created and the original quarks remain confined within the original system. Because quarks cannot be isolated, but are always bound in a baryon or meson, the mass of a quark cannot be accurately known, which is why the masses listed in Table 8-3 are merely educated guesses.

Questions

4. How can you tell whether a particle is a meson or a baryon by looking at its quark content?
5. Are there any quark–antiquark combinations that result in a nonintegral electric charge?

8-5 Field Particles

In addition to the six fundamental leptons and six fundamental quarks, there are other particles called *field particles* or *field quanta* that are associated with the forces exerted by one elementary particle on another. As we have seen, in quantum electrodynamics the electromagnetic field of a single charged particle is described by virtual photons that are continuously being emitted and reabsorbed by the particle. If we put energy into the system by accelerating the charge, some of these virtual photons can be "shaken off" and become real, observable photons. The photon is said to mediate the electromagnetic interaction. Each of the four basic interactions can be described in this way.

The field quantum associated with the gravitational interaction, called the **graviton**, has not yet been observed. The gravitational "charge" analogous to electric charge is mass.

The weak interaction is thought to be mediated by three field quanta called **vector bosons**: W^+, W^-, and Z^0. These particles were predicted by S. Glashow, A. Salam, and S. Weinberg in theory called the *electroweak theory*, which we will discuss in the next section. The W and Z particles were first observed in 1983 by a group of over a hundred scientists lead by C. Rubbia using the high-energy accelerator at CERN in Geneva, Switzerland. The masses of the $W^\pm$ particles (about 80 GeV/c^2) and the Z particle (about 91 GeV/c^2) measured in this experiment were in excellent agreement with those predicted by the electroweak theory. (The W^- particle is the antiparticle of the W^+ particle, so they must have identical masses.)

The field quanta associated with the strong force between quarks are called **gluons**. Isolated gluons have not been observed experimentally. The "charge" responsible for the strong interactions comes in three varieties labeled *red, green,* and *blue* (analogous with the three primary colors), and the strong charge is called the **color charge**. The field theory for strong interactions, analogous to quantum electrodynamics for electromagnetic interactions, is called **quantum chromodynamics (QCD)**.

Table 8-4 lists the bosons responsible for mediating the basic interactions.

Table 8-4 **Bosons that Mediate the Basic Interactions**

Interaction	Boson	Spin	Mass	Electric charge
Strong	g (gluon)	1	0	0
Weak	$W^\pm$	1	79.8 GeV/c^2	$\pm 1e$
	Z^0	1	91.2 GeV/c^2	0
Electromagnetic	γ (photon)	1	0	0
Gravitational	Graviton[†]	2	0	0

[†]Not yet observed

Example 8-5

If a particular weak interaction is mediated by emission of a virtual Z particle, calculate the order of magnitude of the range of the interaction.

From the uncertainty principle, the time that a virtual Z particle can exist without violating the conservation of energy is

$$\Delta t \approx \frac{\hbar}{\Delta E} \approx \frac{\hbar}{m_Z c^2}$$

The range is then given by Equation 8-2:

$$R = c \, \Delta t \approx \frac{\hbar c}{m_Z c^2} = \frac{(6.58 \times 10^{-16} \text{ eV·s})(3 \times 10^8 \text{ m/s})}{91 \times 10^9 \text{ eV}} \approx 2 \times 10^{-18} \text{ m}$$

8-6 The Electroweak Theory

In the **electroweak theory,** the electromagnetic and weak interactions are considered to be two different manifestations of a more fundamental electroweak interaction. At very high energies ($\gg$100 GeV), the electroweak interaction would be mediated by four bosons. From symmetry considerations, these would be a triplet consisting of W^+, W^0, and W^-, all of equal mass, and a singlet B^0 of some other mass. Neither the W^0 nor the B^0 would be observed directly, but one linear combination of the W^0 and the B^0 would be the Z^0 and another would be the photon. At ordinary energies, the symmetry is broken. This leads to the separation of the electromagnetic interaction mediated by the massless photon and the weak interaction mediated by the W^+, W^-, and Z^0 particles. The fact that the photon is massless and that the W and Z particles have masses of the order of 100 GeV/c^2 shows that the symmetry assumed in the electroweak theory does not exist at lower energies.

The symmetry-breaking mechanism is called a **Higgs field,** which requires a new boson, the **Higgs boson,** whose rest energy is expected to be of the order of 1 TeV (1 TeV = 10^{12} eV). The Higgs boson has not yet been observed. Calculations show that Higgs bosons (if they exist) should be produced in head-on collisions between protons of energies of the order of 20 TeV. Such energies are not presently available but will be in a proposed new superconducting super collider (SSC) scheduled to be completed in the mid 1990s.

8-7 The Standard Model

The combination of the quark model, electroweak theory, and quantum chromodynamics is called the **standard model.** In this model, the fundamental particles are the leptons and quarks, each of which comes in six flavors as shown in Table 8-3, and the force carriers are the photon, the $W^{\pm}$ and Z particles, and the gluons (of which there are eight types). The leptons and quarks are all spin-$\frac{1}{2}$ fermions, which obey the Pauli exclusion principle, and the force carriers are integral-spin bosons, which do not obey the Pauli exclusion principle. Every force in nature is due to one of the four basic interactions: strong, electromagnetic, weak, and gravitational. A particle experiences one of the basic interactions if it carries a charge associated with that interaction. Electric charge is the familiar charge that we have studied previously. Weak charge, also called flavor charge, is carried by leptons and quarks. The charge associated with the strong interaction is called color charge and is carried by quarks and gluons but not by leptons. The charge associated with the gravitational force is mass. It is important to note that the photon, which mediates the electromagnetic interaction, does not carry electric charge. Similarly, the $W^{\pm}$ and Z particles, which mediate the weak interaction, do not carry weak charge. However, the gluons, which mediate the strong interaction, do carry color charge. This fact is related to the confinement of quarks as discussed in Section 8-4.

All matter is made up of leptons or quarks. There are no known composite particles consisting of leptons bound together by the weak force. Leptons exist only as isolated particles. Hadrons (baryons and mesons) are composite particles consisting of quarks bound together by the color charge. A result of QCD theory is that only color-neutral combinations of quarks are allowed. Three quarks of different colors can combine to form color-neutral baryons, such as the neutron and proton. Mesons contain a quark and an antiquark and are also color-neutral. Excited states of hadrons are considered to be different particles. For example, the Δ^+ particle is an excited state of the proton. Both are made up of the *uud* quarks, but the proton is in the ground state with spin $\frac{1}{2}$ and a rest energy of 938 MeV, whereas the Δ^+ particle is the first excited state with spin $\frac{3}{2}$ and a rest energy of 1232 MeV. The two u quarks can be in the same spin state in the Δ^+ without violating the exclusion principle because they have different color. All baryons eventually decay to the lightest baryon, the proton. The proton cannot decay because the conservation of energy and baryon number.

The strong interaction has two parts, the fundamental or color interaction and what is called the *residual strong interaction*. The fundamental interaction is responsible for the force exerted by one quark on another and is mediated by gluons. The residual strong interaction is responsible for the force between color-neutral nucleons, such as the neutron and proton. This force is due to the residual strong interactions between the color-charged quarks that make up the nucleons and can be viewed as being mediated by the exchange of mesons. The residual strong interaction between color-neutral nucleons can be thought of as analogous to the residual electromagnetic interaction between neutral atoms that bind them together to form molecules. Table 8-5 lists some of the properties of the basic interactions.

Table 8-5 **Properties of the Basic Interactions**

	Interaction				
				Strong	
	Gravitational	Weak	Electromagnetic	Fundamental	Residual
Acts on	Mass	Flavor	Electric charge	Color charge	
Particles experiencing	All	Quarks, leptons	Electrically charged	Quarks, gluons	Hadrons
Particles mediating	Graviton	$W^{\pm}$, Z	γ	Gluons	Mesons
Strength for two quarks at 10^{-18} m[†]	10^{-41}	0.8	1	25	(Not applicable)
Strength for two protons in nucleus[†]	10^{-36}	10^{-7}	1	(Not applicable)	20

[†]Strengths are relative to electromagnetic strength

For each particle there is an antiparticle. A particle and its antiparticle have identical mass and spin but opposite electric charge. For leptons, the lepton numbers L_e, L_μ, and L_τ of the antiparticles are the negatives of the corresponding numbers for the particles. For example, the lepton number for the electron is $L_e = +1$ and that for the positron is $L_e = -1$. For hadrons, the baryon number, strangeness, charm, topness, and bottomness are the sums of those quantities for the quarks that make up the hadron. The number of each antiparticle is the negative of the number for the corresponding particle. For example, the lambda particle Λ^0, which is made up of the uds quarks, has $B = 1$ and $S = -1$, whereas its antiparticle $\overline{\Lambda}^0$, which is made up of the $\bar{u}\bar{d}\bar{s}$ quarks, has $B = -1$ and $S = +1$. A particle such as the photon γ or the Z^0 particle that has zero electric charge, $B = 0$, $L = 0$, $S = 0$, and zero charm, topness, and bottomness, is its own antiparticle. Note that the K^0 meson ($d\bar{s}$) has a zero value for all of these quantities except strangeness, which is +1. Its antiparticle, the $\overline{K}^0$ meson ($\bar{d}s$) has strangeness −1, which makes it distinct from the K^0. The π^+ ($u\bar{d}$) and π^- ($\bar{u}d$) are somewhat special in that they have electric charge but zero values for L, B, and S. They are antiparticles of each other, but since there is no conservation law for mesons, it is impossible to say which is the particle and which is the antiparticle. Similarly, the W^+ and W^- are antiparticles of each other.

8-8 Grand Unification Theories

With the success of the electroweak theory, attempts have been made to combine the strong, electromagnetic, and weak interactions in various **grand unification theories** known as **GUTs**. In one of these theories, leptons and quarks are considered to be two aspects of a single class of particles. Under certain conditions, a quark could change into a lepton and vice versa, even though this would appear to violate the conservation of lepton number and baryon number. One of the exciting predictions of this theory is that the proton is not stable but merely has a very long lifetime of the order of 10^{31} y. Such a long lifetime makes proton decay difficult to observe.

It was Einstein's dream to be able to describe all the forces in nature through one unified theory. Whether this will ever be accomplished is an open question. There is considerable experimental effort underway to observe the decay of the proton and to test other predictions of the grand unification theories. At the same time, much theoretical work is being done to refine these theories and construct others to further our understanding of our universe.

Summary

1. There are four basic interactions: strong, electromagnetic, weak, and gravitational. All particles with mass experience the force due to the gravitational interaction. All particles with electric charge experience the force due to the electromagnetic interaction. The "charge" associated with the weak interaction is called flavor. Quarks and leptons have flavor and experience the weak interaction. The "charge" associated with the strong interaction is called color. Quarks and gluons have color and experience the strong interaction. Hadrons (baryons and mesons) experience a residual strong interaction resulting from the fundamental strong interaction between the quarks that make up the hadrons.

2. In quantum field theory, each interaction is mediated by the exchange of one or more field particles. The field particle associated with the electromagnetic interaction is the photon, and that associated with the gravitational interaction is the graviton, which has not yet been observed. The field particles associated with the weak interaction are the W^+, W^-, and Z^0. The field particles associated with the strong interaction between quarks are called gluons. All the field particles are bosons. The graviton has spin 2, whereas all of the other field particles have spin 1. The residual strong interaction between hadrons is mediated by mesons, which are also bosons with spin 0 (or 1).

3. There are two families of fundamental particles, leptons and quarks, each containing six members. It is thought that these particles have no size and no internal structure. The leptons are spin-$\frac{1}{2}$ fermions: the electron e and its neutrino ν_e, the muon μ and its neutrino ν_μ, and the tau τ and its neutrino ν_τ. The electron, muon, and tau have mass, electric charge, and flavor but not color, so they participate in the gravitational, electromagnetic, and weak interactions but not the strong interaction. The neutrinos have flavor but no electric charge and no color. They may be massless, but whether or not this is so is still an open question. The quarks are also spin-$\frac{1}{2}$ fermions. The six quarks are called up u, down d, strange s, charmed c, top t, and bottom b. The top quark has not yet been observed. The quarks participate in all of the basic interactions. Because they are always confined in mesons or baryons, their masses can only be estimated.

4. Hadrons are composite particles that are made up of quarks. There are two types, baryons and mesons. Baryons, which include the neutron and proton, are fermions of half-integral spin consisting of three quarks. Mesons, which include pions and kaons, have zero or integral spin. Hadrons interact with each other via the residual strong interaction.

5. Some quantities, such as energy, momentum, electric charge, angular momentum, baryon number, and each of the three lepton numbers, are strictly conserved in all reactions and decays. Others, such as strangeness and charm, are conserved in reactions and decays that proceed via the strong interaction but not in those that proceed via the weak interaction.

6. Particles and their antiparticles have identical masses but opposite values for their other properties, such as charge, lepton number, baryon number, and strangeness. Particle–antiparticle pairs can be produced in various nuclear reactions if the energy available is greater than $2mc^2$, where m is the mass of the particle.

Suggestions for Further Reading

Bloom, Elliott D., and Gary J. Feldman: "Quarkonium," *Scientific American,* May 1982, p. 66.

How quark–antiquark pairs have been observed and are being investigated in order to learn more about the color force binding them together is discussed in this article.

Close, Frank, Michael Martin, and Christine Sutton: *The Particle Explosion,* Oxford University Press, Oxford, England, 1987.

This colorful book describes the discovery of the subatomic particles: the people involved and the instruments they used.

Glashow, Sheldon: "Tangled in Superstring: Some Thoughts on the Predicament Physics Is In," *The Sciences,* May/June 1988, p. 22.

A Nobel Prize–winning physicist raises doubts about a recent theoretical approach to particle physics.

Jackson, J. David, Maury Tigner, and Stanley Wojcicki: "The Superconducting Supercollider," *Scientific American*, March 1986, p. 66.

This article explains why a giant, new colliding-beam particle accelerator 84 km in circumference is needed to test possible extensions of the electroweak theory and new theories such as supersymmetry.

LoSecco, J. M., Frederick Reines, and Daniel Sinclair: "The Search for Proton Decay," *Scientific American*, June 1985, p. 54.

Experimenters who have taken part in the design and operation of the largest of the water Cerenkov proton-decay detectors describe the history of the search and the contribution of their experiment to it.

Quigg, Chris: "Elementary Particles and Forces," *Scientific American*, April 1985, p. 84.

This article provides an overview of leptons, hadrons, and quarks; the fundamental interactions; the unified theory of the electromagnetic and weak interactions; and possible further unification encompassing the strong force. New experimental directions such as the search for proton decay and the Superconducting Supercollider are also discussed.

Sutton, Christine: *The Particle Connection: The Most Exciting Scientific Chase Since DNA and the Double Helix*, Simon and Schuster, New York, 1984.

An excellent 175-page account of the search for the particles that mediate the electroweak interaction: W^+, W^-, and Z^0. The book requires as background only an introductory physics course.

von Baeyer, Hans Christian: "The Voltage Makers," *The Sciences*, January/February 1988, p. 6.

An alternative to the Superconducting Supercollider (SSC), the "beam transformer," may allow further progress in particle physics.

von Baeyer, Hans Christian: "The Atomic Cathedral," *The Sciences*, January/February 1987, p. 8.

Some reflections on the medieval and the modern quest for knowledge, inspired by similarities in the principles of design of cathedrals and particle accelerators.

Weinberg, Steven: "The Decay of the Proton," *Scientific American*, June 1981, p. 64.

One of the three who shared the 1979 Nobel Prize in physics for the theory unifying the electromagnetic and weak interactions describes why attempts at unification of these two forces with the strong force lead to predictions of an extremely small but finite rate of decay for the proton.

Review

A. Objectives: After studying this chapter, you should:

   1. Be able to list the four basic interactions and name some of the particles that participate in each interaction.

   2. Be able to name the particle or particles that mediate each of the basic interactions.

   3. Be able to discuss the difference between hadrons and leptons and between baryons and mesons.

   4. Be able to discuss the quark model of hadrons.

   5. Be able to apply the various conservation laws to tell whether a given decay or reaction can proceed via the strong interaction, the weak interaction, or not at all.

B. Define, explain, or otherwise identify:

   Strong interaction
   Electromagnetic interaction
   Gravitational interaction
   Weak interaction
   Hadron
   Baryon
   Meson
   Quantum electrodynamics (QED)
   Virtual photon
   Pion
   Lepton
   Spin-$\frac{1}{2}$ particles
   Fermion
   Boson
   Baryon number
   Lepton number
   Strangeness
   Strange particle
   Quark
   Flavor
   Charm
   ψ/J particle
   ψ particle
   Y meson
   Bottomonium
   Graviton
   W particle
   Z particle
   Gluon
   Color charge
   Quantum chromo dynamics (QCD)
   Electroweak theory
   Higgs field
   Higgs boson
   Standard model
   Grand Unification theories (GUTs)

C. True or false: If the statement is true, explain why it is true. If it is false, give a counterexample.

   1. Leptons are fermions.

   2. All baryons are hadrons.

   3. All hadrons are baryons.

   4. Mesons are spin-$\frac{1}{2}$ particles.

   5. Leptons consist of three quarks.

   6. The times for decays via the weak interaction are typically much longer than those for decays via the strong interaction.

   7. The electron interacts with the proton via the strong interaction.

   8. Strangeness is not conserved in weak interactions.

   9. Neutrons have no charm.

Problems

In the following problems, use Figure 8-2 to find the strangeness of hadrons.

Level I

8-1 Hadrons and Leptons

1. Suppose the force between two nucleons were mediated by a kaon. What would be the approximate range of the force?

8-2 Spin and Antiparticles

2. Two pions at rest annihilate according to the reaction $\pi^+ + \pi^- \to \gamma + \gamma$. (a) Why must the energies of the two gamma rays be equal? (b) Find the energy of each gamma ray. (c) Find the wavelength of each gamma ray.

3. Find the minimum energy of the photon needed for the following pair-production reactions: (a) $\gamma \to \pi^+ + \pi^-$, (b) $\gamma \to p + p^-$, and (c) $\gamma \to \mu^- + \mu^+$.

8-3 The Conservation Laws

4. State which of the decays or reactions that follow violate one or more of the conservation laws, and give the law or laws violated in each case: (a) $p^+ \to n + e^+ + \bar{\nu}_e$, (b) $n \to p^+ + \pi^-$, (c) $e^+ + e^- \to \gamma$, (d) $p + p^- \to \gamma + \gamma$, and (e) $\nu_e + p \to n + e^+$.

5. Determine the change in strangeness in each reaction that follows, and state whether the reaction can proceed via the strong interaction, the weak interaction, or not at all: (a) $\Omega^- \to \Xi^0 + \pi^-$, (b) $\Xi^0 \to p + \pi^- + \pi^0$, and (c) $\Lambda^0 \to p^+ + \pi^-$.

6. Determine the change in strangeness for each decay, and state whether the decay can proceed via the strong interaction, the weak interaction, or not at all: (a) $\Omega^- \to \Lambda^0 + K^-$ and (b) $\Xi^0 \to p + \pi^-$.

7. Determine the change in strangeness for each decay, and state whether the decay can proceed via the strong interaction, the weak interaction, or not at all: (a) $\Omega^- \to \Lambda^0 + \bar{\nu}_e + e^-$ and (b) $\Sigma^+ \to p + \pi^0$.

8. (a) Which of the following decays of the tau particle is possible?

$$\tau \to \mu^- + \bar{\nu}_\mu + \nu_\tau$$

$$\tau \to \mu^- + \nu_\mu + \bar{\nu}_\tau$$

(b) Calculate the kinetic energy of the decay products for the decay that is possible.

8-4 The Quark Model

9. Find the baryon number, charge, and strangeness for the following quark combinations and identify the hadron: (a) uud, (b) udd, (c) uus, (d) dds, (e) uss, and (f) dss.

10. Repeat Problem 9 for the following quark combinations: (a) $u\bar{d}$, (b) $\bar{u}d$, (c) $u\bar{s}$, and (d) $\bar{u}s$.

11. The Δ^{++} particle is a baryon that decays via the strong interaction. Its strangeness, charm, topness, and bottomness are all zero. What combination of quarks gives a particle with these properties?

12. Find a possible combination of quarks that gives the correct values for electric charge, baryon number, strangeness for (a) K^+ and (b) K^0.

13. The D^+ meson has no strangeness, but it has charm of $+1$. (a) What is a possible quark combination that will give the correct properties for this particle? (b) Repeat (a) for the D^- meson, which is the antiparticle of the D^+.

14. Find a possible combination of quarks that gives the correct values for electric charge, baryon number, strangeness for (a) K^- (the K^- is the antiparticle of the K^+) and (b) $\bar{K}^0$.

8-5 Field Particles

15. What is the approximate range of a weak interaction that is mediated by the W^+ particle?

16. The grand unification theory predicts an X particle that would change a quark into a lepton and vice versa. What would be the approximate range of a virtual X particle if its mass were 10^{15} GeV/c^2?

8-6 The Electroweak Theory

There are no problems for this section.

8-7 The Standard Model

17. (a) What conditions are necessary for a particle and its antiparticle to be the same? Find the antiparticle for (b) π^0 and (c) Ξ^0.

8-8 Grand Unification Theories

There are no problems for this section.

Level II

18. Find a possible quark combination for the following particles: (a) Λ^0, (b) p^-, and (c) Σ^-.

19. Find a possible quark combination for the following particles: (a) $\bar{n}$, (b) Ξ^0, and (c) Σ^+.

20. Find a possible quark combination for the following particles: (a) Ω^- and (b) Ξ^-.

21. State the properties of the particles made up of the following quarks: (a) ddd, (b) $u\bar{c}$, (c) $u\bar{b}$, and (d) $\bar{s}\bar{s}\bar{s}$.

22. Consider the following decay chain:

$$\Xi^0 \to \Lambda^0 + \pi^0$$
$$\Lambda^0 \to p + \pi^-$$
$$\pi^0 \to \gamma + \gamma$$
$$\pi^- \to \mu^- + \bar{\nu}_\mu$$
$$\mu^- \to e^- + \bar{\nu}_e + \nu_\mu$$

(a) Are all the final products shown stable? If not, finish the decay chain. (b) Write the overall decay reaction for Ξ^0 to the final products. (c) Check the overall decay reaction for the conservation of electric charge, baryon number, lepton number, and strangeness. (d) In the first step of the chain, could the Λ^0 have been a Σ^0?

23. Consider the following decay chain:

$$\Omega^- \to \Xi^0 + \pi^-$$
$$\Xi^0 \to \Sigma^+ + e^- + \bar{\nu}_e$$
$$\pi^- \to \mu^- + \bar{\nu}_\mu$$
$$\Sigma^+ \to n + \pi^+$$
$$\pi^+ \to \mu^+ + \nu_\mu$$
$$\mu^+ \to e^+ + \bar{\nu}_\mu + \nu_e$$
$$\mu^- \to e^- + \bar{\nu}_e + \nu_\mu$$

(a) Are all the final products shown stable? If not, finish the decay chain. (b) Write the overall decay reaction for Ω^- to the final products. (c) Check the overall decay reaction for the conservation of electric charge, baryon number, lepton number, and strangeness.

24. Test the following decays for violation of the conservation of energy, electric charge, baryon number, and lepton number:
(a) $n \to \pi^+ + \pi^- + \mu^+ + \mu^-$
(b) $\pi^0 \to e^+ + e^- + \gamma$
Assume that linear and angular momentum are conserved. State which conservation laws (if any) are violated in each decay.

25. Test the following decays for violation of the conservation of energy, electric charge, baryon number, and lepton number:
(a) $\Lambda^0 \to p + \pi^-$
(b) $\Sigma^- \to n + p^-$
(c) $\mu^- \to e^- + \bar{\nu}_e + \nu_\mu$
Assume that linear and angular momentum are conserved. State which conservation laws (if any) are violated in each decay.

Level III

26. (a) Calculate the total kinetic energy of the decay products for the decay

$$\Lambda^0 \to p + \pi^-$$

Assume the Λ^0 is initially at rest. (b) Find the ratio of the kinetic energy of the pion to the kinetic energy of the proton. (c) Find the kinetic energies of the proton and the pion for this decay.

27. A Σ^0 particle at rest decays into a Λ^0 plus a photon. (a) What is the total energy of the decay products? (b) Assuming that the kinetic energy of the Λ^0 is negligible compared with the energy of the photon, calculate the approximate momentum of the photon. (c) Use your result for (b) to calculate the kinetic energy of the Λ^0. (d) Use your result for (c) to obtain a better estimate of the momentum and the energy of the photon.

28. In this problem, you will calculate the difference in the time of arrival of two neutrinos of different energy from a supernova that is 170,000 light-years away. Let the energies of the neutrinos be $E_1 = 20$ MeV and $E_2 = 5$ MeV, and assume that the rest mass of a neutrino is 20 eV/c^2. Because their total energy is so much greater than their rest energy, the neutrinos have speeds that are very nearly equal to c and energies that are approximately $E \approx pc$. (a) If t_1 and t_2 are the times it takes for neutrinos of speeds u_1 and u_2 to travel a distance x, show that

$$\Delta t = t_2 - t_1 = x\frac{u_1 - u_2}{u_1 u_2} \approx \frac{x\,\Delta u}{c^2}$$

(b) The speed of a neutrino of rest mass m_0 and total energy E can be found from Equation 1-32. Show that when $E \gg m_0 c^2$, the speed u is given approximately by

$$\frac{u}{c} \approx 1 - \frac{1}{2}\left(\frac{m_0 c^2}{E}\right)^2$$

(c) Use the results for (b) to calculate $u_1 - u_2$ for the energies and rest mass given, and calculate Δt from the result for (a) for $x = 170{,}000$ c·y. (d) Repeat the calculation in (c) using $m_0 c^2 = 40$ eV for the rest energy of a neutrino.

Chapter 9

Astrophysics and Cosmology

An optical image of galaxy NGC 5128, over which has been superposed a radio image of Centaurus A, a radio source within the galaxy. The optical appearance of the galaxy is very peculiar in that it shows a heavy dust zone surrounding the system in a plane perpendicular to the galaxy's rotation. The intense double radio lobes of Centaurus A extend across 0.12° of the sky, corresponding to a distance of 54,800 $c \cdot y$. The core, or nucleus, of the galaxy is also a source of radio emissions. Not shown are much fainter and larger radio lobes that surround the intense lobes and extend across 10° of the sky. (The full moon covers only 0.5° of the sky.) The thin cone-shaped blue and red emission extending from the nucleus to the northern lobe is a jet of high-velocity gas 15,000 $c \cdot y$ in length. The faint stars surrounding the galaxy lie in shells that computer analyses associate with collisions of galaxies and suggest that Centaurus A is the result of such a collision that occurred about one billion years ago.

Physics is an experimental science. The formulation and acceptance of our current understanding of the physical world, from Newton's laws through Maxwell's equations to relativity theory and quantum mechanics, are based on countless experimental observations. In this chapter, we look outward from the earth into the cosmos and apply the principles and techniques of physics first to the composition and evolution of stars, a branch of physics called **astrophysics,** and then to the large-scale structure of the universe, a field called **cosmology.**

When observing stars and galaxies, astrophysicists and cosmologists are limited to examining the electromagnetic radiation and occasional particles emitted at times past that happen to have traveled to the vicinity of the earth, arriving at the moment of observation. The information thus gained, together with the fundamental assumption that the laws of physics discovered here on the earth are also valid throughout the universe, forms the basis for their work. During most of history, the instrument used for studying the cosmos has been the human eye. Though well adapted to life on earth, the eye is a relatively poor instrument for the scientific examination of the sky because it stores information for only a small fraction of a second before that information is transmitted to the brain for analysis. Today, most of our information about the distant universe is received through telescopes.

9-1 Our Star, the Sun

As we look outward from the earth and beyond the moon, the most obvious object in the sky is, of course, the sun. It is important to us for several reasons: The light that reaches us from the sun is responsible for life on the earth. It sustains a comfortable average temperature on the earth's surface and is the ultimate source of virtually all of our energy. Since the sun contains nearly all of the mass of the solar system, it also provides the gravitational force that binds our planet to the system. But most important for our purposes in this section, the sun is the only star of the 100 billion or so in the Milky Way galaxy that is close enough for us to examine its surface. All of the others are so far away that they appear only as point sources when viewed by even the largest telescopes. What we learn from studies of our star not only provides us with a more complete understanding of the processes taking place on it but surely applies to other stars as well.

The Surface and Atmosphere of the Sun

We can see only the thin layer of the sun, the **photosphere,** which emits the light that makes the sun visible. The photosphere is generally considered to be the surface of the sun. The emitted energy per second per square meter that arrives from the sun at the top of the earth's atmosphere is called the **solar constant** f. It has been measured to be

$$f = 1.36 \times 10^3 \text{ W/m}^2 \qquad 9\text{-}1$$

This quantity for stars other than the sun is called the apparent brightness, as we will see in Section 9-3. Using the solar constant, the earth–sun distance of 1 AU = 1.5×10^8 km, and the conservation of energy, we can calculate the **luminosity** L, which is the total power radiated by the sun or by any star. The area A of a sphere with a radius of 1 AU is

$$A = 4\pi r^2 = 4\pi (1.50 \times 10^{11} \text{ m})^2$$

At that radius, each square meter receives energy from the sun at the rate given by the solar constant. Therefore, the sun's luminosity $L_\odot$ is given by

$$L_\odot = Af = 4\pi(1.50 \times 10^{11} \text{ m})^2 (1.36 \times 10^3 \text{ W/m}^2) = 3.85 \times 10^{26} \text{ W} \qquad 9\text{-}2$$

This is truly enormous power. If we could put a 1000-MW electricity-generating plant on each square meter of the earth's surface, all of them combined would produce only 0.1 percent of the power produced by the sun.

If we assume that the sun radiates as a blackbody, we can use the luminosity of the sun along with its radius (6.96×10^8 m) to calculate the effective temperature at the surface of the sun from the Stefan–Boltzmann law. It states that the intensity I (the power per unit area) radiated by a blackbody in thermal equilibrium is proportional to the fourth power of its surface temperature:

$$I = \sigma T^4 \qquad 9\text{-}3$$

where $\sigma = 5.67 \times 10^{-8}$ W/m^2·K^4 and T is the absolute temperature. If the radius of the sun is $R_\odot$, the intensity radiated at the surface of the sun is

$$I = \frac{L_\odot}{4\pi R_\odot^2} \qquad 9\text{-}4$$

The effective temperature T_e of the surface of the sun is defined as the temperature for which the intensity radiated satisfies the Stefan–Boltzmann law for a blackbody:

$$I = \frac{L_\odot}{4\pi R_\odot^2} = \sigma T_e^4$$

Solving for T_e, we obtain

$$T_e = \left(\frac{I}{\sigma}\right)^{1/4} = \left(\frac{L_\odot}{4\pi R_\odot^2 \sigma}\right)^{1/4} \qquad 9\text{-}5$$

Example 9-1

Use the Stefan–Boltzmann law to calculate the effective temperature of the photosphere.

Using $L_\odot = 3.85 \times 10^{26}$ W in Equation 9-5, we have

$$T_e = \left(\frac{L_\odot}{4\pi R_\odot^2 \sigma}\right)^{1/4} = \left[\frac{3.85 \times 10^{26}\text{ W}}{4\pi (6.96 \times 10^8\text{ m})^2 (5.67 \times 10^{-8}\text{ W/m}^2\cdot\text{K}^4)}\right]^{1/4}$$

$$= 5800\text{ K}$$

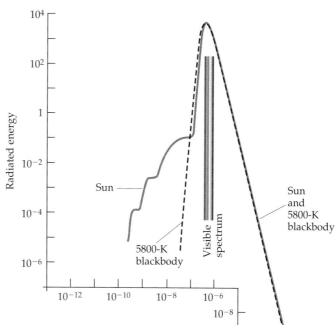

Figure 9-1 The spectral distribution of the energy emitted by the sun closely matches that of a blackbody at 5800 K. The discrepancies are mainly due to the fact that the photosphere is not in thermal equilibrium. The hump at short wavelengths is due to x rays emitted by the corona, which is at a much higher temperature.

The intensity of solar radiation has been measured at wavelengths ranging from about 10^{-13} m in the gamma-ray region to nearly 10 m in the radio region, a range accounting for over 99 percent of the sun's emitted power. Over much of this span, the solar spectrum is quite well predicted by Planck's law of blackbody radiation (see Chapter 2) with $T = 5800$ K as shown in Figure 9-1. The distribution peaks in the yellow region of the visible wavelengths, which is of course why the sun looks yellow. This agreement between the measured and theoretical spectra is very constant and is one of the characteristics of the **quiet sun.**

If we examine the edge of the solar disc, called the **limb**, we see that it is sharply demarcated and darker than the rest of the sun. From the sharpness of the limb, we conclude that the photosphere is very thin from the following reasoning: Atmospheric turbulence during daylight limits the angular resolution of optical telescopes to about 1 arc second (1/3600 of a degree) or more. At the distance of the sun, this corresponds to about 700 km. As we look at the sun, the angle over which the gas of the photosphere changes from rarefied and transparent to optically dense and opaque is smaller than we can resolve. Therefore, the photosphere must be less than 700 km thick, which is only about 0.1 percent of the solar radius.

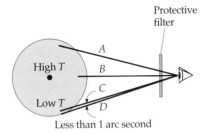

Figure 9-2 Because the photosphere is more transparent when viewed at normal incidence than when viewed at a grazing angle, the light traveling along path B originates deeper in the sun than the light traveling along path A and therefore looks brighter. The limb looks darker. The change in brightness from path C to path D is smaller than we can resolve, so the limb looks sharp.

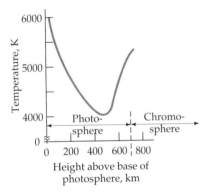

Figure 9-3 The temperature of the sun decreases from the base of the photosphere outward to a minimum at about 500 km. It then increases sharply to an average of about 15,000 K in the chromosphere.

The relatively dark appearance of the limb tells us about the temperature gradient in the sun's atmosphere. Figure 9-2 shows two paths A and B for viewing the sun. Because the photosphere is more transparent when viewed at normal incidence than when viewed at a grazing angle, the light traveling along path B originates deeper in the sun than light traveling along path A. Since the interior is hotter than the outer layers, the light traveling along path B originates in a hotter (brighter) part of the sun than the light traveling path A. Thus, the light from the limb appears darker (cooler). By measuring the change in brightness from path A to path B, we can determine the temperature gradient in the photosphere. It is shown in the left portion of Figure 9-3. Notice in the right portion of Figure 9-3 that the temperature begins to rise sharply, accompanying the transition from the sun's surface, the photosphere, into the solar atmosphere.

Outside the photosphere are two layers of the sun's atmosphere that are not generally seen because of the brightness of the photosphere. The innermost of the two layers of the solar atmosphere, the **chromosphere,** is visible for the first few seconds of totality during a solar eclipse. Under high resolution, the chromosphere resembles a field of burning grass, although each burning "blade" is about 700 km thick and 7000 to 10,000 km high and lasts for only 5 to 15 minutes. Spectral examination indicates that the temperature of the chromosphere *increases* with distance above the photosphere, averaging about 15,000 K.

When the totality of the eclipse blocks out the chromosphere, the outer layer of the sun's atmosphere, the **corona,** becomes visible. It is decidedly nonuniform in thickness, consisting of faint white streamers that extend two to three solar diameters into space, as shown in Figure 9-4. The temperature of the corona is approximately 2,000,000 K. Radiation from the corona would overpower that from the 5800-K photosphere, except for the fact that the gas of the corona is so rarefied that the total energy it emits is miniscule compared to that emitted by the photosphere. It does, however, account for the relatively high intensity of x rays emitted by the sun, which show up in Figure 9-1 as a deviation from the spectral distribution of the blackbody at short wavelengths. It is thought that the extreme temperatures in the corona are produced by acoustic waves generated in the sun's interior that build into shock waves in the corona. These shock waves heat the gases of the outer atmosphere and give the particles so much energy that even the sun's intense gravity cannot confine them. These high-energy particles, mostly electrons and protons, stream outward from the corona continuously. They form the **solar wind** that pervades the entire solar system.

Figure 9-4 The hot but rarefied corona becomes visible during a total solar eclipse.

The Sun's Interior

We cannot see through the photosphere into the interior of the sun. Consequently, our understanding of the processes that occur there is purely theoretical. With the single exception of solar neutrinos, no radiation or particles originating in the interior reach us directly. (As we will see, the single bit of direct observational evidence we do receive is at odds with the current theory.)

For simplicity, theoretical models usually consider the sun to be a nonrotating star in hydrostatic equilibrium. This means that the outward pressure at any point, which is presumed to be due to the energy-conversion processes occurring within the sun, is exactly balanced by the inward pressure of gravity. Although the mean density of the sun (1.4 g/cm³) is not much different from that of the earth (5.5 g/cm³), the enormous pressures that exist in the solar interior substantially exceed those that correspond to the electrodynamic forces that bind electrons to nuclei. Thus, the matter in the interior of the sun—and certainly that within the **core,** the central region in which temperatures are high enough to allow hydrogen fusion—must surely be in the plasma (ionized) state.

Example 9-2

Show that neutral hydrogen is unlikely to exist in the sun's interior.

The pressure at the center of the sun P_c is of the order $P_c = \mu g$, where μ = mass/unit surface area $\approx M_\odot/R_\odot^2$ and $g = \frac{1}{2}GM_\odot/R_\odot^2$ is the average acceleration of gravity in the sun. The pressure turns out to be about 10^{16} N/m². This is the pressure pushing on the surface of a hydrogen atom near the sun's center. The resistance to this gravitational pressure would come from the Coulomb force tending to hold the atom together. That pressure is given by the Coulomb attraction between the proton and electron per unit surface area of the atom. Using the Bohr radius a_0 for hydrogen, we have

$$\frac{F}{A} = \frac{ke^2/a_0^2}{4\pi a_0^2} = \frac{ke^2}{4\pi a_0^4} = \frac{(9 \times 10^9)(1.6 \times 10^{-19})^2}{4\pi(0.5 \times 10^{-10})^4}$$

$$= 2.9 \times 10^{12} \text{ N/m}^2$$

Thus, the gravitational pressure in the sun's interior, at least near the center, exceeds that tending to hold the hydrogen atoms together by about a factor of 10,000—making it unlikely that neutral hydrogen atoms could exist there.

However, given the sun's density, the particles even in the depths of the core are still relatively far apart, so the plasma behaves much like an ideal gas. This allows us to calculate the core temperature from the ideal-gas law. It is found to be 1.5×10^7 K.

The Source of the Sun's Energy

Using the value for the luminosity of the sun that we computed earlier, the present energy content of the sun, as calculated from thermodynamics, would be radiated away in about 3×10^7 y. Since life has existed on earth for approximately a hundred times that long, we can conclude that the sun has been radiating at close to its present luminosity for at least 3×10^9 y. Therefore, the sun must have a supply of energy far larger than that represented by the hot plasma and the observed radiation field.

Figure 9-5 The proton–proton cycle is the primary source of the sun's energy. The neutrino created in the initial reaction escapes from the core. The net energy produced in the cycle is about 26.7 MeV.

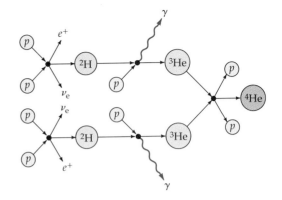

The source of the sun's energy is nuclear fusion. Current theory proposes that as the young sun contracted its temperature rose. Eventually, the temperature of the core reached about 1.5×10^7 K, which is high enough for the hydrogen nuclei (protons) in the plasma to have sufficient energy on the average (about 1 keV) to fuse into helium nuclei. This reaction, actually a chain of reactions, was first proposed by H. A. Bethe in 1938 and is referred to as the **proton–proton cycle** (see Figure 9-5). The first reaction in this chain is

$$^{1}\text{H} + {}^{1}\text{H} \longrightarrow {}^{2}\text{H} + e^{+} + \nu_{e} + 1.44 \text{ MeV} \qquad 9\text{-}6$$

The probability of this reaction occurring is very low except for those protons in the high-energy tail of the Maxwell–Boltzmann distribution. This sets a limit on the rate at which the sun can produce energy and thus ensures a long lifetime for the sun and similar stars. This limit is sometimes called the "bottleneck" in the solar-fusion cycle. Once ^{2}H (deuterium) is formed via Reaction 9-6, the following reaction becomes probable:

$$^{2}\text{H} + {}^{1}\text{H} \longrightarrow {}^{3}\text{He} + \gamma + 5.49 \text{ MeV} \qquad 9\text{-}7$$

It is followed by

$$^{3}\text{He} + {}^{3}\text{He} \longrightarrow {}^{4}\text{He} + 2\,{}^{1}\text{H} + \gamma + 12.86 \text{ MeV} \qquad 9\text{-}8$$

This process by which hydrogen nuclei are "burned" to helium nuclei is shown schematically in Figure 9-5. There are other possible reactions for converting ^{3}He to ^{4}He, all of which have the same net Q-value. Their rates, however, differ depending on the composition and temperature of the interior.

The neutrinos produced in the proton–proton cycle escape from the core, providing our only means for direct observation of the sun's interior. The measured luminosity $L_\odot$ and the known total Q-value of the proton–proton cycle enable a calculation of the total reaction rate. In addition, the alternative reactions for ^{3}He have different neutrino energy spectra, thus providing a way of determining the relative contributions of each reaction and gaining information about the core's composition and temperature. However, the measured rate at which solar neutrinos arrive at the earth is less than half that predicted by theoretical calculations based on the standard solar model. This discrepancy is referred to as the **solar-neutrino problem**.

The as-yet-unresolved solar-neutrino problem has several implications, two of which are particularly important for our purposes. First, there may be a serious gap in our understanding of the properties and behavior of neutrinos. Second, if our theoretical understanding of neutrinos is essentially accurate, then there is a serious error in the current standard solar model. Such an error would have far-reaching ramifications for theories of stellar evolution. For example, Stephen Hawking suggests the possibility that part of the sun's emission of energy may arise from gravitational energy that is released when mass falls into a small black hole at the sun's center. This means that there would be less fusion occurring than current theory suggests and, hence, fewer neutrinos.

The Active Sun

In addition to the relatively stable phenomena that we have discussed, the sun exhibits a number of transient phenomena, most of them associated with its magnetism. We noted earlier that the solar interior must be primarily a plasma composed of protons and electrons. The sun rotates with different angular velocities at different latitudes. At any given latitude, it probably has different angular velocities at different distances from the spin axis as well. The complex motions resulting from this differential rotation and from the rise and fall of charged particles in the convection zone between the core and the photosphere are probably the source of the sun's chaotic magnetic-field structure (see Figure 9-6). This transient structure may have localized magnetic-field strengths exceeding 1 T on occasion.

The transient structure is superimposed onto a general average magnetic field of about 10^{-4} T. The origin of this general field is not known, except that it is not a remnant from the sun's formation since any primeval field would have decayed away by now. Its presence poses formidable problems for any theoretical solar model. Not only must the model explain the origin of the general field, but it must also account for the fact that its polarity reverses every 11 years, in step with the **sunspot cycle.**

Sunspots, dark blemishes on the solar disc, were first reported in pretelescope times and were observed with a telescope by Galileo in 1610. They originate in the following way, according to one of the current models: As shown in Figure 9-7, the sun's magnetic-field lines are distorted into bundles or tubes by the sun's differential rotation. Occasionally, vertical movements in the convection zone may push a bundle through the surface. The area where it leaves the surface and the area where it returns to the surface become the sunspots. They appear darker than the adjacent photosphere, which means that they are cooler, typically around 3800 K. One of the pair of spots will have a magnetic north pole and the other will have a south pole. If the bundle of field lines doesn't protrude completely through the photosphere, only a single sunspot is formed.

The number of sunspots per year varies regularly from about 50 to about 150 in a cycle of 11 years, as can be seen in Figure 9-8. Early in each new cycle, the sunspots form at a latitude of about 30°. As the sun progresses through its 11-year cycle, the spots form progressively closer to the equator. There is an additional cyclical variation in the annual number of sunspots with a period of about 100 years that is also apparent in Figure 9-8. Currently, there is no theoretical explanation for these regularities.

Solar flares are violent, storm-like phenomena that appear to be associated with the large magnetic fields in the vicinity of sunspots. A spectacular flare is shown in the photo at the top of page 304. There is, however, no

Figure 9-6 The magnetic-field lines of the sun show a chaotic structure.

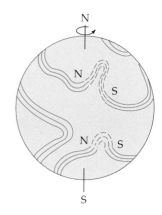

Figure 9-7 The magnetic-field lines of the sun are distorted by the differential rotation of the sun. Sunspots occur where a bundle of field lines leaves and reenters the surface.

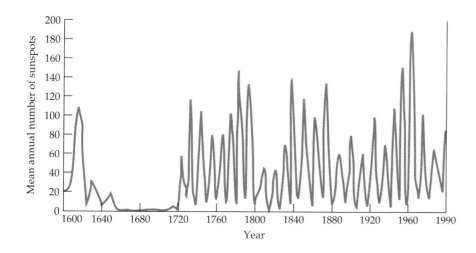

Figure 9-8 Number of sunspots versus year. This number has varied regularly on an 11-year cycle for more than 270 years. The unexplained absence of sunspots between about 1650 and 1700, referred to as the Maunder minimum, coincides approximately with a period of unusually low temperatures in Europe known as the Little Ice Age.

The solar flare shown on the left of this Skylab-4 photograph consists of charged particles confined by the magnetic field of the sun.

generally accepted model to explain them. Solar flares erupt explosively, ejecting particles and emitting radiation ranging from the x-ray through the radio wavelengths of the spectrum. They last anywhere from a few minutes to a few hours and can have temperatures as high as 5×10^6 K. The particles ejected by solar flares reach the earth within a day or so and often produce auroras as they interact with the earth's magnetic field. Solar flares can disrupt some types of radio transmissions and, on rare occasions, can generate surges in high-voltage transmission lines.

Two other transient solar phenomena are plages and filaments. **Plages** are bright (hotter) areas adjacent to the dark sunspots. The evolution of plages suggest that they are areas of increased mass density, resulting perhaps from the movement of the magnetic field bundles generating from the sunspots. **Filaments** are dark, thin lines that thread their way across the disc, sometimes for thousands of kilometers. They do not lie on the surface but extend out into space, sometimes more than 100,000 kilometers, in graceful loops and swirls. Filaments that are seen projecting into space at the sun's edge are called **prominences.** They may erupt and disappear quickly or persist for several weeks. Although prominences appear to be closely related to the shape of the magnetic field, as with other transient features, there is no model that fully accounts for them.

9-2 The Stars

On clear, dark nights, we can see about 6000 stars without the aid of a telescope. The sight is incredibly beautiful and must surely have been just as awesome to our forbearers as it is to us. A cursory glance at the night sky reveals the following features: The distribution of stars is not uniform, the stars do not all have the same brightness, and there is a dim, irregular band of light bisecting the sky. In this section, we will investigate these features.

The hazy band of light that stretches across the entire sky is the Milky Way. With the aid of a small telescope or even binoculars, the band is resolved into a mass of individual stars. It is part of a huge galaxy containing an estimated 10^{11} stars that are bound together gravitationally in our region of the universe. (The term *galaxy* is derived from the Greek word for "milk.") Most of the stars visible to the unaided eye seen in any direction are simply those members of the Milky Way galaxy that are close enough to the earth to be individually resolved by the eye.

Constellations

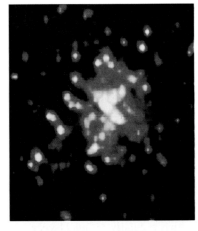

A false-color image of Doradus 30, a star cluster in the Large Magellanic Cloud. The cluster was imaged by the Hubble space telescope, then computer-processed to remove a haze attributed to the telescope's spherical aberration.

Chance groupings in the celestial pattern, usually among the brighter individual stars, are called **constellations.** They were associated by ancient peoples with persons, gods, and objects from their histories, religions, and myths, probably as mnemonic devices. The constellations, as well as several prominent stars, have always had practical uses. For centuries, seafarers have used the Pole Star (in the northern hemisphere) and the Southern Cross (in the southern hemisphere) as aids in navigation. In ancient Egypt, the pharaoh's advisors learned to predict the life-sustaining annual flooding of the Nile by watching for the first appearance of the bright star of Sirius above the horizon in the early spring. Today, eighty-eight constellations (see Figure 9-9 for some of them) are used by astronomers to identify sections of the sky. For example, the center of the Milky Way galaxy is said to be "in Sagittarius," meaning that it is in the direction of the constellation Sagittarius. (The center of the Galaxy is actually more than ten times farther from the sun than are the stars that form the constellation.)

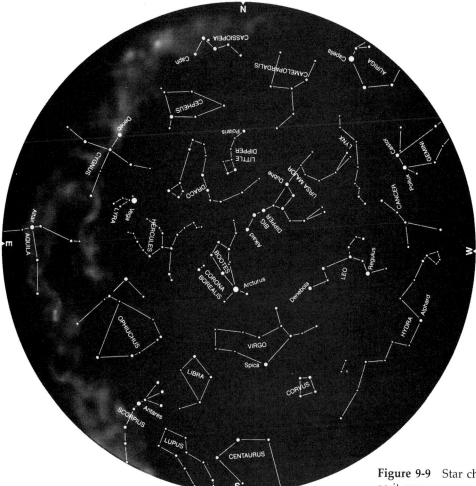

Figure 9-9 Star chart of the sky as it appears on a spring evening at latitude 40° north, showing many of the constellations visible. During the night, the entire pattern revolves about 120° about the Pole Star. To use the chart, hold it (or a copy) in front of you with the S (south) at the bottom while you face south. Match the lower half to the stars that you see. Then rotate the chart, putting the W at the bottom, face west, and again match the lower half to the stars you see, and so on.

Stellar Populations

One characteristic of our Galaxy is that certain regions of it have many more stars than other nearby regions. Such concentrations are called **star clusters**. Clusters are groups of stars that collectively have a small angular diameter, implying that they are all at about the same distance from us. There are two types of star clusters: galactic clusters and globular clusters. **Galactic clusters,** also called **open clusters,** may contain from about 20 to several hundred stars. One such cluster, photographed by the Hubble space telescope, is shown on page 304. All of them appear to have very similar compositions, as inferred from studies of their optical spectra. About 70 percent of their mass is hydrogen, another 28 percent or so is helium, and 2 to 3 percent consists of elements heavier than helium. Stars with this characteristic composition, like our sun, are referred to as **population I stars**. **Globular clusters** may consist of 10^3 to 10^6 stars in a compact, roughly spherical group. Their concentrations of elements heavier than helium are all very similar and are much lower than those of population I stars, typically 0.1 to 0.01 percent. These are called **population II stars.**

Population I stars are thought to be current-generation stars that formed after the gas and dust that exist between these stars had been enriched by the products of ancient fusion reactions in the early universe. The lower concentrations of heavier elements in the population II stars suggest that they are of a previous generation, hence older than those of population I. The fact that they are found in regions of space where there is little dust or gas tends to support this interpretation.

Figure 9-10 A hand drawing of the Milky Way from the perspective of a viewer at the sun.

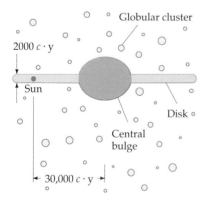

Figure 9-11 A diagram of the presently accepted structure of the Milky Way galaxy based on the work of Harlow Shapley.

The Structure of the Milky Way Galaxy

Figure 9-10 is a map of the Milky Way viewed from the location of the sun. The size and shape of the Galaxy are not at all obvious—hardly surprising from the perspective of an observer inside the Galaxy itself. However, painstaking counts of the number of stars per unit volume in various directions have revealed that the Milky Way galaxy is basically a huge disk. Up until the early 1900s, astronomers thought the sun was located at the disk's center. The true size and shape of the Galaxy (see Figure 9-11) were deduced by Harlow Shapley in 1917 through a brilliant analysis of the distribution of globular clusters. He discovered that 200 or so globular clusters are distributed approximately spherically in space and proposed that the center of that distribution coincided with the center of our Galaxy. The center lies about 30,000 light-years from the sun. It has been said that Shapley dethroned the sun from the center of the Galaxy much as Copernicus had dethroned the earth from the center of the universe.

Following Shapley's work, astronomers studying nearby galaxies with the aid of new, high-resolution telescopes found that the distribution of stars within those systems, many of which have open-spiral structures like Centaurus shown in Figure 9-19b, depends in part upon the ages and compositions of the stars, with open clusters being found mainly in the arms of the spirals. Making the reasonable assumption that such distribution patterns would also hold for the Milky Way galaxy, meticulous measurements of the distances to about 200 open clusters enabled the identification of three spiral arms for the Milky Way galaxy. Thus, if we could look down on the Milky Way galaxy from the galactic north pole, it would look much like Figure 9-12a.

(a)

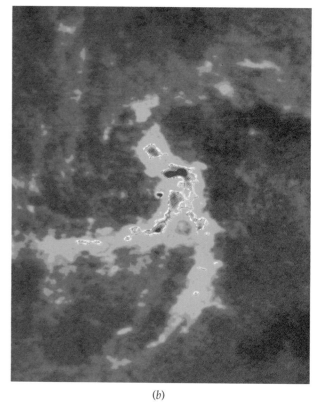
(b)

The Mass (and Missing Mass) of the Milky Way Galaxy

Using the doppler effect, J. Oort and B. Lindblad first demonstrated in 1926 that the Galaxy is rotating. The sun is apparently moving in a circular orbit at a speed of about 2.5×10^5 m/s toward the constellation Cygnus. Assuming that the sun's speed is constant, we can compute the length of the galactic year, that is, the time for the Galaxy to complete one revolution, and the mass of the Galaxy. Since the sun is 30,000 $c \cdot y$ from the galactic center, the galactic year is 2.3×10^8 years (see Problem 3).

Example 9-3

Estimate the mass of that part of the Galaxy that lies inside the sun's orbit from Newton's law of gravity and the sun's orbital speed.

Setting the gravitational force on the sun equal to the mass of the sun $M_\odot$ times its centripetal acceleration, we obtain

$$\frac{GM_\odot M_G}{R^2} = \frac{M_\odot v^2}{R} \qquad 9\text{-}9$$

where M_G is the mass of the Galaxy, v is the sun's orbital speed, R is the distance from the sun to the galactic center, and G is the gravitational constant. Solving for M_G, we obtain

$$M_G = \frac{v^2 R}{G} = \frac{(2.5 \times 10^5 \text{ m/s})^2 (30{,}000 \; c \cdot y)}{6.67 \times 10^{-11} \text{ N} \cdot \text{m}^2/\text{kg}^2}$$
$$= 2.66 \times 10^{41} \text{ kg} = (1.3 \times 10^{11}) \, M_\odot$$

Thus, if the sun's mass is a representative average for the stars of the Milky Way, the Galaxy contains some 1.3×10^{11} stars.

Figure 9-12 (a) The combination of observations in the visible and radio regions of the spectrum reveal a spiral structure for the Milky Way. To an observer looking down on the Galaxy from about a million parsecs, the Milky Way might look like this. The cross marks the position of the sun. (b) Viewed from the earth, the center of the Galaxy is obscured by clouds of dust and gas that prevent visible light from reaching us; however, it contains several areas of strong radio emission, the strongest of which is Sagittarius A, a compact radio source that appears to dominate the large-scale motion of the galactic center. This is a radio image (taken at 6-cm wavelength) of the inner 8 $c \cdot y$ of the Milky Way. The dark red spot at the very center is Sagittarius A, which some astronomers think may contain a huge black hole. This image was made using the Very Large Array, a radio frequency interferometer made of 27 synchronized antennae with an effective diameter of about 20 km, located in New Mexico. Its resolution is better than that of the best ground-based optical telescopes by about a factor of five.

If we add together the masses of all of the visible stars in the Galaxy, including those beyond the sun's orbit, plus all of the dust and gas clouds, even if we confine the examination just to the solar neighborhood where seeing through the interstellar dust and gas is not a serious problem, we can account for only about 10 percent of the gravitational mass necessary to hold the Galaxy together. This discrepancy is referred to as the **missing-mass problem.** It exists for all galaxies and, indeed, for the universe itself. Various solutions to the problem, such as black holes, the possibility that neutrinos have mass, and as yet undiscovered, weakly interacting massive particles called WIMPs, are under intense investigation and debate but as yet there is no clear experimental support for any of them.

9-3 The Evolution of Stars

While no universally accepted theory of stellar formation exists, it is generally agreed that stars are formed from the massive clouds of dust and gas that exist throughout space. At some point in the swirling cloud, gravitational attraction begins to cause aggregations of matter to collect. These contract further due to gravity, attracting still more matter to them and eventually, if the cloud has sufficient mass, increasing the temperature to that necessary to initiate fusion; and a star is born.

In this section, we discuss how stars evolve once they have been formed. Two characteristics of stars are important for this discussion, the luminosity L and the effective temperature T_e. The effective temperature of a star is difficult to measure. It is usually inferred from a comparison of the spectral distribution of its radiation with that of a blackbody or from measurements of the absorption lines of hydrogen and helium in the atmosphere of the star.

The luminosity is the total power radiated by the star. It is determined from the apparent brightness of the star at the earth f (called the solar constant for the sun) and the distance r from the earth to the star (Equation 9-2):

$$L = 4\pi r^2 f \qquad 9\text{-}10$$

Determining the distance to a star is generally a very difficult task. For stars that are relatively close, the distance can be determined from the apparent motion of the star due to the motion of the earth around the sun. During one complete revolution of the earth, a star appears to move in a circle of angular radius θ called the **parallax angle** as shown in Figure 9-13. The parallax angle is given by

$$\theta = \frac{1 \text{ AU}}{r} \qquad 9\text{-}11$$

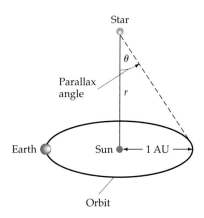

Figure 9-13 The parallax method of finding distances to nearby stars. A parsec is the distance r for which the parallax angle subtended by 1 AU is 1″.

Astronomical distances are often measured in parsecs or light-years. One parsec is that distance at which 1 AU subtends an angle of 1 arc second (1″), which equals 1/3600 of a degree. Setting $\theta = 1″$ in Equation 9-11, we obtain

$$1 \text{ parsec} = \frac{1 \text{ AU}}{1″} \times \frac{3600″}{1°} \times \frac{180°}{\pi \text{ rad}} = 2.0626 \times 10^5 \text{ AU} \qquad 9\text{-}12$$

Using 1 AU = 1.496×10^{11} m and 1 c·y = 9.461×10^{15} m, we can express the parsec in terms of meters or light-years:

$$1 \text{ parsec} = 3.086 \times 10^{16} \text{ m} = 3.26 \text{ c·y} \qquad 9\text{-}13$$

Example 9-4

Proxima Centauri is the star closest to the sun. By measuring the apparent change in the direction of Proxima Centauri between two observations made six months apart, the parallax angle θ is found to be 0.765″. How far is it to Proxima Centauri?

Since 1 AU/1″ equals 1 parsec, we have for $\theta = 0.765″$

$$r = \frac{1 \text{ AU}}{\theta} = \frac{1 \text{ AU}}{0.765″} = \frac{1 \text{ AU}}{1″} \frac{1″}{0.765″} = 1.31 \text{ parsecs} = 4.27 \ c\cdot y$$

The parallax-angle method of Example 9-4 can be used for only about 8000 stars that are relatively close to the sun. For the rest, the parallax angle is immeasurably small. In other situations, more indirect measurements of distance are necessary. These involve complex analyses of intensity variations over time for particular types of pulsating stars found in clusters. Thus, the distances to many clusters can be found. There is as yet no accurate method for determining the distances to individual, nonpulsating stars that are far away.

The various states of stars can be conveniently displayed by plotting the luminosity L versus the effective temperature T_e. The result is called a **Hertzprung–Russell (H–R) diagram.** Figure 9-14 shows an H–R diagram for some stars of representative masses. Each point represents a single star.

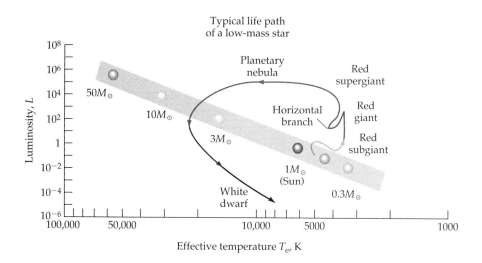

Figure 9-14 A Hertzprung–Russell (H–R) diagram. The points shown are the locations of stars of the indicated masses that have just formed. The shaded band is called the main sequence.

The large majority of stars on an H–R diagram fall in the shaded band called the **main sequence.** Main-sequence stars are normal in that they are homogeneous mixtures except in the core, they have essentially the same chemical composition, and they are fusing hydrogen into helium via one or another of the nuclear reactions discussed earlier. When stars leave the main sequence, they do so by expanding. Thus, stars in the main sequence are often called **main-sequence dwarfs.**

The location of a star along the main sequence in the H–R diagram depends on its luminosity, which is primarily dependent on the mass of the star. The masses of stars range from about $0.08 M_\odot$ to about $60 M_\odot$, where $M_\odot$ is the mass of our sun. Gaseous objects with masses less than about $0.08 M_\odot$ do not have enough gravity for their central cores to be compressed sufficiently to generate the high temperature necessary for sustaining the fusion reactions needed for energy emission. Objects with masses greater than $60 M_\odot$ would generate such enormous internal temperatures that the outward radiation pressure would exceed the gravity-generated inward pressure. Such a system would be very unstable, if indeed it could form at all.

The luminosity of a star is approximately proportional to the fourth power of its mass:

$$L \propto M^4 \qquad 9\text{-}14$$

The lifetime of a star t_L is proportional to the total available energy, which is proportional to the star's mass ($E = Mc^2$), and inversely proportional to the rate of energy emission, which is the luminosity:

$$t_L = \frac{E}{L} \propto \frac{Mc^2}{M^4} \propto M^{-3} \qquad 9\text{-}15$$

Thus, more massive stars burn their hydrogen more quickly than do less massive stars. (Equation 9-15 doesn't work for very small or very large stars because the luminosity–mass relationship of Equation 9-14 is only an average result. The exponent in Equation 9-15 is larger in magnitude for very small stars and smaller for very large stars.)

Considerations of energy balance for stars on the main sequence lead to the approximate proportionality of the radius and the mass:

$$R \propto M \qquad 9\text{-}16$$

Combining this with Equation 9-5, which relates the effective temperature to the luminosity per unit area, we can relate the effective temperature to the mass of the star:

$$T_e = \left(\frac{L}{4\pi R^2 \sigma}\right)^{1/4} \propto \left(\frac{M^4}{M^2}\right)^{1/4} \propto M^{1/2} \qquad 9\text{-}17$$

Thus, stars with larger masses have higher effective temperatures and, hence, higher luminosities than do those with lower masses. It is on the basis of Equations 9-5 and 9-17 that the stellar masses were plotted on the H–R diagram in Figure 9-14.

As a star ages, it consumes its primary fuel, hydrogen. What happens to it as the hydrogen supply in the core becomes exhausted depends on its initial mass. Low-mass and high-mass stars follow somewhat different evolutionary paths. In either case, however, the fundamental processes involved are successive nuclear reactions fueled by the product of the previous cycle. Thus, after the hydrogen in the core has fused to helium, the star must begin fusing helium in a cycle that eventually forms carbon. Before this can occur, the core must heat up still further to the 10^8 K necessary to initiate helium fusion. The chain of events involved in this process is complex and beyond the scope of this book. However, its result for low-mass stars is that the radius (and therefore the surface area) increases while the luminosity remains nearly constant. Thus, the intensity (the luminosity per unit area) and, consequently, the effective temperature decrease and the radiation emitted shifts to longer wavelengths as the star expands to become a **red subgiant**. The photosphere rapidly becomes more transparent as T_e decreases, thus increasing the luminosity and effectively limiting the decrease in temperature. The star is then a **red giant**. The track of a typical evolving low-mass star is shown on the H–R diagram in Figure 9-14.

Helium ignition results in the star again increasing its effective temperature and moving to the **horizontal branch.** When the helium in the core is exhausted, the star begins fusing carbon and ascends the red-giant branch again becoming a red supergiant. What happens after this is not clear.

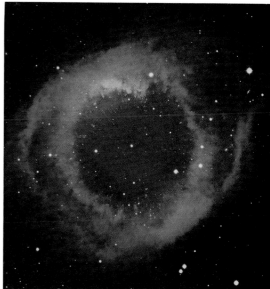

Figure 9-15 The nebula 30 Doradus (left), also known as the Tarantula nebula, is believed older than nebula NGC 7293 (right), also known as the Helix nebula. Its rapidly expanding gas cloud consequently shows a greater degree of diffusion. Located in the Large Magellanic Cloud, the Tarantula contains one of the most massive stars known, as well as supernova SN1987A, the very bright star at the lower right (of the left-hand photo). Ultraviolet radiation from stars heats the gas of nebulae, causing it to radiate.

Through a combination of events that includes the loss of considerable mass and perhaps passing once (or more) through a **planetary-nebula** stage, such as the nebulae shown in Figure 9-15, the star becomes a white dwarf, slowly cooling toward thermal equilibrium with the universe. We will discuss white dwarfs further in Section 9-5.

High-mass stars—those with masses greater than about $6M_\odot$—evolve much more quickly than low-mass ones, as predicted by Equation 9-15. In addition, they have sufficient initial mass to generate gravitationally the high pressures and temperatures necessary to ignite the fusion reactions with oxygen, neon, and then silicon to produce, ultimately, iron. These reactions occur with phenomenal speed and lead to catastrophic events that will be discussed in the next section.

9-4 Cataclysmic Events

Huge explosions and other sorts of cataclysmic events are a natural part of the life cycle of stars. Stars formed in swirling clouds of gas move along the H–R diagram, incorporating such occurrences into their evolution and forming in the process the elements needed to form new stars. Why these events occur is the subject of this section.

Novae

More than half of all stars are estimated to be members of **binary pairs** or even larger associations. These stars orbit their common center of mass as the group moves with the rotation of the galaxy. The periods of binaries vary from a few hours for those with the companions very close to each other to millions of years for those with the companions separated by thousands of astronomical units. Here, we are interested in close binaries.

A complete analysis of the interactions between the two stars forming a close binary is beyond the scope of this book, but a qualitative explanation will suffice. Consider a binary whose stars of masses M_1 and M_2 rotate about their common center of mass in circular orbits. An observer at rest in the rotating system experiences a net force that is the sum of the gravitational forces due to the two stars and the pseudoforces due to rotation. Figure 9-16

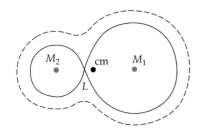

Figure 9-16 Cross sections of two gravitational equipotential surfaces for binary stars of mass M_1 and M_2. The point labeled L is one of five Lagrangian points where the gravitational potential is a minimum and the net force is zero.

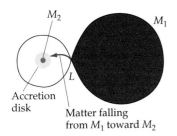

Figure 9-17 Material from M_1 pouring through the Lagrangian point into the Roche lobe of M_2 forms an accretion disk in the equatorial plane of M_2. Material arriving later hits the disk, generating a high-temperature impact area. This causes novae to flicker irregularly.

shows an equipotential surface about a binary pair. It is easy to visualize that there is a point along the line joining the centers of the two stars where the net potential is a minimum. At this point, the net force due to the combined effects of the rotation and the gravitational attraction of the masses M_1 and M_2 is zero. This point is a **Lagrangian point.** The three-dimensional equipotential surface that includes the Lagrangian point forms an envelope around each star called the **Roche lobe.**

Now consider what happens when, through natural evolution, one of the stars, say M_1, begins expanding and fills its Roche lobe. The photosphere of any star "sees" a vacuum outside the surface, the outward pressure at any point being balanced by gravity. But at the Lagrangian point there is no gravity. Thus, material from M_1 pours through the Lagrangian point into the Roche lobe of M_2. Once inside it is gravitationally attracted toward M_2. Since the system is rotating, the material from M_1 doesn't simply move directly toward M_2 because of the Coriolis force, but instead forms a spiralling **accretion disk** (see Figure 9-17).

If M_2 is a normal star, nothing of great consequence occurs, but if it is a white dwarf, then cataclysmic events called **novae** can occur. We will mention two possibilities. Material flowing through the Lagrangian point into the accretion disk is stored there until some instability occurs in the disk that results in the dumping of material onto the surface of the white dwarf. The impact heats the surface, causing a sudden brief increase in intensity by a factor of from 10 to 100. These events recur at intervals of from a few weeks for **dwarf novae** to hundreds or thousands of years for **recurrent novae.** Between these sudden bursts in intensity, the novae flicker as described in the caption to Figure 9-17.

For **classical novae,** which eject substantial amounts of material into space and can brighten by a factor of a million within a few days, astrophysicists suggest that the sudden dumping of material from the accretion disk onto the hot surface of the white dwarf may result in the buildup of sufficient hydrogen to initiate a thermonuclear explosion. After the blast, the system returns to a more quiescent state, pending the accumulation of more hydrogen in the disk. The theoretical problems involved in explaining such an event are formidable, however, and no general agreement on the mechanism exists.

Supernovae

The **supernova**—the catastrophic explosion of an entire star—is, perhaps surprisingly, somewhat more clearly understood than the nova. First, note that supernovae are not just big novae. Their origin is completely different. In Section 9-3, we saw what occurs in a star as it uses up the hydrogen in the core and begins moving off the main sequence of the H–R diagram. The star begins to fuse helium and then carbon. If it is a low-mass star, it has insufficient gravitational energy to ignite the fusion of heavier nuclei in quantity.

For massive stars, however, the situation is different. If the mass is greater than about $8M_\odot$, gravity is strong enough to continue to draw mass from the middle layers into the core as the core uses up fuel. The increasing temperatures, exceeding 10^8 K, are sufficient to ignite fusion in neon and silicon ultimately producing iron. As we saw in Chapter 7, the specific binding energy of iron is the highest in the periodic table. Fusing elements above iron doesn't produce energy; it requires energy. Thus, when the core has been fused to iron, there is nowhere else to go via thermonuclear reactions. With no counteracting outward pressure from nuclear reactions, gravitational contraction continues even more rapidly and the core continues to heat up until it exceeds 10^9 K. At that point, the radiation within the star is intense and the iron nuclei undergo photodisintegation into helium and neutrons, sucking energy from the core and accelerating the gravitational collapse:

$$^{56}_{26}\text{Fe} \longrightarrow 13\, ^{4}_{2}\text{He} + 4n \qquad 9\text{-}18$$

The helium nuclei then begin to photodisintegrate, extracting enormous amounts of energy to overcome the binding energy of helium:

$$^{4}_{2}\text{He} \longrightarrow 2p + 2n \qquad 9\text{-}19$$

The core is now in gravitational free fall, compressing the electrons and protons into neutrons via inverse beta decay:

$$p + e^- \longrightarrow n + \nu_e \qquad 9\text{-}20$$

What happens next is a matter of intense theoretical conjecture that we will explore in Section 9-5.

What happens to the envelope of the star—the material outside of the core—although also unclear theoretically, is certainly apparent visually. The entire envelope is blown away in an incredibly massive explosion. This is a supernova. Supernovae are extremely rare, but scientists were fortunate enough to observe one in 1987 only 170,000 $c \cdot y$ away in the Large Magellanic Cloud, a small, irregular galaxy that is a companion to the Milky Way galaxy. Called SN1987A, it was the first supernova to occur close enough to be visible to the unaided eye since 1604, when both Kepler and Galileo saw one. Two others were recorded earlier, in 1006 and 1054, the latter documented by Chinese astronomers and still visible as the Crab Nebula. Several others have been observed with telescopes.

At its peak light output, a supernova typically shines more brightly than the entire galaxy in which it is located. The spectra of supernovae reveal the presence of elements throughout the entire periodic table. This indicates that some of the energy removed from the core following the production of iron is used to produce elements of even higher atomic numbers. The supernova ejects some of this material into space, where it eventually contributes to the formation of a new generation of stars and their planets via condensation. Such events undoubtedly preceded the birth of the sun and the formation of the earth. We are, as has been said before, "made of the stuff of stars."

There are two kinds of supernovae: *type I* and *type II*. The preceding description applies to type II supernovae. Type I supernovae occur among population II stars, which tend to be of low mass. (Don't be confused by the apparent inconsistency in nomenclature.) This contradiction with our earlier discussion of the evolution of low-mass stars currently has no theoretical explanation.

9-5 Final States of Stars

The cataclysmic events that occur near the end of the life of a star lead to one of only three possible final states: a degenerate dwarf, a neutron star, or a black hole. The mass of the star, particular that of the core, appears to be the primary factor in determining the final state.

Degenerate Dwarfs

Stars whose masses are less than about $6M_\odot$ follow an evolutionary track on the H–R diagram that takes them through one or more periods of substantial mass loss from the envelope. How this occurs is not clear, but the ejected mass, which is heated to a glowing planetary nebula by the hot core, leaves behind a **degenerate dwarf,** also called a **white dwarf** at this stage because it is literally white hot. Its mass is typically about $1M_\odot$ and its radius of the order of 10^7 m, which is about the radius of the earth. Thus, the density of a typical white dwarf is about 5×10^5 g/cm^3. A coin the size of a penny made from the material of a white dwarf would have a mass of over 200 kg.

Thermonuclear reactions have ceased in the white dwarf, so there is no outward pressure due to them from within the star. The star therefore collapses because of the inward gravitational pressure until the exclusion principle prevents the atomic electrons from coming any closer together. This effect is similar to the exclusion-principle repulsion between atoms in a molecule that we discussed in Chapter 5. It results in an outward pressure that is larger even than the thermal pressure of the hot core. It is this **electron-degeneracy pressure** that supports the white dwarf. When the outward electron-degeneracy pressure equals the inward pressure due to gravity, the star stops contracting.

Explicit derivation of the expression for the electron-degeneracy pressure leads to a nonrelativistic relation between the dwarf's radius R and mass M:

$$R = (3.1 \times 10^{17} \text{ m·kg}^{1/3})\left(\frac{Z}{A}\right)^{5/3} M^{-1/3} \qquad 9\text{-}21$$

where Z is the atomic number and A is the atomic mass number of the material of the star. Note the interesting result that the larger the mass, the smaller the radius. For example, a white dwarf with a mass of $1M_\odot$ will have a radius smaller than one with a mass of $0.5M_\odot$. This raises the interesting question of whether, when the electrons become relativistic, the mass might become large enough for the radius of the dwarf to shrink to zero. Although Equation 9-21 does not formally allow that possibility until M approaches infinity, S. Chandrasekhar derived the corresponding relativistic relation and found that the radius would go to zero when the mass reaches about $1.4M_\odot$. This mass is called the **Chandrasekhar limit.** Its validity is strongly supported by the fact that the masses of all the white dwarfs that have been measured are less than that value.

The white dwarf continually loses heat to space and, without a nuclear furnace, it slowly cools off and dims. When it is no longer visible, it has become a **black dwarf** and continues to cool toward thermal equilibrium with the universe. It is likely that no white dwarfs have yet reached this final stage.

Neutron Stars

In the discussion of supernovae, we saw that the enormous pressures in the core forced inverse beta decay to occur, converting the core into neutrons. If the mass of the core after the explosion is greater than the Chandrasekhar limit, what happens? We can get an idea by considering the neutrons to be an ideal gas of fermions and deriving a nonrelativistic expression for the mass–radius relation analogous to Equation 9-21. The result is

$$R = (1.6 \times 10^{14} \text{ m·kg}^{1/3}) M^{-1/3} \qquad 9\text{-}22$$

where M is the mass of the core in kilograms and R is the radius of the core in meters. Such a star is called a **neutron star,** since the envelope was blown away in the supernova and all that is left is the core consisting of neutrons. For $M = 1M_\odot$, Equation 9-22 yields $R = 1.27 \times 10^4$ m = 12.7 km.

The density of the neutron star is about 1.2×10^{14} g/cm^3. This is only slightly less than the density of the neutron itself, which is about 4×10^{14} g/cm^3. Thus, we can conclude that the gravitational pressure of the neutron star is balanced by the repulsive component (due to the exclusion

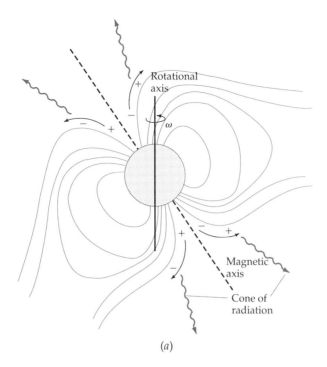
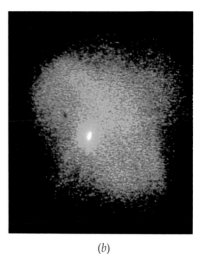

(b)

(c)

principle) of the strong nuclear force between the neutrons. As you might guess from our earlier discussion, gravity can overcome even this resisting pressure. The mass corresponding to the gravity at which that occurs is the maximum mass possible for a neutron star, a mass analogous to the Chandrasekhar limit for white dwarfs. Current theory puts the maximum mass of a neutron star at between $1.7 M_\odot$ and $3 M_\odot$. The few neutron stars that have been tentatively identified and measured all have masses below this limit.

Regularly pulsing radio sources, called **pulsars,** discovered in 1967 in nebulae such as the Crab Nebula that are remnants of supernovae, are thought to be neutron stars. Current theory suggests that the radiation is emitted as the result of the charged particles emitted by the neutron star that are accelerated along the star's magnetic-field lines as a consequence of the star's rapid rotation as illustrated in Figure 9-18. The Crab pulsar also corresponds to an optical variable as illustrated in Figures 9-18b and c. It emits energy at an incredible 3×10^{31} W. Its period is equally incredible, only 0.033 s, one of the shortest known.

As it emits energy into space, the neutron star also slowly cools, approaching thermal equilibrium with the universe.

Figure 9-18 (a) The neutron star acquires much of the original star's angular momentum and magnetic field, causing it to rotate rapidly while dragging along a distorted magnetosphere. Accelerated charged particles radiate in a cone about the rotating magnetic axis like a cosmic lighthouse. (b) As the cone of radiation swings to face the earth, light emitted from accelerated electrons becomes visible (the bright spot in the image). (c) A fraction of a second later, the pulsar has turned and this light is no longer directed toward the earth. Currently rotating about 30 times a second, the pulsar has a period that is increasing by about 10^{-5} s a year.

Black Holes

What happens when the mass of the core remaining after a supernova exceeds the $1.7 M_\odot$ to $3 M_\odot$-upper limit for the formation of a neutron star? The velocity necessary for an object of mass m to escape from a large object of mass M is found by equating the gravitational potential energy at the surface of M to the kinetic energy necessary to escape. This results in

$$v_e = \left(\frac{2GM}{R}\right)^{1/2} \qquad 9\text{-}23$$

For a neutron star with $M = 1.0 M_\odot$, $v_e = 1.3 \times 10^8$ m/s, which is more than 40 percent of the speed of light. If there were no relativistic and quantum-mechanical effects, the escape velocity would equal c when

$$R_S = \frac{2GM}{c^2} \qquad 9\text{-}24$$

where R_S is called the **Schwarzschild radius.** Thus, if an incipient neutron star is so massive that its radius is less than R_S, no object with mass can escape from its surface. In addition, radiation of wavelength λ emitted at some distance R from mass M is shifted to a longer wavelength λ' according to the **gravitational redshift** described in Section 1-11, which is given by

$$\frac{\lambda'}{\lambda} = \left(1 - \frac{v_e^2}{c^2}\right)^{-1/2} = \left(1 - \frac{2GM}{c^2 R}\right)^{-1/2} = \left(1 - \frac{R_S}{R}\right)^{-1/2} \qquad 9\text{-}25$$

If R shrinks to the Schwarzschild radius, then λ' approaches infinity and the energy ($E = hf = hc/\lambda$) approaches zero. Thus, if R is less than R_S, no energy can escape the surface as radiation, either. Such an object is called a **black hole** because it neither emits nor reflects radiation or mass and, hence, appears absolutely black.

The radius of a black hole with a mass of $1M_\odot$, if there is such an object, would be only about 3 km. Thus far, there have been no confirmed observations of black holes, although a number of possible ones are the subject of intense research. Many astrophysicists currently believe that a massive black hole is located at the center of the Milky Way galaxy and may account for part of the "missing mass" of the Galaxy. Unlike degenerate dwarfs and neutron stars, black holes are not cooling toward thermal equilibrium with the universe.

9-6 Galaxies

In Section 9-2, we saw that the Milky Way galaxy is shaped like a spiralled disk with a central bulge located about 30,000 c·y from the sun. The disk is surrounded by a roughly spherical "halo" of globular clusters comprised mostly of population II stars, which are also part of our Galaxy. We will now look at some of the characteristics of galaxies in general.

Material Between the Stars

"Holes in the sky"—regions where no stars are seen—have been observed since the early days of astronomy and were assumed to be empty space. However, studies of open clusters led to the discovery about 60 years ago of a more or less continuous distribution of tiny dust particles called **interstellar dust** between the stars. Consisting of solid specks of silicates and carbides averaging only a few hundred nanometers in diameter (approximately matching the wavelengths of visible light), the interstellar dust both absorbs and scatters some of the starlight striking it. Since blue light scatters more efficiently than red light, starlight is reddened on its trip to us, just as sunlight is reddened at sunset. Although the dust seems to pervade the entire Galaxy, its concentrations are very low. The vacuum in interstellar space is far better than the best vacuum obtainable in the laboratory.

Spectroscopic studies of binaries reveal some absorption lines that are not doppler shifted. In 1904, J. F. Hartmann reasoned correctly, although not to universal acceptance, that the unshifted lines result from the absorption of light from the binary by an intervening gas cloud rather than by gas in the atmosphere of the star. Though still difficult to demonstrate conclusively in all cases, the existence of interstellar gas clouds is now generally accepted. The gas clouds are composed mainly (some exclusively) of hydrogen.

Together, the interstellar dust and the clouds of gas account for an estimated 2 to 3 percent of the mass of our Galaxy. It is nearly certain that there is not enough unseen gas and dust to account for the Galaxy's "missing mass."

Gaseous Nebulae

Though most gas clouds, or nebulae, in interstellar space are irregular in shape, a few are circular, leading to speculation that they are self-gravitating and represent the very early stages of the formation of new stars. Some large hydrogen clouds have spherical regions of ionized hydrogen, with the demarcation between the H and H^+ regions being quite sharp. Astrophysicists believe that the ionized region is maintained by ultraviolet photons with frequencies above the Lyman limit that are emitted by a hot, newly formed star at the center of the region. The view that new stars form in nebulae in an ongoing process is strongly supported by the observation that, although it is of the order of 10^{10} y old, our Galaxy contains main-sequence stars that are no more than 2 to 3×10^6 y old. Furthermore, high-resolution radio-astronomy has in recent years located numerous newly forming stars embedded in clouds of dust and gas that are completely opaque to optical wavelengths.

Classification of Galaxies

Although fuzzy, extended objects, at one time called "nebulae," that were obviously not stars have been observed in the night sky since the 1700s, what and where they were was a matter of active scientific debate until well into the twentieth century. The answer had to await the development of telescopes with sufficient resolution and light-gathering power and a theoretical means of computing distances from observations made with them. These came together in the mid-1920s when Edwin Hubble used the 2.5-m telescope on Mount Wilson, the largest in existence at the time, to measure the intensities of rare stars, called Cepheid variables,* that he discovered in three "nebulae." One of those nebulae, the great spiral Andromeda, he measured to be 2×10^6 c·y away. In one stroke, he was able to demonstrate that the "nebulae" were in fact galaxies much like our own, as had first been suggested by the philosopher Emmanuel Kant 150 years earlier, and that they were far outside the Milky Way galaxy. Exploring Hubble's discovery will take us into the realm of cosmology, the study of the universe.

Following his discovery that "nebulae" were in reality distant galaxies, Hubble conducted a systematic study of the enormous number that were visible. He found that all but a very few fit into four general categories. Most have regular geometrical shapes and occur in two varieties: **ellipticals,** which are roundish, rather like a football; and **disks.** The disks, in turn, have two subgroups: **ordinary spirals** and **barred spirals.** The small percentage that did not have regular shapes he called **irregular galaxies.** Figure 9-19 shows an example of each type of galaxy.

In addition to their geometrical differences, the four types of galaxies have other dissimilarities. A large fraction of the motion of the stars in spirals is rotational about the galactic center, whereas the motion of the stars in ellipticals is generally random with only a relatively small rotational component. Ellipticals also seem to have very little interstellar gas and dust, whereas spirals and many irregular galaxies have substantial amounts. The fact that most ellipticals have no young stars is probably a consequence of that lack. With a few exceptions, ellipticals are much smaller than spirals, typically having only about 20 percent of the diameter of an average spiral like the Milky Way galaxy and only a thousandth of the mass.

---

*Cepheid variables are rare stars for which there exists a relation connecting the period of intensity variation to the brightness and, hence, to the distance from the sun. They are the primary key to our knowledge of astronomical distances. Polaris, the current Pole Star, is a Cepheid variable.

Figure 9-19 Examples of the four types of galaxies in Hubble's classification scheme: (*a*) elliptical, (*b*) ordinary spiral, (*c*) barred spiral, and (*d*) irregular. The Milky Way galaxy is thought to be an ordinary spiral.

Quiet and Active Galaxies

Most of the approximately 10^{10} galaxies in the observable universe appear to be **quiet galaxies;** that is, there is very little activity other than what might be expected for such dynamic systems. The vast majority of these galaxies are so distant that our instruments cannot resolve internal details. Therefore, only the composite spectra and apparent brightness f for the entire galaxy can be observed. The range of velocities Δv among the stars of the regular galaxies, measured by the doppler broadening of the spectral lines, turns out to be related to the total luminosity L by

$$L \propto (\Delta v)^4 \qquad 9\text{-}26$$

Since L is related to f and r, the distance to the galaxy, by Equation 9-10, the distance r can be found from measurements of the doppler shifts and the brightness of the galaxy.

In a very small percentage of galaxies, something extremely violent, even by comparison with stellar supernovae, is occurring. They are called **active galaxies.** There are several distinct types, some of which may not even be galaxies at all. The first discovered were **Seyfert galaxies,** named after Carl Seyfert, their discoverer. They are spirals with extremely bright central cores, or nuclei. In many of them, the light coming from the core exceeds that from all of the stars in the galaxy and, incredibly, may vary in intensity by a factor of two or more in less than a year. Such a rapid variation in the total intensity means that the source must be less than 1 light-year in extent even though it produces as much energy as 10^{11} stars. Even more incredible is the fact that the light emitted by a Seyfert galaxy is an emission line spectrum, not a continuum with absorption lines typical of stars. This suggests that its enormous energy is not coming from thermonuclear reactions. The source is still a mystery.

A similar sort of extreme activity occurs in a few ellipticals called **N galaxies** and **BL Lac objects.** N galaxies are elliptical counterparts of Seyfert galaxies, that is, they have very bright centers. BL Lac objects seem to be like N galaxies, but exhibit substantial short-term intensity variations. In these, an intensity variation of a factor of two can occur within a week and a complete reversal of the polarization of the emitted light can occur within one day, suggesting that the energy source is only 1 light-day in diameter. BL Lac objects are now thought to be giant ellipticals about 10^9 c·y from earth.

Some of the giant ellipticals are also strong emitters in the radio region of the spectrum. Study of these **radio galaxies** has been intense, and the results have been astonishing. For example, the radio source Centaurus A is double-lobed with a small, radio-emitting nucleus midway between the lobes (see page 297). It is one of the largest radio-emitting objects in the universe. Analyses of its spectra indicate that the energy release represented by the radiation we see amounted to 10^{56} J, which is about the equivalent of all the stars in the Milky Way galaxy undergoing supernova explosions simultaneously. The nature of such a colossal event is a mystery.

In a universe of strange phenomena, **quasars,** short for *quasistellar radio sources,* are among the strangest. Their optical images look like stars; that is, they have no resolved structure. Their spectra, however, resemble that of a Seyfert galaxy. Resolved radio images of some quasars show that a few of them are double-lobed, like the radio galaxies, which makes their identification ambiguous. In addition, there is a group of objects about 20 times more numerous than quasars called **quasistellar objects,** or **QSOs.** These are like quasars in every major way, except that they are not radio emitters.

Perhaps the strangest thing about the quasars is the magnitude of the redshift of their spectra, which is very large. *If* it is due to the doppler effect alone, it implies that some quasars are receding directly away from us at speeds greater than $0.9c$, which is larger than the speed of the general expansion of the universe. This would make them the most distant massive objects, of the order of 10^{10} c·y from earth. Their apparent optical brightness f and their great distance imply power outputs of 10^{40} W, greater than that of 10^{12} suns. Not only that, but the intensities of some of them vary over only a few hours, suggesting dimensions of only a few light-hours.

The nature of quasars and QSOs is an unresolved scientific issue. There are two schools of thought. Many astrophysicists think that they are not at any great distance from us but rather are at distances consistent with their brightness. Why they should exhibit such large redshifts is then perplexing because the quasar would violate Hubble's law (discussed next), which is a very simple and very general relation between the velocities of extragalactic systems relative to us and their distances from us. Other astrophysicists feel that the large redshifts are good indicators of the objects' distances from us, which makes their prodigious energy output the issue. The suggestion has been made that the source of the colossal implied power is matter from colliding galaxies falling onto an enormous black hole at the center of one (or both) of the galaxies.

Hubble's Law

E. P. Hubble was the first astronomer to recognize that there is a relation between the redshifts in the spectra of galaxies and their distances from us. This relation is illustrated in Figure 9-20 for a group of spiral galaxies used by astronomers for calibrating distances. Provided that the redshift is due to the doppler effect, the recession velocity v of a galaxy is related to its distance r from us by **Hubble's law:**

$$v = Hr \qquad 9\text{-}27$$

where H is the **Hubble constant.** Figure 9-21 shows the redshifted spectra of five galaxies whose distances from us range from 2.6 megaparsecs to 287.5 megaparsecs.

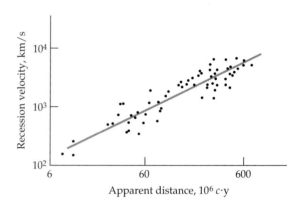

Figure 9-20 A plot of the recession velocities of individual galaxies versus apparent distance illustrates Hubble's law.

In principle, the value of H is easy to obtain since it relies on the direct calculation of v from redshift measurements. However, recall that astronomical distances are very difficult to obtain and that they have been computed for only a fraction of the 10^{10} or so galaxies in the observable universe. Thus, the value of H changes as distance calibration data is refined. The currently accepted value of the Hubble constant is

$$H = \frac{23 \text{ km/s}}{10^6 \ c\cdot y} \qquad 9\text{-}28$$

Notice that the basic dimension of H is reciprocal time. The quantity $1/H$ is called the **Hubble age** and equals about 1.3×10^{10} y. This would correspond to the age of the universe if the gravitational pull on the receding galaxies were ignored.

Exercise
Show that $1/H = 1.3 \times 10^{10}$ y.

Example 9-5

Redshift measurements of a galaxy in the constellation Virgo yields a recession velocity of 1200 km/s. How far is it to that galaxy?

From Hubble's law, we obtain

$$r = \frac{v}{H} = (1200 \text{ km/s}) \frac{10^6 \ c\cdot y}{23 \text{ km/s}} = 52 \times 10^6 \ c\cdot y$$

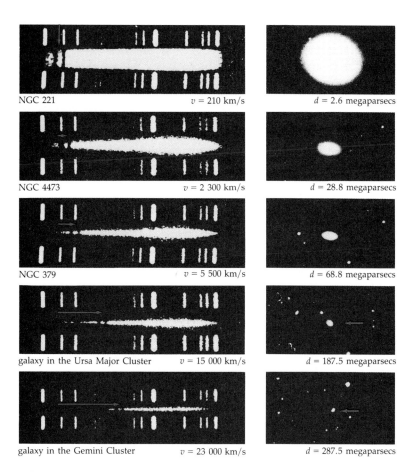

Figure 9-21 The redshift of the Ca, H, and K absorption spectral lines are shown for five galaxies at different distances from us. The line spectra above and below the absorption spectrum are standards used for determining the amount of shift accurately.

Hubble's law tells us that the galaxies are all rushing away from us, with those the farthest away moving the fastest. However, there is no reason why our location should be special. An observer in any galaxy would make the same observations and compute the same Hubble constant (see Problem 20). Thus, Hubble's law suggests that all of the galaxies are receding from each other at an average speed of 23 km/s per 10^6 c·y of separation. In other words, the universe is expanding. This is a profound discovery with enormous theoretical implications.

An obvious question is whether there are other observational results that support Hubble's law. For example, is the observed expansion general, or could it be a statistical accident—a consequence of our having to date measured the redshifts of "only" about 30,000 of the 10^{10} galaxies in the observable universe? Thus, redshift surveys of the universe are an important first step in studying Hubble's expansion. Such surveys have been underway for several years and about 10^{-5} of the volume of the visible universe has now been mapped. These surveys have yielded several unexpected discoveries, but have not yet answered the question conclusively. There are huge voids in space—regions where the density of galaxies is only 20 percent or so of the average for the universe. The galaxies themselves tend to be grouped into local clusters of a dozen or so and the local clusters into superclusters of a few thousand. In addition, the galaxies tend to lie on thin, sheet-like structures. The largest detected thus far is called the "Great Wall" by its discoverers, Margaret J. Geller and John P. Huchra (see Figure 9-22). How such structures might have evolved in the general expansion described by Hubble's law presents a serious challenge to existing theoretical descriptions of the development of the large-scale structure of the universe.

Figure 9-22 These are two views of approximately 4000 galaxies included in the redshift survey of Margaret Geller and John Huchra. The Milky Way is at the vertex. The three layers outlined in white cover latitudes from 26.5° to 44.5° above the celestial equator (called declination by astronomers); the layer outlined in orange covers declinations from 8.5° to 14.5°. Each layer covers the celestial longitude (called right ascension by astronomers) range from 8^h to 17^h, or about 37 percent of the celestial equatorial circle. Each of the 4000 white and yellow points represents a galaxy comparable with the Milky Way. The Great Wall stands out clearly in the lower view, running approximately parallel to the outer boundary of the survey—about halfway from the vertex. Note the huge voids in space, some as large as 10^8 c·y.

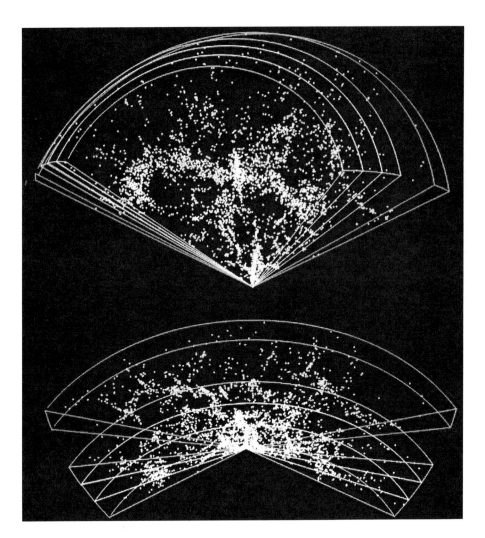

9-7 Gravitation and Cosmology

We have seen that Hubble's law leads inescapably to the conclusion that the universe is expanding, and it provides us with a measure, $1/H$, of how long ago that expansion began. In this section, we will examine the basic theoretical framework that suggests possible tests of that conclusion, in addition to the redshift survey.

The basis for this discussion is the philosophical view that the universe is homogeneous and isotropic at any instant in time. That is, at any given instant, the universe has the same physical properties everywhere and looks the same in all directions from every location. This point of view is called the **cosmological principle.** Note that Hubble's law is consistent with the cosmological principle.

We have already seen that the cosmological principle clearly does *not* hold on a local scale. Galaxies are clustered into local groups. Even on a scale of 10^8 c·y, the dimension typical of galactic superclusters, the universe is neither homogeneous nor isotropic. However, when maps of very distant space are examined (see Figure 9-23), the distribution of galaxies does appear to be homogeneous and isotropic. Whether redshift maps like that shown in Figure 9-22, which extends to about 4×10^8 c·y, will begin to show homogeneity and isotropy as they are extended into deep space remains to be seen.

Figure 9-23 A map showing approximately two million galaxies ranging up to 2×10^9 c·y away. The distribution of the galaxies looks essentially homogeneous and isotropic. This is a composite of 185 contiguous photos taken by the Schmidt telescope at the European Southern Observatory. The south galactic pole is at the bottom center. The four blank squares at the top are covered by photos not yet analyzed.

The Critical Mass Density of the Universe

We noted earlier that the Hubble age $1/H = 1.3 \times 10^{10}$ y ignores the effect of gravity. Our expectation is that gravity tends to slow the expansion over time. Is the gravity in the universe strong enough eventually to reverse the expansion and cause the universe to collapse? Or will the expansion continue forever? The answer depends on the mass density of the universe. We can understand this by considering the motion of a single galaxy of mass m at a very large distance R from the earth. Let M be the total mass of all the galaxies within the spherical volume of radius R. The gravitational potential energy of the galaxy is $-GMm/R$. The total energy of the galaxy is

$$E = K + U = \tfrac{1}{2}mv^2 - \frac{GMm}{R} \qquad 9\text{-}29$$

If we project an object with some speed v from the earth, the object will escape if its total energy is greater than or equal to zero, but if the total energy is negative, the particle will eventually stop and fall back to the earth. Similarly, if the total energy of the galaxy is greater than or equal to zero, it will continue to move away from the earth forever, but if the total energy is negative, the galaxy will eventually stop moving away from the earth and start moving back toward the earth. We can see from Equation 9-29 that the total energy of the galaxy depends on the total mass M within the spherical volume R; that is, it depends on the mass density $\rho = M/(\tfrac{4}{3}\pi R^3)$. We can find the critical mass density of the universe ρ_c by setting the total energy in Equation 9-29 equal to zero:

$$\tfrac{1}{2}mv^2 = \frac{GMm}{R}$$

Substituting $v = HR$ from Hubble's law (Equation 9-27), we obtain

$$\tfrac{1}{2}m(HR)^2 = \frac{GMm}{R}$$

$$\tfrac{1}{2}H^2 = \frac{GM}{R^3}$$

Then

$$\rho_c = \frac{M}{\frac{4}{3}\pi R^3} = \frac{3H^2}{8\pi G} \qquad 9\text{-}30$$

Using present values for H and G, we obtain for the critical mass density of the universe

$$\rho_c \sim 10^{-26} \text{ kg/m}^3$$

This corresponds to about five hydrogen atoms per cubic meter of space.

Determining the present mass density of the universe ρ_0 is thus an important goal. If it is larger than ρ_c, the expansion will reverse and the universe will collapse. If it is smaller, the expansion will continue forever. If it should happen that $\rho_0 = \rho_c$, the universe will coast to a stop but will not begin to contract. It should also be clear that if ρ_0 is greater than ρ_c now, it will always be so because it is actually the conservation of energy that determines whether contraction or continued expansion will occur. Since ρ_0 must decline over time as expansion progresses, the Hubble constant must also decline over time to ensure that ρ_0 remains larger than ρ_c. In other words, the Hubble constant must be a function of time. The value of ρ_0 based on the *visible* universe is only about 4 percent of ρ_c, suggesting that the universe will expand forever. However, the missing mass of the universe discussed earlier affects the value of ρ_0. No generally accepted value of ρ_0 has yet been found, although current estimates put it close to ρ_c.

At this point in time, the emphasis of investigations in cosmology is still centered on developing a basic cosmological model against which to compare the many observational discoveries made by astronomers and astrophysicists, only a few of which we have discussed briefly. To be sure, there is a good candidate for a comprehensive model, the Big Bang. We will review some of its successes and some of the questions it hasn't answered in the final section of the chapter.

9-8 Cosmogenesis

Following his completion of the general relativity theory in 1915, Einstein turned to cosmology. He based his early work on the assumption that the universe was not only homogeneous and isotropic but also constant in time. This is sometimes called the **perfect cosmological principle.** He quickly discovered that such a static universe, like that described by Newton's gravitational theory, is empty; that is, it contains no mass. He accounted for mass by adding a new force of unknown origin via the **cosmological constant,** thereby committing what he later described as the biggest blunder of his life. On learning of Hubble's discovery of the expansion of the universe, he abandoned the cosmological constant. Others, however, were not so quick to give up on the philosophically attractive static model. They argued that the observed expansion would not result in a decrease in the mass density of the universe if new matter were being created in space at a rate sufficient to maintain the density of a steady-state universe.

One difficulty with the steady-state model of the universe is a problem known as Olber's paradox. If there is a uniform distribution of stars throughout an infinite space, then no matter in which direction you look, you will eventually see a star. Therefore, the night sky should look as bright as the surface of the average star. (This is analogous to standing in an infinitely large forest in which all the trees are painted white. Along any line of sight, you will eventually see a white tree, so you should see white in all directions.) Why then is the night sky dark? This dilemma is called Olber's paradox after the nineteenth-century physician–astronomer who publicized it widely. The solution offered by Olber himself was that interstellar dust absorbs the light from distant stars. This is no help since the dust would eventually be heated to glowing and the night sky should still be bright.

The solution to this problem came with Hubble's discovery of the expansion of the universe. Since the velocity of light is finite, looking into space means looking back in time. As we look deeper and deeper into space, we are eventually looking at a time before the stars began to form, that is, at a time greater than the Hubble age. (In terms of our forest analogy, the distant trees have not yet been painted white. Therefore, if the separation of the trees is great enough, many lines of sight will end on dark trees.)

The Big Bang

Two major astrophysical discoveries made in the 1960s were the first of several that have convinced most scientists that the universe is not constant in time but was initiated by a single event at a particular time in the past, the **Big Bang,** and is evolving over time. The first of the two major discoveries that supported the evolving-universe model was Martin Ryle's discovery that there are more distant radio galaxies than nearby ones. Since distant observations correspond to earlier times, this meant that the universe had looked different at earlier times than it does now; that is, it has evolved.

The second discovery was monumental, as important as Hubble's discovery of the expansion of the universe itself. In investigating ways of accounting for the cosmic abundance of elements heavier than hydrogen, cosmologists recognized that nucleosynthesis in stars could explain the abundance of elements heavier than helium but not that of helium. Helium must therefore have been formed during the Big Bang. Synthesizing the amount of helium to account for its present abundance requires that the Big Bang would occur at an extremely high initial temperature to provide the necessary reaction rate before fusion was shut down by the decreasing density of the very rapid initial expansion. The high temperature implies a corresponding thermal (blackbody) radiation field that would cool as the expansion progressed. Theoretical analysis predicted that from the estimated time of the Big Bang to the present, the remnants of the radiation field should have cooled to a temperature of about 3 K, corresponding to a blackbody spectrum with peak wavelength λ_{max} in the microwave region. In 1965, the predicted cosmic background radiation was discovered by Arno Penzias and Robert Wilson at the Bell Labs. Since this landmark discovery, careful analysis has established that the temperature of the background field is 2.7 ± 0.1 K and has shown that it has the isotropic distribution in space that is absolutely essential for a universe that satisfies the cosmological principle.

The Very Early History of the Universe

What was the Big Bang like? The singular event that initiated the expansion of the universe must have been a huge explosion that occurred throughout the entire universe. We don't know whether the universe at the instant of the explosion occupied a small volume (nearly a point) or an infinite one since we don't yet know if the present average density is greater or less than the critical density, that is, whether the universe is infinite or finite. In either case, it is the size of space itself that has been expanding ever since.

Most cosmologists currently favor a theoretical description of the evolution of the universe following the Big Bang called the **standard model**. It relies heavily on recent experimental discoveries and theoretical advances in particle physics and reflects the increasing overlap of frontier research in those areas of physics over the past several years. The standard model's account of how the universe evolved from $t = 0$ to the present, when $t = 10^{10}$ years, is outlined in the following discussion and illustrated in Figure 9-24.

Initially, the four forces of nature (strong, electromagnetic, weak, and gravity) were unified into a single force. Physicists have been successful in developing theoretical descriptions that unify the first three, but a theory of quantum gravity, needed for the extreme densities of the single-force period, does not yet exist. Consequently, until the cooling universe "froze" or "condensed out" the gravitational force at about 10^{-43} s after the Big Bang when the temperature was still 10^{32} K, we have no means of describing what was occurring. At this point, the average energy of the particles created would have been about 10^{19} GeV. As the universe continued to cool below 10^{32} K, the three forces other than gravity remained unified and are described by the grand unification theories (GUTs). Quarks and leptons were indistinguishable and particle quantum numbers were not conserved. It was during this period that a slight excess of quarks over antiquarks occurred, roughly 1 in 10^9, that ultimately resulted in the matter that we now observe in the universe.

At 10^{-35} s, the universe had expanded sufficiently to cool to about 10^{27} K, at which point another phase transition occurred as the strong force condensed out of the GUTs group, leaving only the electromagnetic and weak forces still unified as the **electroweak force**. During this period, the previously free quarks in the dense mixture of roughly equal numbers of quarks, leptons, their antiparticles, and photons began to combine into hadrons and their antiparticles, including the nucleons. By the time the universe had cooled to about 10^{13} K, at about $t = 10^{-6}$ s, the hadrons had mostly disappeared. This is because 10^{13} K corresponds to $kT \sim 1$ GeV, which is the minimum energy needed to create nucleons and antinucleons from the photons present via the reactions

$$\gamma \longrightarrow p^+ + p^- \qquad \text{9-31}a$$

and

$$\gamma \longrightarrow n + \overline{n} \qquad \text{9-31}b$$

The particle–antiparticle pairs annihilated and there was no new production to replace them. Only the slight earlier excess of quarks led to a slight excess of protons and neutrons over their antiparticles. The annihilations resulted in photons and leptons, and after about $t = 10^{-4}$ s, those particles in roughly equal numbers dominated the universe. This was the **lepton era**. At about $t = 10$ s, the temperature had fallen to 10^{10} K ($kT \sim 1$ MeV). Further expansion and cooling dropped the average photon energy below that needed to form an electron–positron pair. Annihilation then removed all of the positrons as it had the antiprotons and antineutrons earlier, leaving only the small excess of electrons arising from charge conservation, and the **radiation era** began. The particles present were primarily photons and neutrinos.

Within a few more minutes, the temperature dropped sufficiently to enable fusing protons and neutrons to form nuclei that were not immediately photodisintegrated. Deuterium, helium, and a bit of lithium were produced in this **nucleosynthesis period**, but the rapid expansion soon dropped the temperature too low for the fusion to continue and the formation of heavier elements had to await the birth of stars.

Section 9-8 Cosmogenesis 327

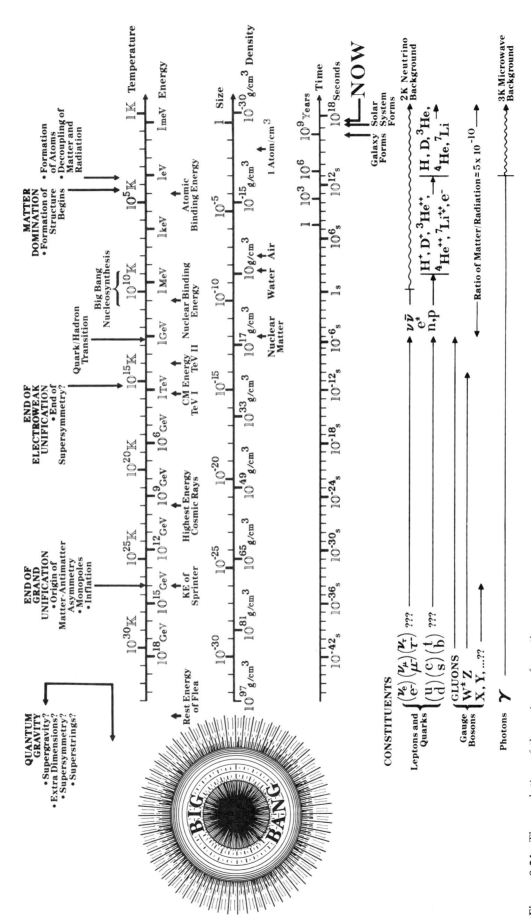

Figure 9-24 The evolution of the universe from time $t = 0$ to the present according to the standard model.

A long time later, when the temperature had dropped to about 3000 K as the universe grew to about 1/1000 of its present size, kT dropped below typical atomic ionization energies and atoms were formed. By then the expansion had redshifted the radiation field so that the total radiation energy was about equal to the energy represented by the remaining mass. As expansion and cooling continued, the energy of the steadily redshifting radiation steadily declined until, at $t = 10^{10}$ y (now), matter came to dominate the universe, with its energy density exceeding that of the 2.7-K radiation remaining from the Big Bang by a factor of about 1000.

Unanswered Questions and the Limits of Knowledge

The standard model of the evolution of the universe and the current theories of stellar and galactic genesis and evolution have been amazingly successful. Still, some fundamental questions that have arisen during our discussions are yet unanswered. Will the universe expand forever or rebound to a point and repeat the Big Bang? The answer depends on whether the present average mass density is greater or less than the critical density of about 10^{-26} kg/m^3. The uncertainty in the current measurement would allow either possibility, but the value is tantalizingly close to the critical value. If it does equal the critical value, an intriguing additional question is, "Why?" We have noted the serious problem of the missing mass of the universe and how it might be explained. The possible explanations—black holes and massive neutrinos—present other major questions. Answering some of them requires that we probe at the current limits of physical knowledge. For example, near a mass m, general relativity prevents our seeing events occurring at dimensions less than L, the **event horizon:**

$$L = \frac{Gm}{c^2} \qquad 9\text{-}32$$

On the other hand, the uncertainty principle in quantum theory places this limit at the Compton wavelength λ_C:

$$\lambda_C = \frac{h}{mc} \qquad 9\text{-}33$$

Equating these yields the **Planck mass** $m = 5.5 \times 10^{-8}$ kg. The length $L = \lambda_C \approx 10^{-35}$ m is called the **Planck length** and the time for light to travel across that length,

$$t = \left(\frac{Gh}{c^5}\right)^{1/2} = 1.35 \times 10^{-43} \text{ s} \qquad 9\text{-}34$$

is called the **Planck time.** Thus, for Planck time the mass density of the universe is such that the mass m is contained within a volume of dimensions $L^3 \sim (10^{-35} \text{ m})^3$. Relativistic space-time is no longer a continuum and even a new theory of gravity—quantum gravity or supergravity—is needed.

Perhaps it really is as some cosmologists have suggested: If the universe had evolved even slightly differently than it has, perhaps due to a slightly different value for h or e or some other fundamental constant, life on earth and maybe the earth itself would be impossible. This is the **anthropic principle,** which holds that the universe looks as it does because we are here to see it.

Summary

1. The luminosity L of a star is the power that it emits. It is related to the power per unit area received at the earth, called the apparent brightness f, and the distance r to the star by

$$L = 4\pi r f$$

   In the case of the sun, where r is accurately known, f is called the solar constant.

2. Computations of stellar surface temperatures are based on Planck's law of blackbody radiation, an assumption well supported by detailed measurements of the sun's electromagnetic radiation spectrum.

3. The source of the sun's energy is the proton–proton fusion cycle which starts with the nuclear reaction

$$^1H + {}^1H \rightarrow {}^2H + e^+ + \nu_e + 1.44 \text{ MeV}$$

4. Stars are classed as population I or population II, based on their compositions. Population I stars have 2 to 3 percent of their mass composed of elements heavier than He and are considered to be the younger of the two classes. Population II stars are nearly devoid of elements heavier than He.

5. The Milky Way is a spiral galaxy consisting of about 10^{10} stars. The sun is located about 30,000 $c \cdot y$ from the center of the Galaxy, which is in the direction of the constellation Sagittarius from us. Approximately 90 percent of the gravitational mass of the Galaxy consists of dark, that is, nonluminous matter.

6. The Hertzprung–Russell diagram relates the luminosity of stars to their effective temperatures. Both quantities are related to stellar masses:

$$L \propto M^4 \qquad T \propto M^{1/2}$$

   Stars that are "burning" hydrogen into helium fall on the main sequence of the diagram.

7. Following the exhaustion of their hydrogen fuel supply, stars evolve along different paths in the H–R diagram that depend primarily on their initial masses, eventually reaching one of three possible final states: white dwarf, neutron star, or black hole. It is in cataclysmic events that precede the latter two final states that elements heavier than Fe are formed.

8. If the mass of an incipient neutron star is so large that its radius is less than the Schwarzchild radius

$$R_S = 2GM/c^2$$

   then no radiation or object with mass can escape its surface. It is a black hole.

9. Galaxies outside the Milky Way were first identified by Edwin Hubble who also showed that, with rare exceptions, they could be grouped into four general classes: spirals, barred spirals, ellipticals, and irregulars.

10. Hubble's law relates the recession velocity of a galaxy, determined from the redshift of its spectrum, to the distance of the galaxy from us:

$$v = Hr$$

where the Hubble constant $H = 23$ km/s per million light-years. From Hubble's law, we conclude that the universe is expanding and that the expansion began approximately $1/H$ years ago.

11. The critical density of the universe is $\rho_c \approx 10^{-26}$ kg/m^3, which is the density at which the kinetic energy of a galaxy equals its gravitational potential energy. If the density of the universe happens to equal this value, the expansion will ultimately coast to a halt. Lower values will result in expansion forever. Higher values will cause an eventual reversal of the present expansion.

12. The model currently used to describe the evolution of the universe is the standard model, in which the universe began with the Big Bang approximately 10^{10} years ago.

13. The standard model is supported by substantial experimental observations, including the isotropic, 2.7-K, background blackbody radiation spectrum. There are also many fundamental questions for which it has not yet provided answers.

Suggestions for Further Reading

Pipkin, Francis M.: "Gravity Up in the Air: Are the Laws of the Universe in Doubt?" *The Sciences,* May/June 1984, p. 42.

This article examines cosmological theories in which the gravitational "constant," G, varies over time.

Schramm, David N. and Gary Steigman: "Particle Accelerators Test Cosmological Theory," *Scientific American,* June 1988, p. 66.

This article describes how cosmological theories have set limits on the possible number of families of elementary particles and how this prediction is being tested.

von Baeyer, Hans Christian: "Creatures of the Deep," *The Sciences,* March/April 1989, p. 2.

The problem of "dark matter," which may make up most of the mass of the universe, puzzles cosmologists, astronomers, and particle physicists.

Review

A. Objectives: After studying this chapter, you should:

1. Know the source of the sun's energy, the general structure of the sun's outer layers, and the approximate value of its surface temperature; and understand the origin of sunspots.

2. Know the approximate size of the Milky Way galaxy in light-years and the sun's approximate location in the Galaxy; and know how the luminosity of a star, its effective temperature, and its mass are related by the Hertzprung–Russell diagram.

3. Know about the missing-mass problem in the Galaxy and in the universe and how it relates to the future expansion.

4. Be able to describe the evolutionary path of a medium size star, like the sun; and know how the mass of a star determines which of the possible final states—white dwarf, neutron star, or black hole—it will reach.

5. Know how galaxies are classified and how they are distributed throughout the universe; and know about the prodigious power emitted by active galaxies.

6. Be able to compute the distances to galaxies with Hubble's law, using the recession velocity determined from the doppler effect.

7. Know about the cosmological principle and the general expansion of the universe.

8. Know how the current theory of cosmogenesis, the standard model, accounts for the evolution of the universe to the present time and suggests what its future may be.

B. Define, explain, or otherwise identify:

Astrophysics
Cosmology
Photosphere
Solar constant
Luminosity
Quiet sun
Limb
Chromosphere
Corona
Solar wind
Core

Proton–proton cycle
Solar-neutrino problem
Sunspot cycle
Sunspots
Solar flares
Plages
Filaments
Prominences
Constellations
Star clusters
Galactic clusters

Open clusters	Neutron star	Electroweak force	Planck mass
Population I stars	Pulsar	Lepton era	Planck length
Globular clusters	Schwarzschild radius	Radiation era	Planck time
Population II stars	Gravitational redshift	Nucleosynthesis period	Anthropic principle
Missing-mass problem	Black hole	Event horizon	
Parallax angle	Interstellar dust		
Hertzprung–Russell (H–R) diagram	Ellipticals		
	Disks		
Main sequence	Ordinary spirals		
Main-sequence dwarfs	Barred spirals		
Red subgiant	Irregular galaxies		
Red giant	Quiet galaxies		
Horizontal branch	Active galaxies		
Planetary nebula	Seyfert galaxies		
Binary pairs	N galaxies		
Lagrangian point	BL Lac objects		
Roche lobe	Radio galaxies		
Accretion disk	Quasars		
Nova	Quasistellar objects (QSOs)		
Dwarf nova			
Recurrent nova	Hubble's law		
Classical nova	Hubble constant		
Supernova	Hubble age		
Degenerate dwarf	Cosmological principle		
White dwarf	Perfect cosmological principle		
Electron-degeneracy pressure			
	Cosmological constant		
Chandrasekhar limit	Big Bang		
Black dwarf	Standard model		

C. True or false: If a statement is true, explain why it is true. If it is false, give a counterexample.

1. The part of the sun that we see, the surface or photosphere, cannot be more than about 700 km thick.

2. A significant portion of the solar energy absorbed by the earth is furnished by neutrinos produced in the sun's proton–proton fusion cycle.

3. The lifetime of a star whose mass is five times that of the sun will be only one-fifth of the sun's lifetime.

4. Supernova SN1987A didn't really occur in 1987.

5. Unlike white dwarfs and neutron stars, black holes are not cooling toward thermal equilibrium with the universe.

6. There is no physical evidence to support the theory that the universe is expanding.

7. The sun is located at the center of the Milky Way.

8. According to the standard model, when the temperature of the expanding universe fell below about ten billion K, the formation of electron–positron pairs ceased.

Problems

Level I

9-1 Our Star, the Sun

1. Measurement of the doppler shift of spectral lines in light from the east and west limbs of the sun at the solar equator reveals that the tangential velocities of the limbs differ by 4 km/s. Use this result to compute the approximate period of the sun's rotation. ($R_\odot = 6.96 \times 10^5$ km)

2. The gravitational potential energy U of a self-gravitating spherical body of mass M and radius R is a function of the details of the mass distribution. For the sun, $U_\odot = -2GM_\odot^2/R_\odot$. What would be the approximate lifetime of the sun, if the source of its emitted energy were entirely derived from gravitational contraction? ($M_\odot = 1.99 \times 10^{30}$ kg)

9-2 The Stars

3. The sun is moving with speed 2.5×10^5 m/s in a circular orbit about the center of the Galaxy. How long (in earth years) does it take to complete one orbit? How many orbits has it completed since it was formed?

4. The reason massive neutrinos are considered a candidate for solving the missing-mass problem is that, at the conclusion of the lepton era, the universe contained about equal numbers of photons and neutrinos. They are still here, for the most part. The photons can be observed and their density is measured to be about 500 photons/cm³; thus, there must be about that number density of neutrinos in the universe, too. If neutrinos have a mass m_ν and if the cosmological expansion has reduced their average speed so that their energy is now primarily rest mass, what would be the individual neutrino mass (in eV/c^2) necessary to account for the missing mass of the universe? Recall that the observed mass accounts for only about 10 percent of that needed to bond the universe.

9-3 The Evolution of Stars

5. A unit of length often used by astronomers to measure distances in nearby space is the parsec, defined as the distance at which a star subtends a parallax angle of one arc second. (See Equation 9-11 and Example 9-4.) The practical limit of such measurements is 0.01 arc second. (a) How many light-years is 1 parsec? (b) If the density of stars in the sun's region of the Milky Way galaxy is 0.08 stars/(parsec)³, how many stars could, in principle, have their distances from us measured by the trigonometric parallax method?

6. Astronomers often use the *apparent magnitude m* as a means to compare the visual brightness of stars and then relate the comparison to the luminosity and distance to standard stars, such as the sun. (See Equation 9-10.) The difference in the apparent magnitudes of two stars m_1 and m_2 is defined as $m_1 - m_2 = 2.5 \log(f_1/f_2)$, a relation based on the logarithmic response of the human eye to the

brightness of objects. Pollux, one of the "twins" in the constellation Gemini, has apparent magnitude 1.16 and is 12 parsecs away. Betelgeuse, the star at Orion's right shoulder, has apparent magnitude 0.41. How far away is Betelgeuse, if they have the same luminosity?

7. Using the H–R diagram (Figure 9-14), determine the effective temperature and the luminosity of a star whose mass is (a) $0.3 M_\odot$ and (b) $3 M_\odot$. (c) Find the radius of each star. (d) Determine their expected lifetimes, relative to that of the sun.

9-4 Cataclysmic Events

8. Compute the energy required (in MeV) to produce each of the photodisintegration reactions in Equations 9-18 and 9-19.

9. The gas shell of a planetary nebula shown in Figure 9-15 is expanding at 24 km/s. Its diameter is 1.5 c·y. (a) How old is the gas shell? (b) If the central star of the planetary nebula is 12 times as luminous as the sun and 1.36 times hotter, what is the radius of the central star in units of $R_\odot$?

9-5 Final States of Stars

10. Calculate the Schwarzschild radius of a star whose mass is equal to that of (a) the sun, (b) Jupiter, (c) the earth. (The mass of Jupiter is approximately 318 times that of the earth.)

11. Consider a neutron star whose mass equals $2 M_\odot$. (a) Compute the star's radius. (b) If the neutron star is rotating at 0.5 rev/s and its density is uniform, what is its rotational kinetic energy? (c) If its rotation slows by 1 part in 10^8 per day and the lost kinetic energy is all radiated, what is the star's luminosity?

12. If the 90 percent of the Milky Way's mass that is "missing" resides entirely in a large black hole at the center of the Galaxy, what would be the black hole's (a) mass and (b) radius?

9-6 Galaxies

13. A particular galaxy is observed to have a recession velocity of 72,000 km/s. (a) Find the distance to the galaxy. (b) What is the upper bound to the present age of the universe according to the Big Bang expansion theory. (c) The value of Hubble's constant depends critically on calibration distance measurements, which are difficult to make. If the calibration distance measurements are in error by 10 percent, by how much is the age calculation in (b) in error?

14. The bright core of a certain Seyfert galaxy had a luminosity of $10^{10} L_\odot$. The luminosity increased by 100 percent in a period of 18 months. Show that this means the energy source of the core is about 9.5×10^4 AU in diameter. How does this compare to the diameter of the Milky Way galaxy?

9-7 Gravitation and Cosmology

15. Evaluate Equation 9-30 for the critical density of the universe.

9-8 Cosmogenesis

16. Cosmological theory suggests that the average distance between galaxies and, hence, the size of the universe is inversely proportional to the absolute temperature. If this is true, how large was the universe relative to its present size (a) 2000 years ago, (b) 10^6 years ago, (c) when $t = 1$ s after the Big Bang, (d) when $t = 10^{-6}$ s, and (e) $t = 10$ s. (See Figure 9-24.)

17. Determine the mass density of the universe for $t =$ Planck time. How does this value compare to the mass density of the proton? Of osmium (density, 2.25×10^4 kg/m^3)?

18. At what wavelength is the blackbody radiation distribution of the cosmic microwave background at a maximum?

19. How long after the Big Bang did it take the universe to cool to the threshold temperature for the formation of muons? What would be the mass of a particle that could be formed by the average energy of the current 2.7-K background radiation?

Level II

20. If Hubble's law is true for an observer in the Milky Way galaxy, prove that it must also be true for observers in other galaxies. (*Hint:* Use the vector property of the velocity.)

21. Supernova SN1987A was first visible at the earth in 1987. (a) How many years before 1987 did the explosion occur? (b) If protons with 100 GeV of kinetic energy were produced in the event, when should they arrive at earth? (See Chapter 8, Problem 28b.)

22. Assume that the sun was composed of 70 percent hydrogen when it first formed. (a) How many hydrogen nuclei were there in the sun at that time? (b) How much energy would ultimately be released if all of the hydrogen nuclei fused into helium? (c) Astrophysicists have predicted that the sun can radiate energy at its current rate until about 23 percent of the hydrogen has been burned. What total lifetime for the sun does that prediction imply?

23. Consider an eclipsing binary, consisting of two stars of mass m_1 and m_2 separated by a distance r, whose orbital plane is parallel to our line of sight. Doppler measurements of the radial velocity of each component of the binary are shown below. Assume that the orbit of each star about the center of mass is circular. (a) What is the period T and angular frequency ω of the binary? (b) Show that $m_1 + m_2 = \omega^2 r^3 / G$. (c) Compute the values of m_1, m_2, and r from the data in the v versus t graph.

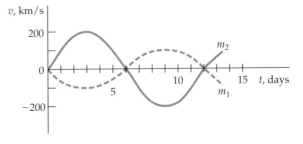

24. Prove that the total energy of the earth's orbital motion $E = (mv^2/2) + (-GM_\odot m/r)$ is equal to one-half of its gravitational potential energy $(-GM_\odot m/r)$, where r is the earth's orbital radius.

25. Given the currently accepted value of the Hubble constant and the fact that the average matter density of the universe is one H-atom/m^3, what creation rate of new H-atoms would be necessary in a steady-state model to maintain the present mass density, even though the universe is expanding? [Give your answer in (H-atoms/m^3)/10^6 y.] Would you expect such a spontaneous creation rate to be readily observable?

Level III

26. The ability of a planet to retain particular gases in an atmosphere depends on the temperature that its atmosphere has (or would have) and the escape velocity for that body. In general, if the average speed of a particular gas molecule exceeds $\frac{1}{6}$ of the escape velocity, that gas will disappear from the atmosphere in about 10^8 years. (a) Graph the average speed of H_2O, CO_2, O_2, CH_4, H_2, and He from 50 K to 1000 K. On the same graph, show the points representing $\frac{1}{6}$ of the escape velocity versus average temperature of the atmosphere for the bodies in the table below. (b) Show that the escape speed from a planet is $v = v_{Earth}\sqrt{\alpha/\beta}$. (c) Of the six gases plotted, above, which would probably be found in the current atmospheres of the solar system bodies in the table, and which would not? Explain *each* answer briefly.

Atmospheric Temperatures

Average T_{Atmo}(K)	Body	α Mass (Earth = 1.0)	β Diameter (Earth = 1.0)
300	Earth	1.00	1.00
390	Venus	0.81	0.95
600	Mercury	0.06	0.38
150	Jupiter	318.00	11.00
60	Neptune	17.00	3.90
290	Mars	0.11	0.53

27. Suppose that the sun's luminosity increases by a factor of 100 as it evolves into a red giant star. Show that the earth's oceans will evaporate, but the water vapor will not escape from the atmosphere.

28. The approximate mass of dust in the Galaxy can be computed from the observed extinction of starlight. Assuming the mean radius of dust grains to be R with a uniform number density of n grains/cm^3, (a) show that the mean free path d_0 of a photon in interstellar dust is given by $d_0 = 1/n\pi R^2$. (b) Starlight traveling toward an earth observer a distance d from the star has intensity $I = I_0 \, e^{-d/d_0}$. In the vicinity of the sun, measurement of I yields $d_0 = 3000$ c·y. If $R = 10^{-5}$ cm, calculate n. (c) The average mass density of solid material in the Galaxy is 2 g/cm^3; and in the disk, the density of stars is about $1M_\odot/300$. Compute the ratio of the mass of dust in 300 (c·y)^3 to $1M_\odot$.

29. Supernova SN1987A certainly produced some heavy elements. To illustrate this, calculate the energy required to fuse two ^{56}Fe-atoms into one ^{112}Cd-atom. Compare this with the energy released in fusing 56 ^{1}H-atoms into one ^{56}Fe-atom, using the proton–proton cycle, and show that more than enough energy is available to produce a ^{112}Cd atom (mass, 111.902760 u).

Appendix A

Summary of Selected Mathematical Relations

In this appendix, we list certain basic results from algebra and trigonometry.

The Binomial Expansion

The binomial theorem is very useful for making approximations. One form of this theorem is

$$(1 + x)^n = 1 + nx + \frac{n(n-1)}{2!} x^2 + \frac{n(n-1)(n-2)}{3!} x^3$$
$$+ \frac{n(n-1)(n-2)(n-3)}{4!} x^4 + \cdots \qquad \text{A-1}$$

If n is a positive integer, there are $n + 1$ terms in this series. If n is a real number other than a positive integer, there are an infinite number of terms. The series is valid for any value of n if x^2 is less than 1. It is also valid for $x^2 = 1$ if n is positive. The series is particularly useful if $|x|$ is much less than 1. Then each term is much smaller than the previous term and we can drop all but the first two or three terms in the equation. If $|x|$ is much less than 1, we have

$$(1 + x)^n \approx 1 + nx \qquad |x| \ll 1 \qquad \text{A-2}$$

Complex Numbers

A general complex number z can be written

$$z = a + bi \qquad \text{A-3}$$

where a and b are real numbers and $i = \sqrt{-1}$. The quantity a is called the real part and the quantity ib is called the imaginary part of z. We can represent a complex number in a plane as shown in Figure A-1, where the x axis is the real axis and the y axis is the imaginary axis. We can use the relations $a = r \cos \theta$ and $b = r \sin \theta$ from Figure A-1 to write the complex number z in polar coordinates:

$$z = r \cos \theta + (r \sin \theta) i \qquad \text{A-4}$$

where $r = \sqrt{a^2 + b^2}$ is called the magnitude of z.

When complex numbers are added or subtracted, the real and imaginary parts are added or subtracted separately:

$$z_1 + z_2 = (a_1 + b_1 i) + (a_2 + b_2 i) = (a_1 + a_2) + (b_1 + b_2) i \qquad \text{A-5}$$

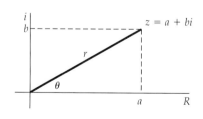

$z = a + bi$
$\quad = r \cos \theta + (r \sin \theta) i$
$\quad = r(\cos \theta + i \sin \theta)$

Figure A-1 Representation of a complex number in a plane. The real part of the complex number is plotted along the horizontal axis, and the imaginary part is plotted along the vertical axis.

However, when two complex numbers are multiplied, each part of one number is multiplied by each part of the other number:

$$z_1 z_2 = (a_1 + b_1 i)(a_2 + b_2 i) = a_1 a_2 + b_1 b_2 i^2 + (a_1 b_2 + a_2 b_1)i \quad \text{A-6}$$
$$= a_1 a_2 - b_1 b_2 + (a_1 b_2 + a_2 b_1)i$$

where we have used $i^2 = -1$.

The complex conjugate of a complex number z^* is that number obtained by replacing i with $-i$:

$$z^* = (a + bi)^* = a - bi \quad \text{A-7}$$

The product of a complex number and its complex conjugate equals the square of the magnitude of the number:

$$zz^* = |z|^2 = (a + bi)(a - bi) = a^2 + b^2 \quad \text{A-8}$$

The notation $|z|^2$ is called the square modulus of z and is just another way of writing zz^*. A particularly useful function of a complex number is the exponential $e^{i\theta}$. Using the expansion for e^x given in Table A-1, we have

$$e^{i\theta} = 1 + i\theta + \frac{(i\theta)^2}{2!} + \frac{(i\theta)^3}{3!} + \frac{(i\theta)^4}{4!} + \cdots$$

Using $i^2 = -1$, $i^3 = -i$, $i^4 = +1$, and so forth and separating the real parts from the imaginary parts, this expansion can be written

$$e^{i\theta} = 1 - \frac{\theta^2}{2!} + \frac{\theta^4}{4!} + \cdots + i\left(\theta - \frac{\theta^3}{3!} + \cdots\right)$$

Comparing this result with the last two equations in Table A-2, we can see that

$$e^{i\theta} = \cos\theta + i\sin\theta \quad \text{A-9}$$

Using this result, we can express a general complex number as an exponential:

$$z = a + bi = r\cos\theta + (r\sin\theta)i = re^{i\theta} \quad \text{A-10}$$

where $r = \sqrt{a^2 + b^2}$.

Table A-1 Exponential and Logarithmic Functions

$e = 2.71828 \quad e^0 = 1$

If $y = e^x$, then $x = \ln y$.

$e^{\ln x} = x$

$e^x e^y = e^{(x+y)}$

$(e^x)^y = e^{xy} = (e^y)^x$

$\ln e = 1 \quad \ln 1 = 0$

$\ln xy = \ln x + \ln y$

$\ln \dfrac{x}{y} = \ln x - \ln y$

$\ln e^x = x \quad \ln a^x = x \ln a$

$\ln x = \ln 10^{\log x} = (\log x) \ln 10$
$\quad\quad = 2.3026 \log x$

$\log x = \log e^{\ln x} = (\ln x) \log e$
$\quad\quad = 0.43429 \ln x$

$e^x = 1 + x + \dfrac{x^2}{2!} + \dfrac{x^3}{3!} + \cdots$

$\ln(1 + x) = x - \dfrac{x^2}{2} + \dfrac{x^3}{3} - \dfrac{x^4}{4} + \cdots$

Table A-2 Trigonometric Formulas

$\sin^2 \theta + \cos^2 \theta = 1 \qquad \sec^2 \theta - \tan^2 \theta = 1 \qquad \csc^2 \theta - \cot^2 \theta = 1$

$\sin 2\theta = 2 \sin \theta \cos \theta$

$\cos 2\theta = \cos^2 \theta - \sin^2 \theta = 2 \cos^2 \theta - 1 = 1 - 2 \sin^2 \theta$

$\tan 2\theta = \dfrac{2 \tan \theta}{1 - \tan^2 \theta}$

$\sin \dfrac{1}{2}\theta = \sqrt{\dfrac{1 - \cos \theta}{2}} \qquad \cos \dfrac{1}{2}\theta = \sqrt{\dfrac{1 + \cos \theta}{2}} \qquad \tan \dfrac{1}{2}\theta = \sqrt{\dfrac{1 - \cos \theta}{1 + \cos \theta}}$

$\sin (A \pm B) = \sin A \cos B \pm \cos A \sin B$

$\cos (A \pm B) = \cos A \cos B \mp \sin A \sin B$

$\tan (A \pm B) = \dfrac{\tan A \pm \tan B}{1 \mp \tan A \tan B}$

$\sin A \pm \sin B = 2 \sin [\tfrac{1}{2}(A \pm B)] \cos [\tfrac{1}{2}(A \mp B)]$

$\cos A + \cos B = 2 \cos [\tfrac{1}{2}(A + B)] \cos [\tfrac{1}{2}(A - B)]$

$\cos A - \cos B = 2 \sin [\tfrac{1}{2}(A + B)] \sin [\tfrac{1}{2}(B - A)]$

$\tan A \pm \tan B = \dfrac{\sin (A \pm B)}{\cos A \cos B}$

$\sin \theta = \theta - \dfrac{\theta^3}{3!} + \dfrac{\theta^5}{5!} - \dfrac{\theta^7}{7!} + \cdots$

$\cos \theta = 1 - \dfrac{\theta^2}{2!} + \dfrac{\theta^4}{4!} - \dfrac{\theta^6}{6!} + \cdots$

Appendix B

Properties of Classical Waves

Despite the immense diversity of wave phenomena such as vibrating strings, sound waves, light waves, water waves, or radio waves, there are many features common to all wave phenomena including the non-classical "electron waves" that arise in quantum theory. A thorough understanding of classical wave properties is therefore a great help in the understanding of modern physics. In fact, much of the difficulty experienced by students in learning quantum theory can be traced to a lack of familiarity with classical wave phenomena. You are therefore urged to review what you have learned about waves in elementary physics. In this appendix, we will give a brief review or outline of some of the properties of waves that are important for the understanding of quantum theory.

The Wave Equation

Consider the familiar example of waves on a very long string, which we assume to be along the x axis. Waves are usually produced on such a string by vibrating one end in some way so that the string has a displacement y that depends on both position x and time t. The acceleration of a point on the string is $\partial^2 y/\partial t^2$, where the partial-derivative notation ∂ is used because y depends on both x and t. A direct application of Newton's second law $\Sigma F = ma$ to a segment of the string results in the **wave equation**

$$\frac{\partial^2 y}{\partial x^2} = \frac{1}{v^2} \frac{\partial^2 y}{\partial t^2} \qquad \text{B-1}$$

where v is the phase velocity, which, for a perfectly flexible string is related to the tension F and the mass per unit length μ by $v = \sqrt{F/\mu}$. A similar application of Newton's laws, along with the gas laws, leads to an identical wave equation for sound waves in a gas in which case the wave function $y(x, t)$ can represent either the displacement of gas particles from their equilibrium position, or the pressure or gas density variation. The same wave equation for light and other electromagnetic waves arises from the laws of electricity and magnetism as expressed in Maxwell's equations. For electromagnetic waves, the wave function $y(x, t)$ represents either an electric field component or a magnetic field component, and the phase velocity v is given by $v = c/n$, where c is the speed of light in vacuum, and n is the index of refraction of the medium. Any function $y(x, t)$ that depends on x and t only in the combination $x - vt$ or $x + vt$ is a solution of the wave equation. For example, if $y = f(x)$ is the shape of the string at time $t = 0$, the function $f(x - vt)$ describes the situation in which this shape is propagated to the right with speed v and $f(x + vt)$ describes propagation to the left.

An important property of the wave equation (and of any linear equation) is that the sum of two solutions is also a solution. This is known as the superposition principle. The superposition of wave functions has important applications in the understanding of standing waves, interference, diffraction, harmonic analysis and synthesis and the propagation of wave pulses.

Harmonic Waves

A particularly useful solution of the wave equation is the **harmonic wave**:

$$y(x, t) = A \sin \frac{2\pi}{\lambda}(x - vt) = A \sin 2\pi\left(\frac{x}{\lambda} - \frac{t}{T}\right) \quad \text{B-2}$$

This function describes a wave traveling in the positive x direction with **amplitude** A, **wavelength** λ, **period** T, **frequency** $f = 1/T$, and phase velocity

$$v = f\lambda \quad \text{B-3}$$

It is convenient to use the **angular frequency** ω and **wave number** k defined by

$$\omega = \frac{2\pi}{T} = 2\pi f \quad \text{B-4}$$

and

$$k = \frac{2\pi}{\lambda} \quad \text{B-5}$$

The harmonic wave function is then written

$$y = A \sin (kx - \omega t) \quad \text{B-6}$$

Other forms of this type of solution are

$$y = A \cos (kx - \omega t) \quad \text{B-7}$$

and

$$y = A e^{i(kx - \omega t)} \quad \text{B-8}$$

Equation B-7 differs from B-6 only in the choice of origin of x or t. Equation B-8 is a linear combination of Equations B-6 and B-7 as

$$e^{i\theta} = \cos \theta + i \sin \theta \quad \text{B-9}$$

Because $y(x, t)$ must be real, it is usually understood that either the real or the imaginary part of the right side of Equation B-8 is to be taken.

Harmonic waves have the somewhat artificial property of extending throughout all space from $x = -\infty$ to $x = +\infty$ and throughout all time. They are very useful for two reasons: (1) they approximate the waves of finite extent that occur in many real situations, for example on a long string by moving one end with simple harmonic motion, and (2) any other type of wave, such as a wave pulse, can be represented by a superposition of harmonic waves of different frequencies and wavelengths.

If we look at a harmonic wave on a string at a certain instant $t = t_0$, the wave function $y(x, t_0)$ describes the shape of the string as it would look in a "snapshot" (Figure B-1). The shape is a sine wave (or a cosine wave depending on the location of the origin). If we look at a particular point x_0, the wave function $y(x_0, t)$ describes the motion of that point on the string. Since y is a sinusoidal function, the motion of any point x_0 is simple harmonic motion of amplitude A and frequency f.

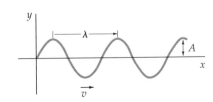

Figure B-1 Harmonic wave at some instant in time. A is the amplitude and λ is the wavelength. For waves on a string, this figure can be obtained by taking a snapshot of the string.

Energy and Intensity

An important property of all waves is that the energy per unit volume or **energy density** η is proportional to the square of the amplitude. For electromagnetic waves the energy density is related to the electric field by

$$\eta = \epsilon_0 E^2 \qquad \text{B-10}$$

The intensity of the wave I, which is the energy transported per unit area per unit time (measured in watts per square meter), is related to the energy density and the wave speed. For a plane wave this relation is

$$I = \eta v \qquad \text{B-11}$$

The intensity is therefore also proportional to the square of the amplitude.

Standing Waves

If a string fixed at both ends is set into oscillation, there are certain resonance frequencies for which standing waves result. The standing-wave condition is that an integral number of half wavelengths fit into the length of the string (Figure B-2)

$$n\frac{\lambda_n}{2} = L \qquad \text{B-12}$$

In terms of the frequency, this condition is

$$f_n = \frac{v}{\lambda_n} = n\frac{v}{2L} \qquad \text{B-13}$$

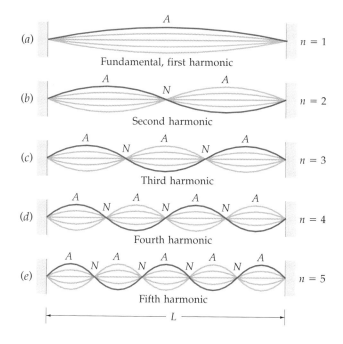

Figure B-2 Standing waves on a string that is fixed at both ends. The length of the string is $\lambda/2$ for the first harmonic, $2\lambda/2$ for the second harmonic, $3\lambda/2$ for the third harmonic, and so on.

Standing waves can be considered as a superposition of two traveling waves of equal frequency and wavelength moving in opposite directions. Standing sound waves can be set up in air columns, as in a flute or organ pipe. Similarly, standing electromagnetic waves can be set up in a cavity such as a microwave oscillator.

The Doppler Effect

When a wave source and a receiver are moving relative to each other, the frequency observed by the receiver is not the same as the source frequency. When they are moving toward each other, the observed frequency is greater than the source frequency; when they are moving away from each other, the observed frequency is less than the source frequency. This is called the doppler effect. A familiar example is the change in pitch of a car horn when the car is approaching or receding.

The change in frequency of a sound wave is slightly different depending on whether the source or the receiver is moving relative to the medium. When the source moves, the wavelength changes, and the new frequency f is found by first finding the new wavelength λ, and then computing $f = v/\lambda$. On the other hand, when the source is stationary and the receiver moves, the frequency is different simply because the receiver moves past more or fewer waves in a given time.

The observed frequency f' is related to the source frequency f_0, the source speed u_s, and the receiver speed u_r by

$$f' = \frac{(1 \pm u_r/v)}{(1 \pm u_s/v)} f_0 \qquad \text{B-14}$$

When the relative speed of the source or receiver, u, is much less than the wave speed, v, the doppler shift is approximately the same whether the source or receiver moves and is given by

$$\frac{\Delta f}{f_0} \approx \pm \frac{u}{v} \qquad \text{B-15}$$

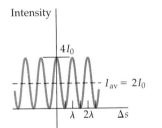

Figure B-3 Intensity versus path difference Δs between two sources that are coherent and in phase. I_0 is the intensity due to each individual source.

Interference and Diffraction

An important example of the superposition of waves is the interference of two waves that differ in phase. Figure B-3 shows the classic interference pattern due to waves from two sources that are coherent and in phase. At points on the screen for which the path difference from the sources is an integral number of wavelengths, the waves are in phase and the amplitude is twice that from either source separately. The intensity is therefore four times that from either source. At points for which the path difference is an integral number of wavelengths $\pm \frac{1}{2}\lambda$, the waves are 180° out of phase and therefore cancel, producing a resultant amplitude and intensity of zero.

We are often interested in determining when the waves from several sources will give complete destructive interference, that is, will add to zero. For N waves of equal amplitude with phase difference δ between each successive pair of waves, complete cancellation occurs when $\delta = 360°/N$ (or any integer times $360°/N$). The phase difference between the first and last waves is then $(N-1)360°/N = 360° - 360°/N$. For a very large number of waves we have the important result that complete destructive interference occurs when the first and last source differ in phase by 360°, that is, they are in phase. We can understand this result by considering 100 waves with a phase difference $360°/100$ between each successive pair. The first wave and the fifty-first wave will then differ in phase by 180° and therefore cancel. Similarly, the second and fifty-second waves will also differ in phase by 180° and cancel, as will the third and fifty-third waves, and so on. The resulting amplitude and the intensity due to the superposition of all of the waves will therefore be zero.

An important wave phenomenon that can be understood in terms of the superposition of many waves is diffraction, the spreading or bending of

(a)

(b)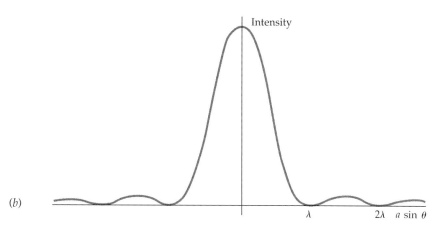

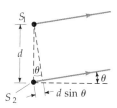

Figure B-4 Single-slit diffraction pattern. The intensity is plotted versus $a \sin \theta$, which is the path difference between the light from the top and bottom of the slit.

Figure B-5 The path difference at an angle θ for two sources a distance d apart is $d \sin \theta$.

waves around corners. The single-slit diffraction pattern is illustrated in Figure B-4.

This pattern can be analyzed using Huygen's principle by replacing the slit opening with a large number of sources, and calculating the resulting interference pattern on the screen. At any angle θ, there will be a phase difference between waves from two sources a distance d apart due to the path difference $d \sin \theta$ as shown in Figure B-5. If a is the width of the slit, the path difference for waves from the top and bottom of the slit is $a \sin \theta$. If this path difference is one wavelength, the first and last waves will differ in phase by 360° and the resulting amplitude due to the superposition of all the waves will be zero. The location of the first minimum in the single-slit diffraction pattern is thus given by

$$a \sin \theta = \lambda \qquad \text{B-16}$$

We see from this result that the larger the width a, the smaller the angle θ to the first minimum, thus the width of the diffraction pattern on the screen varies inversely with the width of the slit.

Figure B-6 shows the diffraction of plane water waves in a ripple tank that meet a barrier with a small opening. The waves to the right of the barrier are not confined to the narrow angle of the rays from the source that can pass through the opening, but are circular waves as if there were a point source at the opening. Figure B-7 shows plane waves hitting a barrier with an opening much larger than the wavelength. The waves beyond the barrier

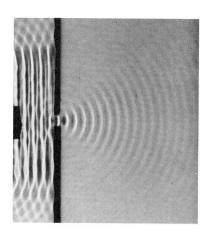

Figure B-6 Plane waves in a ripple tank meeting a barrier with a small opening. The waves to the right of the barrier are circular waves that are concentric about the opening just as if there were a point source at the opening.

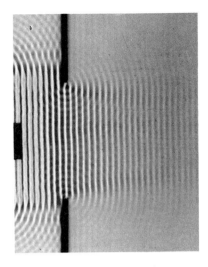

Figure B-7 Plane waves in a ripple tank meeting a barrier with an opening that is much larger than the wavelength. The barrier has a noticeable effect only near the edges of the opening.

are similar to plane waves in the region far from the edges of the opening. (Near the edges the wavefront is distorted, and the wave appears to bend slightly.)

> If the aperture or obstacle is large compared with the wavelength, the bending of the wavefront is not noticeable and the wave propagates in straight lines or rays much as a beam of particles does.

The result is known as the **ray approximation.** Because the wavelength of audible sound ranges from a few centimeters to several meters, and is often large compared with apertures and obstacles (doors or windows, for example), the bending of sound waves around corners is a common everyday phenomenon. On the other hand, the wavelengths of visible light range from about 4×10^{-7} to 7×10^{-7} m. Because these wavelengths are so small compared with the size of ordinary objects and apertures, the diffraction of light is not easily noticed, and light appears to travel in straight lines.

Many important wave phenomena involve both diffraction and inteference.

> Interference and diffraction are the primary wave properties that distinguish wave propagation from the propagation of particles.

Harmonic Analysis and Synthesis

When an oboe and a violin in an orchestra play the same note, for example, concert A, they sound quite different. Both tones have the same "pitch," a psychological sensation as to the highness or lowness of the note that is strongly correlated with frequency—the higher the frequency the higher the pitch. However, the notes differ in what is called **tone quality.** The principle reason for the difference in tone quality is that, although both the violin and oboe are producing vibrations at the fundamental 440 Hz, each instrument is also producing harmonics whose relative intensity depends on the instrument and how it is played. If each instrument produced only the fundamental frequency 440 Hz, the sound would be the same for each. Figure B-8 shows plots of the pressure variation versus time for a tuning fork, a clarinet, and a cornet each playing the same note. These patterns are called **wave forms.** The wave form for the tuning fork is very nearly a pure sine wave, but those for the clarinet and cornet are clearly not simple sine waves. These and other such wave forms can be analyzed in terms of what harmonics make up each wave form. Such analysis is called **harmonic analysis.** (It is

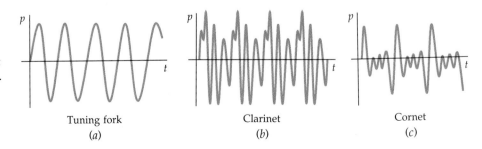

Figure B-8 Waveforms of (*a*) a tuning fork, (*b*) a clarinet, and (*c*) a cornet, each at a fundamental frequency of 440 Hz and all at approximately the same intensity.

Tuning fork (*a*)   Clarinet (*b*)   Cornet (*c*)

also sometimes called **Fourier analysis** after the French mathematician Fourier who developed the mathematics for analyzing periodic functions.) The wave function for a general waveform can be written

$$y(x, t) = \Sigma\, A_i \sin\,(k_i x - \omega_i t) \qquad \text{B-17}$$

where A_i is the amplitude of the ith harmonic wave of wave number k_i and angular frequency ω_i. The calculation of the amplitudes A_i needed to construct a given waveform is a problem in Fourier series.

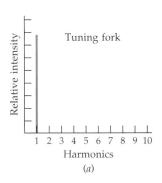
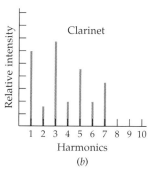
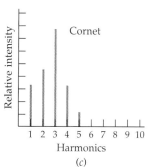

Figure B-9 Relative intensities of the harmonics in the waveforms shown in Figure B-8 for (*a*) the tuning fork, (*b*) the clarinet, and (*c*) the cornet.

Figure B-9 shows a plot of the relative intensity of the harmonics corresponding to the wave forms in Figure B-8. The wave form of the tuning fork contains only the fundamental frequency. That of the clarinet contains the fundamental plus large amounts of the 3rd, 5th, and 7th harmonics in addition to lesser amounts of the 2nd, 4th, and 6th harmonics. For the cornet, there is more energy in the third harmonic than in the fundamental.

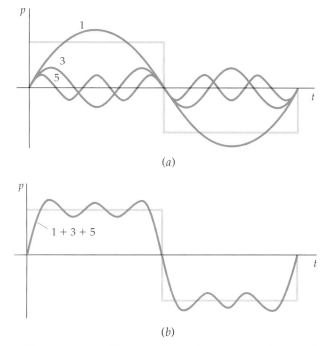

Figure B-10 (*a*) A square wave and the first three odd harmonics used to synthesize it. (*b*) Synthesis of the square wave using the first three odd harmonics.

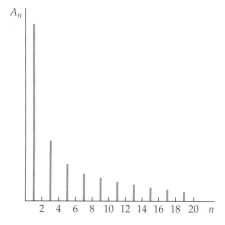

Figure B-11 Relative amplitudes A_n of the harmonics needed to synthesize the square wave shown in Figure B-10. The more harmonics that are used, the closer the approximation is to the square wave.

The inverse of harmonic analysis is **synthesis,** the construction of an arbitrary periodic wave from its harmonic components. Figure B-10 shows a square wave that can be produced by an electronic square-wave generator, and its approximate synthesis using only three harmonics. The relative intensities of the harmonics are shown in Figure B-11. The more harmonics used, the better the approximation to the actual wave form.

Wave Packets

The complicated waveforms discussed in the previous section are periodic in time. Functions that are not periodic, such as pulses, can also be represented by a group of harmonic (sinusoidal) wave functions. To send a signal with a wave, a pulse is needed rather than a periodic wave. The characteristic feature of a wave pulse that distinguishes it from a periodic wave is that the pulse has a beginning and an end, whereas a periodic wave repeats over and over. To describe a pulse that is localized in space, we must construct a group of waves from a continuous distribution of waves. Such a group is called a **wave packet**. We represent this mathematically by replacing A_i in Equation B-17 by $A(k)dk$ and replacing the sum by an integral.

$$y(x, t) = \int A(k) \sin(kx - \omega t)\, dk \qquad \text{B-18}$$

The quantity $A(k)$ is called the distribution function for the wave number k. Either the shape of the wave packet at some fixed time $y(x)$, or the distribution of wave numbers $A(k)$ can be found from the other by methods of Fourier analysis.

There is an important relation between the distribution of frequencies of the harmonic functions that make up a wave packet describing a pulse, and the time duration of the pulse. If the duration Δt is very short, the range of frequencies $\Delta \omega$ is very large. The general relation between Δt and $\Delta \omega$ is

$$\Delta \omega\, \Delta t \sim 1 \qquad \text{B-19}$$

The exact value of this product depends on just how the quantities $\Delta \omega$ and Δt are defined. For any reasonable definition $\Delta \omega$ is of the order of $1/\Delta t$. The wave pulse produced by a source of short duration Δt has a narrow width in space $\Delta x = v\, \Delta t$, where v is the wave speed. Each harmonic wave of frequency ω has a wave number $k = \omega/v$. A range of frequencies $\Delta \omega$ implies a range of wave numbers $\Delta k = \Delta \omega / v$. Substituting $\Delta k\, v$ for $\Delta \omega$ in Equation B-19 gives

$$\Delta k\, v\, \Delta t \sim 1$$

or

$$\Delta k\, \Delta x \sim 1 \qquad \text{B-20}$$

where again the exact value for the product depends on the precise definitions of Δk and Δx.

The relations expressed by Equations B-19 and B-20 are important characteristics of wave packets that apply to all types of waves. Since information cannot be transported by a single harmonic wave which has no beginning or end in time, the transmission of short pulses depends on the ability to transmit a wide range of frequencies.

If a wave packet is to maintain its shape as it travels, all the component harmonic waves that make up the packet must travel with the same speed. This occurs if the speed of harmonic waves does not depend on the wavelength or frequency of the wave. A medium for which the wave speed is independent of wavelength or frequency is called a **nondispersive medium.** Air is a nondispersive medium for sound waves, but solids and liquids generally are not.

If the wave speed in a dispersive medium depends only slightly on the frequency and wavelength, the shape of a wave packet in such a medium will change slightly as the packet travels but the packet will travel a considerable distance as a recognizable packet. However, the speed of the packet—for example, the speed of the center of the packet—is not the same as the (average) speed of the individual component harmonic waves. The speed of the harmonic waves is called the **phase velocity**, and that of the packet is called the **group velocity**.

Appendix C

The Maxwell–Boltzmann Distribution, the Equipartition Theorem, and the Gaussian Distribution

The Maxwell–Boltzmann Velocity Distribution

The Maxwell–Boltzmann velocity distribution describes the distribution of velocities of molecules in a gas that is in thermal equilibrium at some temperature T. In one dimension, the **Maxwell–Boltzmann distribution function** $f(v_x)$ is defined as follows: If N is the total number of molecules, the number that have x components of velocity between v_x and $v_x + dv_x$ is dN given by

$$dN = N f(v_x) \, dv_x \qquad \text{C-1}$$

The fraction of the total number of molecules that have an x component of velocity between v_x and $v_x + dv_x$ is $dN/N = f(v_x) \, dv_x$. Alternatively, $f(v_x) \, dv_x$ is the probability of one molecule having x component of velocity between v_x and $v_x + dv_x$. The distribution function $f(v_x)$ can be derived from statistical mechanics. The result is

$$f(v_x) = \left(\frac{m}{2\pi kT}\right)^{1/2} e^{-mv_x^2/2kT} \qquad \text{C-2}$$

Maxwell–Boltzmann velocity distribution in one dimension

where m is the mass of each molecule, T is the absolute temperature, and k is Boltzmann's constant.

Figure C-1 shows a plot of $f(v_x)$ versus v_x. The fraction $dN/N = f(v_x) \, dv_x$ in a particular range of width dv_x is illustrated by the shaded region in the figure. The most value of v_x is that value for which the distribution is maximum, which is zero as can be seen from the figure.

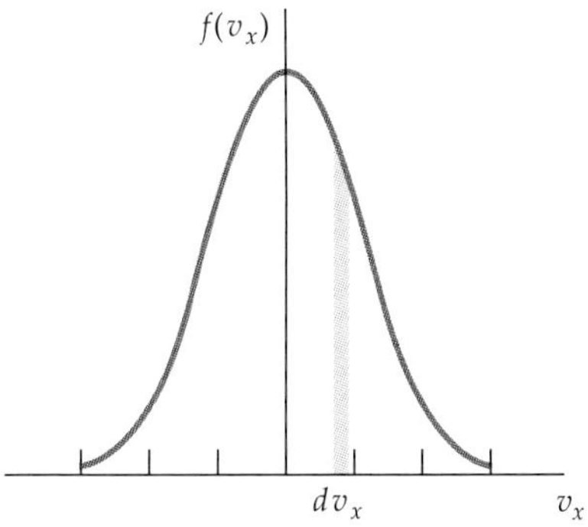

To find the average value of some molecular quantity, we sum the values for each molecule and then divide by the number of molecules. An alternative is to multiply each possible value by the number of molecules having that value, sum them, and divide by the number of molecules. The number of molecules that have their x component of velocity between v_x and $v_x + dv_x$ is $Nf(v_x) \, dv_x$. We obtain the average value of v_x by multiplying v_x by $Nf(v_x) \, dv_x$, summing over all the possible values of v_x, and dividing by the total number of molecules N. The average value of v_x is thus given by

$$(v_x)_{av} = \frac{1}{N} \int_{-\infty}^{\infty} v_x N f(v_x) \, dv_x = \int_{-\infty}^{\infty} v_x f(v_x) \, dv_x$$

$$= \int_{-\infty}^{\infty} v_x \left(\left(\frac{m}{2\pi kT}\right)^{1/2} e^{-mv_x^2/2kT}\right) dv_x = 0 \qquad \text{C-3}$$

Figure C-1 The Maxwell–Boltzmann distribution function $f(v_x)$ for the x component of velocity. The shaded region indicates the probability of finding v_x in the range dv_x.

This result is to be expected from the fact that the distribution function is symmetrical about $v_x = 0$. The average value of v_x^2 is found similarly by integrating $v_x^2 f(v_x)\, dv_x$ from $-\infty$ to $+\infty$.

$$(v_x^2)_{\text{av}} = \int_{-\infty}^{\infty} v_x^2 f(v_x)\, dv_x = \int_{-\infty}^{\infty} v_x^2 \left(\left(\frac{m}{2\pi kT}\right)^{1/2} e^{-mv_x^2/2kT}\right) dv_x$$

$$= \left(\frac{m}{2\pi kT}\right)^{1/2} \int_{-\infty}^{\infty} v_x^2 e^{-\lambda v_x^2}\, dv_x \qquad \text{C-4}$$

where $\lambda = m/2kT$. The integral in Equation C-4 can be found in tables. Its value is

$$\int_{-\infty}^{\infty} v_x^2 e^{-\lambda v_x^2}\, dv_x = \frac{\pi^{1/2}}{2\lambda^{3/2}} = \frac{\pi^{1/2}}{2}\left(\frac{2kT}{m}\right)^{3/2}$$

Thus

$$(v_x^2)_{\text{av}} = \left(\frac{m}{2\pi kT}\right)^{1/2} \frac{\pi^{1/2}}{2}\left(\frac{2kT}{m}\right)^{3/2} = \frac{kT}{m} \qquad \text{C-5}$$

The average kinetic energy associated with motion in the x direction is

$$(\tfrac{1}{2} m v_x^2)_{\text{av}} = \tfrac{1}{2} kT \qquad \text{C-6}$$

The distribution functions for the y and z components of velocity are the same as Equation C-2 with v_y or v_z replacing v_x. The probability that a molecule has x, y, and z velocity components in the ranges v_x to $v_x + dv_x$, v_y to $v_y + dv_y$, and v_z to $v_z + dv_z$ is

$$F(v_x, v_y, v_z)\, dv_x\, dv_y\, dv_z = f(v_x)\, dv_x\, f(v_y)\, dv_y\, f(v_z)\, dv_z$$

$$= \left(\frac{m}{2\pi kT}\right)^{1/2} e^{-mv_x^2/2kT}\, dv_x \left(\frac{m}{2\pi kT}\right)^{1/2} e^{-mv_y^2/2kT}\, dv_y \left(\frac{m}{2\pi kT}\right)^{1/2} e^{-mv_z^2/2kT}\, dv_z$$

or

$$F(v_x, v_y, v_z)\, dv_x\, dv_y\, dv_z = \left(\frac{m}{2\pi kT}\right)^{3/2} e^{-mv^2/kT}\, dv_x\, dv_y\, dv_z$$

where $v^2 = v_x^2 + v_y^2 + v_z^2$. The velocity distribution function is thus given by

Maxwell–Boltzmann velocity distribution

$$F(v_x, v_y, v_z) = \left(\frac{m}{2\pi kT}\right)^{3/2} e^{-mv^2/2kT} \qquad \text{C-7}$$

The Equipartition Theorem

Equation C-6 is an example of the equipartition theorem:

Equipartition theorem

> When a substance is in equilibrium, there is an average energy of $\tfrac{1}{2}kT$ per molecule or $\tfrac{1}{2}RT$ per mole associated with each degree of freedom.

(Each coordinate, velocity component, or angular velocity component that appears as a squared term in the expression for the energy is considered to be a degree of freedom.)

The Maxwell–Boltzmann Speed Distribution

The quantity $dv_x\, dv_y\, dv_z$ is an element of "volume" in a space defined by the rectangular coordinates v_x, v_y, and v_z. Such a space is called velocity space. A point in this space represents a particular velocity v_x, v_y, v_z. The probability that a particular molecule has its velocity in the volume element $dv_x\, dv_y\, dv_z$ is given by $F(v_x, v_y, v_z)\, dv_x\, dv_y\, dv_z$. The points in velocity space corresponding to a given speed v lie on the spherical surface of radius $v = \sqrt{v_x^2 + v_y^2 + v_z^2}$. The probability that a molecule has speed between v and $v + dv$ is found by multiplying $F(v_x, v_y, v_z)$ by the volume of the spherical shell between v and $v + dv$ which is $4\pi v^2\, dv$. The speed distribution function $g(v)$ is thus given by

$$g(v) = 4\pi \left(\frac{m}{2\pi kT}\right)^{3/2} v^2 e^{-mv^2/2kT} \quad \text{C-8}$$

Maxwell–Boltzmann speed distribution

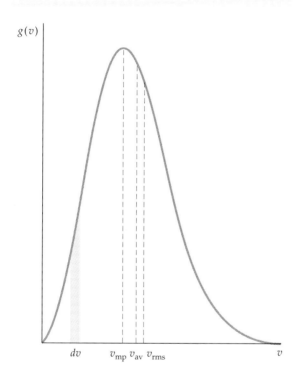

Figure C-2 The Maxwell–Boltzmann speed distribution function $g(v)$. The most probable speed v_{mp}, the average speed v_{av}, and the rms speed v_{rms} are indicated.

Figure C-2 shows a plot of $g(v)$ versus v. The fraction $dN/N = g(v)\, dv$ in a particular range between v and $v + dv$ is illustrated by the shaded region in the figure. The most probable speed v_{mp} is that value at which the distribution is maximum. It is found to be

$$v_{\text{mp}} = \sqrt{\frac{2kT}{m}} \quad \text{C-9}$$

Most probable speed

Since the speeds range from 0 to ∞, the average speed is

$$v_{\text{av}} = \int_0^\infty v g(v)\, dv = \int_0^\infty 4\pi \left(\frac{m}{2\pi kT}\right)^{3/2} v^3 e^{-mv^2/2kT}\, dv$$

$$= 4\pi \left(\frac{m}{2\pi kT}\right)^{3/2} \int_0^\infty v^3 e^{-\lambda v^2}\, dv \quad \text{C-10}$$

where again, $\lambda = m/2kT$. The integral in Equation C-10 can be found in tables. Its value is

$$\int_0^\infty v^3 e^{-\lambda v^2}\, dv = \frac{1}{2\lambda^2} = \frac{1}{2}\left(\frac{2kT}{m}\right)^2$$

The average speed is therefore

Average speed

$$v_{av} = 4\pi \left(\frac{m}{2\pi kT}\right)^{3/2} \frac{1}{2}\left(\frac{2kT}{m}\right)^2 = \sqrt{\frac{8kT}{\pi m}} \qquad \text{C-11}$$

Comparing this with the most probable speed $v_{mp} = \sqrt{2kT/m}$ we see that v_{av} is slightly greater than v_{mp}.

The average value of the square of the speed $(v^2)_{av}$ is found similarly by integrating $v^2 f(v)\, dv$ from 0 to ∞.

$$(v^2)_{av} = \int_0^\infty v^2 f(v)\, dv = \int_0^\infty v^2 4\pi\left(\frac{m}{2\pi kT}\right)^{3/2} v^2 e^{-mv^2/2kT}\, dv$$

$$= 4\pi\left(\frac{m}{2\pi kT}\right)^{3/2} \int_0^\infty v^4 e^{-mv^2/2kT}\, dv$$

This integral can be found in tables. Its value is

$$\int_0^\infty v^4 e^{-mv^2/2kT}\, dv = \frac{3\pi^2}{8}\left(\frac{2kT}{\pi m}\right)^{5/2}$$

Combining these results, we obtain

$$(v^2)_{av} = 4\pi\left(\frac{m}{2\pi kT}\right)^{3/2} \frac{3}{8\pi^2}\left(\frac{2\pi kT}{m}\right)^{5/2} = \frac{3kT}{m} \qquad \text{C-12}$$

The rms speed is thus

rms speed

$$v_{rms} = \sqrt{(v^2)_{av}} = \sqrt{\frac{3kT}{m}} \qquad \text{C-13}$$

The rms speed is slightly greater than the average speed.

The Maxwell–Boltzmann Energy Distribution

The Maxwell–Boltzmann distribution of speed as given by Equation C-8 can also be written as an energy distribution. We write the number of molecules with energy E in the range between E and $E + dE$ as

$$dN = NF(E)\, dE$$

where $F(E)$ is the **energy distribution function**. If we assume that the energy of the molecules is purely kinetic, the energy E will be related to the speed v by $E = \frac{1}{2}mv^2$. Then the number of molecules dN in the relation $dN = NF(E)\, dE$ will be the same as that in $dN = Ng(v)\, dv$ where $g(v)$ is the speed distribution function given by Equation C-8. Thus

$$NF(E)\, dE = Ng(v)\, dv$$

Since the energy E is related to the speed v by $E = \frac{1}{2}mv^2$, the differentials dE and dv are related by $dE = mv\, dv$. We therefore have

$$g(v)\, dv = Cv^2 e^{-mv^2/2kT}\, dv = Cv e^{-E/kT} v\, dv = C\left(\frac{2E}{m}\right)^{1/2} e^{-E/kT} \frac{dE}{m}$$

where $C = 4\pi(m/2\pi kT)^{3/2}$ from Equation C-8. The energy distribution function $F(E)$ is thus given by

$$F(E) = 4\pi \left(\frac{m}{2\pi kT}\right)^{3/2} \left(\frac{2}{m}\right)^{1/2} \frac{1}{m} E^{1/2} e^{-E/kT}$$

Simplifying the constants we obtain

$$F(E) = \frac{2}{\sqrt{\pi}} \left(\frac{1}{kT}\right)^{3/2} E^{1/2} e^{-E/kT} \qquad \text{C-14}$$

Energy distribution function

The energy distribution function is thus proportional to $E^{1/2} e^{-E/kT}$. The quantity $e^{-E/kT}$ is called the **Boltzmann factor**.

The Gaussian Distribution

Equation C-2 for the distribution function for the x component of velocity is a special case of the gaussian distribution function, which occurs in many areas of mathematics and physics. For example, if N measurements are taken and the errors are random, the experimental data are often distributed in a gaussian distribution. The general form of the gaussian distribution function for measurements of some quantity x is

$$f(x) = \frac{1}{\sigma\sqrt{2\pi}} e^{-(x-x_0)^2/2\sigma^2} \qquad \text{C-15}$$

The number of measurements that lie between x and $x + dx$ is $Nf(x)\, dx$. The function $f(x)$ is shown in Figure C-3. It contains two parameters, x_0, which is the average value of x, and σ, which is called the standard deviation. The normalization constant $1/\sigma\sqrt{2\pi}$ is chosen so that the total number of measurements is N

$$\int_0^\infty Nf(x)\, dx = N$$

or equivalently

$$\int_0^\infty f(x)\, dx = 1$$

The width of the curve at half maximum is 2σ as indicated on the figure. The integral of $f(x)\, dx$ from $x = x_0 - \sigma$ to $x_0 + \sigma$ is 0.68

$$\int_{x_0-\sigma}^{x_0+\sigma} f(x)\, dx = 0.68$$

Thus about two thirds of the measurements have values that fall within $x_0 \pm \sigma$.

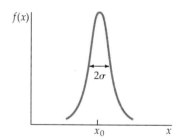

Figure C-3 A gaussian distribution function. The average value is x_0, and the standard deviation is σ. The full width of the curve at half maximum is 2σ.

Appendix D

Numerical Data

Terrestrial Data	
Acceleration of gravity g	
Standard value	9.80665 m/s^2
	32.1740 ft/s^2
At sea level, at equator†	9.7804 m/s^2
At sea level, at poles†	9.8322 m/s^2
Mass of earth M_E	5.98×10^{24} kg
Radius of earth R_E, mean	6.37×10^6 m
	3960 mi
Escape speed $\sqrt{2R_E g}$	1.12×10^4 m/s
	6.95 mi/s
Solar constant‡	1.35 kW/m^2
Standard temperature and pressure (STP):	
Temperature	273.15 K
Pressure	101.325 kPa
	1.00 atm
Molar mass of air	28.97 g/mol
Density of air (STP), ρ_{air}	1.293 kg/m^3
Speed of sound (STP)	331 m/s
Heat of fusion of H$_2$O (0°C, 1 atm)	333.5 kJ/kg
Heat of vaporization of H$_2$O (100°C, 1 atm)	2.257 MJ/kg

†Measured relative to the earth's surface.
‡Average power incident normally on 1 m^2 outside the earth's atmosphere at the mean distance from the earth to the sun.

Astronomical Data	
Earth	
Distance to moon†	3.844×10^8 m
	2.389×10^5 mi
Distance to sun, mean†	1.496×10^{11} m
	9.30×10^7 mi
	1.00 AU
Orbital speed, mean	2.98×10^4 m/s
Moon	
Mass	7.35×10^{22} kg
Radius	1.738×10^6 m
Period	27.32 d
Acceleration of gravity at surface	1.62 m/s^2
Sun	
Mass	1.99×10^{30} kg
Radius	6.96×10^8 m

†Center to center.

Physical Constants

Gravitational constant	G	6.6726×10^{-11} N·m²/kg²
Speed of light	c	2.99792458×10^8 m/s
Electron charge	e	1.602177×10^{-19} C
Avogadro's number	N_A	6.022137×10^{23} particles/mol
Gas constant	R	8.31451 J/mol·K
		1.98722 cal/mol·K
		8.20578×10^{-2} L·atm/mol·K
Boltzmann's constant	$k = R/N_A$	1.380658×10^{-23} J/K
		8.617385×10^{-5} eV/K
Unified mass unit	$u = (1/N_A)$ g	1.660540×10^{-24} g
Coulomb constant	$k = 1/4\pi\epsilon_0$	8.987551788×10^9 N·m²/C²
Permittivity of free space	ϵ_0	$8.854187817 \times 10^{-12}$ C²/N·m²
Permeability of free space	μ_0	$4\pi \times 10^{-7}$ N/A²
		1.256637×10^{-6} N/A²
Planck's constant	h	6.626076×10^{-34} J·s
		4.135669×10^{-15} eV·s
	$\hbar = h/2\pi$	1.054573×10^{-34} J·s
		6.582122×10^{-16} eV·s
Mass of electron	m_e	9.109390×10^{-31} kg
		510.9991 keV/c^2
Mass of proton	m_p	1.672623×10^{-27} kg
		938.2723 MeV/c^2
Mass of neutron	m_n	1.674929×10^{-27} kg
		939.5656 MeV/c^2
Bohr magneton	$m_B = e\hbar/2m_e$	$9.2740154 \times 10^{-24}$ J/T
		$5.78838263 \times 10^{-5}$ eV/T
Nuclear magneton	$m_n = e\hbar/2m_p$	$5.0507866 \times 10^{-27}$ J/T
		$3.15245166 \times 10^{-8}$ eV/T
Magnetic flux quantum	$\phi_0 = h/2e$	$2.0678346 \times 10^{-15}$ T·m²
Quantized Hall resistance	$R_K = h/e^2$	2.5812807×10^4 Ω
Rydberg constant	R_H	1.0973731534×10^7 m⁻¹
Josephson frequency–voltage quotient	$2e/h$	4.835979×10^{14} Hz/V
Compton wavelength	$\lambda_C = h/m_e c$	$2.42631058 \times 10^{-12}$ m

For additional data, see the front and back endpapers and the following tables in the text

Table 1-1   Rest energies of some elementary particles and light nuclei, p. 32

Table 2-1   Approximate dates of some important experiments and theories, 1881–1932, p. 48

Table 4-1   Electron configurations of the atoms in their ground states, pp. 142–143

Table 6-1   Free-electron number densities and Fermi energies at $T = 0$ for selected elements, p. 203

Table 6-2	Work functions for some metals, p. 206	
Table 6-3	Critical temperatures for some superconducting materials, p. 227	
Table 7-1	Atomic masses of the neutron and selected isotopes, p. 242	
Table 7-2	Approximate RBE Factors, p. 267	
Table 7-3	Radiation and dose units, p. 267	
Table 7-4	Average radiation doses received by the United States population, p. 268	
Table 7-5	Recommended dose limits, p. 268	
Table 8-1	Hadrons that are stable against decay via the strong nuclear interaction, p. 278	
Table 8-2	Properties of quarks and antiquarks, p. 287	
Table 8-3	Fundamental particles and their approximate masses, p. 288	
Table 8-4	Bosons that mediate the basic interactions, p. 289	
Table 8-5	Properties of the basic interactions, p. 291	

Appendix E

Conversion Factors

Conversion factors are written as equations for simplicity; relations marked with an asterisk are exact.

Length

1 km = 0.6215 mi

1 mi = 1.609 km

1 m = 1.0936 yd = 3.281 ft = 39.37 in

*1 in = 2.54 cm

*1 ft = 12 in = 30.48 cm

*1 yd = 3 ft = 91.44 cm

1 lightyear = 1 $c \cdot y$ = 9.461 × 10^{15} m

*1 Å = 0.1 nm

Area

*1 m^2 = 10^4 cm^2

1 km^2 = 0.3861 mi^2 = 247.1 acres

*1 in^2 = 6.4516 cm^2

1 ft^2 = 9.29 × 10^{-2} m^2

1 m^2 = 10.76 ft^2

*1 acre = 43,560 ft^2

1 mi^2 = 640 acres = 2.590 km^2

Volume

*1 m^3 = 10^6 cm^3

*1 L = 1000 cm^3 = 10^{-3} m^3

1 gal = 3.786 L

1 gal = 4 qt = 8 pt = 128 oz = 231 in^3

1 in^3 = 16.39 cm^3

1 ft^3 = 1728 in^3 = 28.32 L = 2.832 × 10^4 cm^3

Time

*1 h = 60 min = 3.6 ks

*1 d = 24 h = 1440 min = 86.4 ks

1 y = 365.24 d = 31.56 Ms

Speed

1 km/h = 0.2778 m/s = 0.6215 mi/h

1 mi/h = 0.4470 m/s = 1.609 km/h

1 mi/h = 1.467 ft/s

Angle and Angular Speed

*π rad = 180°

1 rad = 57.30°

1° = 1.745 × 10^{-2} rad

1 rev/min = 0.1047 rad/s

1 rad/s = 9.549 rev/min

Mass

*1 kg = 1000 g

*1 tonne = 1000 kg = 1 Mg

1 u = 1.6606 × 10^{-27} kg

1 kg = 6.022 × 10^{23} u

1 slug = 14.59 kg

1 kg = 6.852 × 10^{-2} slug

1 u = 931.50 MeV/c^2

Density

*1 g/cm^3 = 1000 kg/m^3 = 1 kg/L

(1 g/cm^3)g = 62.4 lb/ft^3

Force

1 N = 0.2248 lb = 10^5 dyn

1 lb = 4.4482 N

(1 kg)g = 2.2046 lb

Pressure

*1 Pa = 1 N/m^2

*1 atm = 101.325 kPa = 1.01325 bars

1 atm = 14.7 lb/in^2 = 760 mmHg
     = 29.9 inHg = 33.8 ftH$_2$O

1 lb/in^2 = 6.895 kPa

1 torr = 1 mmHg = 133.32 Pa

1 bar = 100 kPa

Energy

*1 kW·h = 3.6 MJ

*1 cal = 4.1840 J

1 ft·lb = 1.356 J = 1.286 × 10^{-3} Btu

*1 L·atm = 101.325 J

1 L·atm = 24.217 cal

1 Btu = 778 ft·lb = 252 cal = 1054.35 J

1 eV = 1.602 × 10^{-19} J

1 u·c^2 = 931.50 MeV

*1 erg = 10^{-7} J

Power

1 horsepower = 550 ft·lb/s = 745.7 W

1 Btu/min = 17.58 W

1 W = 1.341 × 10^{-3} horsepower
    = 0.7376 ft·lb/s

Magnetic Field

*1 G = 10^{-4} T

*1 T = 10^4 G

Thermal Conductivity

1 W/m·K = 6.938 Btu·in/h·ft^2·F°

1 Btu·in/h·ft^2·F° = 0.1441 W/m·K

Appendix F Periodic Table

1	2	3	4	5	6	7	8	9	10	11	12	13	14	15	16	17	18
1 H 1.00797																	2 He 4.003
3 Li 6.941	4 Be 9.012											5 B 10.81	6 C 12.011	7 N 14.007	8 O 15.9994	9 F 19.00	10 Ne 20.179
11 Na 22.990	12 Mg 24.31											13 Al 26.98	14 Si 28.09	15 P 30.974	16 S 32.064	17 Cl 35.453	18 Ar 39.948
19 K 39.102	20 Ca 40.08	21 Sc 44.96	22 Ti 47.88	23 V 50.94	24 Cr 52.00	25 Mn 54.94	26 Fe 55.85	27 Co 58.93	28 Ni 58.69	29 Cu 63.55	30 Zn 65.38	31 Ga 69.72	32 Ge 72.59	33 As 74.92	34 Se 78.96	35 Br 79.90	36 Kr 83.80
37 Rb 85.47	38 Sr 87.62	39 Y 88.906	40 Zr 91.22	41 Nb 92.91	42 Mo 95.94	43 Tc (98)	44 Ru 101.1	45 Rh 102.905	46 Pd 106.4	47 Ag 107.870	48 Cd 112.41	49 In 114.82	50 Sn 118.69	51 Sb 121.75	52 Te 127.60	53 I 126.90	54 Xe 131.29
55 Cs 132.905	56 Ba 137.33	57–71 Rare Earths	72 Hf 178.49	73 Ta 180.95	74 W 183.85	75 Re 186.2	76 Os 190.2	77 Ir 192.2	78 Pt 195.09	79 Au 196.97	80 Hg 200.59	81 Tl 204.37	82 Pb 207.19	83 Bi 208.98	84 Po (210)	85 At (210)	86 Rn (222)
87 Fr (223)	88 Ra (226)	89–103 Actinides	104 Rf (261)	105 Ha (260)	106 (263)	107 (262)	108 (265)	109 (266)									

Rare Earths (Lanthanides)

57 La 138.91	58 Ce 140.12	59 Pr 140.91	60 Nd 144.24	61 Pm (147)	62 Sm 150.36	63 Eu 152.0	64 Gd 157.25	65 Tb 158.92	66 Dy 162.50	67 Ho 164.93	68 Er 167.26	69 Tm 168.93	70 Yb 173.04	71 Lu 174.97

Actinides

89 Ac 227.03	90 Th 232.04	91 Pa 231.04	92 U 238.03	93 Np 237.05	94 Pu (244)	95 Am (243)	96 Cm (247)	97 Bk (247)	98 Cf (251)	99 Es (252)	100 Fm (257)	101 Md (258)	102 No (259)	103 Lr (260)

The 1–18 group designation has been recommended by the International Union of Pure and Applied Chemistry (IUPAC).

Appendix G

Properties of Nuclei

(1) Z	(2) Element	(3) Symbol	(4) Chemical Atomic Weight	(5) Mass Number (* Indicates Radioactive) A	(6) Atomic Mass	(7) Percent Abundance	(8) Half-Life (if Radioactive) $t_{1/2}$
0	(Neutron)	n		1*	1.008 665		10.4 m
1	Hydrogen	H	1.0079	1	1.007 825	99.985	
	Deuterium	D		2	2.014 102	0.015	
	Tritium	T		3*	3.016 049		12.33 y
2	Helium	He	4.00260	3	3.016 029	0.00014	
				4	4.002 602	99.99986	
				6*	6.018 886		0.81 s
3	Lithium	Li	6.941	6	6.015 121	7.5	
				7	7.016 003	92.5	
				8*	8.022 486		0.84 s
4	Beryllium	Be	9.0122	7*	7.016 928		53.3 d
				9	9.012 174	100	
				10*	10.013 534		1.5×10^6 y
5	Boron	B	10.81	10	10.012 936	19.9	
				11	11.009 305	80.1	
				12*	12.014 352		0.0202 s
6	Carbon	C	12.011	10*	10.016 854		19.3 s
				11*	11.011 433		20.4 m
				12	12.000 000	98.90	
				13	13.003 355	1.10	
				14*	14.003 242		5730 y
				15*	15.010 599		2.45 s
7	Nitrogen	N	14.0067	12*	12.018 613		0.0110 s
				13*	13.005 738		9.96 m
				14	14.003 074	99.63	
				15	15.000 108	0.37	
				16*	16.006 100		7.13 s
				17*	17.008 450		4.17 s
8	Oxygen	O	15.9994	14*	14.008 595		70.6 s
				15*	15.003 065		122 s
				16	15.994 915	99.761	
				17	16.999 132	0.039	
				18	17.999 160	0.20	
				19*	19.003 577		26.9 s
9	Fluorine	F	18.99840	17*	17.002 094		64.5 s
				18*	18.000 937		109.8 m
				19	18.998 404	100	
				20*	19.999 982		11.0 s
				21*	20.999 950		4.2 s
10	Neon	Ne	20.180	18*	18.005 710		1.67 s
				19*	19.001 880		17.2 s
				20	19.992 435	90.48	

(1) Z	(2) Element	(3) Symbol	(4) Chemical Atomic Weight	(5) Mass Number (* Indicates Radioactive) A	(6) Atomic Mass	(7) Percent Abundance	(8) Half-Life (if Radioactive) $t_{1/2}$
(10)	(Neon)			21	20.993 841	0.27	
				22	21.991 383	9.25	
				23*	22.994 465		37.2 s
11	Sodium	Na	22.98987	21*	20.997 650		22.5 s
				22*	21.994 434		2.61 y
				23	22.989 767	100	
				24*	23.990 961		14.96 h
12	Magnesium	Mg	24.305	23*	22.994 124		11.3 s
				24	23.985 042	78.99	
				25	24.985 838	10.00	
				26	25.982 594	11.01	
				27*	26.984 341		9.46 m
13	Aluminum	Al	26.98154	26*	25.986 892		7.4×10^5 y
				27	26.981 538	100	
				28*	27.981 910		2.24 m
14	Silicon	Si	28.086	28	27.976 927	92.23	
				29	28.976 495	4.67	
				30	29.973 770	3.10	
				31*	30.975 362		2.62 h
				32*	31.974 148		172 y
15	Phosphorus	P	30.97376	30*	29.978 307		2.50 m
				31	30.973 762	100	
				32*	31.973 908		14.26 d
				33*	32.971 725		25.3 d
16	Sulfur	S	32.066	32	31.972 071	95.02	
				33	32.971 459	0.75	
				34	33.967 867	4.21	
				35*	34.969 033		87.5 d
				36	35.967 081	0.02	
17	Chlorine	Cl	35.453	35	34.968 853	75.77	
				36*	35.968 307		3.0×10^5 y
				37	36.965 903	24.23	
18	Argon	Ar	39.948	36	35.967 547	0.337	
				37*	36.966 776		35.04 d
				38	37.962 732	0.063	
				39*	38.964 314		269 y
				40	39.962 384	99.600	
				42*	41.963 049		33 y
19	Potassium	K	39.0983	39	38.963 708	93.2581	
				40*	39.964 000	0.0117	1.28×10^9 y
				41	40.961 827	6.7302	
20	Calcium	Ca	40.08	40	39.962 591	96.941	
				41*	40.962 279		1.0×10^5 y
				42	41.958 618	0.647	
				43	42.958 767	0.135	
				44	43.955 481	2.086	
				46	45.953 687	0.004	
				48	47.952 534	0.187	
21	Scandium	Sc	44.9559	41*	40.969 250		0.596 s
				45	44.955 911	100	
22	Titanium	Ti	47.88	44*	43.959 691		49 y
				46	45.952 630	8.0	
				47	46.951 765	7.3	
				48	47.947 947	73.8	

(1) Z	(2) Element	(3) Symbol	(4) Chemical Atomic Weight	(5) Mass Number (* Indicates Radioactive) A	(6) Atomic Mass	(7) Percent Abundance	(8) Half-Life (if Radioactive) $t_{1/2}$
(22)	(Titanium)			49	48.947 871	5.5	
				50	49.944 792	5.4	
23	Vanadium	V	50.9415	48*	47.952 255		15.97 d
				50*	49.947 161	0.25	1.5×10^{17} y
				51	50.943 962	99.75	
24	Chromium	Cr	51.996	48*	47.954 033		21.6 h
				50	49.946 047	4.345	
				52	51.940 511	83.79	
				53	52.940 652	9.50	
				54	53.938 883	2.365	
25	Manganese	Mn	54.93805	54*	53.940 361		312.1 d
				55	54.938 048	100	
26	Iron	Fe	55.847	54	53.939 613	5.9	
				55*	54.938 297		2.7 y
				56	55.934 940	91.72	
				57	56.935 396	2.1	
				58	57.933 278	0.28	
				60*	59.934 078		1.5×10^6 y
27	Cobalt	Co	58.93320	59	58.933 198	100	
				60*	59.933 820		5.27 y
28	Nickel	Ni	58.693	58	57.935 346	68.077	
				59*	58.934 350		7.5×10^4 y
				60	59.930 789	26.223	
				61	60.931 058	1.140	
				62	61.928 346	3.634	
				63*	62.929 670		100 y
				64	63.927 967	0.926	
29	Copper	Cu	63.54	63	62.929 599	69.17	
				65	64.927 791	30.83	
30	Zinc	Zn	65.39	64	63.929 144	48.6	
				66	65.926 035	27.9	
				67	66.927 129	4.1	
				68	67.924 845	18.8	
				70	69.925 323	0.6	
31	Gallium	Ga	69.723	69	68.925 580	60.108	
				71	70.924 703	39.892	
32	Germanium	Ge	72.61	70	69.924 250	21.23	
				72	71.922 079	27.66	
				73	72.923 462	7.73	
				74	73.921 177	35.94	
				76	75.921 402	7.44	
33	Arsenic	As	74.9216	75	74.921 594	100	
34	Selenium	Se	78.96	74	73.922 474	0.89	
				76	75.919 212	9.36	
				77	76.919 913	7.63	
				78	77.917 307	23.78	
				79*	78.918 497		$\leq 6.5 \times 10^4$ y
				80	79.916 519	49.61	
				82*	81.916 697	8.73	1.4×10^{20} y
35	Bromine	Br	79.904	79	78.918 336	50.69	
				81	80.916 287	49.31	
36	Krypton	Kr	83.80	78	77.920 400	0.35	
				80	79.916 377	2.25	
				81*	80.916 589		2.1×10^5 y

(1) Z	(2) Element	(3) Symbol	(4) Chemical Atomic Weight	(5) Mass Number (* Indicates Radioactive) A	(6) Atomic Mass	(7) Percent Abundance	(8) Half-Life (if Radioactive) $t_{1/2}$
(36)	(Krypton)			82	81.913 481	11.6	
				83	82.914 136	11.5	
				84	83.911 508	57.0	
				85*	84.912 531		10.76 y
				86	85.910 615	17.3	
37	Rubidium	Rb	85.468	85	84.911 793	72.17	
				87*	86.909 186	27.83	4.75×10^{10} y
38	Strontium	Sr	87.62	84	83.913 428	0.56	
				86	85.909 266	9.86	
				87	86.908 883	7.00	
				88	87.905 618	82.58	
				90*	89.907 737		29.1 y
39	Yttrium	Y	88.9058	89	88.905 847	100	
40	Zirconium	Zr	91.224	90	89.904 702	51.45	
				91	90.905 643	11.22	
				92	91.905 038	17.15	
				93*	92.906 473		1.5×10^6 y
				94	93.906 314	17.38	
				96	95.908 274	2.80	
41	Niobium	Nb	92.9064	91*	90.906 988		6.8×10^2 y
				92*	91.907 191		3.5×10^7 y
				93	92.906 376	100	
				94*	93.907 280		2×10^4 y
42	Molybdenum	Mo	95.94	92	91.906 807	14.84	
				93*	92.906 811		3.5×10^3 y
				94	93.905 085	9.25	
				95	94.905 841	15.92	
				96	95.904 678	16.68	
				97	96.906 020	9.55	
				98	97.905 407	24.13	
				100	99.907 476	9.63	
43	Technetium	Tc		97*	96.906 363		2.6×10^6 y
				98*	97.907 215		4.2×10^6 y
				99*	98.906 254		2.1×10^5 y
44	Ruthenium	Ru	101.07	96	95.907 597	5.54	
				98	97.905 287	1.86	
				99	98.905 939	12.7	
				100	99.904 219	12.6	
				101	100.905 558	17.1	
				102	101.904 348	31.6	
				104	103.905 428	18.6	
45	Rhodium	Rh	102.9055	103	102.905 502	100	
46	Palladium	Pd	106.42	102	101.905 616	1.02	
				104	103.904 033	11.14	
				105	104.905 082	22.33	
				106	105.903 481	27.33	
				107*	106.905 126		6.5×10^6 y
				108	107.903 893	26.46	
				110	109.905 158	11.72	
47	Silver	Ag	107.868	107	106.905 091	51.84	
				109	108.904 754	48.16	
48	Cadmium	Cd	112.41	106	105.906 457	1.25	
				108	107.904 183	0.89	
				109*	108.904 984		462 d

(1) Z	(2) Element	(3) Symbol	(4) Chemical Atomic Weight	(5) Mass Number (* Indicates Radioactive) A	(6) Atomic Mass	(7) Percent Abundance	(8) Half-Life (if Radioactive) $t_{1/2}$
(48)	(Cadmium)			110	109.903 004	12.49	
				111	110.904 182	12.80	
				112	111.902 760	24.13	
				113*	112.904 401	12.22	9.3×10^{15} y
				114	113.903 359	28.73	
				116	115.904 755	7.49	
49	Indium	In	114.82	113	112.904 060	4.3	
				115*	114.903 876	95.7	4.4×10^{14} y
50	Tin	Sn	118.71	112	111.904 822	0.97	
				114	113.902 780	0.65	
				115	114.903 345	0.36	
				116	115.901 743	14.53	
				117	116.902 953	7.68	
				118	117.901 605	24.22	
				119	118.903 308	8.58	
				120	119.902 197	32.59	
				121*	120.904 237		55 y
				122	121.903 439	4.63	
				124	123.905 274	5.79	
51	Antimony	Sb	121.76	121	120.903 820	57.36	
				123	122.904 215	42.64	
				125*	124.905 251		2.7 y
52	Tellurium	Te	127.60	120	119.904 040	0.095	
				122	121.903 052	2.59	
				123*	122.904 271	0.905	1.3×10^{13} y
				124	123.902 817	4.79	
				125	124.904 429	7.12	
				126	125.903 309	18.93	
				128*	127.904 463	31.70	$>8 \times 10^{24}$ y
				130*	129.906 228	33.87	$\leq 1.25 \times 10^{21}$ y
53	Iodine	I	126.9045	127	126.904 474	100	
				129*	128.904 984		1.6×10^{7} y
54	Xenon	Xe	131.29	124	123.905 894	0.10	
				126	125.904 268	0.09	
				128	127.903 531	1.91	
				129	128.904 779	26.4	
				130	129.903 509	4.1	
				131	130.905 069	21.2	
				132	131.904 141	26.9	
				134	133.905 394	10.4	
				136*	135.907 215	8.9	$\geq 2.36 \times 10^{21}$ y
55	Cesium	Cs	132.9054	133	132.905 436	100	
				134*	133.906 703		2.1 y
				135*	134.905 891		2×10^{6} y
				137*	136.907 078		30 y
56	Barium	Ba	137.33	130	129.906 289	0.106	
				132	131.905 048	0.101	
				133*	132.905 990		10.5 y
				134	133.904 492	2.42	
				135	134.905 671	6.593	
				136	135.904 559	7.85	
				137	136.905 816	11.23	
				138	137.905 236	71.70	
57	Lanthanum	La	138.905	137*	136.906 462		6×10^{4} y

(1) Z	(2) Element	(3) Symbol	(4) Chemical Atomic Weight	(5) Mass Number (* Indicates Radioactive) A	(6) Atomic Mass	(7) Percent Abundance	(8) Half-Life (if Radioactive) $t_{1/2}$
(57)	(Lanthanum)			138*	137.907 105	0.0902	1.05×10^{11} y
				139	138.906 346	99.9098	
58	Cerium	Ce	140.12	136	135.907 139	0.19	
				138	137.905 986	0.25	
				140	139.905 434	88.43	
				142*	141.909 241	11.13	$>5 \times 10^{16}$ y
59	Praseodymium	Pr	140.9076	141	140.907 647	100	
60	Neodymium	Nd	144.24	142	141.907 718	27.13	
				143	142.909 809	12.18	
				144*	143.910 082	23.80	2.3×10^{15} y
				145	144.912 568	8.30	
				146	145.913 113	17.19	
				148	147.916 888	5.76	
				150*	149.920 887	5.64	$>1 \times 10^{18}$ y
61	Promethium	Pm		143*	142.910 928		265 d
				145*	144.912 745		17.7 y
				146*	145.914 698		5.5 y
				147*	146.915 134		2.623 y
62	Samarium	Sm	150.36	144	143.911 996	3.1	
				146*	145.913 043		1.0×10^{8} y
				147*	146.914 894	15.0	1.06×10^{11} y
				148*	147.914 819	11.3	7×10^{15} y
				149*	148.917 180	13.8	$>2 \times 10^{15}$ y
				150	149.917 273	7.4	
				151*	150.919 928		90 y
				152	151.919 728	26.7	
				154	153.922 206	22.7	
63	Europium	Eu	151.96	151	150.919 846	47.8	
				152*	151.921 740		13.5 y
				153	152.921 226	52.2	
				154*	153.922 975		8.59 y
				155*	154.922 888		4.7 y
64	Gadolinium	Gd	157.25	148*	147.918 112		75 y
				150*	149.918 657		1.8×10^{6} y
				152*	151.919 787	0.20	1.1×10^{14} y
				154	153.920 862	2.18	
				155	154.922 618	14.80	
				156	155.922 119	20.47	
				157	156.923 957	15.65	
				158	157.924 099	24.84	
				160	159.927 050	21.86	
65	Terbium	Tb	158.9253	159	158.925 345	100	
66	Dysprosium	Dy	162.50	156	155.924 277	0.06	
				158	157.924 403	0.10	
				160	159.925 193	2.34	
				161	160.926 930	18.9	
				162	161.926 796	25.5	
				163	162.928 729	24.9	
				164	163.929 172	28.2	
67	Holmium	Ho	164.9303	165	164.930 316	100	
				166*	165.932 282		1.2×10^{3} y
68	Erbium	Er	167.26	162	161.928 775	0.14	
				164	163.929 198	1.61	
				166	165.930 292	33.6	

(1) Z	(2) Element	(3) Symbol	(4) Chemical Atomic Weight	(5) Mass Number (* Indicates Radioactive) A	(6) Atomic Mass	(7) Percent Abundance	(8) Half-Life (if Radioactive) $t_{1/2}$
(68)	(Erbium)			167	166.932 047	22.95	
				168	167.932 369	27.8	
				170	169.935 462	14.9	
69	Thulium	Tm	168.9342	169	168.934 213	100	
				171*	170.936 428		1.92 y
70	Ytterbium	Yb	173.04	168	167.933 897	0.13	
				170	169.934 761	3.05	
				171	170.936 324	14.3	
				172	171.936 380	21.9	
				173	172.938 209	16.12	
				174	173.938 861	31.8	
				176	175.942 564	12.7	
71	Lutecium	Lu	174.967	173*	172.938 930		1.37 y
				175	174.940 772	97.41	
				176*	175.942 679	2.59	3.78×10^{10} y
72	Hafnium	Hf	178.49	174*	173.940 042	0.162	2.0×10^{15} y
				176	175.941 404	5.206	
				177	176.943 218	18.606	
				178	177.943 697	27.297	
				179	178.945 813	13.629	
				180	179.946 547	35.100	
73	Tantalum	Ta	180.9479	180	179.947 542	0.012	
				181	180.947 993	99.988	
74	Tungsten (Wolfram)	W	183.85	180	179.946 702	0.12	
				182	181.948 202	26.3	
				183	182.950 221	14.28	
				184	183.950 929	30.7	
				186	185.954 358	28.6	
75	Rhenium	Re	186.207	185	184.952 951	37.40	
				187*	186.955 746	62.60	4.4×10^{10} y
76	Osmium	Os	190.2	184	183.952 486	0.02	
				186*	185.953 834	1.58	2.0×10^{15} y
				187	186.955 744	1.6	
				188	187.955 832	13.3	
				189	188.958 139	16.1	
				190	189.958 439	26.4	
				192	191.961 468	41.0	
				194*	193.965 172		6.0 y
77	Iridium	Ir	192.2	191	190.960 585	37.3	
				193	192.962 916	62.7	
78	Platinum	Pt	195.08	190*	189.959 926	0.01	6.5×10^{11} y
				192	191.961 027	0.79	
				194	193.962 655	32.9	
				195	194.964 765	33.8	
				196	195.964 926	25.3	
				198	197.967 867	7.2	
79	Gold	Au	196.9665	197	196.966 543	100	
80	Mercury	Hg	200.59	196	195.965 806	0.15	
				198	197.966 743	9.97	
				199	198.968 253	16.87	
				200	199.968 299	23.10	
				201	200.970 276	13.10	
				202	201.970 617	29.86	
				204	203.973 466	6.87	

(1) Z	(2) Element	(3) Symbol	(4) Chemical Atomic Weight	(5) Mass Number (* Indicates Radioactive) A	(6) Atomic Mass	(7) Percent Abundance	(8) Half-Life (if Radioactive) $t_{1/2}$
81	Thallium	Tl	204.383	203	202.972 320	29.524	
				204*	203.973 839		3.78 y
				205	204.974 400	70.476	
		(Ra E″)		206*	205.976 084		4.2 m
		(Ac C″)		207*	206.977 403		4.77 m
		(Th C″)		208*	207.981 992		3.053 m
		(Ra C″)		210*	209.990 057		1.30 m
82	Lead	Pb	207.2	202*	201.972 134		5×10^4 y
				204*	203.973 020	1.4	$\geq 1.4 \times 10^{17}$ y
				205*	204.974 457		1.5×10^7 y
				206	205.974 440	24.1	
				207	206.975 871	22.1	
				208	207.976 627	52.4	
		(Ra D)		210*	209.984 163		22.3 y
		(Ac B)		211*	210.988 734		36.1 m
		(Th B)		212*	211.991 872		10.64 h
		(Ra B)		214*	213.999 798		26.8 m
83	Bismuth	Bi	208.9803	207*	206.978 444		32.2 y
				208*	207.979 717		3.7×10^5 y
				209	208.980 374	100	
		(Ra E)		210*	209.984 096		5.01 d
		(Th C)		211*	210.987 254		2.14 m
				212*	211.991 259		60.6 m
		(Ra C)		214*	213.998 692		19.9 m
				215*	215.001 836		7.4 m
84	Polonium	Po		209*	208.982 405		102 y
		(Ra F)		210*	209.982 848		138.38 d
		(Ac C′)		211*	210.986 627		0.52 s
		(Th C′)		212*	211.988 842		0.30 μs
		(Ra C′)		214*	213.995 177		164 μs
		(Ac A)		215*	214.999 418		0.0018 s
		(Th A)		216*	216.001 889		0.145 s
		(Ra A)		218*	218.008 965		3.10 m
85	Astatine	At		215*	214.998 638		≈100 μs
				218*	218.008 685		1.6 s
				219*	219.011 294		0.9 m
86	Radon	Rn					
		(An)		219*	219.009 477		3.96 s
		(Tn)		220*	220.011 369		55.6 s
		(Rn)		222*	222.017 571		3.823 d
87	Francium	Fr					
		(Ac K)		223*	223.019 733		22 m
88	Radium	Ra					
		(Ac X)		223*	223.018 499		11.43 d
		(Th X)		224*	224.020 187		3.66 d
		(Ra)		226*	226.025 402		1600 y
		(Ms Th$_1$)		228*	228.031 064		5.75 y
89	Actinium	Ac		227*	227.027 749		21.77 y
		(Ms Th$_2$)		228*	228.031 015		6.15 h
90	Thorium	Th	232.0381				
		(Rd Ac)		227*	227.027 701		18.72 d
		(Rd Th)		228*	228.028 716		1.913 y
				229*	229.031 757		7300 y
		(Io)		230*	230.033 127		75,000 y

(1) Z	(2) Element	(3) Symbol	(4) Chemical Atomic Weight	(5) Mass Number (* Indicates Radioactive) A	(6) Atomic Mass	(7) Percent Abundance	(8) Half-Life (if Radioactive) $t_{1/2}$
(90)	(Thorium)	(UY)		231*	231.036 299		25.52 h
		(Th)		232*	232.038 051	100	1.40×10^{10} y
		(UX$_1$)		234*	234.043 593		24.1 d
91	Protactinium	Pa		231*	231.035 880		32,760 y
		(Uz)		234*	234.043 300		6.7 h
92	Uranium	U	238.0289	232*	232.037 131		69 y
				233*	233.039 630		1.59×10^5 y
				234*	234.040 946	0.0055	2.45×10^5 y
		(Ac U)		235*	235.043 924	0.720	7.04×10^8 y
				236*	236.045 562		2.34×10^7 y
		(UI)		238*	238.050 784	99.2745	4.47×10^9 y
93	Neptunium	Np		235*	235.044 057		396 d
				236*	236.046 560		115,000 y
				237*	237.048 168		2.14×10^6 y
94	Plutonium	Pu		236*	236.046 033		2.87 y
				238*	238.049 555		87.7 y
				239*	239.052 157		24,120 y
				240*	240.053 808		6560 y
				241*	241.056 846		14.4 y
				242*	242.058 737		3.73×10^5 y
				244*	244.064 200		8.1×10^7 y

The masses in column 6 are atomic masses, which include the mass of Z electrons. Data are from the National Nuclear Data Center, Brookhaven National Laboratory, prepared by Jagdish K. Tuli, July 1990. The data are based on experimental results reported in *Nuclear Data Sheets* and *Nuclear Physics* and also from *Chart of the Nuclides*, 14th ed. Atomic masses are based on those by A. H. Wapstra, G. Audi, and R. Hoekstra. Isotopic abundances are based on those by N. E. Holden.

Illustration Credits

Introduction p. xvi Courtesy AT&T Archives.

Chapter 1
Opener p. 2 The Hebrew University of Jerusalem/Courtesy of AIP Niels Bohr Library; **p. 3** Courtesy NRAO/AUI; **p. 28** C. Powell, P. Fowler & D. Perkins/Science Photo Library/Photo Researchers; **p. 36 (a)** © Michael Freeman; **(b)** N.A.S.A. (76-HC-612); **p. 37** © Michael Freeman.

Chapter 2
Opener p. 47 Adapted from Eastman Kodak and Wabash Instrument Corporation; **p. 49** Max Planck Institute, Berlin; **p. 53** Courtesy Thorn EMI Electron Tubes Ltd.; **Figure 2-7 (b)** Courtesy General Electric Company; **Figure 2-10** From G. Herzberg, *Annalen de Physick*, Vol. 84, p. 565, 1927; **Figure 2-17 (a,b)** *PSSC Physics*, 2nd ed., 1965. D. C. Heath & Co. and Education Development Center, Newton, Massachusetts; **(c)** C. G. Shull; **(d)** Claus Jönsson.

Chapter 3
Opener p. 74 Lawrence Livermore Laboratory/Science Photo Library/Photo Researchers; **Figure 3-2 (a,b,c)** E. R. Huggins; **(d)** Claus Jönsson; **Figure 3-3** From G. F. Missiroli and G. Pozzi, *American Journal of Physics*, Vol. 44, No. 3, 1976, p. 306; **p. 81** S. Brandt and H. D. Dahmen, from *The Picture Book of Quantum Mechanics*. The color graphics reproduced here were made using the program contained in the book *Quantum Mechanics on the Personal Computer* by S. Brandt and H. D. Dahmen (Springer-Verlag, 1989); **p. 94 (a)** Courtesy of Texas Instruments; **(b)** Courtesy of David Sarnoff Research Center; **p. 95 (c,d,e,f)** Images by Patterson Electronics and computer processing by John Sanford; **p. 101** S. Brandt and H. D. Dahmen, op. cit.; **Figure 3-20 (a,b)** *PSSC Physics*, 2nd ed., 1965. D. C. Heath & Co. and Education Development Center, Newton, Massachusetts; **p. 104 (a,b)** and *(bottom)* S. Brandt and H. D. Dahmen, op. cit.; **Figure 3, p. 113 (a)** R. Tromp/IBM Research; **(b,c)** Courtesy of E. D. Williams; **Figure 4 p. 114** D. M. Eigler, and E. K. Schweizer, IBM Almaden Research Center.

Chapter 4
Opener p. 120 Dr. Bruce Schardt/Courtesy Digital Instruments, Inc.; **p. 127 (a)** Courtesy of IBM Research; **(b)** Courtesy of Park Scientific Instruments, Mountain View, California; **p. 138 (a,b)** A. Jayaraman, AT&T Bell Labs; **p. 140** David Parker/Science Photo Library/Photo Researchers; **p. 141** © Robert Landau/West Light; **p. 147 (a,b)** © 1991 Paul Silverman/Fundamental Photographs; **p. 150 (a,b)** Chuck O'Rear/West Light; **p. 151 (c)** Courtesy of Spectra-Physics Lasers; **(d)** © Bell Labs/Science Source/Photo Researchers; **(e)** Courtesy of Ahmed H. Zewail, California Institute of Technology; **p. 154 (a)** © Philippe Plailly/Science Photo Library; **(b,c)** © Ronald R. Erickson, 1981. Hologram by Nicklaus Phillips, 1978, for Digital Equipment Corporation; **(d)** © 1983 Ronald R. Erickson; **(e)** © Chuck O'Rear/West Light; **Figures 4, 5 p. 159** Courtesy of David Wineland; **Figures 6, 7 p. 162** Courtesy of David Wineland.

Chapter 5
Opener p. 167 Courtesy David S. Lawrence, SUNY Buffalo; **Figure 5-7** © Will and Deni McIntyre/Photo Researchers; **Figure 5-14 (a,b)** Courtesy Dr. J. A. Marquisee.

Chapter 6
Opener p. 190 Courtesy of AT&T Archives; **p. 191** Courtesy of Museum of Modern Art, New York; **p. 192 (a)** Richard Walters/ © Discover Publications; **(b)** Dr. Jeremy Burgess/Science Photo Library/Photo Researchers; **p. 193 (c)** © 1976 Thomas R. Taylor/Photo Researchers; **(d)** Courtesy of AT&T Archives; **p. 195 (a)** Chris Kovach/ © Discover Publications; **(b)** Srinivas Manne, University of California-Santa Barbara; **(c)** Dr. F. A. Quiocho and J. C. Spurlino/Howard Hughes Medical Institute, Baylor College of Medicine; **(d)** W. Krätschmer/Max-Planck Institute for Nuclear Physics; **p. 216** © 1983 C. Falco/Photo Researchers; **p. 218 (a)** Courtesy of AT&T Archives; **(b)** Dr. Jeremy Burgess/Science Photo Library/Photo Researchers; **p. 219 (c)** Courtesy of AT&T Archives; **(d)** Dr. Jeremy Burgess/Science Photo Library/Photo Researchers; **(e,f)** Courtesy of AT&T Archives; **p. 220 (a)** Adapted from "The Photorefractive Effect," by David Pepper, Jack Feinberg, and Nicolai Kukhtarev. © by Scientific American, Inc. 1990; **(b)** Roger S. Cudney/Courtesy of Jack Feinberg, University of Southern California; **p. 221 (c)** Jack Feinberg, "Self-Pumped, Continuous-Wave Phase Conjugator Using Internal Reflection," *Optical Letters*, Vol. 7, 1982, pp. 486–488; **(d)** Courtesy of AT&T Archives; **Figure 6-29 (b)** A. Leitner/Rensselaer Poly-

technic Institute; **p. 225 (a)** Courtesy of IBM Research; **(b)** Akira Tonomura, Hitachi Ltd., Saitama, Japan; **(c)** © Fujifotos/The Image Works; **p. 226** © 1983 C. Falco/Photo Researchers; **p. 230** © Peter Freed/New York University; **Figure 1 (a,b)** Courtesy of IBM Thomas J. Watson Research Center; **Figure 3 p. 233** Courtesy of Samuel J. Williamson.

Chapter 7

Opener p. 238 © Hans Wolf/The Image Bank; **p. 241** Courtesy of Ronald Y. Cusson; **p. 245** Courtesy of Mallinckrodt Institute of Radiology; **p. 254 (a)** © 1991 by the Metropolitan Museum of Art, New York; **(b,c)** Courtesy of Paintings Conservation Dept., Metropolitan Museum of Art, New York; **p. 260** © Jerry Mason/Science Photo Library/Photo Researchers; **p. 262 (a,b,c)** Courtesy of Princeton Plasma Physics Laboratory; **p. 263 (a,b)** Courtesy of Lawrence Livermore National Laboratory, U.S. Department of Energy.

Chapter 8

Opener p. 277 Lawrence Berkeley Laboratory/Science Photo Library/Photo Researchers; **p. 279 (a)** Science Photo Library/Photo Researchers; **(b)** Lawrence Berkeley Laboratory/Science Photo Library/Photo Researchers; **p. 281 (a)** From *The Particle Explosion*, by Frank Close, Michael Marten, and Christine Sutton, Oxford University Press, 1987, p. 130; **(b)** Lawrence Berkeley Laboratory/Science Photo Library/Photo Researchers; **Figure 8-1** Richard Ehrlich; **p. 282** Courtesy of CERN; **p. 283** Courtesy of CERN; **p. 285** Lawrence Berkeley Laboratory/Science Photo Library/Photo Researchers.

Chapter 9

Opener p. 297 Optical image: © Anglo-Australian Telescope Board; radio image: Jack O. Burns/University of Southern California; **Figure 9-4** Patrick Wiggins/© Hansen Planetarium; **p. 304** (*top*) N.A.S.A. (74-HC-260); (*bottom*) N.A.S.A. (90-HC-508); **Figure 9-9** Helmut Wimmer/Courtesy of *Natural History*; **Figure 9-10** Lund Observatory, Sweden; **Figure 9-12 (a)** From *Foundations of Astronomy*, 1986 ed., by Michael A. Seeds, © 1986 by Wadsworth, Inc., reprinted by permission of the publisher. Adapted from a study by G. De Vaucouleurs and W. D. Pence, *The Astronomical Journal 83*, No. 10, October 1978, p. 1163; **(b)** NRAO/AUI and Farhad Yusef-Zadeh/Northwestern University; **Figure 9-15** (*left*) © Anglo-Australian Telescope Board; (*right*) © David F. Malin/Anglo-Australian Telescope Board; **Figure 9-18 (b,c)** Harvard-Smithsonian Center for Astrophysics **Figure 9-19 (a)** U.S. Naval Observatories; **(b)** © 1980 Anglo-Australian Telescope Board; **(c)** David Malin/© Anglo-Australian Telescope Board; **(d)** California Institute of Technology; **Figure 9-21** California Institute of Technology; **Figure 9-22** M. J. Geller and J. P. Huchra, Harvard-Smithsonian Center for Astrophysics/© 1989 by the American Association for the Advancement of Science. From "Mapping the Universe," by Margaret J. Geller and John P. Huchra, *Science*, Vol. 246, November 17, 1989, pp. 897–903; **Figure 9-23** S. Maddox, W. Sutherland, G. Efstathiou, and J. Loveday/University of Oxford; **Figure 9-24** © 1984 Fermilab.

Answers

These answers are calculated using $g = 9.81$ m/s^2 unless otherwise specified in the exercise or problem. The results are usually rounded to three significant figures. Differences in the last figure can easily result from differences in rounding the input data and are not important.

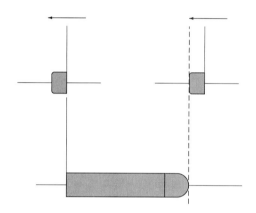

Chapter 1

True or False 1. True 2. True 3. False 4. True 5. False
6. False 7. True

Problems

1. (*a*) 0.183 ms (*b*) 1.83×10^{-12} s (*c*) No
3. (*a*) 4.94×10^{-8} s (*b*) 12.6 m (*c*) 6.63 m
5. (*a*) 44.7 μs (*b*) 13.4 km
7. $0.527c$
9. (*a*) 130 y (*b*) 88.1 y
11. $0.9991c$
13. 2.60×10^8 m/s
15. (*a*) 4.5×10^{-10} percent (*b*) The time elapsed on the pilot's clock is 3.15×10^7 s $- 1.42 \times 10^{-4}$ s; the time lost in minutes is 2.37×10^{-6} min.
17. 80 $c \cdot$min
19. answer given in problem
21. $L_p V/c^2 = 60$ min
23. 0.0637
25. $0.6c$
27. $0.696c$
29. (*a*) $-0.882c$ (*b*) $-60{,}000$ m/s $+ 6 \times 10^{-4}$ m/s
31. $-0.994c$
33. (*a*) 1.11×10^{-17} kg (*b*) 0.351 mg
35. (*a*) 9×10^{13} J (*b*) $2.5 million (*c*) 28,571 y
37. $E = 0.522$ MeV, $K = 1.05 \times 10^{-2}$ MeV, $p = 0.104$ MeV/c
39. (*a*) 2.23×10^8 m/s (*b*) 1039 MeV/c
41. 3.55×10^{14} reactions/s
43. 0.782 MeV
45. 50 percent
47. $0.8c$
49. 1.85×10^4 y
51. 9.61 ms
53. The required speed is $0.4c$; event B precedes event A for an observer moving with a speed $0.4c < v < c$.
55. (*a*) $0.625c$ (*b*) 31.2 y
57. (*a*) 52.7 m (*b*) $-0.987c$ (*c*) 16.1 m (*d*) 2.07×10^{-7} s (*e*)
59. (*a*) 4.97 MeV/c (*b*) 0.995
61. (*a*) 0.75 percent (*b*) 68.7 percent
63. (*a*) 630 m/c (*b*) 777 m/c (*c*) 148 m/c (*d*) 778 m/c (*e*) 4.36 h (*f*) 19 h
65. (*a*) To the left with speed $0.5c$ (*b*) 1.73 y
67. 3.84×10^{14} kg/day
69. (*a*) $0.333c$ (*b*) 20 m in the $+x$ direction (*c*) 60 m/c
71. (*a*) 290 MeV (*b*) 625 MeV
73. (*a*) 1.30 m (*b*) 0.825 m
75. (*a*) E/Mc (*b*) EL/Mc^2
77. answer given in problem
79. answer given in problem
81. answer given in problem
83. answers given in problem

Chapter 2

True or False 1. True 2. True 3. False 4. True 5. True
6. False 7. True 8. True 9. True 10. True
11. True 12. False

Problems

1. $E = 6.626 \times 10^{-26}$ J $= 4.14 \times 10^{-7}$ eV
3. (*a*) 2.41×10^{14} Hz (*b*) 2.41×10^{17} Hz (*c*) 2.41×10^{20} Hz
5. $E_{400} = 3.11$ eV, $E_{700} = 1.77$ eV
7. (*a*) $f_t = 1.11 \times 10^{15}$ Hz, $\lambda_t = 271$ nm (*b*) 1.63 V (*c*) 0.39 V
9. (*a*) 4.74 eV (*b*) 2.36 V
11. (*a*) 4.97×10^{-19} J (*b*) 0.01 J (*c*) 2.01×10^{16} photons/s
13. 9.27×10^3 V
15. 1.215 pm

17. (a) $p = 1.66 \times 10^{-27}$ kg·m/s = 3.11 eV/c (b) $p = 3.32 \times 10^{-25}$ kg·m/s = 621 eV/c (c) $p = 6.63 \times 10^{-24}$ kg·m/s = 12.4 keV/c (d) $p = 2.21 \times 10^{-32}$ kg·m/s = 4.14×10^{-5} eV/c
19. (a) 17.5 keV (b) 76.0 pm (c) 16.3 keV
21. answer given in problem
23. $\Delta E_{3 \to 2} = 1.89$ eV, $\lambda_{3 \to 2} = 656$ nm; $\Delta E_{4 \to 2} = 2.55$ eV, $\lambda_{4 \to 2} = 486$ nm; $\Delta E_{5 \to 2} = 2.86$ eV, $\lambda_{5 \to 2} = 434$ nm
25. (a) $\Delta E_{\infty \to 4} = 0.850$ eV, $\lambda_{\infty \to 4} = 1459$ nm (b) $\lambda_{5 \to 4} = 4052$ nm, $\lambda_{6 \to 4} = 2627$ nm, $\lambda_{7 \to 4} = 2168$ nm

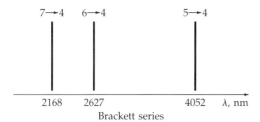

27. (a) 0.775 nm (b) 0.0775 nm (c) 0.0245 nm (d) 0.00775 nm
29. (a) 3.313×10^{-27} kg·m/s (b) 6.024×10^{-24} J
31. 0.203 nm
33. 4.40×10^{-13} m
35. 1.52×10^{-34} m
37. 22.8 eV
39. 6.31 keV
41. 6.80×10^6 m = 4225 mi
43. 1.69×10^{-14} W/m²
45. (a) 3.18 W/m² (b) 1.04×10^{15} photons/s
47. (a) 4.86 pm (b) 92.7 keV (c) 92.7 keV
49. (a) 13.6 eV (b) 54.4 eV (c) 122.4 eV
51. (b) $E_n = n^2(37.6 \text{ eV})$

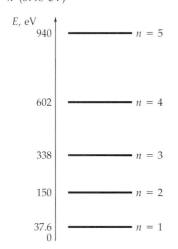

(c) 11.0 nm (d) 6.60 nm (e) 1.37 nm
53. (a) $E_1 = 5.13 \times 10^{-3}$ eV, $E_2 = 2.05 \times 10^{-2}$ eV, $E_3 = 4.61 \times 10^{-2}$ eV (b) 80.8 μm (c) 48.5 μm (d) 30.3 μm
55. (b) $R_\infty = 10.97373$ μm^{-1}; $R_H = 10.96776$ μm^{-1} (c) 0.0545 percent
57. (a) 10^{-22} W (b) 53 min
59. (a) 2.42×10^{-10} m (b) 0.512 MeV (c) 2.73×10^{-22} kg·m/s (d) 2.42 pm
61. answer given in problem

Chapter 3

True or False **1.** False; the electron velocity equals the group velocity of the wave packet describing it **2.** False; the position can be known precisely if the momentum is completely undetermined **3.** True **4.** False; energy exchanges, such as in the photoelectric effect, require a particle model **5.** False **6.** True **7.** False; the expectation value is the *average* of the measurement for a large number of identical systems. It may, or may not, be one of the possible values measured **8.** False; α decay is a physical example of the penetration of a barrier by a wave **9.** True

Problems
1. (a) 0 (b) 0.00417
3. (a) $|A|^2 \, dx$ (b) $|A|^2 e^{-1/2} \, dx$ (c) $|A|^2 e^{-2} \, dx$ (d) The electron is most likely to be found in a region centered at $x = 0$.
5. $\Delta x \, \Delta k \approx 4\pi$
7. 2.64×10^{-19} m
9. 5.79 m
11. (a) $f_0 = N/\Delta t$ (b) $\Delta k \approx 1/\Delta x = (v \, \Delta t)^{-1}$
13. 1.66×10^{-33} m
15. $\lambda = 8.68 \times 10^{-11}$ m, which is roughly the size of an atom
17. answer given in problem
19. (a) $K(x) = (\hbar^2/2mL^2)(1 - x^2/L^2)$
21. (a) $2L$ (b) $2L/3$ (c) $h/2L$ (d) $p^2/2m = h^2/8mL^2$
23. (a) 0.0205 eV (b) 205 MeV
25. (a) 0 (b) 0.003 (c) 0
27. 3.02×10^{19}
29. (a)

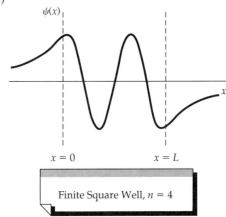

(b)

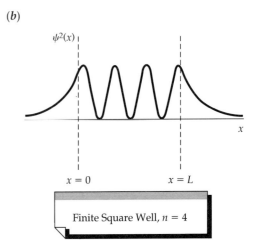

31. (a) $L/2$ (b) $0.321L^2$
33. (a) answer given in problem (b) $<x> = L/2$, $<x^2> = L^2/3$
35. (a) $k_2 = (6mU_0)^{1/2}/\hbar = \sqrt{3/2}k_1$ (b) 0.0102
(c) 0.990 (d) 9.90×10^5 particles are expected to continue past the potential step. Classically, 100 percent of the particles are transmitted.
37.

n_1	n_2	n_3	$E/(h^2/288mL_1^2)$
1	1	1	49
1	1	2	61
1	2	1	76
1	1	3	81
1	2	2	88
1	2	3	108
1	1	4	109
1	3	1	121
1	3	2	133
1	2	4	136

39. (a)

n_1	n_2	n_3	$E/(h^2/128mL_1^2)$
1	1	1	21
1	1	2	24
1	1	3	29
1	2	1	33
1	1	4	36
1	2	2	36
1	2	3	41
1	1	5	45
1	2	4	48
1	3	1	53
1	1	6	56
1	3	2	56

(b) $n_1 = 1$, $n_2 = 1$, $n_3 = 4$ and $n_1 = 1$, $n_2 = 2$, $n_3 = 2$ both give 36; $n_1 = 1$, $n_2 = 1$, $n_3 = 6$ and $n_1 = 1$, $n_2 = 3$, $n_3 = 2$ both give 56
41. (a) $\psi(x, y, z) = A \cos(\pi x/L) \sin(\pi y/L) \sin(\pi z/L)$
(b) The energy levels are the same.
43. $10E_1$
45. (a) 205 MeV (b)

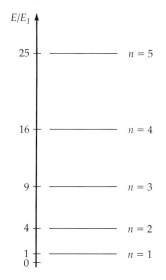

(c) 2.02×10^{-15} m (d) 1.21×10^{-15} m (e) 7.57×10^{-16} m
47. (a) 0.5 (b) 0.196 (c) 0.909
49. (a)

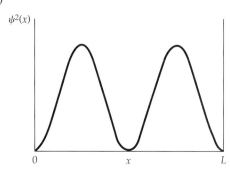

(b) $L/2$ (c) 0 (d) There is no contradiction.
51. (a) 0.5 (b) 0.402 (c) 0.75
53. (a) $T \approx 10^{-18}$ (b) $T \approx 10^{-2}$
55. $0.289L$
57. $T \approx 10^{-1}$
59. (a) 1.66×10^{-26} m/s (b) $T = 1.21 \times 10^{21}$ s = 3.82×10^{13} y
61. $<x> = 0$, $<x^2> = L^2(1/12 - 1/2\pi^2)$
63. $<x> = 0$, $<x^2> = L^2(1/12 - 1/8\pi^2)$
65. $E = 65h^2/8mL^2$ is four-fold degenerate, with quantum numbers $n = 7$, $m = 4$; $n = 4$, $m = 7$; $n = 8$, $m = 1$; $n = 1$, $m = 8$.
67. answer given in problem
69. answer given in problem
71. (b) $T = 0.823$, $R = 0.177$ (c) $T = 0.971$, $R = 0.0294$
(d) $T = 0.9993$, $R = 6.93 \times 10^{-4}$
73. answers given in problem

Chapter 4
True or False 1. True 2. True 3. True 4. True
Problems
1. (a) $\sqrt{2}\hbar$ (b) $-1, 0, +1$ (c)

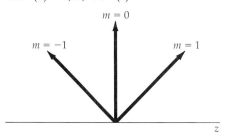

3. (a) $\ell = 0, 1, 2$ (b) For $\ell = 0$, $m = 0$; for $\ell = 1$, $m = -1, 0, 1$; for $\ell = 2$, $m = -2, -1, 0, 1, 2$ (c) 18
5. (a) 3.49×10^{-3} kg · m^2/s (b) 3.31×10^{31}
7. (a) $n \geq 4$, $m = -3, -2, -1, 0, 1, 2, 3$ (b) $n \geq 5$, $m = -4, -3, -2, -1, 0, 1, 2, 3, 4$ (c) $n \geq 1$, $m = 0$
9. (a) $\psi = (1/\sqrt{\pi})(1/a_0)^{3/2}e^{-1}$ (b) $\psi^2 = (1/\pi a_0^3)e^{-2}$
(c) $P(r) = (4/a_0)e^{-2}$
11. (a) 0.0162 (b) 0.00879
13. (a) $\psi = (1/4\sqrt{2\pi})(1/a_0)^{3/2}(2 - Z)e^{-Z/2}$ (b) $\psi^2 = (1/32\pi a_0^3)(2 - Z)^2 e^{-Z}$ (c) $P(r) = (1/8a_0)(2 - Z)^2 e^{-Z}$
15. (a) 6.95×10^{-5} eV (b) 1.79 cm
17. 3

19. (a) 5.05×10^{-27} J/T (b) 3.15×10^{-12} eV/G
21. $\ell = 1$ or $\ell = 2$
23. (a) $j = 1$ or $j = 0$ (b) $\sqrt{2}\hbar$
(c)

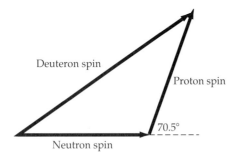

25. (a) $1s^2 2s^2 2p^2$ (b) $1s^2 2s^2 2p^4$
27. (a) Silicon (b) Calcium
29. Both gallium and indium have a single p-shell electron shielded from the nucleus by the inner closed shells, making it relatively easy to remove.
31. 1.59
33. Potassium, chromium and copper each have a single 4s electron as the outer electron; calcium, titanium, and manganese each have two 4s electrons as the outer electrons.
35. (a) 2s or 2p (b) $1s^2 2s^2 2p^6 3p$ (c) $1s2s$
37. (a) 0.0611 nm, 0.0580 nm (b) 0.0543 nm
39. Zirconium
41. (a) 1.01 nm (b) 0.155 nm
43. (a) 15 mJ (b) 5.24×10^{16}
45. 2.81 km
47. (a) 3p (b) $K = 6.4$ eV
49. (a) 0.305 nm (b) 0.715 nm
51. Tin
53. (a) $E = 2.10505$ eV and $E = 2.10291$ eV (b) $\Delta E = 2.14 \times 10^{-3}$ eV (c) $B = 18.5$ T
55. (a) $E = 1.61774$ eV and $E = 1.61041$ eV (b) $\Delta E = 7.33 \times 10^{-3}$ eV (c) $B = 63.3$ T
57. (a) 3.32×10^6 m/s^2 (b) $\Delta z = (a\,\Delta x_1/v_0^2)(\Delta x_1 + 2\,\Delta x_2)$, with $\Delta x_1 = 0.75$ m and $\Delta x_2 = 1.25$ m
59. (a) 2, 1, 0 (b) 1, 0 (c)

ℓ	s	j
2	1	3, 2, 1
2	0	2
1	1	2, 1, 0
1	0	1
0	1	1
0	0	0

(d) 3/2, 1/2 (e)

j_1	j_2	j
3/2	3/2	3, 2, 1, 0
3/2	1/2	2, 1
1/2	3/2	2, 1
1/2	1/2	1, 0

The same values of j occur here as in part (c).
61. $P = 9.01 \times 10^{-15}$
63. answer given in problem
65. answer given in problem
67. $r = (3 \pm \sqrt{5})a_0/Z$
69. 0.323

Chapter 5

True or False 1. False 2. False; it is due to exclusion-principle repulsion 3. False; the dissociation energy is the energy to separate the ions *and* form neutral atoms from the ions. For NaCl, the dissociation energy is 1.53 eV less than the energy needed to remove the ions to an infinite separation 4. False; it is very common 5. False; they are much less than the energy of electronic excitation

Problems
1. 0.940 nm
3. (a) Ionic bonding (b) Covalent bonding (c) Metallic bonding
5. (a) 1 eV/molecule = 23.0 kcal/mole (b) 98.1 kcal/mole
7. 43.6 percent
9. 2.63×10^{-29} Cm
11. 0.110 nm
13. answer given in problem
15. 0.121 nm
17. (a) -5.39 eV (b) 4.66 eV (c) 0.230 eV
19. (a) $U = -6.09$ eV. From the figure the dissociation energy is 5.79 eV. (b) 0.305 eV
21. (a) -6.64 eV (b) 5.75 eV (c) 0.68 eV
23. 1.25×10^{14} Hz
25. 1.58×10^{-5} eV
27. 477 N/m
29. The Lenard-Jones potential appropriate for an H_2 molecule is shown in the figure as curve (1).

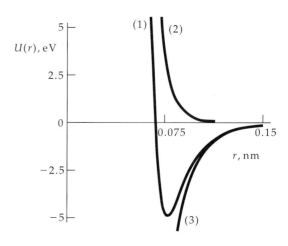

Curve (2) shows the $U_0(a/r)^{12}$ term, and curve (3) shows the $-2U_0(a/r)^6$ term.
31. $F \sim x^{-4}$
33. (a) $I = 1.45 \times 10^{-46}$ kg·m^2, $E_{0r} = 0.000239$ eV (b) Energy levels for $\ell = 0$ to $\ell = 5$ are shown in the figure. Choosing $E = 0$ for the $\ell = 0$ state yields $E = \ell(\ell + 1)E_{0r}$ for general ℓ. (c) $\ell = 1 \to \ell = 0$, $E = 0.000478$ eV; $\ell = 2 \to \ell = 1$, $E = 0.000956$ eV; $\ell = 3 \to \ell = 2$, $E = 0.00144$ eV; $\ell = 4 \to \ell = 3$, $E = 0.00191$ eV; $\ell = 5 \to \ell = 4$, $E = 0.00239$ eV. (d) $\ell = 1 \to \ell = 0$, $\lambda = 2.60$ mm; $\ell = 2 \to \ell = 1$, $\lambda = 1.30$ mm; $\ell = 3 \to \ell = 2$, $\lambda = 0.865$ mm; $\ell = 4 \to \ell = 3$, $\lambda = 0.649$ mm; $\ell = 5 \to \ell = 4$, $\lambda = 0.519$ mm. These photons fall in the

microwave and short radio wave portion of the electromagnetic spectrum.
35. (*a*) $\mu = 0.9722$ u for H^{35}Cl, and $\mu = 0.9737$ for H^{37}Cl; $\Delta\mu/\mu = 0.00153$ (*b*) answer given in problem (*c*) $\Delta f/f = -0.00153$; According to Figure 38-17, the major peak is at approximately $f_{\text{major}} = 9.21 \times 10^{15}$ Hz. The associated minor peak should therefore be at $f_{\text{minor}} = f_{\text{major}} + \Delta f = 9.21 \times 10^{15}$ Hz $- 0.00153(9.21 \times 10^{15}$ Hz$) = 9.196 \times 10^{15}$ Hz, which is in agreement with the location of the minor peak in the figure.

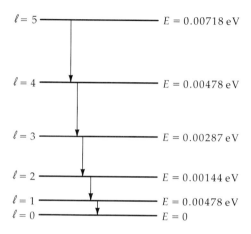

Chapter 6
True or False 1. True **2.** False; the heat capacity due to the free electrons is much less than that predicted by the classical theory **3.** True **4.** False **5.** True **6.** True **7.** False

Problems
1. (*a*) The unit cell is a cube with sides of length $2R$. (*b*) 0.524
3. 2.07 g/cm^3
5. (*b*) 0.141 nm
7. (*a*) 5.86×10^{22} electrons/cm^3 (*b*) 5.90×10^{22} electrons/cm^3
9. 3.00
11. 4.00
13. (*a*) 1.36×10^5 K (*b*) 2.46×10^4 K (*c*) 1.18×10^5 K
15. (*a*) 11.2 eV (*b*) 1.30×10^5 K
17. (*a*) Silver and gold (*b*) 0.1 V
19. At $T = 0$, $U/N = 4.2240$ eV; at $T = 300$ K, $U/N = 4.2242$ eV. The classical prediction is $U/N = 0.0388$ eV
21. (*a*) 3.42×10^{-8} m (*b*) 4.11×10^{-8} m (*c*) 4.29×10^{-9} m
23. 1.09×10^{-6} m
25. 1.77×10^{-7} m
27. (*a*) *p*-type semiconductor (*b*) *n*-type semiconductor
29. 116 K
31. (*a*) 0.0596 V (*b*) -0.0596 V
33. (*a*) 2.17×10^{-3} eV (*b*) 0.455 mm
35. (*a*) 3.67×10^{-3} (*b*) 1.22×10^{-2}
37. 9.46 eV
39. (*a*) 9.48×10^{25} atoms (*b*) 7.58×10^{26} states (*c*) 9.29×10^{-27} eV/state (*d*) The thermal energy is 0.0259 eV, which is 2.79×10^{24} greater than the average spacing between conduction states.
41. 2.19 eV
43. 250
45. (*b*) 500 MΩ (*c*) $R = 1.03$ Ω, $I = 0.485$ A (*d*) 0.0515 Ω
47. (*a*) 0.0189 eV (*b*) 0.00531 eV (*c*) 48.2 percent
49. $P = 3.82 \times 10^{10}$ N/m^2 = 3.78×10^5 atm
51. (*a*) 66.1 nm (*b*) 2.27×10^{-20} m^2
53. (*a*) 1.28×10^{13} Hz (*b*) $\lambda = 23.4$ μm, which is of the same order of magnitude as the absorption bands at 61 μm
55. (*a*) $v_d = -eE\tau/m$ (*b*) $\rho = m/ne^2\tau$

Chapter 7
True or False 1. False; it does not contain electrons **2.** True **3.** False; after 2 half-lives, $\frac{3}{4}$ of the nuclei in a sample have decayed and $\frac{1}{4}$ remain **4.** True **5.** True **6.** False; it is a measure of the energy deposited per unit mass

Problems
1. (*a*) ^{15}N, ^{13}N (*b*) ^{57}Fe, ^{58}Fe (*c*) ^{117}Sn, ^{119}Sn
3. (*a*) 31.99 MeV, 5.332 MeV/nucleon (*b*) 333.7 MeV, 8.557 MeV/nucleon (*c*) 1636 MeV, 7.868 MeV/nucleon
5. (*a*) 20.58 MeV (*b*) 7.253 MeV
7. answer given in problem
9. (*a*) 5 s (*b*) 250 counts/s
11.

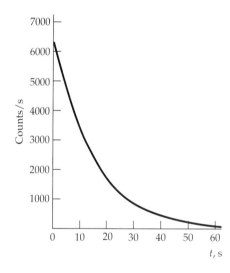

$\lambda = 0.0693$ s^{-1}
13. 3.61×10^{10} decays/s, which is approximately 1 Ci = 3.7×10^{10} decays/s
15. (*a*) ^{22}Na can decay by β^+ emission to become ^{22}Ne. (*b*) ^{24}Na can decay by emission of a β^- to become ^{24}Mg.
17. 3352 y
19. 13,940 y
21. (*a*) -0.7638 MeV (*b*) 3.269 MeV
23. 1.56×10^{19} fissions/s
25. 1.16×10^8 K
27. (*a*) 0.0581 cm (*b*) 0.00518 cm
29. 2.72 barns
31. 21.0 barns
33. (*a*) 22.96 MeV (*b*) 4.178 MeV (*c*) 1.286 MeV
35. $\lambda = 1.05 \times 10^{-23}$ s^{-1}, $\tau_{1/2} = 2.10 \times 10^{15}$ y
37. $\lambda = 4.16 \times 10^{-9}$ s^{-1}, $\tau_{1/2} = 5.27$ y

39. (a) 156 h (b) 551 h
41. (a) 1.20 MeV (b) The mass of ^{13}C does not include the mass of the seventh electron on the atom and the mass of the emitted β^+.
43. (a) For ^{141}Ba, $R = 7.81$ fm; for ^{92}Kr, $R = 6.77$ fm. (b) 200 MeV
45. 7.84×10^7 AU
47. (a) 3.42×10^{12} neutrons/s (b) 493 rem/h (c) 1.01 h
49. 7.03×10^8 y
51. (a) 2.47×10^5 atoms (b) 26.6 mrem (c) 1.177
53. $E_n = 14.1$ MeV, $E_{He} = 3.56$ MeV
55. (a) 1.19 MeV/c (b) 0.752 keV (c) 0.0962 percent
57. (b) 55
59. (c) The rate of proton consumption is 3.74×10^{38} protons/s. At this rate the sun will last 50.5×10^9 y.
61. 1.77×10^{-3} rem

Chapter 8

True or False **1.** True **2.** True **3.** False; hadrons include both baryons and mesons **4.** False **5.** False **6.** True **7.** False; the electron interacts with the proton via the weak interaction (as well as the electromagnetic interaction and the gravitational interaction) **8.** True; strangeness can change by ± 1 in weak interactions **9.** True

Problems
1. 0.397 fm
3. (a) 279.2 MeV (b) 1878 MeV (c) 212 MeV
5. (a) $\Delta S = +1$, the reaction can proceed via the weak interaction (b) $\Delta S = +2$, the reaction is not allowed (c) $\Delta S = +1$, the reaction can proceed via the weak interaction
7. (a) $\Delta S = +2$, the reaction is not allowed (b) $\Delta S = +1$, the reaction can proceed via the weak interaction
9.

	Quark structure	Baryon number	Charge, e	Strangeness	Hadron identity
(a)	uud	+1	+1	0	p^+
(b)	udd	+1	0	0	n
(c)	uus	+1	+1	−1	Σ^+
(d)	dds	+1	−1	−1	Σ^-
(e)	uss	+1	0	−2	Ξ^0
(f)	dss	+1	−1	−2	Ξ^-

11. uuu
13. (a) $c\bar{d}$ (b) $\bar{c}d$
15. 2.44×10^{-18} m
17. (a) It is necessary that its charge, lepton number, baryon number, strangeness, charm, topness, and bottomness all be zero. (b) π^0 (c) Ξ^0
19. (a) $\bar{u}d\bar{d}$ (b) uss (c) uus
21. (a) Charge $-1e$, baryon number 1, and strangeness 0 (b) Charge 0, baryon number 0, strangeness 0, and charm -1 (c) Charge $+1e$, baryon number 0, strangeness 0, and bottomness -1 (d) charge $+1e$, baryon number -1, and strangeness 3
23. (a) The final products shown are not all stable. The neutron decays to yield $p + e + \bar{\nu}_e$. (b) $\Omega^- \rightarrow p + 3e^- + e^+ + 3\bar{\nu}_e + \nu_e + 2\bar{\nu}_\mu + 2\nu_\mu$ (c) All quantities are conserved except for strangeness, which changes by $+3$
25. (a) All conservation laws satisfied (b) Energy and baryon number conservation are violated (c) All conservation laws satisfied
27. (a) 1193 MeV (b) 77 MeV/c (c) 2.66 MeV (d) $E = 74.3$ MeV, $p = 74.3$ MeV/c

Chapter 9

True or False **1.** True **2.** False **3.** False; since the lifetime varies as $M_\odot^{-3}$, it will be 1/125 the sun's lifetime **4.** True; it occurred about 170,000 years ago **5.** True **6.** False **7.** False **8.** True

Problems
1. 25.3 d
3. The time for one orbit is 2.26×10^8 y. The sun has completed about 44 orbits.
5. (a) $3.26 \, c \cdot y$ (b) 3.35×10^5 stars
7. (a) $T_e \approx 3300$ K, $L \approx 5 \times 10^{-2} L_\odot$ (b) $T_e \approx 13{,}500$ K, $L \approx 10^2 L_\odot$ (c) $R_{0.3} = 0.3 R_\odot$, $R_3 = 3 R_\odot$ (d) For $M = 0.3 M_\odot$, $t_L = 37 t_{L\odot}$; for $M = 3 M_\odot$; $t_L = 0.037 t_{L\odot}$
9. (a) 9360 y (b) $1.86 R_\odot$
11. (a) 10.1 km (b) 8.01×10^{38} J (c) 1.85×10^{26} W
13. (a) $3.13 \times 10^9 \, c \cdot y$ (b) 1.30×10^{10} y (c) 10 percent
15. 1.06×10^{-26} kg/m^3
17. Planck density = 5.5×10^{97} kg/m^3, proton density = 1.67×10^{18} kg/m^3, and osmium density = 2.45×10^4 kg/m^3.
19. The muon threshold was at roughly 10^{-3} s after the Big Bang. At the current temperature, the maximum rest mass that could be created is of order of magnitude 10^{-40} kg.
21. (a) 170,000 y (b) 7.34 y after the light, in 1994
23. (a) $T = 12$ d, $\omega = 6.06 \times 10^{-6}$ rad/s (c) $m_1 = 4.45 \times 10^{31}$ kg, $m_2 = 2.23 \times 10^{31}$ kg, $r = 4.95 \times 10^{10}$ m
25. 9.68×10^{-4} H atoms/m^3. It is unlikely this rate would be observable.
27. The earth's new temperature would be 949 K, enough to boil water but not to remove it from the atmosphere.
29. 26.42 MeV needed to form ^{112}Cd, 462 MeV released in forming ^{56}Fe

Index

Absorption spectrum, 146, 156, 184–185
Acceptor levels, 213
AC Josephson effect, 226, 230
Accretion disk, 312
Aharonov–Bohm effect, 225
Alpha (α) decay, 246, 250–252
    energies of α particles, 251
    four decay chains, 251
    tunneling through Coulomb barrier, 252
    variation in half-lives, 103, 251, 252
Amorphous solids, 191
Amplitude (A), 75, AP-6, AP-11
Anderson, Carl, 239, 280
Andromeda galaxy, 317
Angular frequency, 78–80 (see also Frequency)
    definition, AP-6
Angular momentum
    magnetic moments and, 130
    of nucleus, 244
    orbital, 123, 130
    quantized, 61, 66, 123, 130
    rule for addition of, 134
    spin, 130
    total, 133–136
Anthropic principle, 328
Antileptons, 280, 283
Antineutrino, 249, 280, 283–284
Antiparticles, 249 (see also specific antiparticles)
    particle–antiparticle pair, 281
    spin and, 280–282
Antiproton, 281–282
Antiquark, 286, 326
    properties, *table*, 287
Antisymmetric wave functions, 107–108, 137, 169–171
Apparent brightness, 308, 318
Astronomical data, *table*, AP-18
Astronomical unit (AU), 308–309
Astrophysics, 297–321 (see also specific topics)
    cataclysmic events, 311–313
    galaxies, 316–321
    stars, 304–316
    sun, 298–304
Atomic clocks, 161
Atomic force microscope (AFM), *photo*, 127
Atomic magnetic moments, 129–131

Atomic masses, AP-24–32
    of neutron and some isotopes, *table*, 242
Atomic number (Z), 120, 239, 240, AP-24–32
Atomic orbitals, 175–176
Atom manipulation, 161
Atoms, 120–166
    absorption, scattering, stimulated emission, 146–147
    addition of angular momenta, 133–136
    chemical and physical properties of elements, 120–121
    hydrogen-atom wave functions, 125–129
    laser, 147–154
    magnetic moments and electron spin, 129–131
    optical and X-ray spectra, 140–145
    periodic table, 136–140
    "plum pudding" model of, 59
    quantum theory of hydrogen atom, 121–125
    shell structure of electrons, 120
    spin–orbit effect, 133–136
    Stern–Gerlach experiment, 131–133
Avalanche breakdown, 215
Avogadro's number, 180

Balmer, Johann, 58
Balmer formula, 58
Balmer series, 21, 58, 62
Band, 210
Band theory of solids, 209–212
    conduction band, 210–212
    conductors versus insulators, 209–211
    dielectric breakdown, 210
    forbidden energy band, 211
    intrinsic semiconductor, 211–212
    valence band, 210–211
Bardeen, John, 216, 223
Barn, unit of nuclear cross section, 253–254
Barrier penetration, 98, 102–104, 224 (see also Tunneling)
    in α decay, 103, 252
    scanning tunneling electron microscope, 104, 111–114

    total reflection and, 102
    by water waves, 103
    of wave packet, *photo*, 104
Baryon, 277
    strangeness, 284
Baryon number, 283, 286–287, 292
BCS theory of superconductivity, 223–224
Beats, 78
Becquerel, Antoine Henri, 238
Becquerel (Bq), unit of radioactive decay, 248
Bednorz, 227
Bennet, W. R., Jr., 149
Beryllium, electron configuration, 139
Beta (β) decay, 246, 248–250
    daughter nucleus, 248
    decay of free neutron, 248, 283
    energy released, 248
    neutrino and, 248–249
    radioactive carbon dating, 249–250
Bethe, Hans, 302
Bevatron (Berkeley), 281
Big Bang model, 324–328
    cosmic background radiation and, 325
    radio galaxies and, 325
Binary pairs (stars), 311–312, 316
Binding energy, 32–33, 137, 241–243
    total nuclear, 242–243
Binnig, Gert, 111, 113
Blackbody radiation, 49–50, 299, 325
Black dwarf, 314
Black holes, 37, 315–316
Bohr, Niels, 48, 51, 59–61, 67, 256
Bohr formula, 125, 141
Bohr magneton, 130, 244
Bohr model of atom, 58–62, 121, 124, 125, 145, 276
Bohr orbit, 66, 138
Bohr radius, 61, 125, 128
Bohr's correspondence principle, 91
Bohr's postulates, 60–61
Boltzmann factor, 185, AP-17
Boltzmann's constant, 49, AP-13
Bose–Einstein condensation, 162
Bosons, 108, 280
    Higgs, 290
    mediating fundamental interactions, 290, *table*, 289
Bottleneck, in solar-fusion cycle, 302
Bottomness, 276, 292

Bottomonium, 288
Boundary conditions for standing-wave functions, 96, 121
　　particle in a box, 88–90
Bound state, 124
Brackett, F., 62
Bragg peak, 264
Brattain, Walter H., 216
Breeder reactor, 259–260
Bremsstrahlung, 264
Bremsstrahlung spectrum, 55
Bubble chamber tracks, *photos*, 276, 279, 281, 285

Carbon
　　bonding of atoms, 176
　　crystalline forms, *photo*, 195
Carbon dating, 249–250
Central-field problem, 122
Cepheid variables (stars), 317
CERN accelerator, 289
　　photo, 282, 283
Chadwick, J., 239
Chamberlain, O., 281
Chandrasekhar, 314
Chandrasekhar limit, 314–315
Characteristic rotational energy, 180, 181
Characteristic spectrum, 55
Characteristic x-ray spectrum, 145
　　frequencies of, 145
　　measurement of wavelengths, 145
Charge-coupled devices (CCDs), *photo*, 94–95
Charm, 288, 292
Chernobyl accident, 260
Chromosphere, 300
Classically forbidden region, 97–98, 101
Classical particle, 85
Classical wave, 85
Clock synchronization, 15–20
Cockcroft, J. D., 239, 252
Color charge, 289, 290
Complex conjugate, 128, AP-2
Complex functions, 87, 128
Compton scattering, 56–57, 147, 266
Compton wavelength, 57, 328
Conduction, electric (*see* Electric conduction)
Conduction band, 210–212, 213, 215
Conduction electrons, 211
Conductivity, 195
Conductor, electric, 195–199
　　free electrons in, 195–199
Conservation of angular momentum, 283
Conservation of baryon number, 283
Conservation of charge, 283
Conservation of charm, 288
Conservation of energy, 31, 60, 283
Conservation of lepton number, 283–284
Conservation of momentum, 283
Conservation of strangeness, 285
Constellations, 304, *photo*, 305

Contact potential, 205–206
Control rods, 258–259
Conversion factors, AP-18–20
Cooper, Leon, 223
Cooper pairs, 222–224, 226, 230
Copernicus, 306
Corona, 300
Correspondence principle of Bohr, 91
Cosmic background radiation, 325
Cosmic ray, *photo*, 28
Cosmogenesis, 324–328
　　Big Bang, 324–328
　　unanswered questions and limits of knowledge, 328
　　very early universe, 325–328
Cosmological constant, 324
Cosmology, 3, 322–328
　　Big Bang model, 324–328
　　cosmogenesis, 324–328
　　cosmological principle, 322, 325
　　critical mass density of universe, 323–324, 328
　　gravitation and, 322–324
　　Olber's paradox, 324–325
　　perfect cosmological principle, 324
　　steady-state model of universe, 324
Coulomb barrier, 252, 261
Coulomb energy, 103, 169
Covalent bond, 170–173
　　antisymmetric wave function, 170–171
　　electric dipole moment and, 172–173
Crab Nebula, 313, 315
Cross section (σ), 253–254, 265–266
Crystal
　　photorefractive, *photo*, 220–221
　　quartz, *photo*, 193
　　silicon, *photo*, 191
　　snowflake, *photo*, 192
　　sodium chloride, *photo*, 192
Crystal structures
　　body-centered-cubic (bcc), 194
　　face-centered-cubic (fcc), 191–192, 194
　　hexagonal close-packed (hcp), 195
Curie (Ci), rate of radiation emission, 248
Current gain (transistors), 217
Cyclotron, 157, 252
Cygnus A, 319

Daughter nucleus, 248, 250, 251
Davisson, C. J., 64, 74
DC Josephson effect, 226, 230
De Broglie, Louis Victor, 63, 66, 74
De Broglie equations, 80, 86, 121
De Broglie wavelength, 63–64, 66, 161
Decay constant, 246, 248
Decay rate, 246, 248
Deep inelastic scattering experiments, 287
Degeneracy of energy levels, 105–106, 125, 185
Degenerate dwarfs, 313–314
Density of states, 202–204

Depletion region, 214
Detection efficiency, 247
Deuterium, 239, 258
Deuteron, 32
Dielectric breakdown, 210
Diffraction, AP-8–10
　　of electrons, 64–65
　　intensity for single-slit, AP-9
　　and interference for two slits, AP-8
　　single-slit, AP-9
Diode, 213–214
　　forward-biased, 214, 217
　　light-emitting (LEDs), 216
　　pn junction, 215
　　reverse biased, 214–215, 217
　　tunnel, 215–216
　　Zener, 215
Diode lasers, 152
Dipole moment (*see* Magnetic moment)
Dirac, P., 130, 280
Dirac equation, 130, 244, 280
Dispersive medium, AP-12
Dissociation energy, 169–170
　　of a crystal, 193–194
Distribution function, for wave number, 79–80, 82
DNA, *tunneling micrograph*, 74, 174
Donor levels, 213
Doping, of semiconductor, 213, 217
Doppler cooling, 160
Doppler effect, 8, 156, AP-8 (*see also* Redshift)
　　applications, 161
　　Hubble's law, 319–321
　　relativistic, 20–21, 23
Dosage (radiation)
　　biological effects, 267–268
　　radiation received by U.S. population, *table*, 268
　　recommended dose limits, *table*, 268
　　studies of atomic-bomb victims, 267
　　units, 267, *table*, 267
Doublet energy states, 141, 144
Doublet spectral lines, 141
Drude, P., 195

Einstein, Albert, 2, 8, 31, 59, 160
　　absorption vs. stimulated emission, 147
　　energy quantization, 48, 50–51
　　general relativity, 2–3, 34–37, 47
　　photo, 2
　　photoelectric effect, 50–54
　　special relativity, 2–3, 8, 47
　　train illustration of simultaneity, 16–18
　　work in cosmology, 292
　　work on unified theory, 292
Einstein relation, 50
Einstein's photoelectric equation, 52
Einstein's postulates of special relativity, 8–12, 25–26
Elastic scattering, 146

Electric charge, 196–199
Electric conduction
 classical model, 195–196
 microscopic picture of, 195–199
 quantum mechanical theory, 199–200
Electric current, 196–197
 free electrons in conducting wire, 196–198
 motion of charges and, 196–197
 resistance and Ohm's law, 197–199
Electromagnetic force, 277
Electromagnetic waves
 energy density in, AP-7
 intensity of, AP-7
 wave equation for, AP-5
Electron
 diffraction, 64–65, 74
 discovery of, 276
 free, in conductor, 195–199
 kinetic energy of, 80
 magnetic moment, 129–130
 microscope, 66
 orbital angular momentum, 123, 133–134
 scanning tunneling microscope, 104
 spin angular momentum, 129–130, 133–134
 velocity of, 81
 wave function, 74–77
 wavelengths, 64
 wave–particle duality, 85
Electron charge density, 128
Electron-cloud model, 91, 127, 138
Electron configuration, 136
 of atoms in ground state, *table*, 142–143
 schematic, 140
Electron-degeneracy pressure, 314
Electron diffraction, 64–65
Electron microscope, 66, 111
 scanning tunneling, 104, 111–114
Electron spin, 129–131
Electron volt (eV), unit of energy, 31
Electron wave function, 74–78
Electron wave packets, 78–81
 group velocity of, 80–81
 phase velocity, 81
 uncertainty principle and, 82
Electron waves, 63–67, 74
 barrier penetration, 102–104
 diffraction and interference of, 64–65, 74
 reflection and transmission, 100–104
 standing-wave patterns of, 74
 step potential, 100–101
 wavelength of, 80
Electroweak force, 326
Electroweak theory, 289, 290, 292
Elementary particles, 276–292
 conservation laws, 283–286
 field particles, 289–290
 grand unification theories, 292
 hadrons and leptons, 277–281
 quark model, 286–289
 spin and antiparticles, 280–282
 standard model, 277, 290–292

Emission spectra, 146, 156, 182–183
Endothermic reaction, 252–253
Energy
 binding, 32–33
 conservation of, 31, 60
 distribution function, AP-16–17
 kinetic, 28–34
 quantization, 48, 50–51, 66–67, 87–88, 95, 121, 181
 relativistic, 28–34
 rest, 29–34
 transmitted by waves, AP-7
Energy barrier, 112
Energy density
 in electromagnetic waves, 75, AP-7
Energy-level degeneracy, 105–106
Energy-level diagram, 61–62, 92, 111
 from absorption spectrum, 146
 for optical transitions in sodium, 141, 144
Energy of interaction, 137
Envelope (of the group), 78–79
Equipartition theorem, 51, 200, AP-14
Equivalence principle, 34–35
Escape speed, 37
Ether, 5
European Laboratory for Particle Physics (CERN), *photo*, 282, 283
Event horizon, 328
Excitation energy, 82, 141
Excited states, 90, 92, 97, 141
Exclusion principle (*see* Pauli exclusion principle)
Exclusion-principle repulsion, 168, 192, 314
Exothermic reaction, 252–253
Expansion of universe, 321–328
 effect of gravity, 323
Expectation values, 98–100
 of $f(x)$ defined, 99
 of x defined, 98
 of x for time-independent state, 99
Exponential decrease, 246, 265, 266

Fermi–Dirac distribution, 204–206, 208–209
Fermi electron gas
 average energy at $T = 0$, 204
 contact potential, 205–206
 energy distribution at $T = 0$, 201–204
 energy distribution at temperature T, 204–205
 energy distribution function, 202
 Fermi–Dirac distribution, 204–205
 one-dimensional model, 201–202
 in three dimensions, 202–203
Fermi energy, 112, 201, 204–205
 table, 203
 in three dimensions, 203
Fermi factor, 202–207
Fermion, 108, 280, 290
Fermi speed, 206–207
Fermi temperature (T_F), 204
Field particles, 289–290
Field quanta, 289–290

Filaments (in sun), 304
Fine-structure splitting, 129, 131, 135–136, 183
Finite square-well potential, 94–98, 112
 allowed energies, 96
 behavior of wave functions, 96–97
 classically forbidden region, 97–98
 ground-state energy, 97
Fission, 254, 255–257 (*see also* Nuclear fission reactors)
 critical energy for, 256
 emission of neutrons, 257
 energy released, 256
 induced, 256
 spontaneous, 243, 255
 of uranium, 256
FitzGerald, George Francis, 14
Flavor charge, 290
Flavors, 286
Fluorescence, 146
 photo, 147
Fluorescent dyes, *photo*, 150
Fluxon, 224
 photo, 225
Flux quantization, 224
Forbidden energy band, 211
Force (*see* Interactions)
Fourier analysis, 79–80
Free-electron lasers, 152
Free-electron number densities, *table*, 203
Frequency (*see also* Angular frequency)
 of absorption spectra, 184–185
 Bohr, 162
 De Broglie, 63
 definition, AP-6
 Doppler shift, 21, AP-8
 in Einstein relation, 50
 Josephson, 226
 nuclear magnetic resonance, 245
 of standing waves, 67, 78, AP-7
 threshold, 52
Friedrich, W., 55
Frisch, Otto, 160
Fullerene crystals, *photo*, 195
Fundamental elementary particles, 288, *table*, 288
Fusion, nuclear, 261–263
 break-even point, 262–263
 confinement time, 261
 controlled fusion, problems of, 261
 energy released, 261
 inertial confinement, 263
 magnetic confinement, 262–263
 source of sun's energy, 302
 in stars, 309–310

Galaxies, 316–321
 Andromeda, 317
 barred spirals, 317
 BL Lac objects, 319
 Centaurus A, *photo*, 297
 classification of, 317
 disks, 317
 ellipticals, 317

gaseous nebulae, 317
Great Wall, 321
Hubble's law, 319–321
irregular, 317
material between stars, 316
Milky Way, 306
N galaxies, 319
ordinary spirals, 317
quiet and active, 318–319
radio galaxies, 319, 325
Seyfert galaxies, 319
total luminosity, 318
The Galaxy (see Milky Way)
Galilean transformation, 9–10
Galileo, Galilei, 4, 303, 313
Gamma decay, 246
metastable states, 250
wavelength of emitted photon, 250
Gamow, George, 103, 251–252
Gaussian distribution, AP-17
Geiger, H. W., 59, 238, 240
Geller, Margaret J., 321
Gell-Mann, M., 284, 286
General relativity, 2–3, 34–37, 47
black holes, 37
deflection of light in gravitational field, 35–36
frequencies of light in gravitational field, 36–37
perihelion of orbit of Mercury, 36
Gerlach, W., 131, 133
Germer, L., 64, 74
Glashow, S., 289
Glass, 191
Gluons, 289–290
Goudsmit, S., 129
Gradiometer, 232
Grand unification theories (GUTs), 292, 326
Gravitational lensing, photo, 3
Gravitational potential, 36
Gravitational red shift, 37, 316
Graviton, 289
Gravity waves, 233
antenna for, 37
Gray (Gy), unit of radiation energy, 267
"Great Wall," 321
Ground state energy, 89–90, 92, 93
Group velocity, 79–81, AP-12
Gyromagnetic ratio (g), 130
Gyroscope, 36

Hadronic force, 239 (see also Strong nuclear force)
Hadrons, 277–278 (see also Baryon; Meson)
table, 278
Hahn, 256
Half-life ($t_{1/2}$), 247
Harmonic analysis, AP-11
Harmonic synthesis, AP-11
Harmonic waves, AP-6, AP-12
superposition, AP-6
Hartmann, J. F., 316
Hawking, Stephen, 302

Heat capacity, of metals, 200, 208–209
Heat conduction of metals, 200, 208–209
Heavy water (see Deuterium)
Heisenberg, Werner, 82
Helium atom
electron configuration, 137
Schrödinger equation for, 106–108
Herriott, D. R., 149
Hertz, Heinrich, 51
Hertzprung–Russell (H–R) diagram, 309–313
horizontal branch, 310
main sequence, 309–310
red-giant branch, 310
Higgs boson, 290
Higgs field, 290
Holograms, 152–154
electron, 225
Hubble, Edwin, 317, 320, 324–325
Hubble age, 320, 323, 325
Hubble constant, 320–321, 324
Hubble's law, 319–321
expanding universe, 324–328
Huchra, John P., 324
Huygens' principle, AP-9
Hybridization, 176–177, 195
Hybrid orbitals, 176–177
Hydrogen atom
Bohr model, 58–62, 121, 124
energy-level diagram, 124
energy levels, 61–62, 67, 123
quantum numbers for, 122–123
quantum theory of, 121–125
Schrödinger equation for, 87, 106, 121–122
standing-wave problem, 67
wave functions, 125–129
Hydrogen bond, 174
Hydrogen spectrum, 58
Balmer series, 21, 58, 62
Lyman series, 59, 62
Paschen series, 59, 62

Impulse (I), 100
Impurity semiconductor, 213–214
acceptor levels, 213
diode, 213–214
donor levels, 213
doping, 213, 217
transistor, 213–214
Inelastic scattering, 146
Inert element, 137
Inertial confinement, 263
Inertial reference frame, 4
Infinite square-well potential, 88, 93, 99
allowed energies, 89
energy-level diagram for, 90
ground-state energy, 89
wave functions for, 90
Infrared absorption spectroscopy, 184
Integrated circuit, photo, 218–219
Intensity
amplitude and, 75, AP-7

definition, 75
diffraction, AP-8–9
of sun's radiation, 298–299
of waves, AP-7
Interaction of particles with matter
charged particles, 263
dosage, 267–268
energy loss rate, 264
minimum ionizing particles, 264
neutrons, 265–266
nuclear radiation therapy, 264
photons, 266–267
range, 264
Interactions
basic properties, table, 291
electromagnetic, 277
field particles associated with, 289–290, table, 289
gravitational, 277
strong nuclear, 277–278, table, 278
unification of electromagnetic and weak interaction, 290
Interference
of harmonic waves, AP-8
two-slit pattern, 75–76, AP-8
Interferometer (see Michelson interferometer)
Interstellar dust, 316
Intrinsic semiconductor, 211–212
Ionic bond, 168–170, 173
Ionization, 62
Ionization energy, 137, 139, 140, 168
Ions, 121
Ion trapping, 157–159
Isotope, 239

Javan, Ali, 149
Josephson, Brian, 226, 230
Josephson frequency, 226, 230
Josephson junction, 226
Junction diode, 214
Junction lasers, 152
Junctions (semiconductors), 214–221

Kant, Emmanuel, 317
Kepler, Johannes, 313
Kinetic energy (K)
definition, 28
of electrons, 80
relativity and, 28–34
Kinetic theory of gases
distribution of molecular speeds, AP-15–16
kinetic energy and absolute temperature, AP-14
Knipping, P., 55

Lagrangian point, 312
Laser (light amplification by stimulated emission of radiation), 147–154
dyes for tunable, photo, 150

femtosecond pulsed, photo, 151
free-electron laser, 152
helium–neon laser, 149–150, 152
holograms, 152
liquid laser, 152
population inversion, 148, 150, 152
range of wavelengths, 152
ruby laser, 148–152
semiconductor laser, 152, photo, 151
titanium–sapphire, photo, 151
tunable, 152, 160
Laser cooling, 160–161
Laue, M., 55
Lawrence, E. O., 252
Lawson, J. D., 261
Lawson's criterion, 261–262
Lenard, P. E., 51
Length contraction, 13–15, 17
Lepton era, 326
Lepton number, 283–284, 292
Leptons, 279–280
heavy lepton, 279 (see also Tau particle)
six leptons and antiparticles, 280
Light (see also Diffraction; Interference)
diffraction, AP-8–10
interaction with film, 75–76
interference, AP-8–10
quantization of light waves, 75
scattering, 146
speed of, 4–5
wave versus particle, 74, 85
Light-emitting diodes (LEDs), 216
Light-year ($c\cdot y$), unit of distance, 308
Lindblad, B., 307
Line spectrum, photos, 47
Liquid-drop model of nucleus, 240, 256
Lithium, electron configuration, 138–139
Little Ice Age, 303
Livingston, M. S., 252
Lorentz, Hendrik, A., 14
Lorentz–FitzGerald contraction, 14
Lorentz transformation, 9–15, 18, 24
length contraction, 13–15, 17
time dilation, 11–13, 18
Luminosity, 298, 308, 310, 318
Lyman, T., 62
Lyman limit, 317
Lyman series, 59, 62, 146

Madelung constant, 192, 194
Magnetic bubbles, photo, 219
Magnetic confinement, 262–263
Magnetic dipole moment (see Magnetic moment)
Magnetic moment
angular momentum and, 130–131
atomic, (see also Atomic magnetic moments)
of electron, 129–131, 244
of neutron, 244
of nucleus, 133n, 244
of proton, 244
quantization of, 130
Magnetic quantum number (m), 123

Magnetron motion, 158
Magnetron tube, 158
Maiman, Theodore, 148
Marsden, E., 59, 238, 240
Mass (m)
binding energy and, 241–243
energy and, 31
gravitational versus inertial, 34
of neutrino, 249
of nucleus, 241
Planck mass, 328
reduced, 180, 182
relativistic, 27, 241
Mass number (A), 239
Mass spectrometer, 242
Mathematics, review of, AP-1–3
Maunder minimum, 303
Maxwell, James Clerk, 160 (see also Maxwell's equations)
Maxwell–Boltzmann distribution, 197, 200–201, 302, AP-13–14, AP-15–17
definition, AP-13
energy distribution function, AP-16–17
most probable speed, AP-15–16
Maxwell's equations, 5
Mean free path, 198–199, 207
Mean lifetime, 247
Mega electron volt (MeV), unit of energy, 31
Meissner, 221
Meissner effect, 221–222
Meltdown, 260
Meson, 280 (see also Pion)
prediction of, 277–278
Y meson, 288
Metallic bond, 174
Metals, classical free-electron theory of, 195–200
energy distribution, 195–200
heat capacity, 200
heat conduction, 200
Metastable state, 147, 148–149, 250
Michelson, Albert, 5–7
Michelson interferometer, 6
Michelson–Morley experiment, 5–8, 9, 14
Microchip, photo, 219
Microcurie (μCi), rate of radiation emission, 248
Microscope
atomic force (AFM), 127
electron, 66
scanning tunneling electron, 111–114
Milky Way, 304
galactic year, 307
mass of, 307
missing-mass problem, 308, 316, 328
photo, 306
Sagittarius A, 307
structure of, 306
total energy of, 323
Millicurie (mCi), rate of radiation emission, 248
Millikan, Robert A., 52

Missing-mass problem, 308, 316, 328
Moderator, 258
Molecular bonding, 128
covalent bond, 170–173
hydrogen bond, 174
ionic bond, 168–170
metallic bond, 174
for polyatomic molecules, 175–176
van der Waals bond, 172, 173
Molecular orbitals, 168
Molecular wave functions, 168
Molecules, 167–186
absorption spectra, 184–185
characteristic rotational energy, 180
electronic energy, 177
emission spectra, 182–183
molecular bonding, 168–174
polyatomic, 175–177
rotational energy levels, 177, 180–181
rotational quantum number, 177
structures of common molecules, table, 178–179
successes of molecular physics, 167
two views of, 167–168
vibrational energy levels, 181–182
vibrational quantum number, 181
Momentum
relativistic, 26–27, 29–30, 33–34
Morley, Edward M., 5, 7
Moseley, H., 145
Muller, 227
Muon, 13–14, 276

National Council on Radiation Protection, 268
Nebulae
Crab, 313, 315
gaseous, 317
planetary, 311, 314
Tarantula, photo, 311
Neutrino, 30
antineutrinos, 249
in beta decay, 248–249
mass of, 249
solar, 301–302
from supernova, 280
three kinds of, 249, 284
Neutron
capture cross section, 254, 265
cross section for scattering, 265–266
delayed, 259
discovery of, 239, 276
magnetic moment of, 244
pair production, 266–267
photoelectric effect, 266
reactions with nuclei, 253–254
thermal, 253–254
Neutron diffraction, 65
Neutron stars, 314–315
Newton, Isaac, 4
Newtonian relativity, 3–4, 8
Newton's first law of motion, law of inertia, 3
Newton's second law of motion, 60

Nishijima, K., 284
Nobel prize, 51, 54, 111
Normalization condition, 88, 90, 125–126
Novae, 311–312
    classical, 312
    dwarf, 312
    recurrent, 312
Nuclear fission reactors, 255, 257–261 (*see also* Fission)
    breeder reactor, 259–260
    control rods, 258–259
    delayed neutrons, 259
    moderator, 258
    pressurized-water reactor, 258
    problem of radioactive waste, 261
    reproduction constant (k), 257–259
    safety of, 260–261
Nuclear fusion (*see* Fusion, nuclear)
Nuclear magnetic resonance, 244–245
    medical use for imaging, 245
    resonance frequency of free protons, 245
    RF radiation, 245
Nuclear magneton, 244
Nuclear power plant
    fission, *photo*, 238, 260
    fission, *schematic*, 258
    fusion, *photo*, 262–263
Nuclear radiation therapy, 264
Nuclear reactions, 252–254
    cross section (σ), 253–254
    reactions with neutrons, 253–254
Nuclear spin I, 244
Nuclear weapons, 255, 257, 267–268
Nuclei, 238–268
    angular momentum of, 244
    artificial nuclear disintegration, 238–239
    discovery of radioactivity, 238
    fission, 255–257
    interaction of particles with matter, 263–268
    liquid-drop model, 240, 256
    mass and binding energy, 241–243
    N and Z numbers, 240 (*see also* Atomic number)
    nuclear magnetic resonance, 244–245
    nuclear reactions, 252–254
    nuclear reactors, 255, 257–261
    nuclear size and shape, 240
    nuclear weapons, 255, 257, 267–268
    properties of, 239–244
    radioactivity, 246–252
    spin and magnetic moment, 244
    uniform density, 240, 243
Nucleon, 239
Nucleosynthesis period, 326
Nuclide, 239

Ochsenfeld, 221
Ohm's law, 197–199
Olber, 325
Olber's paradox, 324–325
Oort, J., 307

Oppenheimer, J. Robert, 37
"Optical molasses," 161
Optical pumping, 148
Optical spectra, 141–144, 183
    selection rules, 141, 144
Orbital angular momentum (**L**), 123, 133–134
Orbital quantum number (l), 123, 136

Pair production, 266–267
Parallax angle, 308–309
Parsec, unit of distance, 308
Particle accelerators, 239, 276, 281–282, 289
Particle in a box, 88–95
    allowed energies, 89
    boundary conditions, 88–89
    finite square-well potential, 94–98
    ground-state energy, 89–90
    infinite square-well potential, 88–93
Paschen series, 59, 62
Pauli, Wolfgang, 108, 129, 159, 248
Pauli exclusion principle, 108, 136–137, 144, 168, 172, 204, 211, 314
Paul trap, 159
Penning, F. M., 157
Penning trap, 157–159
Penzias, Arno, 325
Perihelion, 36
Periodic table, 120, 136–140, AP-23
    beryllium ($Z = 4$), 139
    boron to neon ($Z = 5$ to 10), 139
    electron configuration, 136
    helium ($Z = 2$), 137–138
    lithium ($Z = 3$), 138–139
    Pauli exclusion principle, 136–137
    quantum numbers and, 136–137
    sodium to argon ($Z = 11$ to 18), 139
    transition elements, 140
Perl, 279
Permeability of free space, 5
Permittivity of free space, 5
Pfund, H. A., 62
Phase velocity, 78, 81
Phipps, 133
Phosphorescent materials, 147
    photo, 147
Photoelectric effect, 50–54, 147, 266
    Einstein's equation, 52, 54
    Millikan experiments, 52
    stopping potential, 51, 54
    threshold frequency, 52
    threshold wavelength, 52, 54, 206
    work function, 52, 54
Photomultiplier, *photos*, 17
Photon, 30, 52
    Compton scattering, 56–57, 147, 266
    cross section for absorption and scattering, 266
    energy, 51, 53–54
    intrinsic angular momentum, 124–125
    speed of light, 25

    virtual, 277
    wave equation for, 86
Photorefractive material, *photo*, 220–221
Photosphere, 298–299, 312
Physical constants, *table*, AP-19
Piezoelectric ceramic, 113
Pion, 276, 280 (*see also* Meson)
Plages (in sun), 304
Planck, Max, 48, 50–51, 59, 299, *photo*, 49
Planck length, 328
Planck mass, 328
Planck's constant, 50, 60, 82–83, 141
Planck time, 328
Plasma, 261
"Plum pudding" model of atom, 59
Pole Star, 304
Population inversion, 148, 150, 152
Positron, 238
    annihilation by electrons, 281
    discovery of, 239, 276, 280
Potential, square-well (*see* Finite square-well potential; Infinite square-well potential)
Potential energy (U), step function, 100–101
Principal quantum number (n), 123, 136
Probability
    of finding electron in unit volume, 77, 82, 93, 121, 126
    of finding photon in unit volume, 76
    of reflection, 101
    of transmission, 101–102
Probability density, 77, 126
    for complex functions, 128
    radial, 127, 129
Probability distribution, 77, 98–99, 128, 171
Prominences (in sun), 304
Proper length, 13–14
Proper time, 11, 13, 15
Proton
    diffraction, 65
    magnetic moment, 244
    in NMR, 245
    spin quantum number, 244
Proton–proton cycle, 302
Proxima Centauri, 309
Pseudowork, particle, 288
Pulsars, 315

Quanta, 50
Quantization
    of angular momentum, 61, 66
    of atomic energies, 58–62
    of energy, 48, 50–51, 66–67, 87–88
    standing-wave condition, 66–67
Quantum chromodynamics (QCD), 289, 291
Quantum electrodynamics (QED), 277
Quantum mechanics (*see also* Quantum theory)

barrier penetration, 102–104
electron wave function, 75–77
electron wave packets, 78–81
expectation values, 98–100
particle in a box, 88–95
particle in finite square well, 95–98
reflection and transmission of electron waves, 100–104
Schrödinger equation, 74–77, 86–88, 105–108
uncertainty principle, 82–85
wave–particle duality, 85
Quantum numbers, 90, 105
for hydrogen atom, 122–123
magnetic, 123
orbital, 123
principal, 123
for total angular momentum, 134–135
Quantum theory, 2, 74–114 (see also Quantum mechanics)
blackbody radiation, 49–50
Bohr model, 58–62
Compton scattering, 56–57
dates of experiments and theories, table, 48
electron waves, 63–67
of hydrogen atom, 121–125
photoelectric effect, 50–54
quantization of atomic energies, 58–62
Schrödinger equation, 74–77, 86–88, 105–108
Quantum theory of electrical conduction
Fermi–Dirac distribution for electron gas, 206, 208–209
wave properties of electrons in scattering, 206–207
Quark, 244, 278
baryon number, 286–287
charm, 288
excess over antiquarks, 326
flavors, 286
fractional charges, 286
properties, table, 287
strangeness, 286
Quark model, 286–289
deep inelastic scattering experiments, 287
fundamental elementary particles, 288, table, 288
quark confinement, 288, 291
Quartz crystal, photo, 193
Quasars (quasistellar radio sources), 319
Quasistellar objects (QSOs), 319
Q value, 252–253, 302

Rad (radiation absorbed dose), 267
Radial equation, 122
Radial probability density, 127, 129
Radiation, nuclear
biological effects of, 267–268
dosage, 267–268
radiation therapy, 264

rates of emission, 248
units, table, 267
Radiation, blackbody, 49–50
Radioactive carbon dating, 249–250
Radioactivity, 2
alpha decay, 246, 250–252
beta decay, 246, 248–250
decay constant, 246
decay rate, 246
exponential time dependence, 246
gamma decay, 246, 250
half-life ($t_{1/2}$), 247
mean lifetime, 247
Radio galaxies, 319, 325
Radio waves, RF radiation, 245
Raman, C. V., 146
Raman scattering, 146
Range, 264
Ray approximation, AP-10
Rayleigh, Lord, 146
Rayleigh–Jeans law, 49–50
Rayleigh scattering, 146
RBE (relative biological effectiveness) factor, 267, table, 267
Red giant, 310
Redshift, 21
gravitational, 37, 316
Hubble's law, 320
maps, 322
in quasars, 321
Red subgiant, 310
Red supergiant, 310
Reduced mass, 180, 182
Reflection
of electron waves, 100–102
total internal, 101, 102
Reflection coefficient (R), 101
Relativistic energy, 28–34
Relativistic mass, 27
Relativistic momentum, 26–27, 29–30, 33–34
Relativistic transformation equation, 10–11 (see also Lorentz transformation)
Relativity, 2–37
clock synchronization and simultaneity, 15–20
doppler effect for light, 20–21, 23
Einstein's postulates, 8–12
general theory, 2–3, 34–37
Lorentz transformation, 9–15, 18
Michelson–Morley experiment, 5–9, 14
Newtonian, 3–4
relativistic energy, 28–34
relativistic momentum, 26–27, 29–30, 33–34
special theory, 2–3, 8, 23
twin paradox, 21–23
velocity transformation, 24–26
Rem (roentgen equivalent in man), 267
Reproduction constant (k), 257–259
Residual strong interaction, 291
Resistivity, 195–199
of insulators and conductors, 209
of semiconductors, 212
Resonance absorption, 146

Rest energy, 29–34
table, 32
of unified mass unit, 241
Rest mass, 27, 31, 33
RF (radiofrequency) radiation, 245
Ring nebula, 311
Rms speed, AP-16
Roche lobe, 312
Roentgen (R), radiation measure, 267
Rohrer, Heinrich, 111, 113
Röntgen, W., 54
Rotational energy levels, 177, 180–181
Rotational quantum number for molecules, 177
Rubbia, C., 289
Ruska, Ernst, 111
Rutherford, Ernest, 59, 238, 246
Ryle, 325

Sagittarius A, 307
Salam, A., 289
Saturated bond, 172
Scanning tunneling electron microscope (STM), 104
essay, 111–114
Scattering
Compton, 147
elastic, 146, 252
inelastic, 146
of neutron, 253
Raman, 146
Rayleigh, 146
Schrieffer, Robert, 223
Schrödinger, Erwin, 67, 74, 122
Schrödinger equation, 74–77, 86–88, 101
applied to hydrogen atom, 87, 106, 121–122
applied to simple harmonic oscillator, 87, 181
complex solutions, 87, 128
energy quantization and, 87–88, 95, 121, 181
normalization condition and, 88
for rotation, 177
solutions of, 125–127
in spherical coordinates, 122
in three dimensions, 105–106, 121–122
time-dependent, 87
time-independent, 87, 96, 107 (see also Particle in a box)
for two identical particles, 106–108
Schwarzschild radius, 37, 316
Segré, E., 281
Seivert (Sv), unit for dose equivalent, 267
Selection rules, 124, 141, 144, 182
Semiconductor (see also Semiconductor junctions)
depletion region, 214
hole, 211–212
impurity, 213–214
intrinsic, 211–212
junctions and devices, 214–221
n-type, 213–216

p-type, 214–216
    temperature and resistivity, 212
Semiconductor junctions
    avalanche breakdown, 215
    depletion region, 214
    forward bias, 214, 217
    light-emitting diodes, 216
    pn junction, 214–216
    reverse bias, 214–215, 217
    solar cell, 216
    transistors, 213–214, 216–218
    tunneling current, 215–216
    Zener diode, 215
Semiconductor lasers, 152
Seyfert, Carl, 319
Seyfert galaxies, 319
Shapley, Harlow, 306
Shockley, William, 216
Silicon, *photo*, xvi
    crystal, *photo*, 191
Simultaneity, 15–20
Single crystal, 191
SI units, 248
Snowflake, *photo*, 192
Snyder, 37
Sodium chloride crystal, *photo*, 192
Solar cell, 216
Solar constant *f*, 298, 308
Solar corona, 300
Solar eclipse, *photo*, 300
Solar flares, 303–304
Solar-neutrino problem, 302
Solar wind, 300
Solids, 190–233
    amorphous, 191
    application of quantum mechanics to, 190
    band theory of, 209–212
    classical free-electron theory of metals, 195–200
    crystal structures, 191, 194–195
    equilibrium separation in crystal, 193–194
    Fermi electron gas, 201–206
    impurity semiconductors, 213–214
    potential energy of ion in crystal, 191–193
    quantum theory of electrical conduction, 206–208
    semiconductor junctions and devices, 214–221
    structure of, 190–195
    superconductivity, 221–227
Sound waves (*see* Waves)
Southern Cross, 304
Space quantization, 133
Special theory of relativity, 2–3, 8, 23, 47, 130
Spectral distribution curves, 49
Spectral distribution function, 49
Spectroscope, 58
Spectrum, line, *photos*, 47
Speed (*see also* Velocity)
    escape, 37
    of light, 4–5
    of molecules, distribution, AP-15–16

root-mean-square (rms), AP-15–16
    of waves on a string, 78, AP-5–6
Spin
    angular momentum, 130, 133–134
    and antiparticles, 280–282
    magnetic moment and, 244
Spin angular momentum (**S**), 133–134
    intrinsic, 124–125, 280
Spin-orbit effect, 135–136
Spin quantum number
    of electron, 129–131
    of neutron and proton, 244
Splitting, of spectral lines, 129, 131
Spontaneous emission, 146, 149
Square modulus, A-2
Square-well potential (*see* Finite square-well potential; Infinite square-well potential)
SQUID (Superconducting Quantum Interference Device), 226, 230–233
    basic principles, 230–231
    biomagnetic research, 231–233
    essay, 230–233
    foundations, 230
    magnetic fields in heart and brain, 226, 230
    other applications, 233
    sensitivity, 233
Standard deviation, 80, 82, AP-17
Standard model, 277, 290–292
    for evolution of universe, 326, 328
    properties of basic interactions, *table*, 291
    residual strong interaction, 291
Standing waves, AP-7
    condition for, AP-7
    of electron waves, 74
    energy quantization and, 66–67
    for hydrogen atom, 67
    solutions of Schrödinger equation, 121
Star chart, 305
Star clusters
    galactic (open), 305–306
    globular, 305–306
Stars
    accretion disk, 312
    apparent brightness, 308
    binary pairs, 311–312
    black holes, 315–316
    cataclysmic events, 311–313
    constellations, 304, *photo*, 305
    degenerate dwarfs, 313–314
    distance to, 308–309
    effective temperature, 308, 310
    evolution of, 308–311
    final states of, 313–316
    helium fusion in, 310
    Hertzprung–Russell (H–R) diagram, 309–313
    lifetime of, 310
    luminosity, 308, 310
    main-sequence, 309–310
    main-sequence dwarfs, 309
    mass–radius relation, 310, 314
    Milky Way, 306–308

neutron stars, 314–315
novae, 311–313
planetary-nebula stage, 311
Pole Star, 304
population I, 305
population II, 305
Roche lobe, 312
stellar populations, 305
supernovae, 312–313
Stationary states, 60
Stefan–Boltzmann law, 49, 298
Step potential, 100–101
    classical vs. quantum-mechanical results, 100–101
Stern, O., 131, 133
Stern–Gerlach experiment, 131–133
Stimulated emission, 147, 150
Stopping potential, 51, 54
Strangeness, 284–285, 286, 292
    and center of charge, 285
    conservation of, 285
Strange particles, 284–285
Strassmann, 256
Strong nuclear force, 239, 277–278, 289 (*see also* Hadronic force)
Sun
    active sun, 303–304
    chromosphere, 300
    core temperature, 301
    corona, 300
    effective temperature, 298–299
    intensity radiated, 298–299
    interior, 301
    limb, 299–300
    luminosity, 298, 302
    mean density, 301
    photosphere, 298–299
    polarity reversals, 303
    rotation of, 303
    solar spectrum, 299
    source of its energy, 301–302
    surface and atmosphere, 298–300
    surface temperature, 298
Sunspot cycle, 303
    Little Ice Age, 303
Superconducting supercollider (SSC), 290
Superconductivity
    BCS theory of, 222–224, 227
    Cooper pair, 222–224
    critical field, 221
    critical temperature, 221, *tables*, 221, 227
    critical voltage, 224
    flux quantization, 224
    high-temperature superconductors, 227
    Josephson junction, 226
    levitation, 222–223
    magnetic levitation, 222
    Meissner effect, 221–222
    quantum mechanics and, 224, 226
    SQUID, 226, 230–233
    superconducting energy gap, 223–224
    tunneling, 224, 226
    type I superconductors, 222

type II superconductors, 222, 227
vortex state, 222
Supernova explosion, 312–313
electron neutrinos from, 280
gravitational collapse, 313
SN1987A, 313
spectra of, 313
type I, 313
type II, 313
Surface-barrier detectors, 216
Symmetric wave functions, 107–108
Synchronization of clocks, 15–20
Synchrotron, 157

Tables, list of, AP-19–20
Tau (τ) particle, 249, 279, 284
Taylor, 133
Terms of energy states, 124
Terrestrial data, *table*, AP-18
Thompson, G. P., 65
Thomson, J. J., 59, 85, 276
Three Mile Island accident, 260
Threshold frequency for photoelectric effect, 52
Threshold wavelength for photoelectric effect, 52, 54
Time dilation, 11–13, 18
Tokamak (U.S.S.R.), 262
Tone quality, AP-10–11
Topness, 276, 292
Total internal reflection, barrier penetration and, 102
Transistors, 213–214, 216–218
base, 216
collector, 216
current amplifier, 218
current gain, 217
emitter, 216
npn, 216–217
pnp, 216–217
voltage gain, 218
Transition elements, 140
Transitions
photon wavelengths for, 92
selection rules for, 124
spectral lines and, 124
Transmission coefficient (T), 101–102
Trapped atoms and laser cooling, *essay*
atomic clocks, 161
atom manipulation, 161
collision studies, 161
condensed matter, 161–162
high-accuracy atomic spectroscopy, 160
ion trapping, 157–159
Triode vacuum tube, 217
Triplet spectral lines, 144
Tritium, 239
Tunnel diode, 215–216
Tunneling, 224, 226, 252, 261 (*see also* Barrier penetration)
Tunneling current, 104, 215–216
Tunnel junction, 230
Twin paradox, 21–23
Two-body problems, 106

Two-slit interference pattern, 75–76, AP-9
for electrons, 76

Uhlenbeck, G., 129
Ultraviolet catastrophe, 50
Uncertainty principle, 82–85, 98, 106
energy and time, 82–83, 156
momentum and position, 82
Planck's constant and, 82–83
Unified mass unit (u)
definition, 180, 241
rest energy of, 241
Unit cell, 191

Vacuum tubes, 214, 217
triode, 217
Valence band, 210–211
Valence electrons, 121
Van de Graaff, R., 252
Van de Graaff generator, 252
Van der Waals bonds, 168, 172, 173
Vector bosons, 289
Velocity (*see also* Speed)
group, 79–81
transformation, 24–26
Velocity transformation, 24–26
Very Large Array (VLA), 307
Vibrational energy levels, 181–182
Vibrational quantum number, 181
Virtual photons, 277
Voltage gain (transistors), 218
Vortex state, of superconductor, 222

Walton, E. T. S., 239, 252
Wave equation
for electrons, 74
for string waves, 78, AP-5
Wave function, 78, 86
for electrons, 74–77
for general wave packet, 79
molecular, 168
normalized, 90
symmetric, antisymmetric, 107–108, 137
well-behaved, 97, 122
Wavelength, AP-6
Compton, 57
cutoff, for x-ray spectrum, 55–56
de Broglie, 63–64, 66
electron, 64
threshold, for photoelectric effect, 52, 54
wave number and, 78
Wave mechanics (*see* Quantum theory)
Wave number, 78–80, 82, 89
Wave-number distribution function, 79–80
Wave packet, AP-11–13
barrier penetration of, 104
electron, 78–81

motion of, *photo*, 81
properties of, 78–80
wave function for, 79
Wave–particle duality, 74, 85
Waves
diffraction, AP-8–10
in dispersive medium, AP-12
electromagnetic, 75, AP-5–7
electron, 63–67
energy, AP-7
equation, AP-5
group velocity, 79
harmonic, AP-6
interference, AP-8–10
phase velocity, AP-5
standing, AP-7
on a string, 75–78, AP-5
superposition, AP-5, AP-7
wave function, 74–79, 107–108, 137
Wave theory of electrons, 63–67
Schrödinger equation and, 76–80
Weak charge, 290
Weak nuclear force, 277
Weinberg, S., 289
Wheeler, 256
White dwarf, 311, 314
Wien's displacement law, 49
Williams, Ellen D., 111–114
Williamson, Samuel J., 230–233
Wilson, Robert, 325
WIMPs (weakly interacting massive particles), 308
Wineland, D. J., 156–162
Work function, 52, 54, 112
for some metals, *table*, 206
W particles, 289–292

X-ray diffraction, 55
X rays, 54–56
Compton scattering, 56–57
diffraction, 55
X-ray spectrum, 144–145
bremsstrahlung spectrum, 55
characteristic spectrum, 55
cutoff wavelength, 55–56

Y meson, 288
Young, Thomas
two-slit interference experiment, 65, 75, 85
wave theory of light, 85
Yukawa, H., 277–278

Zeeman, P., 136
Zeeman effect, 136, 244
Zener diode, 215
Zero-point energy, 83, 90
Z particles, 289–292
Zweig, G., 286

A Pictorial Tribute to an American Icon

ALEX ROGGERO

WITH TONY BEADLE

BARNES & NOBLE BOOKS
NEW YORK

This edition published by Barnes & Noble, Inc. by arrangement with Osprey, an Imprint of Reed Books.

© 1996 Barnes & Noble Books
© Reed International Books 1995

CIP Cataloging Data Available Upon Request.

All rights reserved. Apart from any fair dealing for the purposes of private study, research, criticism or review, as permitted under the Copyright, Designs and Patents Act, 1988, no part of this publication may be reproduced, stored in a retrieval system, or transmitted in any form or by any means electronic, electrical, chemical, mechanical, optical, photocopying, recording or otherwise, without prior written permission. All enquiries should be addressed to the Publisher.

ISBN 0-7607-0236-5

Editor: Shaun Barrington
Art Editor: Mike Moule
Design: the Black Spot

M 10 9 8 7 6 5 4 3 2 1

Printed in Hong Kong

OVERLEAF *The inimitable Clark Gable and Claudette Colbert in Frank Capra's 1934 Oscar winner,* It Happened One Night. *This comedy about a runaway heiress and an opportunistic journalist exploited the romantic plot device of an escape across country, courtesy of Greyhound, to the full.* (Columbia: Courtesy Kobal)

The montage pictured on page 9 was created by Andy White.

All contemporary photographs by Alex Roggero apart from page 47, courtesy Hyman Myers. Captions by Alex Roggero. Main text by Tony Beadle.

to Carlo and Louise

ACKNOWLEDGEMENTS

This book would not have been possible without the knowledgeble assistance of many dedicated and professional Greyhound employees. My warmest thanks go to all the drivers, station managers and service-crew members I had the pleasure of meeting during my time on the road. Special thanks must go to Amy Engler and everyone at the Dallas depot, as well as the managers of Evansville IN, Cleveland Ohio, Uvalde Texas and Jackson MS terminals.

As anyone who has taken a Greyhound trip will know, a bus journey is as much about meeting people as it is about going from A to B. During my travels I got to know hundreds of fellow passengers. I thank each and every one of them for their time, and wish them all a happy journey.

Of the many people who helped during the course of my research, I am indebted to architect Hyman Myers, of Vitetta Group in Philadelphia, for providing the pictures of the restored Washington terminal.

I am also grateful for the support of Karen and Richard Spector, who provided invaluable archival information and helped organize several bus trips. American Airlines' generous assistance made a number of research trips possible, therefore making a huge contribution to the project.

Many thanks also to everyone at AutoCapital magazine in Milan, especially Gilberto Milano and former editors Luca Grandori and Filippo Piazzi, for supporting my 'road' articles and photography over a period of several years.

In London, Rosie Thomas, Steve Brookes, Andy White and Matt Prince gave leads and ideas and supplied all kinds of materials, whilst Ceta West End processing lab took care of the colour transparencies with their usual effortless competence. My gratitude also goes to my editor, Shaun Barrington, for giving me the time and freedom to represent Greyhound's past, present and future exactly as I wanted.

Thanks are insufficient to acknowledge the role played by Jacquie Spector Roggero, who was instrumental during the picture selection process and provided tremendous support throughout.

Finally, and for reasons that have to do with photography, design and the romance of the road, I would like to acknowledge the influences of Russell Lee, Alan Hess, Jack Duluoz and the old favourite, my 'American' globe-roaming grandpa.

Alex Roggero

I would like to thank the following people who provided extremely useful help and information for this book: Ed Stauss of *Bus World*, Woodland Hills, California; Nick Georgano; Kit Foster; Nicky Wright; Amy Engler and Betty Haynes in the Public Relations Department of Greyhound Lines Inc, Dallas, Texas.

Tony Beadle

CONTENTS

FOREWORD 8

INTRODUCTION 12

A COLD START 14

ART DECO JEWELS 32

KINGS OF THE ROAD 48

PANORAMIC VIEWS 74

GO GREYHOUND 100

FOREWORD
Spirits of the Road

It all started on a rutted mud track, a single lane road between two small mining towns in Minnesota. In a sense, it could not have happened any other way. Even then, in 1914, what was to become the Greyhound empire was based on a simple but visionary idea: the necessity to link America's myriad communities with an efficient and inexpensive transportation network. In the country where to be without wheels was a symbol of second class citizenship, there were millions of people, living in thousands of towns, who had no wheels. The Greyhound bus gave them mobility, that quintessentially American commodity, and turned them into coast to coast travellers.

Over the course of many years the bus, with its conspicuous aluminium and stainless steel body criss-crossed the nation, knitting together those communities, embedding itself deeper and deeper in the American psyche. So much so that in the end it became another symbol for the country itself, like a Coke bottle or a pair of Levi's. But unlike these objects, the Greyhound was more than a typically American product. It was a mobile icon, a travelling receptacle of people's dreams and people's problems. Its silver sides reflected, like a mirror, many big and small events that shaped American history – from the pioneering days of road building, when the Greyhound routes were inaugurated almost as quickly as the new highways were opened, to the engineering advances of the thirties, when Greyhound buses were the first with rear-mounted engines, the first to use diesel, the first to have on board air-conditioning. And then there were the buildings, the network of impressive streamlined terminals, with massive towers and multicoloured neon signs.

Greyhound was one of the first US companies to realize the importance of what is now described as "the corporate image". Its impact on American society was such that it came to be reflected in all facets of popular culture. In 1934 Frank Capra's Oscar winning movie, *It Happened One Night*, romanticized bus travel for the newly affluent middle classes, while at the same time blues singers such as Robert Johnson and Lee Brown dedicated entire songs to the Greyhound, the vehicle of escape for many Southern blacks on the way to a better future up North. In the post-war years 50% of all intercity travellers went by Greyhound, and the Silversides became a symbol of the nation's unprecedented wealth and desire for travel. The Beats discovered the bus and immediately fell in love with it. Jack Kerouac's *On the Road* became an inspiration for generation after generation of incurable road romantics, for whom the great American landscape, geographical and spiritual, was best observed from the widescreen windows of a Scenicruiser. During those golden

years the Greyhound bus even managed to accommodate two very different sets of people: those who went searching for the American Dream, and those who rejected it.

But it could not last forever. Inevitably, the years that followed were not so golden. Competition from airlines and privately owned cars as well as periods of economic recession made life increasingly difficult for the old Greyhound. In 1983 the deregulation of the airline sector forced the company to implement a massive programme of pay cuts, which the drivers reiected. The resulting industrial dispute turned out to be one of the most intractable in US history, and has only recently been fully settled. The Greyhound image has also changed. What was once a symbol of the future appears to have turned into a photograph from the past. The bus, redolent of a gentler, more relaxed America, now has to survive in a much harsher environment. Some of the magnificent terminals have become stranded in the middle of deserted and unsafe downtown areas, while economic pressures are forcing the company to close many secondary routes.

In this climate of striking contrasts, where the old Greyhound myths are constantly buffeted by the realities of modern American life, I decided to board a coach and, like many photographers, writers and movie makers before me, take a ride that lasted several thousand miles. It was a journey of rediscovery, in the sense that I went looking for the old buses, stations and signs that helped create the Greyhound myth, as well as a contemporary exploration, in that I wanted to document bus travel in its present form. To my surprise, I found much of Greyhound's glorious past still intact, and much loved by both staff and passengers. But perhaps the most striking feature of the trip was the realisation that the bus is one of the last public spaces in America where people from very different backgrounds are still willing to interact. In a country where people's attitudes are often polarised by class, colour, gender and language, this seems to me to be a major achievement. The ritual begins with a simple question: "Where ya headed?" The answer can last twelve hours. The fact is that in the last 80 years some things at least haven't changed. If you want to get to know America's soul, to put your finger on its pulse, and experience the sense of community that makes total strangers talk for hours on end there is, still, only one way to go.

Alex Roggero

VACATION PLEASURE in DOUBLE Measure!

CROSS the continent twice—by different routes, if you wish—and see BOTH the New York and San Francisco World's Fairs... all for $69.95 total transportation cost! Your ticket is good for three months, so you can stay as long as you like or stop off en route to spend the night, visit friends, or go sightseeing. With scientifically planned air conditioning to keep you comfortably cool... deeply cushioned, reclining chairs... and wide, clear windows, you'll find your trip the most pleasant ever. Friendly fellow-passengers add to the interest, and frequent schedules provide maximum travel convenience. See your Union Pacific or Interstate agent for information!

Visit **BOTH** Fairs *for only* **$69.95**

MAIL COUPON FOR FREE FOLDERS

Paste this coupon on a penny postal card and mail to Travel Dept., Interstate Transit Lines, Omaha, Nebr. (Check information desired.)

☐ San Francisco World's Fair ☐ New York World's Fair
☐ Western Wonderlands ☐ Other points of interest

Name_____
Address_____
City_____

HT-639

Cool and Fresh as a mountain top

Cool vacations begin the moment you step inside an air-conditioned Interstate or Union Pacific Super-Coach. Gentle no-draft circulation keeps the air clean, fresh and invigorating throughout the trip.

INTERSTATE TRANSIT LINES • UNION PACIFIC STAGES

INTRODUCTION
America's 'Silver Bullet'

My first experience of travelling by Greyhound was in October 1974. In those days the Toronto bus depot was in the heart of downtown, only a block or two away from the Holiday Inn where I had squandered far too many of my pitifully small roll of precious dollars on a bed for the night.

I had tried to catch the bus the previous night and sleep on board to avoid the hotel bill, but my flight from England had been delayed and I arrived too late to catch the daily departure heading west to Winnipeg. Clutching my 'Ameripass' in my hand – good for one month's unlimited travel in the USA and Canada – I entered the vault-like building and approached the dark wooden ticket desk in one corner. As it was fairly early in the morning (a combination of jet lag and trepidation at being alone in a strange country had prevented much sleep) the queues at the windows were pretty short and soon I was able to exchange the pass for a sheaf of tickets.

There were a few hours to wait before my bus was called and I spent some of the time just watching the comings and goings as the depot grew busy – the hustle of the passengers, the bark of the diesel engines, the announcements of buses arriving from, and departing to, cities I had only dreamed of: New York, Montreal, Detroit, Chicago – but most of all I marvelled at those wonderful, gleaming, stainless steel and aluminium machines painted in red, white and blue with the sleek greyhound along the sides.

I climbed aboard eventually, and as instructed by friends back home who had done this sort of thing before, took the window seat in the second row on the opposite side to the driver. The theory was that during the day you got to see out of the front, but at night could duck down and sleep without being bothered by the headlights of the oncoming traffic (it works pretty good, too!).

As we headed north out of the city I still wasn't sure what to expect. My destination was over 1,300 miles away – an amazing distance to travel by road for someone brought up on an island less than 600 miles from top to bottom – I would be on the bus for more than 30 hours, and yet I still wouldn't have reached the middle of the country!

Long distance travel by Greyhound soon develops its own rhythm as the road disappears beneath the big wheels at a steady pace, the scenery slipping effortlessly by like a never-ending travelogue, and the small towns passed through quickly fading to be forgotten. You look out of the window, read a few pages of a large book bought especially for the journey (in my case Tolkein's 'Lord of the Rings') or talk to fellow travellers. Then, every three hours or so, there's a longer stop to change drivers which gives the passengers a chance to stretch their legs and grab some refreshment. I remember my surprise at the first such stop being told by the driver that I could reserve my seat by simply leaving my jacket on it, and my bag could safely be left in the overhead rack while the bus was cleaned and refuelled. Is that still true today I wonder? When the new driver took over, he was more in the mood to talk than his predecessor and I learnt that Greyhound pilots spent their time driving back and forth over the same stretch of highway, passing the bus along as if it were a giant baton in a relay race lasting for thousands of miles.

It was on the second day that I also began to understand just how important the Greyhound bus could be in such a huge country. Passing along the shores of Lake Superior the bus was flagged down by a young couple whose car had broken down – we were miles from the nearest small town and if it hadn't have been for the Greyhound they would have had to wait (probably for several hours as traffic was very sparse) for a passing motorist to rescue them.

My journey to Winnipeg was completed without any problems and after a few days staying with friends I rejoined the Americruiser heading west. Following the Trans Canada Highway (little more than a two lane blacktop in places) I visited Regina, Calgary, Edmonton and Vancouver before heading south to my ultimate destination – Los Angeles, California.

Riding the Greyhound bus in Canada in '74 was a great experience. Air travel was still relatively expensive and the railways had somehow lost their glamour, so it seemed that everybody used the bus: students, families, old people, military personnel, children – it was a cosmopolitan and popular form of transport.

The contrast when we crossed over the border into the United States was immediate. The bus was now for those who couldn't afford a car. The first person who sat next to me as we headed out of Portland, Oregon, hadn't shaved for a while and his opening gambit was to try to 'borrow' some money for a phone call so's he could contact his relatives when we arrived at Salem in an hour's time! Looking back, it is easy to see the problems that were mounting up for Greyhound in the 1970s. Their depots were often run-down and in undesirable locations in the poorer areas of cities and on long distance routes they were facing increasing competition from the airlines. For local travel, people were using their own automobiles and the bus company was undoubtedly feeling the pinch.

On reflection, my journey across Canada was probably as close as I could have got to travelling by Greyhound in its heyday of the '50s and '60s. And the importance of the bus to rural communities in those days cannot be overstressed – Greyhound shipped parcels and freight as well as people across the country. For thousands of small towns across America, the bus was an essential lifeline.

I vividly remember one stop in the middle of the prairie. It was nothing more than a crossroads with a small diner on one corner, and no other building in sight as far as the eye could see in any direction. The main highway stretched from horizon to horizon in an unbroken line and it was over an hour since we had passed the last signs of habitation. As the bus pulled up I counted six pick-up trucks parked outside the cafe. Four passengers got off the bus and were driven away, two new passengers joined and the last remaining pair of pick-ups disappeared. After a few minutes, when the dust had settled, it was hard to believe there had been anyone there at all.

Other images, just as striking, remain with me over twenty years later. Sunrise in the Rocky Mountains; toast, jelly and coffee in the diners (my staple diet); sentry-like grain silos standing alongside the railroad tracks; the driver arguing with a pedestrian who had walked in front of the bus as he pulled out of the depot; a row of tennis balls dangling from the roof of one depot to guide the bus drivers into their parking bays; and many, many other delights of that journey of discovery between 13 October and 7 November 1974.

There isn't a single photograph of my trip; I had deliberately not taken a camera with me because I didn't want to be looked upon as a tourist. Such a vain and foolish youth…

Tony Beadle

A COLD START
Hupmobile to Greyhound

No one man invented long distance bus travel, and the idea of transporting passengers for a profit in America predates the invention of the automobile by over 150 years. The first regularly scheduled public stagecoach line opening in New Jersey in March 1732, when Solomon Smith and James Moore advertised that they were going to operate two wagons between Burlington and Amboy, a distance of fifty miles. Although some local citizens expressed fears that the speed of the eight-miles-per-hour wagons would make them a danger to other road users, it is not recorded how much discomfort the passengers endured on the journey.

In 1756, a stagecoach line ran for the first time between New York and Philadelphia, taking three days to complete the route, travelling for 18 hours a day. Commissions from Congress to carry mail helped boost stagecoach lines, but it was still an arduous means of long distance travel – in 1785 Weddale Stage Lines took six days to go from New York to

Left In 1914 Carl Eric Wickman, a Minnesota car dealer, bought for $600 a seven passenger Hupmobile, like this one, and changed the course of US transportation history. He used the car to transport miners from the firehouse in Hibbing to the town of Alice, ten miles away. The ride cost 15c one way and 25c round trip, but it was worth it: Alice had the best saloon in the area. The little bus was an instant succes, and eventually it evolved into the last great American whales seen in the background.

ABOVE *Hibbing to Alice: the first seeds of an American icon.*

Boston, along rutted and dusty tracks. Journey times began to come down however, and by 1831 Washington to Philadelphia could be done in five days; but on the eastern seaboard the stagecoach was facing increasing competition from the burgeoning railroad systems with their steam powered locomotives.

Out west nevertheless, the stagecoach was still the only method of passenger travel available, and on October 7, 1858, a big Concord coach (of the type seen in the western movies) drawn by six horses, arrived in Los Angeles, California, 20 days after leaving St. Louis. Carrying only five passengers and a small amount of letter mail, the Overland Mail Company's stagecoach did the 2,600 mile marathon non-stop – running day and night – crossing deserts, the vast plains and even going through hostile Comanche Indian territory along the way.

The legendary stagecoach company of Wells Fargo was formed in 1852 by John Butterfield, Henry Wells and William Fargo and in 1868 they were granted a government subsidy for a daily, cross-country mail service to California. But the completion of the transcontinental railroad the following year foretold the end for stagecoach travel over long distance routes.

While the railroad was infinitely better than the stagecoach, it did have one major drawback – the trains could only travel where there were tracks laid down for them. Horse-drawn coaches were still in demand for local journeys and there was even a special type of vehicle developed for getting people to the railroad – called a depot hack or station wagon.

The demise of the horse as transport didn't happen overnight, and although Charles and Frank Duryea produced the first US-built gasoline powered motorcar in 1892, it wasn't until 1903 that an automobile crossed the continent. That car, a two-cylinder, six horsepower Winton, took nearly two months for the drive from coast-to-coast, but had demonstrated the capability of the new fangled 'horseless carriage'.

Development of the automobile in America proceeded at a frantic pace, and soon there were hundreds of companies in the business of constructing motorised vehicles. Most of these were hand-built, and therefore expensive, and it was only when Henry Ford introduced his famous Model T in October 1908 that owning a car started to come within the reach of the working man.

It was around this time that a young Swedish immigrant

ABOVE *Aerodynamics not a priority in 1914.*

Thrill to ALL of CALIFORNIA

Above left: *Fabulous Treasure Island is now brilliant with color and excitement as San Francisco's Golden Gate Exposition thrills thousands of gay visitors.*

Above right: *There is always something exciting going on in fascinating Hollywood. Here is a movie premiere!*

ENJOY GREYHOUND'S
Nite Coach Sleeper!

Here is the pleasant and enjoyable way to travel to and from California...the Greyhound Nite-Coach sleeper. By day you can relax in a soft, comfortable seat, at night, you can stretch out in a full-length berth. Each compartment of the Nite-Coach contains a wash basin, running water, space for clothes and a radio. Two lavatories and a ladies' lounge are likewise available. Beginning June 1 (Eastbound) and June 3 (Westbound) this service will operate daily between Kansas City and Los Angeles. (Extra berth fare $4.50 single and $6.00 double.) Make your trip to California, or between Los Angeles and San Francisco ($1.00 Berth Fare), this luxurious convenient way.

FOR COMPLETE INFORMATION WRITE NEAREST GREYHOUND TRAVEL BUREAU

920 Superior Avenue, Cleveland, Ohio	12th and Wabash, Chicago, Illinois	917 McGee Street, Kansas City, Mo.
905 Commerce Street, Fort Worth, Texas	560 S. Los Angeles St., Los Angeles, Cal.	Pine & Battery Streets, San Francisco, Cal.

GREYHOUND

arrived in the United States seeking, as did millions of fellow travellers, his fortune in this fabled 'land of opportunity'. Carl Eric Wickman ventured as far west as Arizona, where he worked for a short time in a saw mill; he then moved north to Hibbing, Minnesota, where he was employed as a drill operator in an iron mine. Job security in the Mesabi open-pit mining business was non-existent, and lay-offs occurred quite often. When available, the work was dirty and hard, and the hours were long, but the pay must have been pretty reasonable because, when he left the mines in 1913, Wickman was able to buy his own business – a dealership for Hupmobile cars and Goodyear tyres – located in Hibbing.

As any automobile dealer would have done in those days, Wickman also undoubtedly ran a taxi service in and around Hibbing. The story goes that, in 1914, when a new Model 32 Hupmobile touring car was proving difficult to sell, Wickman paid $600 to buy the automobile for himself and began using it to ferry miners from Hibbing to Alice, a distance of ten miles. Seven miners could be crammed into the Hupmobile,

From little acorns...Following the success of the Hupmobile (above) the Mesaba Transportation Co. was formed in 1916 when Wickman was joined by Ralph Bogan, Arvid Heed and Dominic Bretto. By 1918 the company had a fleet of 18 buses and a route system that covered most of Minnesota.

Above By the early 1920s federal and state governments were spending over 1 billion dollars a year on new roads, and as a result bus networks grew enormously. Competition was fierce. In 1928 a Yelloway Bus completed the first coast to coast trip: SF to NY in five days and fourteen hours. Wickman and his partners responded by improving quality and demanding that all drivers wore smart military-style uniforms. They kept buying out smaller firms and in 1929 managed to take over Yelloway itself.

each paying a fare of 15 cents for a one way ride, or 25 cents for the round trip between the firehouse in Hibbing and the Alice saloon over rough and rutted roads. From this tiny beginning was to grow the huge Greyhound bus network that eventually covered the entire North American continent.

However, the notion (largely promoted by Greyhound) that Wickman's fledgling enterprise was the first intercity bus route in America seems to be inaccurate. As early as October 1907, W.B. Chenoweth was operating a service between Colorado City and Snyder, Texas, using his own design, six-cylinder 'motor driven stage coach'. Chenoweth's venture was short-lived, mainly due to passengers being wary of his unusual vehicle and, when a local cattleman purchased four Buicks and went into direct competition, his fate was sealed.

As far as can be ascertained, the first successful, regularly maintained, scheduled intercity bus service began on 1 March 1912 between Luling and San Marcos, Texas. The operator was Josh Merritt who converted a 1906 Packard to carry up to seven passengers and their luggage.

In these early days of course, there was no such thing as a bus. The vehicles used were, like Wickman's Hupmobile and Merritt's Packard, ordinary cars that were sometimes modified to carry more people. Local taxi services expanded the territory they covered and developed routes between towns, forming into small intercity transport companies – but many of these soon fell by the wayside.

Meanwhile Wickman's Minnesota operation was booming, thanks to the iron mines working at full capacity to meet the demand created by the war in Europe, and he soon brought in a partner to help share the driving chores, A.G. 'Bus Andy' Anderson. In 1916, the Mesaba Transportation Company was formed by a merger between Wickman's company and Ralph Bogan's taxi company that plied the route between Hibbing and the 'big city' of Duluth. The newly formed organisation had a fleet of five cars and five drivers on the payroll. In that first year of 1916, the Mesaba Transportation Co posted a profit of $16,000 – a tidy sum in those days.

An important step in the history of intercity bus travel also came about in 1916, with the introduction of the Federal Highway Act in July of that year. President Woodrow Wilson gave his full backing to the legislation which sought to bring

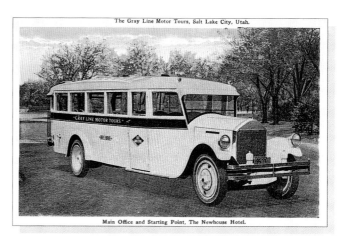

In 1928 and 1929 Wickman's organisation acquired several bus companies that bore the name "Greyhound": Northern Greyhound, Southern Greyhound and Eastern Greyhound Lines used comfortable, reliable coaches such as this Z-250 (below) built by The Yellow Motor Coach Co. of Chicago, Illinois, a subsidiary of General Motors.

about the much needed improvements required to the U.S. road system and, in so doing, helped speed the development of bus travel.

By 1918, Mesaba Transportation had expanded to an 18 vehicle fleet with a network of routes that covered a large part of Minnesota. Then, in 1922, Carl Wickman sold his interest in the company and moved his family to Duluth, where he took on the task of organising bus lines to complement the train services of the Great Northern Railway rather than compete with them. Forming the Northland Transportation Company, in four short years he helped build up a system that he was able to sell to Great Northern for $240,000.

This sale had two effects. First, it created an interest in the idea of the railway companies owning bus lines and, more importantly, it gave Wickman and his associates the wherewithal to begin creating their own system – one based on the

ABOVE *The Mack bus was one of the best of its time. The state-of-the-art six-cylinder gasoline engine produced plenty of power and the bullet-style shock absorbers efficiently dealt with the bumpy roads of the day.*

BELOW *The first Greyhound Company logo was a blue shield that incorporated the name, a stylised dog and a drawing of the Mack. In later years, as different models were introduced, the bus drawing was dropped.*

ABOVE *An American institution was officialy born in 1930, when Greyhound Lines Inc. was formed. The newly named company had over 1800 buses travelling the nation's highways. The magnificent, specially designed Mack bus arrived in 1931, having been vastly improved since its inception in 1926. Built by the famous Mack Truck Company, it offered unprecedented speed and comfort and was used primarily – not surprising in the light of the cost – on the newly opened transcontinental routes.*

notion of small, independently owned bus companies getting together in large, regional affiliations. In September 1926, a holding company called the Motor Transit Corporation was formed by Wickman, Ralph Bogan, Orville Caesar and E.C. Eckstrom. Basically the new enterprise came from the partnership between Mesaba Transportation and Eckstrom's Safety Motor Coach Lines which operated out of Chicago and also served much of Michigan.

It is from E. C. Eckstrom's Safety Motor Coach Lines that the name 'Greyhound' seems to have originated, although his company was not alone in using this descriptive identification. Eckstrom apparently settled on the name after hearing a passenger say that his buses ran "like greyhounds". Coming at a time when bus companies were looking for more zappy trade names to create the idea of speed and comfort, and in so doing promote the advantages of travelling with them, it was a natural. After a short while, Safety Motor Coach Lines started using 'Greyhound Lines' as their operating name.

As soon as it was up and running, the Motor Transit Corporation started making acquisitions, the first in 1926 being Interstate Stages which ran buses between Chicago and Detroit. The division of responsibilities between the four men at the top of MTC seems to have been fairly clear-cut: Wickman and Caesar provided the day-to-day operations

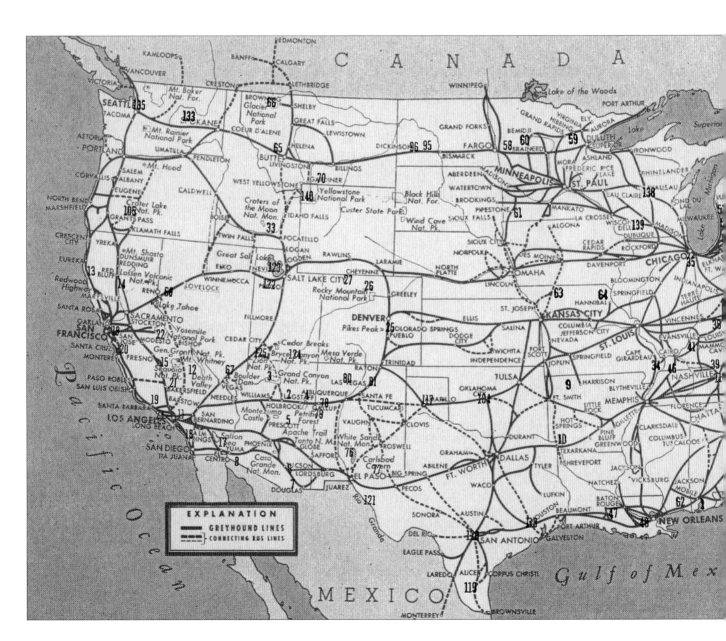

LEFT *The typical 1930s Deco interior lights on the Mack allowed night-time reading and the thick cloth side curtains kept sun and dust at bay. In those days there was no air conditioning.*

BELOW *The 1931 Mack bus was amazingly comfortable, even by today's standards. The seats were covered in pure Massachusetts mohair, with horsehair stuffing. For kids or last minute passengers there were special flip-up seats that made intelligent use of the aisle space.*

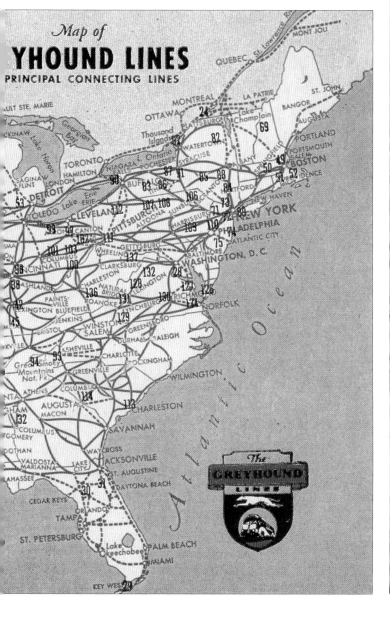

management skills, Eckstrom handled the promotions and advertising, while Bogan searched for potential take-over candidates. However, cash was still in relatively short supply and it was only the arrival of Glenn W. Traer, an investment banker from Chicago, that got things properly in motion. Traer had purchased a large tranche of MTC's first stock issue and resold it at a profit, and he was therefore able to arrange loans and establish lines of credit to support the growing company.

By 1928, a dozen different companies were amalgamated in the MTC 'Greyhound' system, with some 325 buses on the road. Aggressive marketing of the period using the Greyhound name is typified by a newspaper advertisement of the Southland-Red Ball Motorbus Company exhorting businesses to get their salesman to use the bus instead of driving a car. Touted as being a third of the cost of a private automobile, the 'Greyhounds of the Highway' were said to be 'the most economical means of transportation available'. The slogan 'Ride The Greyhounds' is repeated in many other similar advertisements, firmly establishing the public's awareness of the name.

Above right It's difficult to imagine now just how snappy this rig must have looked back in 1930; and it doesn't look all that 'quaint' today.

Above Early Greyhound coaches advertised on their flanks the main cities covered by the network. A coast to coast trip cost around $55.

Left The twenties was all about 'rationalisation' for the MTC Greyhound group. Logic dictated that the myriad tiny bus companies across the country would begin to amalgamate.

1928 was also the year that a bus completed the trip from San Francisco to New York in 5 days and 14 hours. Leaving the west coast on September 5, the bright yellow Pioneer Stage bus, belonging to the Yelloway System of W.E. Travis, was the forerunner of a transcontinental service operated by the American Motor Transporation Company.

East of St Louis the route was run in conjunction with other independent companies, as AMT had yet to reach that part of the country.

Northland Transportation, the company started by Carl Wickman in 1922, became Northland-Greyhound Lines in 1929 when Motor Transit Corporation bought

In the days before streamlining (or more accurately, before it was taken seriously for everyday, workhorse machines) coaches looked more like trains than buses. The Mack's brakelights were indeed railroad lamps. Baggage was stowed on the roof and had to be secured very firmly in order to avoid dangerous shifts when the bus had to negotiate tight bends.

a part interest. But this was only a tiny part of the Greyhound expansion going on in that year. In February 1929, MTC paid a reported $6 million for American Motor Transportation and in so doing, not only removed the threat of Yelloway encroaching on their territory in the mid-west, but also opened up the California routes to Greyhound.

At the same time, the Greyhound network was spreading rapidly in other directions too, with MTC working in co-operation with the wealthy railroad companies to add more bus companies and more routes to their system. The eastern sector, taking in Pittsburg, Philadelphia and New York came into the fold via a link-up with the Pennsylvania Railroad which evolved into a jointly owned operation, Pennsylvania Greyhound Lines.

With the name Greyhound proliferating in such a rapid manner, it came as little surprise when, in 1930, the Motor Transit Corporation officially became Greyhound Lines Inc. That same year, the corporate headquarters were moved from Duluth to Chicago.

Altogether, there were 3,520 intercity bus companies operating in North America in 1930, with 14,090 vehicles covering 318,715 route miles and totalling up an incredible 7.5 billion passenger miles. These were the Depression years of course, and Greyhound advertisements stressed the economical benefits of bus travel more than the (apparent) convenience and (arguable) comfort aspects.

During the process of expanding by buying up a number of small companies, the Greyhound 'fleet' inevitably consisted of a variety of different makes and types of buses. Famous names such as White, Fageol, Mack, Yellow Coach and others produced what were known in the 1920s as 'parlor cars' – which looked like stretched versions of the sedans available at the time – even automobile makers like Studebaker and Pierce-Arrow were in on the act. Often the bus companies themselves would produce their own bus designs by taking a truck chassis and adding a hand-built body that fitted their particular local requirements.

However, it is the Fageol Safety Coach which really set the early standards for intercity bus construction. Founded by brothers Frank R. and William B. Fageol in 1917, the Fageol

Motors Company of Oakland, California, started out making expensive luxury cars that were guaranteed to do eighty miles an hour – thanks to a six-cylinder, 125 horsepower aviation engine supplied by the Hall-Scott Motor Company. Demand for the big Fageol automobiles was extremely limited, and the company sensibly turned its attention to the manufacture of trucks and buses.

In 1921, Fageol produced what is probably the first ever American vehicle designed specifically as a bus. Low-slung, with four wheels and four passenger doors down the side (it would be a few years before the centre aisle configuration caught on) the 22-seater 'Safety Coach' proved immediately popular with operators and passengers alike. Competitors were quick to see the advantages of the design, and brought out their own versions, notably White in 1922. Fageol responded by stretching the wheelbase further and increasing the width of the body to accommodate 29 passengers.

Wickman and Bogan's Mesaba Transportation Co. used Fageol Safety Coaches on their busy route between Hibbing and Duluth in 1922, and the ever-widening Greyhound organisation obviously had many Fageol buses on its roster in the latter part of the decade.

By 1927, Greyhound itself was in the bus building business, thanks to the acquisiton of the C.H. Will Motors Corporation of Minneapolis, Minnesota. Early Will buses looked almost identical to the Fageol Safety Coach and were obviously copied from them, but gradually they became bigger, heavier and provided more luggage space. The final bus from Will was delivered to Greyhound in January 1931, thereafter they sourced a large number of their new buses from the GMC division of Yellow Coach.

It was during 1926 that Mack introduced a six-cylinder bus designed for long distance, high-speed use. By 1929 this bus had been considerably improved and the BC Model as it was

Right *The details show the high quality of design and manufacture of the mighty Mack. A chrome greyhound sits on top of perfectly spaced engine air vents.*

called, now carried 29 passengers in some style. The central aisle did away with the need for a door for each row of seats, but interior luggage space was still fairly limited, and most of it had to travel on the roof rack. Mack also produced a six cylinder, 126 hp BK version until 1931, but thereafter the company cut back on interstate building, although production of the Mack BC continued until 1937 in small quantities.

At the start of the 1930s, the vast majority of intercity travellers who didn't use their own car went by train – over 70% in fact – while the bus companies handled around 25% of the passenger traffic. This was to change gradually throughout the period leading up to World War Two, with the bus taking passengers from the railroad and reaching a peak of a 35 per cent share of the market.

ABOVE The practice of putting a shining chrome greyhound on the coach's front started with the Mack. This went on for many years and was one of the little touches that made Greyhound buses special. Sadly, the chrome emblem is missing from current models. Design is in the detail...

LEFT Going Greyhound: in the 1930s travelling became the new national pastime. People from every corner of the country flocked to the 1933 Century of Progress Exposition in Chicago, thanks to all-expense-covered Greyhound bus tours. Magazine ads proudly proclaimed "There is no better way to see America than the Greyhound way".

ART DECO JEWELS
Station to Station

Survival was the name of the game for Greyhound in the early part of the 1930s and although its route coverage continued to expand as it bought up more bus companies it wasn't until the 1933 Chicago World's Fair – officially named the Century of Progress Exposition – that the tide began to turn.

Thanks to the amazing exhibition, demand for travel to the Windy City boomed, and Greyhound was ideally placed to benefit from the increase in passenger traffic to its centre of operations. Having been granted the exclusive on providing transportation inside the fairgrounds, Greyhound then took a huge gamble and reserved 2,000 hotel rooms for the duration of the Exposition.

In what was probably one of the first examples of the cheap 'package holiday', Greyhound then advertised all-in bus tours to Chicago. Anywhere in the USA, passengers could buy a ticket that would provide bus travel, admission to the World's Fair and hotel accommodation in one. The idea caught on and the tours were a sell-out, with many people making more than one visit to the event.

But if leisure travel provided a much needed boost to the income of Greyhound and other bus companies, it was the regular daily services which produced the 'bread and butter' revenue. Exactly how important the bus had become in rural America can be seen by two letters of support sent to the Texas Bus Owners Association in 1929.

J.E. Guthrie, Secretary of the Salado Chamber of Commerce, wrote: 'Years ago, Salado being an inland town, felt great need of a railroad. We no longer need one. We have six Greyhound buses each way in 24 hours. Our people like the service – the bus line fills a long felt want.'

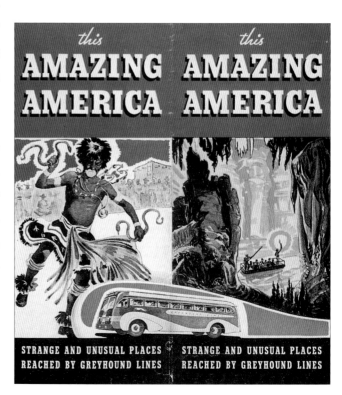

SWING AROUND AMERICA!
THIS SUMMER.. by GREYHOUND

Call it what you will—a fortunate conjunction of the planets, or the inscrutable march of events, *or just plain good luck*—but 1939 has brought to America the most amazing cycle of fun, excitement and thrills in its fast-moving history!

Shining stars in this galaxy are the New York World's Fair and the Golden Gate Exposition. But sprinkled between, on this giant coast-to-coast orbit, are the scarcely lesser lights of vacation enchantment—cool and wondrous national parks, northern lakes and mountains, surf-swept beaches, dude ranches, every summer scene on the map.

"I can't see them all in my short vacation, on my small budget," say you. But you can! That's where Greyhound steps into the picture, with the amazing rate of $69.95 to both Fairs and a hundred places of thrilling interest in between—over any route.

"But I simply can't take time to see both Fairs," you come back (wistfully). Well then—visit whichever Fair you've set your heart upon, throw in a cool vacation resort or two—and still save dollars over any other type of transportation that rolls, swims, or flies.

A great fleet of Greyhound Super-Coaches—streamlined, smooth-rolling, efficiently ventilated or completely air-conditioned—awaits your pleasure!

GRANDEST CIRCLE TOUR IN TRAVEL HISTORY—
visiting both Fairs ...

This amazingly low rate includes transportation from your home, across the continent to one Fair, then back to the other, and return to your home—following your choice of scenic routes. You can take as much as ninety days—or the trip can easily be made in two weeks. It's an all-time bargain, no matter how you plan it.

for only
$69.95

Ask about Expense-Paid Tours ... they save time and money, add pleasure, assure hotel reservations.

PRINCIPAL GREYHOUND INFORMATION OFFICES

Cleveland, O. . . . . East 9th & Superior
Philadelphia, Pa. . . . Broad St. Station
New York City . . . . 245 W. 50th Street
Chicago, Ill. . . . . . . 12th & Wabash
Boston, Mass. . . . . . . 60 Park Square
Washington, D.C.,1403 New York Ave.,N.W.
Detroit, Mich., Wash. Blvd. at Grand River
St. Louis, Mo. . Broadway & Delmar Blvd.
Charleston, W. Va. . 155 Summers Street
Minneapolis, Minn. . 509 Sixth Ave., N.

San Francisco, Cal. . Pine & Battery Sts.
Ft. Worth, Tex. . . . 905 Commerce Street
Memphis, Tenn. . . . 527 N. Main Street
New Orleans, La. . . 400 N. Rampart St.
Lexington, Ky. . . . . 801 N. Limestone
Cincinnati, O. . . . . 630 Walnut Street
Richmond, Va. . . . 412 E. Broad Street
Windsor, Ont. . . . 403 Ouellette Ave.
London, England . . . .
. A. B. Reynoldson, 49 Leadenhall Street

THIS BRINGS PICTORIAL WORLD'S FAIR BOOKLETS

Mail this coupon to nearest Greyhound information office, listed at left, for bright, informational folders on NEW YORK WORLD'S FAIR ☐, or SAN FRANCISCO'S GOLDEN GATE EXPOSITION ☐. Please check the one desired. For information on any other trip, jot down place on line below.

Information on trip to _____

Name _____

Address _____ HT-6

1937 GREYHOUND SUPER COACH
NOSTALGIA ON WHEELS!

Yes, 1937 was a very good year. The San Francisco Golden Gate Bridge was completed and was the longest in the world at that time. The Greyhound Super Coach was also introduced, and was considered the "Ultimate in bus design."

Some of its features included…first rear engine bus — first with baggage compartment underneath — jump seat for the porter — first to be air-conditioned.

This bus was recently located in South Dakota where it was being used by a church. Greyhound decided to restore the Super Coach to its original condition in keeping with the current nostalgia trend and to preserve a little history.

Today, the 1937 Greyhound Super Coach is a vivid example of how much Greyhound Motor Coach travel has changed for the better over the decades — changes that have made Greyhound the number 1 motorcoach carrier.

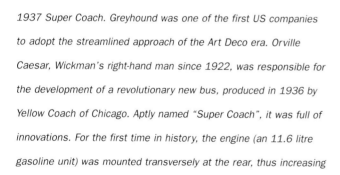

1937 Super Coach. Greyhound was one of the first US companies to adopt the streamlined approach of the Art Deco era. Orville Caesar, Wickman's right-hand man since 1922, was responsible for the development of a revolutionary new bus, produced in 1936 by Yellow Coach of Chicago. Aptly named "Super Coach", it was full of innovations. For the first time in history, the engine (an 11.6 litre gasoline unit) was mounted transversely at the rear, thus increasing the number of seats to 37. This also allowed the creation of a luggage compartment underneath the passanger area, doing away once and for all with the fiddly and dangerous roof racks. The blue and white colour scheme emphasised the streamlined design and promoted an image of speed and efficiency. The significance of the Super Coach for Greyhound is indicated by this later advert, a perfect example of what Willie Davidson called "The New Nostalgia".

Jasper Chamber of Commerce Secretary, Ed C. Burris, was even more fulsome in his praise: '… it is not possible for an individual or group of individuals to estimate the importance of Bus transportation in Texas and the United States.' Burris went on to explain in meticulous detail: 'You take this town for an example, and I feel there are numerous other communities in which similiar conditions exist. We can depend upon the Bus service for more frequent means of getting about than the Railroad. This condition is not brought about by neglect of the Railroad but is made possible by six Highways radiating out of our little town. Each of these Highways accommodate from one to three Buses each way every day …The Buses assist the surrounding communities in reaching our little city, for shopping and other business

ABOVE Evansville, Indiana. As the Super Coaches criss-crossed the country and revenue came on tap, Greyhound went on a building spree. This resulted in a network of terminals of unprecedented beauty. The one in Evansville is internationally renowned as an Art Deco masterpiece. Designed by W.S. Arrasmith, it was completed in 1938, at a cost of $150,000. Over 100 buses were scheduled in and out of this terminal each day.

RIGHT A hand-painted 1939 view of Evansville Terminal.

ABOVE When Greyhound adopted Streamline Moderne as its company image, it did not fail to exploit the power of neon. All over America, terminals designed by the Louisvillle firm of Wischmeyer, Arrasmith & Elswick featured signs that would not have looked out of place on the Las Vegas Strip. In small towns like Evansville the Greyhound terminals lit up the sky, shining like beacons of modernity, harbingers of adventure.

LEFT The typically Deco central tower advertises the Greyhound name in vibrant red, while at the top the blue neon exercises the running dog. Although many terminals of the period featured a 'running dog' sign, this is the last working example left in the country.

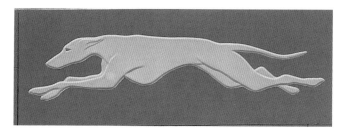

ABOVE *The blue and white colour scheme adopted for the Super Coaches was also used for the terminals. In this way Greyhound was able to enhance its corporate identity in city after city. The large glazed tiles and emblems are typical of these stations throughout the 1930s.*

purposes ... This manner of service could not be rendered by any other means of transportation.'

But it wasn't all plain sailing: whenever there was a big demand for transport (such as the Chicago World's Fair) then unscrupulous 'wildcat' operators would try to get in on the act. These rogue companies rarely published a list of fares or a timetable, and often gathered a vehicle load of passengers together for a trip on a 'cost sharing' basis. Some wildcat drivers even resorted to snatching luggage away from a rival bus line and loading it on their vehicles – a guaranteed method of getting the passenger to switch carriers. As the situation worsened, complaints from customers grew louder, bus companies and state regulatory bodies protested in unison, and this eventually resulted in the US Congress passing the Motor Carrier Act of 1935.

This act brought all bus companies engaged in running interstate routes under federal control, via the Interstate Commerce Commission. From now on, all bus lines operating across state borders were required to adhere to fixed schedules and tariffs. And, while existing carriers who were operating 'regular' bus services when the Act came into force were granted 'grandfather' certificates to continue in business, afterwards any company wishing to start up a new service had to apply to the ICC for a certificate.

Because existing bus lines could protest these applications, one effect of the Act was to limit the number of new companies able to enter the system. Another result was that, in certain cases, an existing bus line could gain a virtual monopoly over a route. In turn, the granting or holding of an ICC certificate became a valuable commodity, one that could be bought and sold.

Overall, while the ICC regulation was welcomed, it tended to accelerate the process of the preceding years of reducing the number of bus lines in operation – mainly by mergers. Although Greyhound's expansion programme had been relatively dormant in the early '30s as the company looked to modernize its bus fleet and upgrade the facilities at bus terminals and garages.

However, in 1935 Greyhound set up an agreement with the New York Central railroad which led to the creation of the

ABOVE *More Evansville decoration. In those days attention to detail was such that even the door handles were something special. Each and every one was solid brass and decorated with the shield-type logo of the period.*

ABOVE *The Evansville Greyhound station is no museum. It still welcomes hundreds of buses every week, and each time Station Manager Joe Kaylor is there to help weary passangers down the steps, give directions and even unload baggage. Which makes this one of the nicest terminals to arrive at in all the network.*

RIGHT *Glass block – the construction material of the Deco era – is featured in most terminals of the period.*

RIGHT *A period "linen" postcard of the Minneapolis Greyhound depot in 1937, with a brand new Super Coach on the right.*

ABOVE *This postcard of the Norfolk, Virginia terminal (1943) shows the departure from the blue and white colour scheme of the thirties. Most new terminals were muted, in harmony with the Silversides.*

Central Greyhound Lines of New York which operated between New York, Albany, Buffalo and Cleveland as well as many other routes. As a consequence, the existing Central Greyhound Lines that covered routes from Detroit to Columbus, Cinncinati and Louisville had its name changed to Ohio Greyhound Lines to prevent any confusion.

Because the Greyhound network had been created over a number of years by pulling together a wide variety of different companies, for any long distance traveller there were still some obstacles to be overcome en route. When changing from one line's area to another, often a new ticket had to be issued, and sometimes it was necessary to change buses – inconveniences that the passenger could do without.

In addition to its own extensive advertising in the press, Greyhound bus travel received some big boosts from Hollywood during the depressed years of the '30s. The first

RIGHT *Hopkinsville, Kentucky. Compare the Evansville Terminal with one of today's anonymous bus stops. In some extreme cases, such as the one pictured here, the Greyhound "station" is no more than a small sign lost in a sea of other signs.*

film to prominently feature a bus as part of the plot was *Fugitive Lovers* from MGM in 1934. This was followed by *Cross Country Cruise* from Universal Pictures and then, arguably the most famous bus movie of all time, *It Happened One Night* starring Clark Gable and Claudette Colbert. This Columbia Pictures release created the impression that riding the bus could be exciting and might even lead to romance. Some twenty years later, the 1956 remake, entitled *You Can't Run Away With It,* starring June Allyson might have failed to

America's "OUT-OF-TOWN CAR"

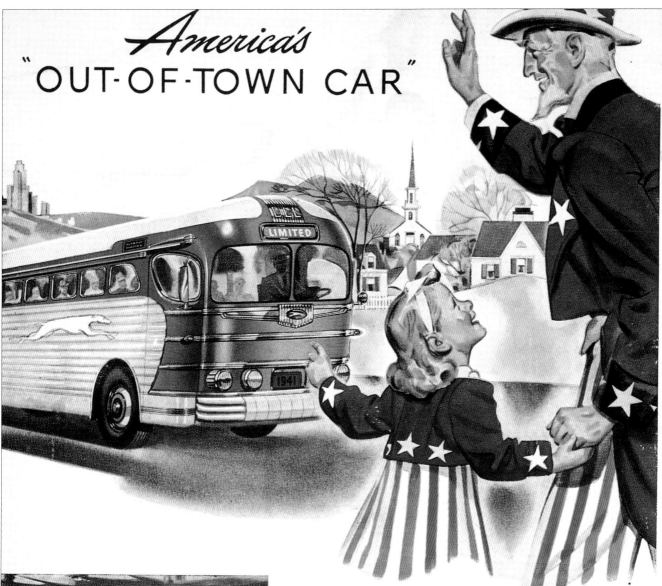

is the finest on the highways

—yet it costs less to ride, has the world's safest driver

●Millions of Uncle Sam's nieces and nephews are making their trips out of town these days in their *second cars*—we mean Greyhound Super-Coaches, of course! Why not you? You'll save a lot of wear and tear on your automobile—on your pocketbook—and on your nerves, too! These big "out-of-town cars" are warmed and ventilated like a pleasant living room—they ride as smooth as silk—and the men at the wheel are 14 times safer than the average driver! It's the pleasant way to travel on business—and you'll enjoy a business-like saving on pleasure trips! Fares are only a fraction of the cost of driving even a small car. *Try Greyhound—next trip!*

Principal Greyhound information offices are located at: New York City • Cleveland, Ohio • Philadelphia, Pa. • Boston, Mass. • Washington, D. C. • Detroit, Michigan • St. Louis, Mo. • San Francisco, California • Fort Worth, Texas • Minneapolis, Minn. • Lexington, Kentucky • Charleston, W. Va. • Cincinnati, Ohio • Richmond, Va. • Memphis, Tenn. • New Orleans, La. • Windsor, Ontario, (44 London Street, East) • Montreal, Quebec, (1188 Dorchester Street, West)

You won't find all these Super-Coach features in even the costliest private car: Four-position easy chairs with sponge-rubber cushions; massive rubber footrests; translucent pull-down shades; directed-beam reading lights; perfected air-conditioning.

FREE! NEW CARTOON MAP

A brand new "Amazing America" Cartoon Map, with more than 100 entertaining pictures and stories, in full color. Simply mail this coupon to the nearest Greyhound Information Bureau listed above (no local address necessary).

Name _____

Address _____

City _____ HT-2

GREYHOUND

ABOVE *Jackson, Mississipi. Another of the "blue" 1930s stations, this terminal was also designed by W. M. Arrasmith, who completed it in 1937. Decommissioned several years ago, it would surely have been demolished if local architect R. Parker Adams, who used to play pinball in the station every day as a child, had not bought it and restored it to perfection.*

RIGHT *The Jackson Depot before it was decommissioned.*

have the same impact, but it still provided Greyhound with some very welcome free advertising.

When the Revenue Act of 1936 came into being, although aimed at restricting public utility holding companies, it had a direct effect on Greyhound whereby some 20 operating subsidiaries were merged and brought under the control of The Greyhound Corporation. Indeed, this can be seen to be the start of a lengthy consolidation policy which ultimately brought about the unified Greyhound system in the 1970s.

1936 also saw the beginnings of the National Trailways Bus System which, in contrast to Greyhound, was an association of independent bus companies rather than a holding company with a number of subsidiaries. Trailways, under the prompting of people like Aaron E. Greenleaf and Paul J. Neff, rapidly developed into a truly national organisation, quickly adopted a standard red and cream colour scheme for its buses and began operating a through ticketing system which the public favoured.

But without doubt, the most important event of 1936 as far as Greyhound was concerned was the arrival of the Model 719 'Super Coach' built by GM division Yellow Coach. This radical new streamlined design was to change the face of intercity bus travel forever. Thanks to the rear-mounted engine, the driver now sat ahead of the front wheels and enjoyed an unobstructed forward view. Greater passenger capacity (now up to 36) in the same overall length of vehicle came about by raising the floor above the engine level, which also allowed a capacious baggage compartment to be created in between the front and rear wheels. Extensive used of aluminium in the construction of this bus reduced the weight by two tons over previous models.

Although the Super Coach was an innovation, the principles of its design had been around for while, and some of them had already been used on earlier Greyhound buses. The outline shape was first seen on the 1927 Fageol Twin Coach, and the underfloor stowage of luggage had also been seen before a decade earlier on buses built by the American Car & Foundry Co (ACF) and on some other makes too. Even the transverse rear engine wasn't new – Dwight Austin had patented an 'angle drive' and used it on a series of Nite Coach buses he constructed for Greyhound in 1933-34 using the old Pickwick Corporation factory in Los Angeles, California.

Yellow Coach brought Dwight Austin to their Pontiac, Michigan, headquarters in 1934 and immediately took his patented drivetrain and adapted it to their own buses. Greyhound contributed towards some of the cost of the engineering development and design work, and also made

BELOW *Jackson, Tennessee. One of the most incredible examples of the old Greyhound terminals is to be found in the plantation state of Tennessee. The quiet, sleepy town of Jackson harbours a huge Art Deco station, which seems to have hardly changed since the day it was inaugurated. Completely unrestored, this is a fully working station, to which the patina of time has added a wonderful sense of character.*

Inevitably, such a successful concept as the Super Coach was quickly taken up by other bus builders who produced their own versions. ACF, Kenworth, and the Fort Garry Motor Body Co. of Winnipeg, Canada (later to become Motor Coach Industries in 1940), were some of the first to copy the ideas – the rest soon followed.

The New York World's Fair, opened by President Franklin D. Roosevelt on April 30th 1939, not only generated extra passenger traffic through bus tours and charters, it also provided Greyhound and Yellow Coach with an ideal platform to introduce an improved and restyled Super Coach that became known as the 'Silversides'.

The prototype bus displayed at the World's Fair was fitted with the same 707 cu.in. (11.6 litre) gasoline engine as had been used in most of the earlier Super Coaches, but the vast majority of Silverside production buses were powered by six cylinder Detroit Diesel engines as Greyhound led the industry

ABOVE *Ever since the days when the bus went only from Hibbing to Alice, Greyhound travel has always meant much more than just getting from A to B. More than a mere transportation vehicle, the bus is ultimately a container of people, a melting pot on wheels where people of all backgrounds and colours can get to know each other. It may seem surprising, but in present day America, where everybody sometimes seems to be afraid of everybody else, the Greyhound bus is one of the last public places where total strangers will still strike up a conversation. Lonnie, photographed in Jackson at the end of a ten hour trip, told stories of life in the L.A. Projects and gave first-hand descriptions of its dangers and lack of opportunities. By the end of the trip, the whole bus was listening.*

suggestions based on their experiences as an operator. Dwight Austin is also credited with being involved with the exterior styling of the Super Coach, together with James J. St.Croix, and Orville Caesar seems to have had a major influence on behalf of Greyhound. Air conditioning was another innovation that first appeared in the Model 743 Super Coach introduced in 1937 (right), using a system called Tropic-Aire manufactured by a Greyhound subsidiary. Production of the Super Coach totalled nearly 1600 by 1939 when the design was superseded.

in switching to this more economical method of propulsion. In 1938 it is estimated that there were only something like 200 diesel-powered buses operating in the USA; by 1948 this number had multiplied to over 18,000 – a fantastic revolution.

But, before that revolution could get into full swing, a much more cataclysmic event was to take place. On December 7, 1941, the Japanese attacked Pearl Harbor, Hawaii, and the USA entered World War Two. Production of large intercity buses was halted as the factories were turned over to armament manufacture. However, smaller urban buses were still being made under the auspices of the War Production Board, and many saw use on long distance routes.

Hundreds of old buses that had been scheduled for retirement were dragged out of the boneyards, refurbished and put back to work. Even so, overcrowding became an everyday occurrence as military personnel and civilians on essential

ABOVE *Uvalde Texas. Throughout the 1930s the Greyhound Co. was affiliated with several smaller organisations, which serviced secondary routes and operated their own terminals. This Painter terminal in Southern Texas has remained untouched for 60 years. Inside, the layout is very simple, and as you can see, the notice boards are still homemade.*

ABOVE *The Uvalde Depot Cafe is a rare surviving example of the coffee shops and restaurants that were often attached to Greyhound stations. In the early '40s there were almost 100 such restaurants, called Post Houses. Nowadays most have been replaced by fast food chains, but this little diner has managed to keep its original identity against all the odds.*

journeys filled every seat and stood packed in the aisles. Despite a vigorous camapign using the slogan 'Is This Trip Necessary?' bus passenger mileage went from 10.1 million miles in 1940, to more than double in 1942 at 21.5 million. It peaked at 27.4 million miles in 1943.

It was a difficult time for the bus operator; spare parts were hard to obtain, gasoline and tyres were rationed, and staff shortages meant longer hours. In some cases, transporting workers to shipyards and aircraft factories was taken care of by the use of converted trailers, and competing companies were forbidden from duplicating bus routes by the government for the duration of the war. The war years also saw the first female bus drivers being hired by Greyhound.

Towards the end of the conflict, in 1945, a one-time bus station ticket agent named M. E. Moore merged the Tri-State Transit Co and Bowen Motor Coaches to form the Continental Bus System Inc. based in Dallas, Texas. This marked the first time two formerly independent member companies of the Trailways system had been bought out and combined, and was the beginning of the creation of a second strong national bus system with central management control.

Washington DC. Thanks to the efforts of the Art Deco Society of Washington, this marvellous terminal has been saved and has been turned into the main lobby of a modern tower, built in 1991 by a Canadian insurance company. The original Greyhound terminal was opened in 1939 and was perhaps the most flamboyant of all the ones built by W.S.Arrasmith. The interior was even more outré than the outside, with a domed skylit ceiling, copper-edged walnut benches and a greyhound motif in the middle of the terrazzo floor. Restoration architect Hyman Myers went out of his way to use traditional materials such as flexwood, a canvas backed veneer which was used for the curved balcony. Artist John Grazier was commissioned to paint murals of buses outside national landmarks, in the spirit of the lost originals.

KINGS OF THE ROAD
Caesar, Silversides & Scott

Although Greyhound had plans to introduce a completely new type of bus after the war, it never made it into production and in 1946, Orville Caesar announced that the company had ordered a new fleet of Silversides from the GMC Truck & Coach Division of General Motors (formerly Yellow Coach – the name was changed at the end of 1942). Eventually 2,000 buses of this type with either 37 or 41 seat capacity were delivered between 1947 and 1948 and these formed the basis of the first large fleet of diesel-powered buses in the USA.

Instantly recognisable by their fluted aluminium sides, the classic Silversides (in effect, a Super Coach with different styling) featured some mechanical improvements over the pre-war buses, but were basically similar in appearance. One alteration noticed by the driver was the adoption of a column-mounted gear shifter, following the trend found in automobiles of the period.

An aborted project, a 50-seater double-decker bus known as the Highway Traveler, began in 1945 when Greyhound issued design contracts to GMC and Convair. Also referred to as the GX-1, this bus featured many ideas that were to become standard on later bus designs: air suspension, lavatory compartment, thermostatically controlled air conditioning and high-level seating to give better visibility.

Another unique idea incorporated in the design of the Highway Traveler (apparently at the insistence of Orville Caesar) was the use of two air-cooled gasoline engines. It was intended that one engine would run all the time, and the second engine could be brought into use for climbing hills, overtaking, or whenever extra power was needed. This was supposed to be more economical than running one big engine constantly, but the mechanical complication of the arrangement proved unworkable. Styled by famous designer Raymond Loewy, only one prototype Highway Traveler was built – by Greyhound itself, after GMC and Convair pulled out of the project – and it was eventually finished in the early part of 1947.

In 1948 Greyhound, who had been purchasing almost the entire output of MCI buses over the years, acquired the Winnipeg based bus and coach building company, thereby returning to the situation of having an in-house supplier for the first time since 1930.

Although the bulk of the Greyhound fleet in the late 1940s consisted of Sliversides built by GMC, it was by no means completely standardised and the company still used many

Right *1947 Silversides. From a design point of view, the new GMC bus was full of surprises. The huge rear stop light was as much an advertisment as it was an effective stop signal.*

Hardly surprisingly, the war put a stop to Greyhound expansion plans, but only temporarily. Even during the course of the conflict the company kept building new buses – aptly named Victory buses – which carried thousands of military personnel to their induction centres. For obvious reasons, these were relatively basic, unreliable machines. In 1946, however, Orville Caesar was able to announce that the company had ordered 1,200 "Silversides" from the GMC Truck & Coach Division of General Motors. Once again, the new coaches were state-of-the-art in terms of design, engineering and comfort. The Silversides was a 41-seat fully air-conditioned, centrally heated bus with transverse rear-mounted 6-71 Detroit Diesel engine. (Heating and air conditioning take up a lot of space in the 'Troubleshooting Guide' issued to all drivers today: "Blowers not on – blower reset tripped – reset, left front baggage bin.") The nation's new affluence and sense of optimism was duly reflected in the Silversides' gleaming stainless steel and aluminium design.

other makes of buses. An outstanding example of these other buses was the ACF Brill, built in Philadelphia. Powered by a six cylinder 707 cu.in. Hall-Scott gasoline engine mid-mounted between the axles, ACF Brill buses were very fast and used by other lines including Trailways. One disadvantage with the ACF Brill was that, because of the engine postion, the only luggage carrying space was underneath the seats. The first ACF Brill was delivered to Virginia Stage Lines in December, 1945 and Pennsylvania Greyhound received their intial batch in January 1946. The bulk of ACF Brill buses were built between 1945

and 1950, although the bus stayed in production until 1953 and the later models came with a Cummins diesel engine.

By 1950 the bus lines carried almost 37% of intercity travellers who used public transport, and the railroads share had declined to less than 47%. The airlines were taking an increasing percentage of the market, up from 3.2% in 1942 to 14.6% in 1950, showing how they were on course to dominate the US travel scene in the decades to come.

Greyhound increased the number of all-expense tours on offer to over 200, and the selection continued to grow. The company also provided a large choice of charter services that could be used for business travel, sports events or pleasure excursions. In some advertisements, the company liked to imply that a charter bus was just like having a luxury limousine and personal chauffeur at your disposal! But the pressure was certainly already on...

As always, the little chrome greyhound was placed on the coach's nose, this time enclosed in an oval shaped motif.

YOUR VACATION WILL BE *Easy to Take*

when you take it by GREYHOUND

It's *Easy to Find* the right schedules—Greyhound buses leave so frequently, at such convenient hours. It's *Easy to Reach* the choice vacation places, since Greyhound serves all 48 States and Canada. ● It's *Easy to Relax* in deeply-cushioned, adjustable chairs. *Easy on the Eyes* are the beauty spots, found only by Greyhound. ● *Easy to Buy* are Greyhound tickets, offering big savings on almost any trip... *Easy to Plan* is your outing, with the help of a friendly Greyhound agent.

Only by Greyhound you meet the Real America

 **GREYHOUND**

William Wyan, a Greyhound driver for 30 years, at the wheel of an old Silversides.

"The seven mile bridge between Miami-Key West"

"The bus that goes to sea"

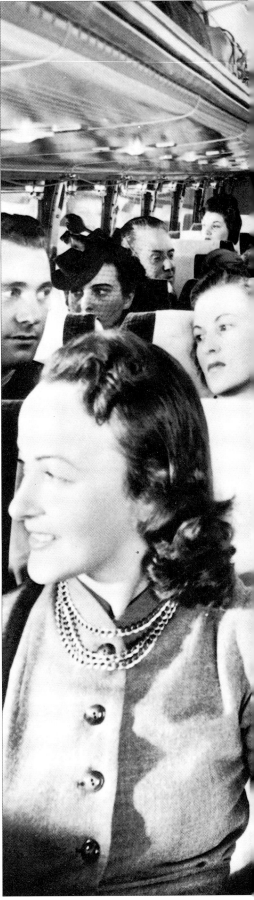

Above "The bus that goes to sea," a Silversides on its way to Key West on the famous Overseas Highway Bahia-Honda bridge.

Right The Silversides had a new modern interior which provided the passengers with unprecedented levels of comfort. They could relax in four-position adjustable chairs stuffed with the all new "sponge rubber" foam, stretch their legs thanks to retractable rubber footrests and read their Highway Traveler magazines with the help of directional reading lights. Not surprisingly, during and after the war buses continued to be filled to capacity, enabling Greyhound to earn almost $20 million in 1946.

Above left *Dallas, Texas. A vintage Silversides parked in front of the beautifully restored Dallas depot.*

Below left *Cleveland, Ohio. As the number of passengers increased, Greyhound built bigger and bigger terminals. When the one in Cleveland was completed (1943) it was claimed to be the biggest bus station in the world. It cost a record $1,250,000 and served over three million passengers every year, thanks to its 21 loading platforms and huge curved lobby. After the War, terminals similar to the one pictured here kept being built at an ever faster rate. In 1949 alone 21 terminals, 9 garages and 7 combinations of the two were constructed.*

Right *Cleveland Terminal's crowning glory. The hands may have fallen off the clock at the top of the tower, but time has yet to run out for this typical 1940s station, which remains a favourite among Greyhound travellers.*

Below *The Super Coaches of the 1930s and early '40s had a small, discreet greyhound painted on their flank. The new Silversides, as if to underline the company's new aggressive attitude, carried much more visible, larger and sleeker dogs running along their sides.*

LEFT *Bowie, Arizona. The only tepee-shaped Greyhound terminal in existence, and surely the most bizarre bus station in the world. Extravagantly named Geronimo's Castle (the great Apache chief was apparently captured nearby) it was built in the 1940s, at a time when such vernacular buildings were springing up all over the Western United States. In its heyday the Tepee attracted many thirsty soldiers on their way to training camps in California.*

INSET *A 1945 view of the Bowie, Arizona terminal. Buses to and from Tucson, Arizona, stop there every day.*

ABOVE *Inside Geronimo's Castle. Known these days simply as the Tepee Bar, the place hasn't changed a bit. Customers are still served by a lady dressed as an Indian, and dollar bills left as tips are still used as wallpaper.*

ABOVE Back in the 1940s, the Greyhound fleet was not completely standardised. While most coaches came from GMC, the company also bought equipment from other manufacturers. One such firm was ACF-Brill, a subsidiary of the American Car & Foundry Motors Co. of Detroit, Michigan. In 1945 they started making for Greyhound a beautifully streamlined coach, known simply as the ACF Brill. It had a six-cylinder Hall-Scott underfloor engine with optional Spicer hydraulic transmission, and was a favourite among the drivers, who loved its reliability and smooth power.

LEFT The rear (naturally) featured the hound.

RIGHT Like their GMC counterparts, the Brills had aluminium logos attached to their sides. The southeastern logo, with the added compass rose symbolizing the company's pan-continental policy, was perhaps the most beautiful of all.

63

LEFT *Aerodynamic and fast, the Brill was also extremely comfortable. Thanks to fans mounted inside the ceiling ducts, its air-conditioning system was extremely effective at keeping the interior cool, even when temperatures outside were unbearably high. If the weather turned, there were heating units under the floor to convey warm air to the new, ergonomically contoured seats.*

The Highway Traveler magazine was distributed free of charge to Greyhound passengers and was the equivalent of today's "in-flight" publications. Full of (pretty good) articles on history and travel, it was published bi-monthly and had a circulation of about 500,000.

ABOVE *Kingman, Arizona. Travelling through Arizona, the bus follows old Route 66, one of the oldest – and the most famous – road in America. The Silversides painted on a wall near Kingman's railway station is a reminder of the halcyon days of the "Mother Road".*

RIGHT *Benson, Arizona. The Quarter Horse Motel, another relic from the 1940s, welcomes travellers passing through this tiny Arizona town with one of the best vintage neons of the area.*

Left Old US Highway 86, going through Bowie, Arizona.

Below Highway 66 East of Kingman, Arizona.

Above Greyhound recycling at its best: a heavily customised GMC Silversides 4104 bus patiently awaits a new owner in Bracketville, Texas.

LEFT *Situated close to the old Greyhound terminal, the Yucca Lodge Motel in Bowie is typical of the mid 1940s. All over America motels like this one are disappearing fast, leaving the roadside to nationwide hotel chains and fast food restaurants.*

TOP *Dinosaur, Colorado. The Miner's Cafe, not far from the Dinosaur National Park, is a favourite stop for travellers. Inside this tiny diner passengers and cowboys often gather in front of the woodstove with a much needed cup of coffee. Outside, the bitter winter winds can bring temperatures down to -20C.*

ABOVE *Aluminium had of course been a reserved resource in the war years; the aluminium and steel are the essence of the Greyhound look.*

Top Gallup, New Mexico. The Indian Capital of the USA has some of the best preserved 1940s diners and cafes. The abundance of period signs makes this one of those US towns where time seems to have stood still.

Above Chief Yellow Horse Trading Post, 10 miles west of the city, still attracts plenty of business. Fifty years ago it was a major Indian tourist attraction and the site of several Hollywood movie productions.

Right Greyhound logo, inevitably, on the Silversides rear brake light.

PANORAMIC VIEWS
This Beautiful America...

B arely seven months of the Fifties had elapsed before events took place that would lead to the outbreak of the Korean War. President Truman declared a state of emergency on December 16, 1950 which, among other things, restricted the availability of essential materials like steel, nickel and chromium. And for the first three months of 1951 the US auto industry suffered a compulsory 20% restriction of output to conserve metal stocks for military use.

Dwight D. Eisenhower won the Republican nomination and then the 1952 presidential election and, with his running mate Richard Nixon, took over in the White House in January 1953. On July 12 1954, in Washington DC, Vice President Nixon announced a major new road building programme that would cost $50 billion, spread over the next decade. Concentrating on the construction of a network of interstate highways, in its own way this initiative would have as profound effect on the bus industry as the 'Good Roads' legislation introduced in the Federal Highway Act of 1916. Greyhound used the ever-improving freeways to introduce a number of limited and express runs which allowed cross-country passengers to complete their journey without changing buses.

But it was the unveiling of the split-level Scenicruiser bus that really put Greyhound ahead of its competitors. Taking the

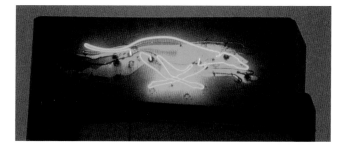

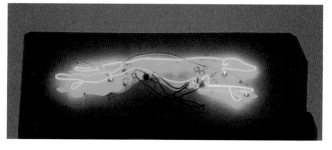

ABOVE *The last "running dog" in the country, at Evansville, Indiana.*

ill-fated Highway Traveler concept a stage further, in 1949 a prototype bus named the GX-2 was constructed by GMC in co-operation with Greyhound, using a Raymond Loewy design (Loewy was employed as a consultant by GM's head of styling Harley Earl). A three-axle configuration, stretched to 40 feet in length (instead of the normal 35 feet) with 10 passengers at the same level as the driver and 33 on the rear upper deck, the Scenicruiser was a masterpiece. Before the Scenicruiser could take to the freeways in earnest Greyhound had to set

ABOVE In the early 1950s, as President Eisenhower's massive interstate highway programme was beginning to transform the American landscape, the Greyhound empire was at its peak. With 6,280 buses and 90,000 miles of scheduled routes, Greyhound dominated the market and was able to think big. Very big. Its contribution to the new age of technological optimism was the world-famous Scenicruiser, a totally new kind of coach which, like the cars of the period, looked capable of inter-planetary travel. Despite the tragedy of the Korean War, the fifties was America's decade.

OVERLEAF 1954 Scenicruiser. The revolutionary new coach was built by GMC specifically for Greyhound. It was one of the first ever bi-level buses, with 10 passengers seated on the driver level and 33 up above. The new arrangement provided increased luggage space and, most important, a fully furbished rear toilet compartment. Originally the bus was powered by two rear-mounted 4-71 Detroit Diesels with a fluid coupling connection and common propeller shaft. It suffered from unreliability until the introduction of the 220 bhp 8V-71 Detroit engines in 1961.

out on a countrywide lobbying campaign with the GX-2 prototype to get approval for this larger bus and make certain it would be legal for use in all states. The GX-2 also saw service on regular routes as various ideas were tried and modifications carried out. In 1954, Greyhound ordered 1,000 Scenicruisers (GMC Model No. PD-4501) and the first was delivered in July of that year. In addition to air suspension and a lavatory to make travel more comfortable for the passengers, the extra luggage capacity of the Scenicruiser allowed Greyhound to promote an express parcel service – which proved a major boon for outlying communities and gave the company welcome additional revenue.

Originally, all 1,000 Scenicruisers were built with dual rear-mounted four cylinder Detroit Diesel engines driving

Above *The slanted windows gave the impression of speed, enhancing the sleek design of the Scenicruiser. The 90% glare-resistant glass used for the panoramic windows was developed especially for Greyhound, and there were even skylights in the roof. Teenagers could now lean back and gaze at the stars.*

Right *A true child of the fifties, this was possibly the last really innovative, anything-is-possible American bus. When it covered the 3,000-mile-long intercontinental "Through Bus Schedules" the Scenicruiser was a kind of "road-plane", with reserved seat service, free pillows, radio and on board hostess. The driver used the new public address system to point out places of interest and generally keep the customers entertained. Thus the tradition of Greyhound drivers' fantastic – notorious? – sense of humour was born.*

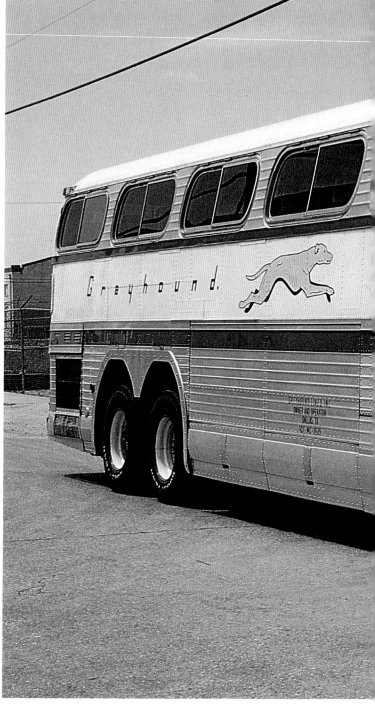

through a fluid coupling linked to a single transmission. Although complicated, this system functioned reasonably well, but was unreliable in service, not liked and consquently the surviving 979 buses were re-equipped with a single eight cylinder Detroit Diesel engine and a Spicer 4-speed manual transmission by the Marmon-Herrington Corporation at Indianapolis, Indiana, between 1961 and 1962.

Because the Scenicruiser was exclusive to Greyhound, competitors had to look elsewhere for a bus of similar design. Continental Trailways went to the Flexible Company of Loudonville, Ohio, who came up with a split-level design called the Vistaliner. Seating either 37 or 41 passengers, the Vistaliner was a two-axle bus and only 35 feet in length and, although sold to other lines as well as Trailways, only 208 were built between 1955 and 1958.

In 1957 Trailways imported 50 Golden Eagle buses from Germany. Built by Karl Kassbohrer AG, these 40 foot long, three-axle buses seated 41 passengers at high-level and had a lavatory plus lounge area at the rear. The following year the first Silver Eagle buses – 45 seaters without any luxury amenities – were delivered, and became the standard Trailways bus. These were the first of thousands of Eagle buses bought by

Trailways over the years. In 1962 the company set up its own factory in Belgium (operated in partnership with La Brugeoise) and Bus & Car N.V. used many American components (in particular Detroit Diesel engines and transmissions) in the Eagle buses it produced. Then, in 1974, Eagle International Inc was established in Brownsville, Texas, and started assembling buses using parts imported from Belgium.

Meanwhile, Greyhound was also very much in the business of building its own buses. Late in 1963, it established an assembly plant at Pembina, North Dakota, for its subsidiary Motor Coach Industries. The MCI MC-5 bus was the first to be produced at Pembina, with running gear and interiors being added to bodies shipped from the Winnipeg factory 70 miles north across the USA/Canada border. *(Cont. p 90)*

Thanks to a unique three-axle alignment (above) and springs mounted in tension rubber shackles, the Scenicruiser ride was smoother than ever. The new air suspension system was so effective that advertisments claimed the bus "floated entirely on cushions of air". The styling retained the traditional white and blue colours, together with the stainless steel and aluminium finish. Incredibly, it wasn't until 1957 that Greyhound adopted a real live dog as a promotional device. Incongruously named 'Steverino' (because Greyhound sponsored the Steve Allen TV show at the time), her name was later changed to the far more suitable "Lady Greyhound'. Wearing a jewelled tiara and collar, she made hundreds of public appearances for the company and for charity.

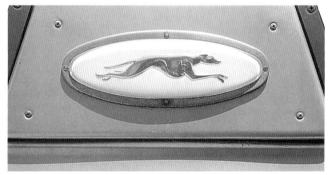

ABOVE Plastics were everywhere. The seats, often upholstered with the mosaic patterns typical of the 50s, were adjustable and resembled those used on airplanes. The fifties would see Greyhound drivers begin making on-board announcements to passengers thanks to the installation of a speaker system, and bus hostesses were introduced to assist travellers. The decade also produced a classic slogan, the world-famous: 'Go Greyhound – And Leave The Driving To Us.'

LEFT San Antonio, Texas. Casual and striking at the same time, the sign advertising the San Antonio terminal is pure fifties.

Every mile a Magnificent Mile...
every highway a strip of velvet...

when you travel in the amazing new
Scenicruiser

Get ready to experience the smoothest, most thrilling travel in highway history, when you step aboard Greyhound's luxurious new *Scenicruiser!* This is the revolutionary motor coach that floats entirely on cushions of air, to give you the gentlest ride ever known — that provides panoramic sightseeing on two observation levels — offers washroom convenience, many, many other luxury features.

A great fleet of 500 *Scenicruisers* is scheduled to serve all America — scores are already in operation. And *only by Greyhound* can you travel in this motor coach of tomorrow!

GREYHOUND

RAISED LEVEL SIGHTSEEING — Relax in a body-contoured easychair, enjoy panoramic sightseeing on four sides and overhead.

GENTLE AIR SUSPENSION RIDE

Entire coach floats on cushions of air, contained in flexible rubber-nylon bellows. Road shock and vibration are magically absorbed.

Hundreds of modern Greyhound "Highway Traveler" coaches also offer Air Suspension, huge picture windows, air conditioning.

COMPLETE WASHROOM
The *Scenicruiser* has a completely equipped washroom — with wash basin, running water, toilet, mirror, other features so convenient on longer trips.

FREE! Pleasure Map of America
Mail to Greyhound Tour Dept., 71 W. Lake Street, Chicago 1, Ill., for full-color trip-planning map — with details on 50 Greyhound Expense-Paid Tours.

Name_____

Address_____

City & State_____

Send me special information on a tour to:

_____

L-9-54

The Scenicruiser design details (above) display all the confidence and bravura that marked so much of US manufacturing in the 1950s. "It's such a comfort to go by bus – and leave the driving to us." Apart from forming an integral part of the cultural fabric of "This Wonderful America", the Greyhound company also inspired one of the best forced rhymes in popular music: "She shook his hustle, A Greyhound bus'll – take the one that got away." (Tom Waits)

JOIN the GREYHOUND TRAVEL CLUB

Here's what you'll receive as a member:

One year's subscription to The Highway Traveler magazine (six issues).

Greyhound's color folders and booklets: information on request relative to trips, tours, resorts, hotels, etc.

Membership fee is $1.00 a year. Mail coupon below with check or money order.

●

"Trip of the Month"
A 10-DAY CIRCLE TOUR OF FLORIDA FOR $106.80
(double rates lower)

(Transportation from Jacksonville around State and return. Hotel accommodations with both. Sightseeing trips at Bok Tower, Marine Studios, Silver Springs, St. Petersburg, Miami, Miami Beach and Key West.)

THE HIGHWAY TRAVELER
2341 Carnegie Ave., Cleveland, Ohio.

Please enter my name as a full paid member of the Greyhound Travel Club and as a subscriber to The Highway Traveler magazine, for one year. $1.00, enclosed herewith, covers membership for one year. (Residents in Canada, $1.50 for one year.)

NAME_____

STREET & NUMBER_____

CITY & STATE_____

RIGHT *San Antonio, Texas. One of the friendliest cities in the USA, San Antonio has one of the busiest terminals. Situated close to the famous Alamo, the depot is always teeming with people travelling to and from the Mexican city of Nuevo Laredo, only a few hours away.*

LEFT Dallas, Texas. When the Greyhound Corporation moved its headquarters to Dallas, the city's terminal was restored and modernised. One of the most beautiful modern terminals, Dallas retains many of the elements which over the years have been used to enhance the company image.

ABOVE Vernal, Utah. A pink brontosaurus (nicknamed Dino) welcomes travellers to the DineAville Motel. Vernal claims to be the dinosaur capital of America, due to its proximity with the Dinosaur National Monument. Many similarly outrageous motel signs built in the post-war years have sadly disappeared.

RIGHT Period views of the Akron, Ohio, and Pittsburgh, Pennsylvania, terminals.

For many people, the magical country that lies between New York and Los Angeles can still only be experienced directly from the front seat of a Greyhound bus. The Greyhound experience is de rigueur for any self-respecting student backpacker from Europe.

The MC-5 was very popular, particularly with operators wanting a shorter 35 foot coach. 2,255 were built, with later versions carrying A, B and C suffixes to the basic model number.

Greyhound's 50th anniversary was celebrated in 1964; to mark the event, every Greyhound bus had a broad gold stripe added to the traditional blue and white colour scheme. Once the commemoration was over, the gold stripe was replaced by a red one, and the now familiar red, white and blue design was established, still to be seen now, thirty years on.

In an effort to build an improved replacement for the spliit-level Scenicruser (by now well over ten years old) in 1967 MCI put a new prototype on the road. Called the MC-6, it was 40 feet in length with three axles and all the passengers sat at high level. The most controversial aspect of the MC-6 though, was its width. Greyhound had added six inches to the normal bus width of 96 inches to provide more passenger room and comfort – outside the legal limit in most states.

As with the Scenicruser in 1954, Greyhound tried to get the 102 inch wide bus accepted but in this instance the company failed – Congress refused to alter the maximum legal width allowed on the Interstate highway system. *(Cont. p 96)*

Above The Scenicruiser may have gone, but the views are still there. The imposing New Mexico landscape unravels outside the tinted windows of a modern Americruiser 2 on its way to Truth or Consequences, the town named after a 1950s TV show.

Left Tucson, Arizona. The pleasant terminal interior, always busy with passengers 24 hours a day.

Right When the sixties arrived, the oval logo used for years on the front of the buses became particularly popular. In Tucson it was featured at the top of the terminal's flagstaff style sign.

Above Amarillo, Texas. The Cadillac Ranch, famous celebration of the "chrome and fin" car-culture of America. It can be seen easily from buses travelling on Interstate 40 between Wildorado and Amarillo.

Left Colombus, Ohio. A driver checks his schedule while refuelling. Greyhound drivers are rightly considered to be the safest bus drivers in America.

ABOVE Horseshoe Café, US 80, Arizona.

RIGHT Pictures of horses, blueberry pies and the old Wurlitzer are what make the Horseshoe Café a favourite among Greyhound travellers.

ABOVE Bus sign at Columbus, Ohio.

RIGHT In 1948 Greyhound acquired Motor Coach Industries, manufacturers of bus body shells, and in 1963 started making their own buses at plants in Pembina, North Dakota and Winnipeg, Canada. The first Greyhound MCI bus was the 39-passenger MC-5, followed by the experimental MC6 and then, in 1968, the 6 wheeled 47-passenger MC-7, shown here.

Having already built 100 of these so-called 'Supercruiser" buses, powered by a huge V12 diesel engine in place of the usual V8, Greyhound was forced to limit their use to routes where state laws permitted wider buses or where special permission was granted. Fifteen MC-6 buses spent all their working life in Canada, the remaining 85 were used for a time on the east coast, but ended up in California.

The twelve cylinder engines in the west coast buses were removed in 1977 and replaced by Detroit Diesel V8s and Allison automatic transmissions (originally the MC-6 had a manual transmission) and were sold off from Greyhound service in 1980 – an expensive gamble for the company to lose.

Facing these difficulties with the MC-6, MCI introduced the MC-7 in 1968. Of standard 96 inches width, this forty foot long, three axle bus was based on the shorter MC-5 and equipped with a conventional drivetrain. The MC-7, sometimes called the "Super 7 Scenicruiser", remained in production until 1973, by when 2,550 had been produced.

Above Guess what's on the front?

Below Back to basics with the MC-7, following MC-6 problems, with conventional drivetrain.

Above and left 1968 MC-7. Even though it looked a little bit like a Scenicruiser, and was even called Super-Scenicruiser, the MC-7 was a single-deck coach. In 1964 a gold stripe was added to the blue and white colour scheme to celebrate the company's 50th anniversary. This was changed to red two years later, when Greyhounds started wearing the colours of the American flag. It is also around that time that buses started sporting the new slogan "Go Greyhound – and leave the driving to us" on their flanks. It proved to be amazingly successful and is still in use today.

GOING GREYHOUND
The Red, White & Blue

For America, the 1970s were of course a turbulent decade: as if you needed reminding of some of the grimmer highlights, the Vietnam War, Watergate, gasoline shortages caused by the oil embargo in 1973, and the nuclear disaster at Three Mile Island in 1979. In the ten years from 1970 to 1980 the population of the USA rose from 203 million to 226 million, but the bus companies were facing increasing competition for passenger traffic – mainly from the private automobile, but also from the airlines over the longer distances.

At the start of the Seventies, in 1971 to be precise, Greyhound moved its corporate headquarters from Chicago to Phoenix, Arizona, citing in its company literature 'the need for more spacious facilities' as the reason for the relocation. This was also the year that Amtrak, the National Railroad Passenger Corporation began operations, but the day of the train as the premier mass passenger carrier had long passed. Mostly, headlines about buses in the '70s were associated with 'busing' students in order to integrate American schools. Violence and rioting were commonplace when busing took place and the buses usually required armed police escorts. The segregation in education debate continued right through the 1970s and beyond, but the busing issue gradually faded from prominence.

On a more positive note, Greyhound introduced the 'Ameripass' in 1972. This scheme offered unlimited travel by bus in the USA and Canada over a weekly or monthly period for a single special payment which equated to a huge discount

Tucson, Arizona. Another stop, another Coke. Getting ready for yet another night on the bus.

Above *In 1973 Greyhound introduced the MC-8, the last coach with Scenicruiser-type slanting windows. Six years later came the MC9. This no-nonsense, comfortable coach proved to be one of the best workhorses the company ever had, so much so that it is still widely used today.*

on the regular fares charged. The 1970s also saw the phasing out the use of on-board hostesses that had been introduced by Greyhound some twenty years earlier.

Towards the end of the decade, Greyhound introduced the MCI MC-9 bus, in 1979. In the automobile industry, the MC-9 would probably be referred to as a face-lift of the MC-8, as it retained most of the mechanical components and layout of the earlier model but with a revised body design. Increased glass area, plus the elimination of the slanting window panels on either side, and a higher windshield as the dip in the roof line above the driver was done away with, are the main identifying features of the MC-9 when compared to the MC-8. Greyhound mainly used the MC-9 in a 43-passenger configuration, and the overall dimensions of 40 feet in length and width of 96 inches remained. Many other bus operators also bought MC-9 buses and well over 9,000 were produced as manufacturing continued into the late 1980s.

In fact, the MC-9 was so popular that it was continually being re-ordered – even by Greyhound – long after it would normally have been replaced by a newer version. A new MCI bus (and a new model designation), the 102A3, would become available in October 1985, but we're getting a little ahead of ourselves.

By 1979, through acquisitions, mergers and diversification, The Greyhound Corporation listed its activities as: meat and poultry packers, bus transportation, soap products, food, financial services and pharmaceuticals. One of its most famous brand names – Dial soap – would turn out to have an inordinately significant part to play in the future of the Greyhound organisation.

Problems started for Greyhound in 1978 with the deregulation of the airlines, followed by freight in 1980 and buses in 1982. Then came a bitterly divisive strike by Greyhound workers in 1983 – by the end of all that the company had lost

ABOVE *Greyhound drivers do a lot more than deliver passengers to their destinations. On the road they point out places of interest, historical markers and even the local wildlife. Their jokes are legendary, if not always new. Many experienced drivers like Jim Burgos, photographed on the road in Colorado, have rejoined the company after the bitterly divisive strike. The badges on this driver's uniform are proof of thirty years of accident-free driving.*

ABOVE RIGHT *Lordsburgh, New Mexico.*

nearly half of its customers and shed the same percentage of employees. Troubled times indeed.

Over the next few years things were to get even worse. In December 1986, a group of Dallas, Texas, investors offered a $350 million leveraged buyout deal to purchase Greyhound

Lines from The Greyhound Corporation of Phoenix, Arizona, and created a new company. The GLI Holding Co, headed by Fred Currey, assumed control of Greyhound Lines in March 1987 and moved the company's headquarters to Dallas. For a time, the company's fortunes seemed to be heading in the right direction, particularly when only four months later, in July 1987, Greyhound bought up its great rival Trailways Lines Inc, the largest member of the National Trailways Bus System. With this move Greyhound established itself as the only nationwide provider of intercity bus transportation services.

Above Hope, Arkansas. Outside the station.

Below Reflections in the windows of a California station.

Right Ozona, Texas. Passengers look for shelter from the midday sun.

Overleaf Water towers silhouetted against a bloodred Sonoran sunset indicate the imminent arrival at yet another small town station.

ABOVE *Greyhound buses wait in line at the huge depot near downtown Dallas, where the company's headquarters were relocated in 1987.*

RIGHT *Isla Morada, Florida.*

Meanwhile, The Greyhound Corporation acquired the production rights of the RTS transit bus from General Motors in January 1987. Using its subsidiary, Transportation Manufacturing Corporation in Roswell, New Mexico, Greyhound started producing the RTS by obtaining bodies from GM's factory in Pontiac, Michigan, but by September the buses were being built entirely in-house.

By now, the MCI 102A3 bus was in service and this vehicle is notable because of its unusual shape, which is not always apparent from most photographs. It is 102 inches in width (harking back to the MC-6), but only across the passenger section – the front and windshield are still at the old 96 inch width and the driver's compartment body sides taper outwards at a 5.5 degree angle. The wider 102A3 was immediately successful and thanks to the increasing number of states allowing the bigger bus access to the Interstate highways (102 inch wide buses and trucks were legal in all US states by 1990) it soon became an industry standard.

But all was not well at Greyhound, and in March 1990, following unsuccessful contract renewal negotiations, just about all the company's bus drivers, clerical workers and mechanics represented by the Amalgamated Transit Union (ATU) went on strike. Although Greyhound continued to operate by hiring replacement drivers as quickly as possible, and most of the

ABOVE *Memphis, Tennessee. The station is very close to the Sun Records studios, where Elvis cut his first singles. The Denny's restaurant across the road advertises itself with a cautious "The King probably ate here" neon.*

striking workers (apart from the drivers) returned to work, the loss of revenue and extra expenses incurred by the strike exhausted the company's cash resources. In June 1990, Greyhound filed a voluntary petition under Chapter 11 of the United States Bankruptcy Code.

A reorganisation plan was formulated and Greyhound emerged from bankruptcy on October 31, 1991 with Frank J. Schmieder as President and Chief Executive Officer. However, it seems that Frank Schmieder regarded the operation of a bus company as being the same as running an airline and tried to use the same management techniques. Critics also said he employed people in key positions who did not understand the bus industry, getting rid of experienced personnel who had spent years working for Greyhound. At this time Greyhound withdrew from the National Trailways Bus System, thereby giving up the right to use the Trailways name.

It was also in 1991 that The Greyhound Corporation changed its name to The Dial Corporation – reasoning that the familiar and profitable soap brand provided a better corporate image than the troubled bus company. However, despite the change of name, Dial continued to own much of the property used by Greyhound Lines Inc. which it leased to the bus company and kept control of Greyhound of Canada. Dial also retained Motor Coach Industries as a subsidiary (renamed Motor Coach Industries International in 1993) with a contract which stated that Greyhound Lines must buy at least 75% of its new buses from MCII until March 1997.

The 1990 strike had seen the number of passengers carried by Greyhound in that year drop to 15.7 million from 22 million in 1989. By the end of 1992 that figure had increased slightly to 16.2 million, but it was still well below the '89 level.

1993 turned out to be another traumatic year for Greyhound as management tried to implement some radical changes to the organisation. The most notorious of these was the introduction of seat reservations in the same manner as the airlines. This was something completely new to Greyhound customers – ever since intercity bus travel had begun, passengers just turned up at the terminal and paid their fare on the day they wanted to travel. Traditionally, if the scheduled bus was full, the company would lay on another bus to make sure the extra passengers got to their destination. The idea of booking a seat two weeks in advance took the spontaneity out of bus travel.

The software for the new computerised ticket reservation system had been developed internally (at great expense) by Greyhound and was given the grandiose name of Transportation Reservation Itinerary Planning System (TRIPS). But, when introduced in August 1993, the system failed to cope with the demand, resulting in long queues and aggravating delays at the terminals. Not surprisingly, business suffered and the number of passengers carried by Greyhound that year dropped to 15.4 million with the company recording another hefty loss.

There were a couple of bright spots amid the gloom. In order to upgrade its fleet Greyhound ordered 518 new buses in 1993 – most of these being MC-12 models, an improved version of the MC-9. And, in May 1993, the company reached a settlement with the ATU – ending the three-year-long dispute – agreeing to recall 800 striking drivers. There was still the matter of back pay and the union had launched a law suit with the National Labor Relations Board claiming $143 million, which the company contested.

In order to attempt to reverse the trend initiated by TRIPS,

BELOW *Memphis, Tennessee. Thanks to an extremely efficient maintenance system, Greyhound buses are kept on the road longer than ever. On average each bus covers 125,000 miles every year.*

ABOVE *El Paso, Texas. This modern terminal, always packed with people waiting to cross the border, retains the traditional adobe look of the Southwest.*

Greyhound mounted an aggressive marketing campaign and offered substantial discounts on fares. It also revised its ideas on seat reservations and stopped taking them over the phone while the system was upgraded (TRIPS was reopened in January 1994). Passengers returned, it's true, but the discounted ticket prices meant revenue was still well down on expectations. Saddled with long-term debts of $291 million, Greyhound offered 4.7 million shares of common stock to the public in May 1993, generating $93 million in proceeds which was primarily used to purchase new buses. In March 1994, 151 new buses were ordered from MCI at a cost of $34.8 million; delivery was completed by September the same year.

Beginning in 1994, virtually all of Greyhound's dispatching activities were managed from the Central Dispatch Office using an automated scheduling system called Bus Operations Support System (BOSS). Designed to eliminate certain operational problems, BOSS enabled drivers to be informed of their assignments several days in advance, instead of when they arrived at the depot.

Other factors were also combining to create further problems at this time. The Greyhound management decided to compete against, rather than co-operate with, regional bus lines and started placing passengers on Greyhound buses where before they had used the local services which were often more efficient. In addition, where other bus companies shared the use of the Greyhound depot, they suddenly found their lease payments increased significantly. As a result, many of these bus companies stopped using the Greyhound facilities and, once again, the travelling public suffered more inconvenience.

Its no wonder then, that Greyhound's fortunes continued to decline and Frank Schmieder was to resign in August 1994. In September the company was yet again teetering on the brink of bankruptcy. Following some intense negotiations, on 13 October 1994 Greyhound entered into a new $35 million credit facility with the Foothill Capital Corporation, which replaced the company's previous bank facility.

On November 11 1994, the financial restructuring of Greyhound was agreed to by bond holders and Craig R. Lentzsch took over as President and Chief Executive Officer four days later. Lentzsch had previously been the Executive Vice President and Chief Financial Officer of Motor Coach Industries – at last, a bus person was back in the driving seat!

The number of passengers carried by Greyhound in 1994 fell to 14.9 million, but the signs for the future look promising. Lentzsch stated recently that there would be no profit in

RIGHT *Fort Stockton, Texas. Coaches travelling through the border country near Mexico are often stopped by the US Border Patrol, always on the lookout for illegal immigrants. In the great melting pot, the officers who carry out these searches are often of course the sons of immigrants themselves.*

Above Sleep is a valuable commodity on the long intercontinental routes. Here, a man from Southern Mexico tries to sleep through a fuel stop in Texas, while a relief driver (top) takes a well earned rest in Southern California.

Above Despite the inevitable contraction of the bus industry under the onslaught of domestic flights and auto ownership, Greyhound is still an important part of the infrastructure of America today, just as it was in the 1940s. What is remarkable is not the amount that has been lost, but just how much has survived. The street furniture and cars date the picture, not the station itself, nor, if you look quickly, the bus.

1995, but prospects for '96 were looking good. Today Greyhound operates approximately 215 million miles of regularly scheduled services to more than 2,500 destinations across 48 US states. The company also has two subsidiary bus lines – Texas, New Mexico, and Oklahoma Coaches Inc and Vermont Transit Co Inc.

Greyhound has something like 10,500 employees, including 3,500 drivers, who operate a fleet of nearly 2,000 buses. The average age of a 47-seat Greyhound bus is approximately 7 years and it will cover 125,000 miles in a year.

The immediate survival of Greyhound seems to be reasonably secure, but what of the future? The lucrative parcel delivery service has been hit in recent years by the upsurge in competition from other carriers offering an express or overnight service and Greyhound is now concentrating on local deliveries of under 400 miles. The package express part of the business now contributes less than 7% of Greyhound's annual revenue (charters account for only 1%). Competition for passengers is fiercer than ever with regional airlines offering big discounts to tempt people on board *(cont p 122)*

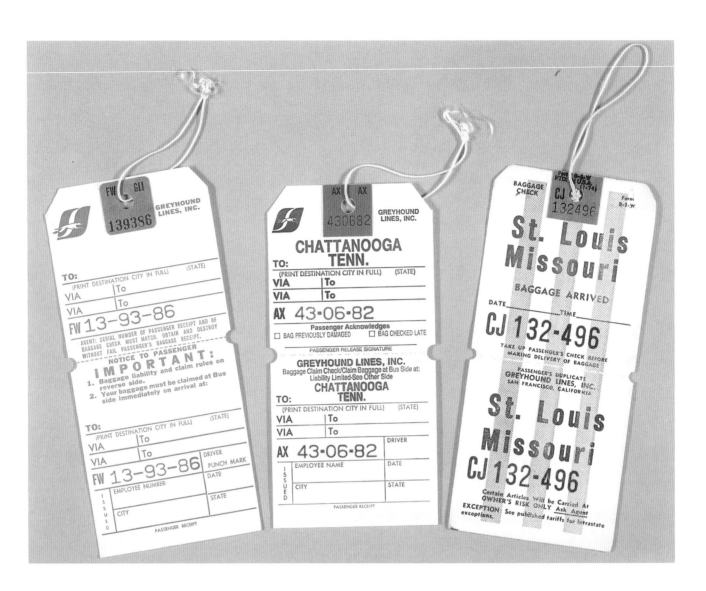

Above As can be imagined, this book involved an awful lot of baggage checks.
Left New Mexico.
Right Two-hundred-year-old saguaros stand guard on the verges of a secondary Arizona Highway.

ABOVE Homestead, Florida. Still recovering from the devastation caused by Hurricane Andrew in 1992.

RIGHT A late winter blizzard welcomes an Americruiser 2 to Steamboat Springs, Colorado.

ABOVE Baby Cody, 4 months old, enjoys the view from the front seat of an Americruiser travelling through the Ohio plains. Many babies have actually been born on Greyhound buses, with drivers often having to assist with the delivery.

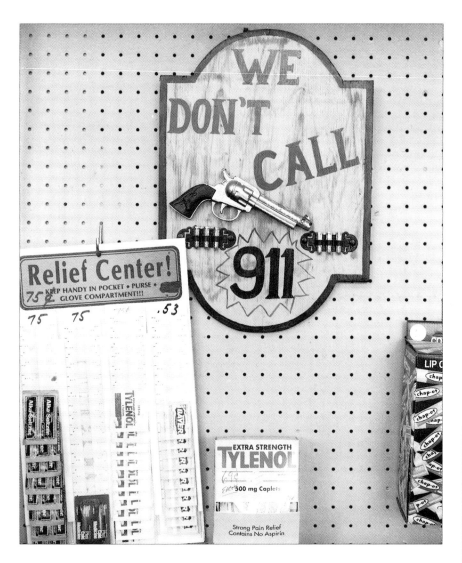

Above Self-defence inside the Sonora, Texas terminal.

Right Carl Eric Wickman's dream finally became a reality in 1987, when Greyhound Lines Inc. purchased Trailways Lines and became the only intercity bus carrier in the country. As part of the deal the company inherited many Eagles A3 buses, which were quickly converted to Greyhound colours. Some Eagles are still in use, providing efficient and comfortable service thanks to their aerodynamic body and Torsilastic suspension system. But the Eagle has never really found a place in the hearts of Greyhound aficionados. Maybe it's the large glass area, maybe it's the inner-city style doors, but for most people the Eagle somehow just isn't a Greyhound.

Above Miami, Florida. Since 1972, when the Ameripass was introduced, foreign tourists have been an important source of income for Greyhound. Yukiko Kubo, from Hokaido in Japan, celebrates at the end of a long vacation on the road.

Right An Americruiser 2 picks up passengers in Utah. On rural roads like this, flagging down the bus is a pleasant antidote to creeping agoraphobia!

and private automobiles are being used for longer journeys more than ever before. It isn't going to be an easy ride into the next century.

Even so, it seems unthinkable that the famous 'running dog' logo and those gleaming silver-sided Americruisers could ever disappear from the USA highways – the Greyhound bus has rightly become established as an essential part of the American way of life; long may it remain so.

ABOVE *"Board a Greyhound in Pittsburgh"* sang Paul Simon, and leave the past behind. For Eddie Conner from Virginia, a new life in Arizona is only two days and two nights away.

RIGHT The MC-12 is the latest coach made by Motor Coach Industries for Greyhound. Mechanically similar to the MC-9, but much more aerodynamic. The new bus retains all the traditional Greyhound trademarks: small glass area, silver sides and white, blue and red paintwork. The image that has turned the Greyhound into an American classic.

VEHICLE SPECIFICATIONS

HUPMOBILE
Year: 1914
Model: Model 32 Touring Car
Engine: 32 hp, 4 cylinder gasoline
Wheelbase: 106 inches
Seating: 7 (or as many as could be crammed aboard!)

MACK
Year: 1931
Model: BC
Engine: 6 cylinder gasoline
Wheelbase: 233 inches
Seating: 29

YELLOW COACH
Year: 1937
Model: Model 743 Super Coach
Engine: 6 cylinder GMC 707 cu.in. gasoline
Wheelbase: 245 inches
Length: 33 feet
Width: 96 inches
Height: 117 inches
Seating: 36

GMC TRUCK & COACH DIVISION
Year: 1947
Model: PD-3751 'Silversides'
Engine: 6 cylinder, 200 hp, Detroit Diesel 6-71
Wheelbase: 264 inches
Length: 35 feet
Width: 96 inches
Height: 118 inches
Seating: 37

ACF BRILL
Year: 1948
Model: IC-41
Engine: 6 cylinder, Hall-Scott 190-2 gasoline
Wheelbase: 270 inches
Length: 35 feet
Width: 96 inches
Seating: 37
Aisle width: 14 inches

GMC TRUCK & COACH DIVISION
Year: 1954
Model: PD-4501 Scenicruiser
Engine: Two Detroit Diesel 4-71 (rebuilt with single Detroit Diesel 6-71)
Wheelbase: 261 inches
Length: 40 feet
Width: 96 inches
Height: 134 inches
Seating: 43
Luggage capacity: 344 cu.ft.
Aisle width: 14 inches

MOTOR COACH INDUSTRIES
Year: 1964
Model: MC-5
Engine: 8 cylinder Detroit Diesel 8V-71
Wheelbase: 261 inches
Length: 35 feet
Width: 96 inches
Height: 120 inches
Seating: 39-45
Luggage capacity: 212 cu.ft.
Aisle width: 14 inches

MOTOR COACH INDUSTRIES
Year: 1968
Model: MC-7 'Super 7 Scenicruiser'
Engine: 8 cylinder, 252 hp Detroit Diesel 8V-71
Wheelbase: 285 inches
Length: 40 feet
Width: 96 inches
Height: 129 inches
Seating: 43-49
Luggage capacity: 325 cu.ft.
Aisle width: 14 inches

MOTOR COACH INDUSTRIES
Built by Transportation Manufacturing Corporation
Year: 1973
Model: MC-8 'Americruiser'
Engine: 8 cylinder, Detroit Diesel 8V-71C
Wheelbase: 285 inches
Length: 40 feet
Width: 96 inches
Height: 132 inches
Seating: 43/47
Luggage capacity: 300 cu.ft.
Aisle width: 14 inches
Gross Vehicle Weight: 36,500 lbs
Safe Operating Range: 750 miles

MOTOR COACH INDUSTRIES (TMC)
Year: 1980
Model: MC-9 'Americruiser 2'
Engine: 6 cylinder, Detroit Diesel 6V-92TA
Wheelbase: 285 inches
Length: 40 feet
Width: 96 inches
Height: 132 inches
Seating: 43/47
Luggage capacity: 300 cu.ft.
Aisle width: 14 inches
Gross Vehicle Weight: 36,500 lbs
Safe Operating Range: 750 miles

MOTOR COACH INDUSTRIES (TMC)
Year: 1985
Model: TMC-102A3
Engine: 6 cylinder, Detroit Diesel 6V-92TA
Wheelbase: 285 inches
Length: 40 feet, 4.25 inches
Width: 102 inches
Height: 135 inches
Seating: 47
Luggage capacity: 362 cu.ft.
Aisle width: 14 inches
Gross Vehicle Weight: 37,800 lbs
Safe Operating Range: 700 miles

MOTOR COACH INDUSTRIES
Year: 1993
Model: MC-12
Engine: 6 cylinder, Detroit Diesel
Wheelbase: 285 inches
Length: 40 feet
Width: 96 inches
Height: 132 inches
Seating: 47
Aisle width: 14 inches

GREYHOUND BUS PASSENGER PROFILE

(Taken from a 1992 On-Board Passenger Survey conducted by Greyhound)

- 56.5% of passengers are female, 43.5% male.
- Over 41% of passengers are aged between 18 and 34.
- Singles predominate – over 38% of passengers have never married, and nearly 29% are either divorced, separated or widowed.
- Most people travel alone (almost 73%) with couples the next biggest group (25%).
- 21% are retired, 13% are students and there are 2% military personnel.
- Most passengers make between one and three trips by Greyhound of over 50 miles each year (53%).
- 23% of passengers have never taken an airline flight.
- The majority of passengers get to the bus terminal by car (58%), while 33% arrive on another bus, the subway or a taxi, and only 6% walk.

Left A Greyhound approaches a small town station. This image, conjured up time and time again by advertisers selling a feel for old time America, is becoming rarer. Market pressures are forcing the closure of many secondary routes and the famous Greyhound web that kept America's communities together is embattled – but fighting back.

Overleaf As another scorching day nears its end Raul Pena, driver with some 37 years of Greyhound experience, salutes a colleague going in the opposite direction on US 90 between Laredo and El Paso.

*Bury my body
down the Highway side*

*So my ole evil spirit
can get a Greyhound bus
and ride.*

The blues of Robert Johnson, 1936.